Conservation Agriculture

Jaideep Kumar Bisht • Vijay Singh Meena
Pankaj Kumar Mishra • Arunava Pattanayak
Editors

Conservation Agriculture

An Approach to Combat Climate Change in Indian Himalaya

Springer

Editors
Jaideep Kumar Bisht
ICAR-Vivekananda Institute of Hill Agriculture
Almora, Uttarakhand
India

Vijay Singh Meena
ICAR-Vivekananda Institute of Hill Agriculture
Almora, Uttarakhand
India

Pankaj Kumar Mishra
ICAR-Vivekananda Institute of Hill Agriculture
Almora, Uttarakhand
India

Arunava Pattanayak
ICAR-Vivekananda Institute of Hill Agriculture
Almora, Uttarakhand
India

ISBN 978-981-10-2557-0 ISBN 978-981-10-2558-7 (eBook)
DOI 10.1007/978-981-10-2558-7

Library of Congress Control Number: 2016959265

Printed on acid-free paper

This Springer imprint is published by Springer Nature
The registered company is Springer Nature Singapore Pte Ltd.
The registered company address is: 152 Beach Road, #22-06/08 Gateway East, Singapore 189721, Singapore

Foreword

The Himalayan region with distinct ecosystem and climatic conditions is an ideal habitat of numerous unique flora and fauna, which also provides a natural resource base to a large part of the Indian subcontinent since the time immemorial. Recognition as one of the global biodiversity hotspots and predominance of diverse agricultural forms in the region reflect the prevalence of rich diversity all across the Himalayan ecosystem. But nowadays, the climate change (CC), faced worldwide, has become a major global concern. And, the Indian Himalayan region (IHR) is also not unperturbed by CC impacts. In fact, it has become highly vulnerable to the globally changing climate. These CC impacts are reflected in terms of general consequences such as unpredictable or erratic rainfall patterns, drying up of local springs and streams, species migrating to different elevations, shifts in sowing and harvesting periods of crops, emergence of invasive species, increasing evidences of disease/pest in crops as well as fodder species and many more unpredictable threats to the ecological security.

To combat with these CC challenges, an urgent need has been felt time and again to mitigate the challenges by sufficient food production, which is feasible only by adopting eco-friendly conservation agriculture practices in the IHR. Agriculture can help not only in ensuring food security for the growing global population but also in reducing the greenhouse gas (GHG) emission. Observing the high vulnerability and dependency of the agriculture-based food production on CC, a new time-tested approach/technique is the need of the present time, which should be adaptable and resilient to CC.

Nowadays, under the changing climatic scenario the conservation agriculture is the proven best, operational, integrated management approach of agroecology to manage agroecosystems, which helps in mitigating the CC impacts, improving C-sequestration, ensuring sustainable food security with increased profits and resilience to cope with extreme weather, protecting land degradation, restoring carbon in the soil and thus reducing global carbon footprint while preserving the resource base and the environment in the IHR.

On the ground of old, historic paradigm of knowledge and methodologies, short-term projects, inadequate conservation agriculture equipment, outdated experimental set-ups, etc. are among the major constraints faced by the research and development (R&D) as well as farming communities alike. To bring agriculture at the forefront, farmers seek advices from the researchers and government and are ready to adopt both improved farming systems and state-of-the-art technologies that can allow them to adapt to the changing climate, be it gradual or aberrant emergent change.

In the wake of CC, this is an opportune time to bring out the book entitled *Conservation Agriculture*: *An Approach to Combat Climate Change in Indian Himalaya*, edited by Drs. J.K. Bisht, V.S. Meena, P.K. Mishra and A. Pattanayak, which presents innovative practices and acts as an essential reference and knowledge base on sustainable conservation agriculture for all, i.e. the scholars, scientists and professors on one hand and development officials and policy-makers on the other. At this outset, I congratulate them all for their great endeavours in the direction of "conservation agriculture" vis-à-vis "climate change (CC)" and appreciate the editors and authors for guiding the Himalayan communities towards an environmentally secured and flourishing Himalayan region.

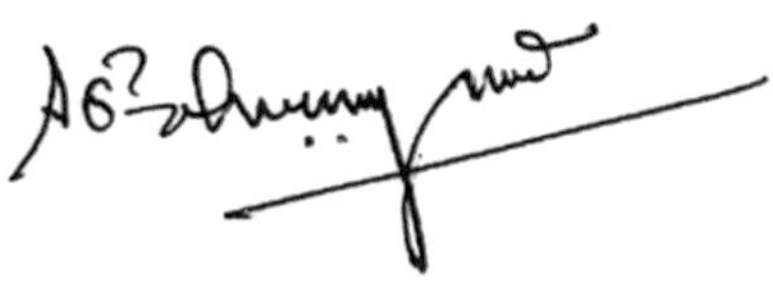

(P.P. Dhyani)
Director

G.B. Pant National Institute of Himalayan
Environment and Sustainable Development
Kosi-Katarmal, Almora
Uttarakhand, India

Contents

Contributors

Shahid Ahmed Crop Improvement Division, ICAR-Indian Grassland and Fodder Research Institute, Jhansi, India

Sanjay Arora Regional Research Station, Central Soil Salinity Research Institute (ICAR), Lucknow, Uttar Pradesh, India

Debarati Bhaduri ICAR-Directorate of Groundnut Research, Junagadh, Gujarat, India

Rajan Bhatt Farm Advisory Service Scheme, Tarn Taran, Punjab Agricultural University, Ludhiana, Punjab, India

Indu Bhushan Department of Biotechnology, Shri Mata Vaishno Devi University, Katra, Jammu & Kashmir, India

Jaideep Kumar Bisht ICAR-Vivekananda Institute of Hill Agriculture, Almora, Uttarakhand, India

A.K. Biswas ICAR-Indian Institute of Soil Science, Bhopal, India

R. Roy Burman ICAR-Indian Agricultural Research Institute, New Delhi, India

Nirmal Chandra ICAR-Vivekananda Institute of Hill Agriculture, Almora, Uttarakhand, India

D. Chatterjee ICAR-RC for NEH Region Nagaland Centre, Jharnapani, Medziphema, India

Mahipal Choudhary ICAR-Vivekananda Institute of Hill Agriculture, Almora, Uttarakhand, India

Bidyut C. Deka ICAR-RC for NEH Region Nagaland Centre, Jharnapani, Medziphema, India

C.K. Dotaniya College of Agriculture, SKRAU, Bikaner, India

M.L. Dotaniya ICAR-Indian Institute of Soil Science, Bhopal, India

S.K. Dwivedi ICAR-Research Complex for Eastern Region, Patna, India

Prakash Chand Ghasal ICAR-Indian Institute of Farming Systems Research. Modipuram, Meerut, Uttar Pradesh, India

P.K. Ghosh ICAR-Indian Grassland and Fodder Research Institute, Jhansi, Uttar Pradesh, India

B. Gupta Dr. Y S Parmar University of Horticulture and Forestry, Nauni, Solan, India

R.L. Jat ICAR-Indian Institute of Pulses Research, Kanpur, India

Renu Jethi ICAR-Vivekananda Institute of Hill Agriculture, Almora, Uttarakhand, India

Pramod Jha ICAR-Indian Institute of Soil Science, Bhopal, India

Kushagra Joshi ICAR-Vivekananda Institute of Hill Agriculture, Almora, Uttarakhand, India

H.L. Kharbikar ICAR-Vivekananda Institute of Hill Agriculture, Almora, Uttarakhand, India

Shrikaant Kulkarni Department of Chemical Engineering, Vishwakarma Institute of Technology, Pune, India

A. Kumar International Rice Research Institute, Los Baños, Philippines

Ashok Kumar Department of Botany, MMV, Banaras Hindu University, Varanasi, India

Rakesh Kumar ICAR-Reseach Complex for Eastern Region, Patna, Bihar, India

ICAR-RC for NEH Region Nagaland Centre, Jharnapani, Medziphema, India

Sandeep Kumar ICAR-Centre for Environment Science and Climate Resilient Agriculture, Indian Agricultural Research Institute, New Delhi, India

Santosh Kumar ICAR-Research Complex for Eastern Region, Patna, India

Amrit Lamichaney ICAR-Indian Institute of Pulses Research, Kanpur, Uttar Pradesh, India

Manju Lata University of Rajasthan, Jaipur, India

Aniruddha Maity ICAR-Indian Grassland and Fodder Research Institute, Jhansi, Uttar Pradesh, India

A.L. Meena ICAR-Indian Institute of Farming Systems Research, Modipuram, India

B.L. Meena ICAR-Central Soil Salinity Research Institute, Karnal, India

B.P. Meena ICAR-Indian Institute of Soil Science, Bhopal, India

Ramawatar Meena Department of Soil Science and Agricultural Chemistry, Institute of Agricultural Sciences, Banaras Hindu University, Varanasi, India

Sunita Kumari Meena Division of Soil Science and Agricultural Chemistry, ICAR-Indian Agriculture Research Institute, New Delhi, India

Vijay Singh Meena ICAR-Vivekananda Institute of Hill Agriculture, Almora, Uttarakhand, India

Amit Mishra International Rice Research Institute, Bhubaneswar, Odisha, India

J.S. Mishra ICAR-Research Complex for Eastern Region, Patna, India

T. Mondal ICAR-Vivekananda Institute of Hill Agriculture, Almora, Uttarakhand, India

Anirban Mukherjee ICAR-Indian Agricultural Research Institute, New Delhi, India

ICAR-Vivekananda Institute of Hill Agriculture, Almora, Uttarakhand, India

Arindam Nag Bihar Agricultural University, Sabour, Bihar, India

Seva Nayak Crop Improvement Division, ICAR-Indian Grassland and Fodder Research Institute, Jhansi, India

D.R. Palsaniya ICAR-Indian Grassland and Fodder Research Institute, Jhansi, Uttar Pradesh, India

Jai Singh Patel Department of Botany, Banaras Hindu University, Varanasi, India

A.K. Patra ICAR-Indian Institute of Soil Science, Bhopal, India

M.K. Patra ICAR-RC for NEH Region Nagaland Centre, Jharnapani, Medziphema, India

Sudipta Paul ICAR-Indian Agricultural Research Institute, New Delhi, India

Ved Prakash ICAR-Research Complex for Eastern Region, Patna, India

Pragati Pramanik Center for Environment Science and Climate Resilient Agriculture, ICAR-Indian Agricultural Research Institute, New Delhi, India

S. Rajendiran ICAR-Indian Institute of Soil Science, Bhopal, India

Shantanu Rakshit ICAR-Indian Agricultural Research Institute, New Delhi, India

K.K. Rao ICAR-Research Complex for Eastern Region, Patna, India

Mrinmoy Ray ICAR-Indian Agricultural Statistics Research Institute, New Delhi, India

K.L. Regar Department of Soil Science and Agricultural Chemistry, Institute of Agricultural Sciences, BHU, Varanasi, India

Subhodeep Roy ICAR-Indian Institute of Vegetable Research, Varanasi, Uttar Pradesh, India

J.K. Saha ICAR-Indian Institute of Soil Science, Bhopal, India

Gitanjali Sahay Crop Improvement Division, ICAR-Indian Grassland and Fodder Research Institute, Jhansi, India

S.K. Samal ICAR-Research Complex for Eastern Region, Patna, India

Sujit Sarkar ICAR-Indian Agricultural Research Institute, New Delhi, India

Pradeep Saxena Crop Improvement Division, ICAR-Indian Grassland and Fodder Research Institute, Jhansi, India

A.O. Shirale ICAR-Indian Institute of Soil Science, Bhopal, India

Kumari Shubha ICAR-Indian Agricultural Research Institute, New Delhi, India

Gurjant S. Sidhu ICAR-National Bureau of Soil Survey and Land Use Planning, New Delhi, India

Mahendra Singh Bihar Agricultural University, Sabour, India

Tejveer Singh Crop Improvement Division, ICAR-Indian Grassland and Fodder Research Institute, Jhansi, India

Sher Singh ICAR-Vivekananda Institute of Hill Agriculture, Almora, Uttarakhand, India

A. Thirugnanavel ICAR-RC for NEH Region Nagaland Centre, Jharnapani, Medziphema, India

Dunna Vijay ICAR-Indian Grassland and Fodder Research Institute, Jhansi, Uttar Pradesh, India

R.P. Yadav ICAR-Vivekananda Parvatiya Krishi Anusandhan Sansthan, Almora, India

ICAR-National Bureau of Soil Survey and Land Use Planning, New Delhi, India

About the Authors

Dr. Jaideep Kumar Bisht is currently working as Principal Scientist & Head of Crop Production Division in ICAR-Vivekananda Institute of Hill Agriculture, Almora, Uttarakhand, India

Dr. Vijay Singh Meena is currently working as Scientist in ICAR-Vivekananda Institute of Hill Agriculture, Almora, Uttarakhand, India

Dr. Pankaj Kumar Mishra is currently working as Principal Scientist in ICAR-Vivekananda Institute of Hill Agriculture, Almora, Uttarakhand, India

Dr. Arunava Pattanayak is currently working as Director in ICAR-Vivekananda Institute of Hill Agriculture, Almora, Uttarakhand, India

Conservation Agriculture and Climate Change: An Overview

1

Mahipal Choudhary, Prakash Chand Ghasal, Sandeep Kumar, R.P. Yadav, Sher Singh, Vijay Singh Meena, and Jaideep Kumar Bisht

Abstract

Conservation agriculture (CA) is the integrated management of the available natural resources such as soil, water, flora, and fauna with partial outside inputs which increases the efficiency of natural resource use. It provides sustainability in farming production through maintaining the quality of natural resources by stable or semi-stable organic cover to soil. Zero or minimum tillage or no-till (NT) and minimum disturbance of soil along with varying rotation of crops are a must for future prospects. CA is an integrated approach to agriculture cultivation that helps enhance food security, allay poverty, conserve biological diversity, and preserve ecosystem services. CA practices are also helpful in making farming systems more resilient to recent climatic changes. CA can comprise wide-ranging practices such as management of forage and farm animals, fallows improvement, combined cultivation of agricultural crops and trees as agroforestry, management of watershed, and management of areas which are reserved for village and community people. In this chapter, climate change predictions for Indian Himalayan Region (IHR) will be discussed. Then the potential of CA as a source to alleviate and acclimatize to climate change will be examined for climatically affected environments.

M. Choudhary (✉) • S. Singh • V.S. Meena (✉) • J.K. Bisht
ICAR-Vivekananda Institute of Hill Agriculture, Almora 263601, Uttarakhand, India
e-mail: mahipal.babal@gmail.com; vijay.meena@icar.gov.in

P.C. Ghasal
ICAR-Indian Institute of Farming Systems Research, Modipuram, Meerut 250110, Uttar Pradesh, India

S. Kumar
ICAR-Centre for Environment Science and Climate Resilient Agriculture, Indian Agricultural Research Institute, New Delhi 110012, India

R.P. Yadav
ICAR-Vivekananda Parvatiya Krishi Anusandhan Sansthan, Almora 263601, India

J.K. Bisht et al. (eds.), *Conservation Agriculture*,
DOI 10.1007/978-981-10-2558-7_1

Keywords
Conservation agriculture • Climate change • Zero tillage • Crop rotation • Indian Himalayan Region (IHR)

1.1 Introduction

Agriculture is the mainstay of people in the Indian Himalayan Region (IHR). However, with climate variability and the growing capriciousness in precipitation and rising temperatures, the traditional farming systems and cropping patterns are increasingly under threat. The ecological fragility and vulnerability of the IHR to climatic aberrations and increasing demand for land to grow more food have been issues of concern for quite some time. The resource-poor rural farming people are most pregnable to the risks of changing climate due to poor adaptation capacity. Furthermore, the physiographic and environmental constraints associated with the wide variability in altitude, slope, and aspect limit the adoption of modern agricultural technologies in the IHR. With regard to this ecosystem, extreme variations in growing conditions, very small holdings, remoteness of villages, degraded watershed, and poverty are serious challenges to sustainable hill agriculture. This calls for urgent extensive investigation on the hill agriculture to develop resilience in this vulnerable sector. The focal geographical areas of this task force are Eastern (NEH region), Central (Uttarakhand), and Western Himalayas (Leh, Ladakh, HP, and J&K) (NMSHE 2016) (http://knowledgeportal-nmshe.in/). Indian Himalayan region covers about 53.8 Mha (~16 % of country's total geographical area). Densely populated areas of South Asia are particularly pregnable to climate change and are envisaged to suffer crop productivity reduces of at least ~20 % by 2050 with ~40 % risk of crop failure for wheat in a given season across large areas (Cairns et al. 2013). Instead of abovementioned challenges, the major factors responsible for unsustainable farming systems include soil and water erosion, desertification, reduction in organic matter (OM) content of soil, and accumulation of acids and salts in the soil. The principal causes of these problems are (i) intensive tillage causing deterioration in soil structure, reduction in organic matter content of soil, reduced water infiltration, soil erosion, crusting, and compaction of soil surface, (ii) inadequate OM supply, and (iii) monocropping (FAO 2012). Moreover, there is incriminated degradation of land resources, reduction in biological diversity, reduced efficiency of energy, and involvement to worldwide warming trouble. Therefore, it is an important tool for diversification in farming techniques to increase future productivity and prospects while nourishing the natural wealth. The various crop management practices termed "conservation agriculture" (CA) are globally promulgated to enhance crop yield, conservation of the soil, and development of resilient systems to stresses which are induced by weather including those created due to irregularity and change in climate. CA is an important agronomic practice which is concerned about agriculture sustainability and has progressively augmented globally to cover ~11 % of the globe's cultivable land

(157.8 Mha) (FAO 2016). CA is also showing a new, very attractive advantage in the mitigation of global climate change by sequestering soil organic carbon (Lal 2013; UNEP 2013). Hence, CA is an well-organized substitute to conventional agricultural systems, as it improves soil properties like water retention, infiltration, nutrient dynamics, CEC, soil biota, and efficient in terms of money, time, and fossil fuel.

1.2 What Is Conservation Agriculture?

It is an agronomic practice that comprises reduced tillage (RT) or no-tillage (NT) or minimum tillage along with stable cover to soil with organic materials or by retaining residue of crops or growing green manure crops as cover crop and rotation of crops with pulses and legumes. According to FAO guideline, "CA is an approach to managing agricultural ecosystems for enhanced and sustained productivity, improved returns, and food security while preserving and enhancing the resource base and the environment."

The terms "conservation agriculture" (CA) and "resource-conserving technologies" (RCTs) are not synonyms, but sometimes both words are used for a similar meaning. Therefore, the practices which improve resource or input-use efficiency are termed as RCTs. RCT is a wider term than CA which is used for nitrogen-efficient varieties; fuel-, money-, and time-efficient techniques; and improvement in plot-level water productivity including CA. CA practices will include only those RCTs with the following features:

- Crop residue retention on the soil surface (Fig. 1.1) or soil covering with crop organic materials
- Rational, gainful crop rotations
- Minimum soil disturbance, e.g., minimum or zero tillage

The abovementioned principles are said to be common to systems of CA. However, components which are specific in nature such as selection of implements for farm, methods of establishment, rotation of crops with pulses or legumes, management of fertility in soil, management of mulch and residue of crops, etc., are varied according to environment. The potential of CA is that it can be applied in all types of agricultural ecological zones and helpful in enhancing food security for a large number of smallholder residents of developing countries (Sayre 2000; Derpsch and Friedrich 2009).

CA is the best management of natural resources like soil, water, vegetation, and biodiversity for sustaining the future prospects. CA have potential to decrease the effects of changing climate by optimizing crop productivity and advantages while maintaining a coordination among agricultural, monetary, and ecological benefits (FAO 2011a; Giller et al. 2009).

Fig. 1.1 The various conservation agricultural management practices to maximize the soil nutrient water-retention capacity. As depicted in the figure, crop residue cultivation, crop rotations (cereal–legumes–fallow), cover crops, and minimum soil disturbances

1.3 The Chronicle of Conservation Agriculture (CA)

Soil ploughing or tilling began for the first time at Mesopotamia ~3000 BC (Hillel 1998). Tillage was an important component of agronomic practices in agricultural cultivation (Table 1.1). It was used first time by Edward H. Faulkner in the 1930s in a manuscript called *Plowman's Folly* (Faulkner 1943). With time, the concept of conserving the soil with the help of minimum tillage or zero tillage and keeping the soil covered with crop residues gained popularity. This system of the soil and water conservation was called conservation tillage (Friedrich et al. 2012). The zero tillage (ZT) was introduced in Brazil in the early 1970s to reduce severe water erosion problems (Derpsch 2001). Moreover, escalating fuel prices and severe erosion problems during the 1970s forced the farmers to move in the direction of

Table 1.1 The chronicle list of conservation agriculture worldwide

Year	Development	References
1930	Great dust bowl and start of CA in the USA	Hobbs et al. (2008)
1940	Development of direct seeding machinery, first no-tillage sowing	Friedrich et al. (2012)
1950	First successful demonstration of no-tillage, crops direct sowing in the USA	Harrington (2008)
1960	No-tillage commercial adopted in the USA	Lindwall and Sonntag (2010)
1962	Registration of paraquat as first herbicide for broad-spectrum control of weed	Lindwall and Sonntag (2010)
1962	Long-term experiment on no-till was started in Ohio, USA	Perszewski (2005)
1964	First no-tillage experiments in Australia	Barret et al. (1972)
1966	Demonstration trails on direct seeding system in Germany	Baumer (1970)
1968	First no-tillage trail in Italy	Sartori and Peruzzi (1994)
1970	First time reported about herbicide resistance development in weeds	Ryan (1970)
1974	First time no-tillage demonstrated in Brazil and Argentina	Friedrich et al. (2012)
1975	Book on CA entitled "One straw revolution" by Fukuoka	Fukuoka (1975)
1980	CA introduced and demonstrated on farms in subcontinent	Harrington (2008)
1981	The first national-level conference on no-tillage held in Ponta Grossa, Parana, Brazil	Derpsch (2007)
1990	Development and commercial release of reliable seeding machines	Lindwall and Sonntag (2010)
1990	CA introduced in India, Pakistan, and Bangladesh	Friedrich et al. (2012)
1992	CA research started in China	Derpsch and Friedrich (2009)
1996	Launching of commercial transgenic glyphosate-resistant soybean	Dill (2005)
1998	Identified resistant weed (rigid rye grass) to glyphosate	Powles et al. (1998)
2001	First World Congress on Conservation Agriculture, Madrid, Spain	www.ecaf.org
2002	Introduction of no-tillage systems in Kazakhstan	Derpsch and Friedrich (2009)
2003	2nd World Congress on Conservation Agriculture, Iguassu Falls, Brazil	www.febrapdp.org.br
2005	3rd World Congress on Conservation Agriculture, Nairobi, Kenya	www.act.org.zw/congress Derpsch and Friedrich (2009)
2008	A technical workshop on investing in sustainable crop intensification: the case of improving soil health. FAO, Rome (Italy)	www.fao.org

(continued)

Table 1.1 (continued)

Year	Development	References
2009	4th World Congress on Conservation Agriculture, New Delhi, India	www.icar.org.in
2011	5th World Congress on Conservation Agriculture and 3rd Farming Systems Design Conference (WCCA5 and FSD3), Brisbane Australia	www.wcca2011.org
2014	6th World Congress on Conservation Agriculture (WCCA 6), Winnipeg, Canada	www.ctic.org/wcca
2015	The limit lessons learned from long-term conservation agriculture	Thierfelder et al. (2015)
2016	Role of conservation agriculture to mitigate the climate change	Powlson et al. (2016)

resource-saving agricultural systems (Fig. 1.1). In this scenario, CA was adopted by commercial or large farmers to reduce the drought-driven soil degradation jointly with saving of fuel (Haggblade and Tenbo 2003).

1.4 Objectives of Conservation Agriculture

1.4.1 Minimum Mechanical Soil Disturbance or No-Tillage (NT) or Reduced Tillage or Minimum Tillage

It refers to opening the slot with the help of khurpi or any other equipment and putting the seed in it or direct seeding or broadcasting of seed which leads to minimum soil disturbance. The disturbed area must be less than 15 cm wide or less than 25 % of the cropped area (whichever is lower) (FAO 2014b).

1.4.2 Soil Cover with Organic Materials

Three groups are well known: (1) 30–60 %, (2) >60–90 %, and (3) >90 % cover of ground, calculated without delay following the direct seeding process. CA does not include area <30 % soil cover (Kassam et al. 2015).

1.4.3 Crop Rotation with Pulses or Legumes

Minimum of three different agricultural crops should be included in rotation. However, cyclical cropping of wheat, maize, or rice is not an omission factor for collection of data in this purpose, but rotation is recorded where practiced (Kassam et al. 2015).

1.5 Why Adopt CA Practices?

Excessive/meager rainfall, relatively low temperature, poor and shallow soils, and soil erosion in terms of loss of fertile top soil are the major natural problems that plague the sustained agriculture in the hills. In the IHR, soil and water erosion leads to disastrous loss of soil and severe degradation of farmland. In response to these problems, CA is one of the best management practices (BMPs) to solve all these problems of hill agriculture. There are some important reasons for adoption of CA.

- It is very economical for farmers which helps in cost reduction of fuel and machinery and saving time for operations which allow development of other farming and non-farming harmonizing activities.
- It is very resilient to technical possibilities for sowing, fertilizer application, and weed control (helping in carrying out of operations within time).
- It is important to increase productivity and greater stabilization of productivity in long-term effect.
- It conserves the soil and water against water and wind erosion.
- It increases the agronomic efficiency (AE) and production efficiency (PE).
- It increases the nutrient use efficiency (NUE).
- It increases water use efficiency (WUE) or water productivity and water economy in dryland areas where water scarcity is a serious problem.

The CA aims for improving the yearly input of new organic matter (OM), controlling losses of organic matter in the soil through soil erosion, and reducing organic material decomposition (mineralization) rate in soil or increases immobilization of organic matter. The CA promotes and potential of most soils which have a higher biological activity and diversity of microbes, improved moisture availability in soil and a very high resistance against splash erosion by raindrops (Fig. 1.2). Soil erosion and deforestation are major problems in the Himalayan region, which needs immediate attention and important steps should be taken toward conservation of soil, water, and biodiversity (they are chief agent of soil erosion) and to enhance soil agronomic inputs and protect the land from degradation.

Farmers reassess the agricultural practices to make it more sustainable farming system due to adverse circumstances, and this journey is earmarked by a long period or several years. This long journey period comprises of successive phases, each phase characterized by making use of special practices that progressively integrate more practices and mastery of the three principles (Corbeels et al. 2014).

Climate variability and change pose a substantial risk to livelihoods and economic activities in the hills. Many households that are vulnerable to food insecurity are also highly susceptible to weather shocks and climatic hazards. The CA practices can also add to making farming systems more flexible to climate change. In several cases, CA has been demonstrated to reduce agricultural systems' greenhouse gas (GHG) emissions and increase their job as carbon sequestration. CA can boost the capability of smallholder farmers toward adaption to climate change by decreasing susceptibility to drought and enrichment of natural resource base of the

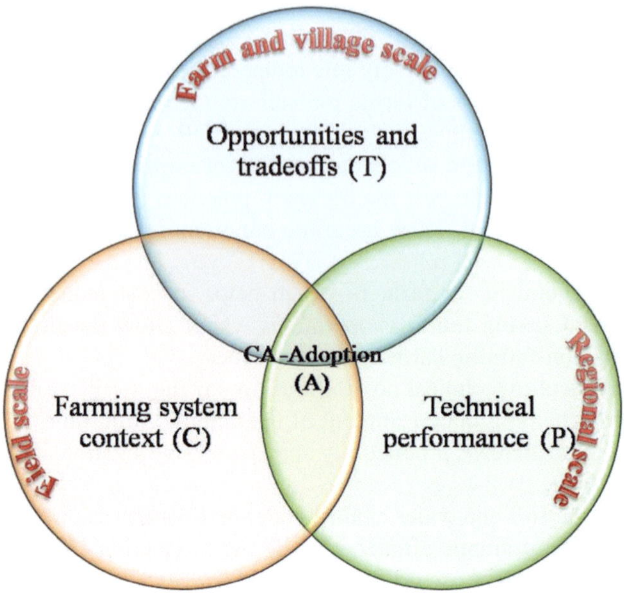

Fig. 1.2 Model delineation of the determinants of adoption of CA (Adopted and modified from Corbeels et al. 2014)

locality on which farm productivity depends. The CA practices such as no-tillage and covering of soil with crops can also used for perennial crops which includes olives, nut crops, vines of grapes, or fruit crop orchards. The CA can also be used for crops in winter, and for conventional rotations with leguminous crops, sunflower, and canola, and in farm crops with irrigation where CA can help optimize irrigation system management for conservation of water, energy, and quality of soil, decrease salinity and alkalinity problems, and increase fertilizer-use efficiency (FEU), water infiltration, and resistance against drought (Colmenero et al. 2013).

The IHR is distinguished with light- to medium-textured soils, undulating topographical features, high rainfall during summer, and physically highly eroded soils due to slopy land. This meticulous grouping of type of weather, land, and tillage features made this key farming area particularly susceptible to soil degradation (Bhattacharyya et al. 2009). The CA is emphasized on enhancing resource-use efficiency (RUE), and production efficiency (PE) is the need of the hour as a sturdy tool for management of natural resources and to gain sustainability in agriculture. The conventional methods through intensive agricultural practices were successful in achieving goals of production for the increasing populations but simultaneously led to degradation of natural resources (Bhattacharyya et al. 2012). The conventional method of cultivation, involving intensive tillage and crop establishment practices, is not only resource (labor, water, and energy) intensive but also emits significant amounts of GHGs (Grace et al. 2003). To meet these challenges, scientists have developed BMPs, including CAPs and new crop rotations with legumes that are more productive and profitable and assumed to be economically feasible and environmental friendly (Gathala et al. 2013; Laik et al. 2014). Different components of CA such as zero tillage, minimum tillage, no-tillage, conservation tillage, dry drill seeding of rice (DSR), and residue retention have been evaluated in cereal systems (Gathala et al. 2011a, b; Kumar et al. 2013; Ghasal et al. 2014; Bhatt and Kukal 2015).

1.6 Difference Between Conventional and Conservation Agriculture

In the conventional agriculture, management practices are extensively used of various tillage operations for plowing of the land for preparation of seedbed and to keep weed down, i.e., moldboard or animal drawn plow or harrowing, drilling, cultivator, etc. These tillage operations are repeated many times; due to this conditions break down the soil structure and destroy pore and soil becomes prone to erosion and leads to heavy cost of time, fuel, and labor. However, conventional tillage exposed soil to air and sunlight which causes oxidation of organic matter and leads to low carbon content in soil which affects soil structure. The oxidation of organic matter releases CO_2 into environment which causes global warming or climate change (Grace et al. 2003). It was observed that soil organic matter decomposes more rapidly in the tropics compared to subtropical and moderate climates because of the higher temperature (Steiner 2002).

The conservation agriculture system involves specific agronomic field operations such as minimum soil disturbance or use of zero tillage or NT, soil cover with green manure or crop residue (mulching), and crop rotation; these management practices are highly beneficial for farm community because it saves fuel, time, and labor as well as it provides good soil structure, porosity, more accumulation of the organic matter in soil which provides better soil aggregation (Table 1.2), water-holding capacity, soil moisture for the long term, nutrient recycling, and transformation. Meanwhile, the conservation tillage improves soil fertility, water, and crop productivity; the no-tillage gives better soil protection than conventional tillage. This happens as the conventional tillage system leaves ~1–5 % of the soil surface covered with crop residues (Hussain et al. 1998).

1.7 Global and Indian Scenario of Conservation Agriculture

Globally, CA is being practiced on 157.8 Mha (Table 1.3). The major CAP countries are the USA (35.61 Mha), Brazil (31.81 Mha), Argentina (29.18 Mha), Canada (18.31 Mha), and Australia (17.69 Mha). In India, adoption of CA is still in the preliminary phase. Over the precedent small number of years, adoption of ZT and CA is slowly increasing and has expanded to cover ~1.5 Mha (Jat et al. 2012).

Table 1.2 Characteristic features between conventional and CA

Conventional agriculture	Conservation agriculture
Cultivation of land with the help of machinery, using science and technology	Minimal disturbances to nature during crop grown processes
Excessive motorized tillage and leads to deterioration of soil pores or structure	No-tillage or considerably reduced tillage (bio-tillage) increases porosity
Higher wind speed and water erosion	Lower wind speed and water erosion
Removal of crop residue from field or burning or uncovered surface	Soil covered with crop residue permanently covered
Low water infiltration	High rate of infiltration of water
Decrease water-holding capacity	Enhance water-holding capacity
FYM/composts added from outside or green manuring (incorporated)	Use of in situ organics/composts or brown manuring/cover crops (stubble retention)
Due plowing established weeds are kills but also stimulates more weed seeds to germinate	Weeds create problem during early stage of implementation but reduce with time
Soil become very compacted due to more use of heavy traffic or machinery	Control of traffic, there is less compaction

Table 1.3 The global area covered under conservation agriculture

Country	Area under CA (10^3 ha)	Country	Area under CA (10^3 ha)
Argentina	29,181	Malawi	65
Australia	17,695	Mexico	41
Azerbaijan	1.3	Morocco	4
Belgium	0.268	Mozambique	152
Bolivia (Plurinational State of)	706	Namibia	0.34
Brazil	31,811	Netherlands	0.5
Canada	18,313	New Zealand	162
Chile	180	Paraguay	3000
China	6670	Portugal	32
Colombia	127	Republic of Moldova	40
Democratic People's Republic of Korea	23	Russian Federation	4500
Finland	200	Slovakia	35
France	200	South Africa	368
Germany	200	Spain	792
Ghana	30	Switzerland	17
Greece	24	Syrian Arab Republic	30
Hungary	5	Tunisia	8
India	1500	Turkey	45
Iraq	15	Ukraine	700
Ireland	0.2	United Kingdom	150

(continued)

Table 1.3 (continued)

Country	Area under CA (10^3 ha)	Country	Area under CA (10^3 ha)
Italy	380	United Republic of Tanzania	25
Kazakhstan	2000	USA	35,613
Kenya	33.1	Uruguay	1072
Kyrgyzstan	0.7	Uzbekistan	2.45
Lebanon	1.2	Venezuela	300
Lesotho	2	Zambia	200
Madagascar	6	Zimbabwe	332
Total	156,991.1		

Source: www.fao.org/ag/ca/6c.html

The main CA-based technologies being adopted is ZT wheat in the rice–wheat (R-W) cropping system in the Indo-Gangetic plains (IGP). It has been seen that the area under wheat (Table 1.3) adopting the ZT drill has been rising fast, and presently ~25–30 % of wheat is ZT in rice–wheat-growing areas of the IGP of India. In addition, planting with raised-bed and laser land leveling are progressively enhancing adoption by the farmers of the northwestern region (Bhatt and Kukal 2015; Bhan and Behera 2014).

1.8 Principles of Conservation Agriculture (CA)

The CA principles are globally applicable to all agricultural landscapes and land uses with locally adapted practices. CA enhances biodiversity and natural biological processes above and below the ground surface (Fig. 1.3). Soil interventions such as mechanical soil disturbance are reduced to an absolute minimum or avoided, and external inputs such as agrochemicals and plant nutrients of mineral or organic origin are applied optimally and in ways and quantities that do not interfere with, or disrupt, the biological processes (FAO 2014a).

CA facilitates good agronomy, such as timely operations, and improves overall land husbandry for rain-fed and irrigated production. Complemented by other known good practices, including the use of quality seeds; integrated pest, nutrient, weed, and water management; etc., CA is a base for sustainable agricultural production. It opens increased options for integration of production sectors, such as crop–livestock integration and the integration of trees and pastures into agricultural landscapes (FAO 2016).

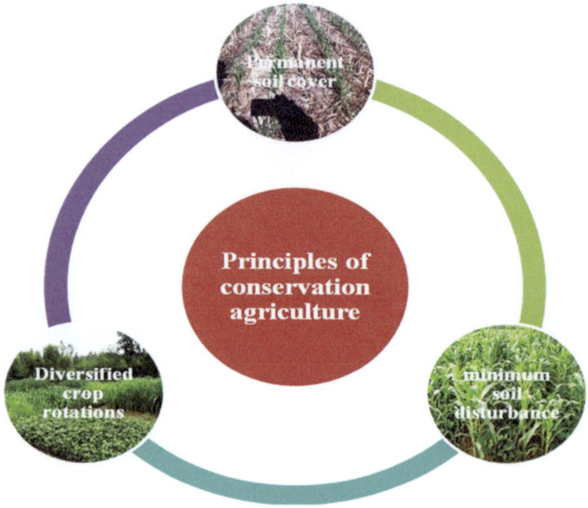

Fig. 1.3 The conservation agriculture covers three basic principles: (i) direct seeding or planting or minimum soil disturbance; (ii) minimum soil disturbance or direct seeding, maintaining soil cover permanently or semi-permanently; and (iii) diversified crop rotation

1.8.1 Direct Seeding or Planting or Minimum Soil Disturbance (Just Enough to Get the Seed into the Ground)

Direct seeding deals with farming systems in which seeds and fertilizer are put directly into undisturbed soil in a single field operation or two separate operations of fertilizing and seeding. Only narrow strips or slots of soil are opened by the equipment openers used to place fertilizer and seed in the soil. Ideally the seed slot is completely covered by mulch or green manure crops again after seeding, and no loose soil should be visible on the surface (Fig. 1.4). No-tillage (NT) or minimum tillage is very important agronomic practice which follows such criteria of minimum soil disturbance, and we can put seed and fertilizer into the soil without loosening it which maintains various physical properties such as soil structure, aggregation, aggregate stability, and porosity. It eases permeability of water and good aeration (exchange of gases) to the roots. It also provides niche of diverse microbial populations. Thus, CA involves the use of minimum or no-tillage along with crop residue retention as this addresses soil physical degradation problems by reducing subsurface compaction (Sayre and Hobbs 2004). Residue retention or incorporation generally increases soil carbon content in the soil (Das et al. 2014).

The stable aggregation is formed by the activity of soil fauna and flora when they are high in number or ample amount is present in the soil. They improve porosity, air permeability, and water infiltration in the soil; this process is known as "biological tillage" and it is not compatible with mechanical tillage. With mechanical tillage soil becomes disturbed, and the biological soil structuring processes vanish. Minimum soil disturbance maintains optimum proportions of respiration gases in the rooting-zone, moderate organic matter oxidation, porosity for water movement, retention, and release and limits the re-exposure of weed seeds and their

Fig. 1.4 As depicted in the figure, three basic principles of conservation agriculture

germination (Kassam and Friedrich 2009). Land preparation for seeding or planting under no-tillage involves precise land leveling with laser land leveler; slashing or rolling the weeds, previous crop residues, or cover crops; or spraying herbicides for weed control and seeding directly through the mulch. Crop residues are retained either completely or to a suitable amount to guarantee the complete soil cover, and fertilizer and amendments are either broadcasted on the soil surface or applied during seeding (Roper et al. 2013).

1.8.2 Maintaining Soil Cover Permanently or Semi-permanently (Using Either Previous Crop Residue or Specifically Growing a Cover Crop for This Purpose)

Maintaining soil cover permanently or semi-permanently with crop residue is essential to conserve the soil and moisture, to provide enough moisture during adverse or drought conditions. Soil cover with crop residue protects the soil from erosion and runoff. It is important for providing the micro- and macroorganisms in

Fig. 1.5 Residue mulch on surface, crop residues can either be incorporated or retained on the surface to protect the soil against erosion and disturbance

the soil with a constant supply of "food" as organic matter which is known as source of energy (Fig. 1.4). It is also important for maintaining and altering the microclimate in the soil for optimal growth and development of soil organisms, including plant roots (FAO 2014a, b). Overall, it improves soil aggregation, soil biological activity and soil biodiversity, and carbon sequestration (Ghosh et al. 2010).

Crop residues on the soil surface protect the soil and reduce its erosion (Boulal et al. 2011; Brouder and Gomez-Macpherson 2014) and runoff (Thomas et al. 2007). Plants are left growing or killed and their residues left to decompose in situ. The primary function of this is to protect organic matter-enriched topsoil against chemical and physical weathering. Plant residues intercept energy from falling raindrops, provide a barrier from strong winds, and moderate temperatures, improving water infiltration and decreasing surface evaporation from sunlight (Fig. 1.5). Surface cover also favors enhanced levels of biological activity by providing food for soil microbes, especially in tropical and subtropical areas. Retaining crop residue and eliminating tillage improved infiltration and soil moisture content where potential evapotranspiration was high and water-holding capacity was low (TerAvest et al. 2015). Soil cover should ideally be above 100 % measured immediately after planting operation; ground cover of less than 30 % is not considered as a conservation agriculture practice (Friedrich et al. 2009).

1.8.2.1 Advantage of Soil Cover

- Ease of infiltration in the soil and soil moisture retention during adverse or harsh conditions increase availability of plant nutrients.

- It provides energy and habitat for diverse soil biota, creation of channels for air and water, biological tillage and substrate for biological activity through the recycling of organic matter, and plant nutrients.
- Increases humus formation.
- Reduces the direct impacts of raindrops which causes soil crusting and splash erosion.
- Resultant reduction of runoff and erosion.
- Soil regeneration is higher than soil degradation.
- It provides better penetration, development of roots, and seedling growth.

1.8.2.2 Means and Practices

- Use of high-quality seeds for high yields as well as high residue production and good root development.
- Integrated management and reduced competition with livestock or other uses, e.g., grow cover crops as green manure which also serves as feed for animals and fodder crops in the rotation.
- Use of various cover crops, especially multipurpose crops, like biological N_2 fixing, soil porosity restoring, pest repellent, etc.
- Optimization of crop rotations in spatial, timing, and economic terms.
- "Targeted" use of herbicides for controlling cover crop and weed development.

1.8.3 Diversified Crop Rotations

Crop rotation refers to growing of different types of crops in the same area in sequenced season. It is important for soil health which reduces allelopathic effect of crops and increases crop productivity and soil fertility and quality. Growing of same crop in same place for many years or monocropping causes adverse effect on soil health, and soil becomes sick. Balanced crop rotation or addition of legume or pulse crops in crop rotation reduces the many dominant pest and disease problems, allelopathic effect of various crops which is harmful for beneficial crops, including the proliferation of insect pests and other harmful bacteria, viruses, and fungi, by increasing the diversity and abundance of beneficial soil microorganism that can help keep pest and disease problems in check. Rotating crops also interrupts the life cycle of many weeds, thereby leading to a reduction in overall weed population. These benefits translate to a typical yield increase of ~10 % of crops grown in rotation, as compared to those grown in monoculture (Vanlauwe et al. 2014). Rotation should involve at least three different crops. The rotation of crops is not only necessary to offer a diverse "diet" to the soil microorganisms, but as they root at different soil depths, they are capable of exploring different soil layers for nutrients and water. Nutrients that have been leached to deeper layers and that are no longer available for the commercial crop can be "recycled" by the crops in rotation. This way the rotation crops functions as biological pumps. Furthermore, a diversity of crops in rotation leads to a diverse soil flora and fauna, as the roots excrete different organic substances that attract different types of bacteria and fungi, which in turn play an important role in the transformation of these substances into available plant nutrients (FAO 2015).

1.8.3.1 Effects of Crop Rotation

- Higher diversity in plant production and thus in human and livestock nutrition.
- It reduces the risk of pest, diseases, and weed infestations.
- Diversified crops are grown which enhance root penetration, porosity, moisture, and nutrient recycling equal transformation in soil profile.
- Enhance biological N_2 fixation through certain plant–soil biota symbionts and improved balance of N/P/K from both organic and mineral sources.
- Increased humus formation.

1.8.3.2 Other Important Features of CA

Ridge Till

The soil remains undisturbed from harvest to planting except for nutrient injection. Planting takes place in a seedbed prepared on ridges with sweeps, disk openers, coulters, or row cleaners. Residue is left on the surface between ridges (Fig. 1.6). Weed control is by herbicides and/or cultivation. Ridges are rebuilt during cultivation.

Fig. 1.6 Ridge furrow plantation under conservation agriculture

Mulch Till/Reduced Tillage/Minimum Tillage

The soil is disturbed prior to planting. Tillage tools such as chisels, field cultivators, disks, sweeps, or blades are used. Weed control is by herbicides and/or cultivation. In non-inversion tillage, soil is disturbed (but not inverted) immediately after harvest to partially incorporate crop residues and promote weed seed germination to provide soil cover during the intercrop period. These weeds are later chemically destroyed (using herbicides) and incorporated at sowing, in one pass, with non-inversion drills (FAO 2015).

Utilization of Green Manures in CA

Green manure cover crops can have significant long-term soil fertility benefits, but smallholder farmers prefer crops that increase food production in the short-term (Bezner-Kerr et al. 2007). Green manure crops (GMC) are grown specifically to help maintain soil fertility, productivity, as well as quality of the soil. Green manure crops secretes some acids (humic acid, acetic acid) which causes as acidifying act to decrease the pH of the soil. GMCs increase soil organic matter (SOM) levels in at least one of two ways – by conserving soil or by adding fresh plant residues to the soil which improves water infiltration and retention, aeration, and other soil characteristics (Fig. 1.7). Leguminous cover crops offer the added advantage of being able to fix N_2 from the atmosphere and add it to the soil, thereby increasing overall nitrogen availability for other crops. Cover crops are usually mowed, sprayed with chemical herbicides, or otherwise killed before or during soil preparation for the next economic crop. It is generally recommended that you leave a week or two between the killing of the cover crop and the planting of a primary crop in order to allow for some decomposition to occur as well as to lessen the effects of nitrogen immobilization and allelopathic effects.

No Burning of Crop Residues

Since crop residues are the principal element of permanent soil cover, they must never be burned or otherwise removed from the soil surface. Rather, plant residues

Fig. 1.7 Sesbania – an excellent green manure crop

Fig. 1.8 Biomass burning of agricultural field residue (stalks and stubble) during harvesting periods, in the Indo-Gangetic plains, has led to substantial emission of trace gases and particles

are left on the soil surface in order to protect organic matter-enriched topsoil from erosion while also adding fresh organic matter upon decomposition (Fig. 1.8). Burning not only creates significant air pollution but also dramatically increases mineralization rates, leading to the rapid depletion of soil organic matter and nutrients from the soil (Kumar et al. 2015). However, in some situations farmers need to think of the tradeoff between removing residues to feed their animals and leaving them to feed the soil.

Integrated Disease and Pest Management

CA depends heavily on enhanced biological activity to help control insect pests and other disease-causing soil microorganisms. Integrated pest management (IPM) entails the judicious use of crop rotations and other beneficial plant substances as well as chemical pesticides, herbicides, and fungicides to control insect pest and disease problems. Over time, the enhanced biological activity and abundance brought on by no-tillage and other CA technologies results in decreased applications of agrochemicals.

Integrated Nutrient Management (INM)

Integrated nutrient management deals with application of organic and inorganic fertilizers, along with farmyard manure (FYM), crop residue, and vermicompost for sustaining soil health for the long term. Feed the soil rather than fertilizing the crop. This will reduce chemical pollution, improve water quality, and maintain the natural ecological integrity of the soil while optimizing crop productivity and economic returns (Vanlauwe et al. 2014).

Adoption of Zero tillage or No-tillage Reduces Use of Fossil Fuel and Greenhouse Gas (GHG) Emissions

Compared to conventional tillage, which often requires four to eight tractor passes in a typical growing season, no-tillage or zero tillage is significantly reducing the

Fig. 1.9 The surface crusting due to heavy machinery load under conventional tillage system

use of tractors and other heavy farm machinery and thus alleviating pollution through tractors and machinery. Furthermore, the increased levels of soil organic matter (SOM) and plant-available nitrogen typically found in CA soils greatly reduce the need for applying large amounts of chemical fertilizers, which causes pollution through N_2 emission in the environment (Fig. 1.9). Thus, overall fossil fuel use and greenhouse gas emissions are greatly reduced under the CA (Niggli et al. 2009).

Decreases Soil Compaction Through Controlled Traffic over Land

According to CA principle, minimum soil disturbance occurs when the number of tractor that passes over a given field is significantly reduced, as compared to conventional tillage systems. However, increased bulk densities have been reported under CA, though this can be corrected by limiting the use of heavy farm machinery when soils are wet and most prone to compaction by converting to a permanent raised-bed system.

Adoption of Agroforestry–Horticulture System

Promoting agroforestry-cum-horticulture system is judicious use of land for fiber, fruit, fuel, and medicinal purposes. Agroforestry (trees on farms) and horticulture crops provide many opportunities for value-added production, particularly in tropical regions, but these technologies are also used as living contour hedges for erosion control, to conserve and enhance biodiversity, and to promote soil carbon sequestration (Hanif et al. 2015).

1.9 Conservation Agriculture and Climate Change Mitigation

In India, more than 600 MT crop residues are produced every year, out of which ~16 % crop residues are being burned in farm in situ. Out of total crop residue being burned, ~40 % residue belongs to rice and ~22 % from wheat. Burning of residue is mainly practiced in Indo-Gangetic Plain region where harvesting of rice–wheat cropping system is done by a combine harvester. Burning of crop residue leads to release of greenhouse gases and loss of nutrients. As an estimate, nearly 1.5 MT plant nutrient are being lost due to burning annually (Jain et al. 2014). Residue burning also kills the beneficial microorganisms present in the soil required for different biological processes. Release of different greenhouse gases leads to increase in atmospheric CO_2 which is the ultimate cause of the changing climate (Chivenge et al. 2007). The CAPs is a powerful mechanism to adapt to climate change by increasing resilience to drought and increasing water and nutrients use efficiencies (WUE and NUE). Climate change is believed to have a great impact on soils. Increasing temperature would increase oxidation of the organic carbon in soil. Its levels will go down further. Incidence of runoff or wind erosion may increase due to increase in extreme events. These changes may reflect themselves in poorer soil fertility, loss of soil biota, water stress, and groundwater depletion (FAO 2011a, b). The ecological benefits from CA are operationally defined as "increase in soil fertility," "retention of soil moisture," "long-term yield increase," "decreasing yield variations," "greater food security," "increased growing time of an area," and "production per hectare" (Pittelkow et al. 2015).

Small-scale farmers are among the first to feel the impacts of climate change because of their greater dependence on the natural environment. They have very limited resources even if they do sell produce, their ability is limited, and any disturbance in production, including climate change, can result in starvation or poverty. Extreme climate variability (drought, floods, sudden rainfall, cloud burst, cyclones, and frost) can spoil the economies and welfare of poor rural small-farm families because they lack technologies, social protection mechanisms (such as benefits, insurance, and savings), and adequate protection of their crops and animals. Adverse effects of climate change continue to be a major threat to rural livelihoods (IPCC 2007). This poses a challenge of developing innovative technologies to improve rural livelihoods and environmental conservation and ensuring adoption of such technologies. Climate change refers to any change in climate over time, whether due to natural variability or as a result of human activity (IPCC 2007). Climate change imposes constraints to development especially among smallholder farmers whose livelihoods mostly depend on rain-fed agriculture (IPCC 2007; Tanner and Mitchell 2008).

Negative impacts of extreme events such as floods and droughts are expected to be high in developing countries especially in rural areas (IPCC 2007). In the situation, where climate change is central in development policies and practices, CA potentially contribute in addressing the challenge of adapting agricultural practices to climate change (Govaerts et al. 2009; FAO 2011a). Climate variability is expected to unfavorably affect agricultural production in IHR. Concerns that climate variability will have an adverse impact on the livelihoods of the rural poor in developing countries have been raised (Below et al. 2010). Due to heavy dependence on rain-fed agriculture (Lybbert and Sumner 2012), the effects of climate variability are expected to be particularly prominent (Lybbert and Sumner 2012). Agricultural production, including access to food in Himalayan region, is projected to become severely compromised by climate change. Therefore, adaptation is a key factor that will shape the future severity of climate change impacts on food production (Lobell et al. 2008). CA takes time to build up enough organic matter and have soils become their own fertilizer; the process does not start to work overnight. But if producers make it through the first few years of production, results will start to become more satisfactory. Improved soil quality and improved nutrient cycling with conservation agriculture will improve the resilience of crops to adapt to changes in local climate change, while drought tolerance can be increased in some areas with CA (Hobbs and Govaerts 2010).

Several studies showed that the climate change and variability is occurring in IHR in various forms, for example, dry spells occurring in drought-prone areas, floods in high-rainfall areas, and delayed onset or early cessation of rainfall. Considering topographic variability which correlates with rainfall pattern, climate variability and change is expected to continue to compromise agricultural production if no proper adaptation and mitigation strategies are put in place. Tillage is claimed to be one of main sources of GHGs emitted in the atmosphere resulting in climate change which, in turn, negatively affects agricultural production (Six et al. 2004); thus, promoting agricultural practices that mitigate climate change by reducing GHGs emissions is important (Hobbs and Govaerts 2010).

The hill agriculture is facing a shift from subsistence agriculture to profitable crop production. The IHR have a great potential of conservation of natural resources and environmental preservation; adoption of CA would face multiple tradeoffs. CA require a long-time planning and commitment to resources protection, while agriculture in Himalayas is still of subsistence; thus, to adopt CA requires an alternative source of food while waiting for the increased yield as results of effect of CA. Since the diverse agroclimatic conditions of hills impart a unique advantage and competitive edge over a plain region, conservation agriculture (CA) in the hill farming systems has to be different compared to plains in managing the different production systems. Thus, in view of widespread resource degradation and to reduce production costs, increase profitability, and make agriculture more competitive, CA can prove a vital tool in addressing these issues directly.

From this point of view, it is of great importance to research to translate CA into field practices attractive to farmers. Smallholder rain-fed crop production in semi-arid areas is characterized by low residue production levels (Mashingaidze et al. 2012). Smallholder crop–livestock systems face tradeoffs among various options for crop residue use (Valbuena et al. 2012).

The lack of adoption of CA in many countries has been attributed to farmers conservative mind-set (Sanginga and Woomer 2009), but as mentioned earlier, farmers are most interested in agricultural technologies that can directly increase yield for food security. In addition, weed control in CA is a greater challenge than in conventional agriculture because there is no weed seed burial by tillage operations, and soil-applied herbicides are not incorporated, resulting in reduced efficacy (Chauhan et al. 2012). CA has been, as well, reported as complex and requires intensive community-based extension; thus, adoption rates have been low (Li et al. 2011). To encourage adoption of CA in Himalayan regions requires increasing awareness to farmers with stronger and wider demonstrations.

1.10 Advantage of Conservation Agriculture

Conservation agriculture (CA) improves following parameters:

1.10.1 Soil Properties: Soil Aggregation, Aggregate Stability, and Structure

Soil aggregate stability deals with the susceptibility of soil to change under natural or anthropogenic activities. It is important for conservation of soil due to water erosion because erodibility of soils is directly related to aggregate stability (Fig. 1.10). There is medium to high aggregate correlation between aggregate stability in water, aggregate size, and total organic carbon content. The conventional tillage promotes loss of soil organic matter, which leads to disruption of soil aggregates leading to soil erosion. CA approaches that leave more crop residues on the soil surface generally allow improvements in soil aggregation and aggregate stability. It also protects surface aggregates against the effects of raindrops or splash erosion. Research on CA showed that no-tillage with stubbles retained treatment and had more water-stable aggregation (Fig. 1.11). CA characterized by no-tillage,

Fig. 1.10 A better soil aggregation can result in good water infiltration and nutrients storage

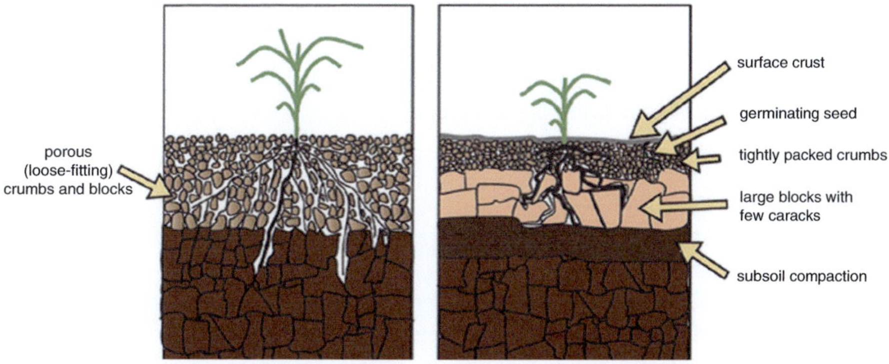

Fig. 1.11 The plants growing is soil with good soil structure and right picher soil having three types of compaction: surface crusting, plow layer/surface compaction, and subsoil compaction

and crop residue retention is helpful for soil aggregating and the aggregate stability (Li et al. 2011). Soil structure research showed that the lack of annual tillage, as provided under continuous no-tillage management, encouraged the development and persistence of a soil surface horizon rich in stable aggregates in semiarid Morocco (Mrabet and El Brahli 2005). Mrabet (2002) observed higher mean weight diameter (MWD) and aggregation index (AI) at the surface (0–7 cm) of a self-mulching swelling clay soil under no-tillage than under conventional tillage. CA requires adequate soil cover to perform critical functions including improving infiltration, plant-available water, and aggregate stability (Palm et al. 2013). It also enhances the proportion of micropores in the soil, increases water-holding capacity, and reduces evaporation from the soil surface and upper levels of soil (Kassam et al. 2009; Palm et al. 2013).

1.10.2 Soil Moisture

The CA emphasizes on soil cover with crop residues and mulches help in conserving moisture and soil. It generally requires significantly less water use due to increased infiltration and enhanced water-holding capacity from crop residues left on the soil surface. Mulches also protect the soil surface from extreme temperatures and greatly reduce surface evaporation, which is particularly important in tropical and subtropical climates. From the several studies, it has been found that conservation agriculture saves 20–30 % of irrigation water because of lower evaporation losses from surface as surface is covered with residues (Jat et al. 2012). In CA, the soil moisture is conserved and more water is available for crops (Verhulst et al. 2010). Tillage can influence the evaporation process because of its effects on the physical properties of the soil surface (albedo, roughness, mulch) and on the hydraulic properties (Mrabet 2007). CA can provide instant advantages to small farmers by increasing rainwater harvesting and serving as a alleviation strategy against climate change and late or variable rainfall (Rockström et al. 2009; Thierfelder and Wall 2010; Pittelkow et al. 2014). In, IHR, the advantages of CA are most salient during drought, when the risk of total crop failure is significantly reduced due to increased water use efficiency.

1.10.3 Increases Water Infiltration and Hydraulic Conductivity

The CA provide ease of infiltration because minimum disturbing of soil which have better soil pores structure or porosity and enhances hydraulic conductivity. Hence, the effect of tillage and residue cover on water infiltration is probably due to changes in soil structure (Mrabet 2007). Long-term studies in Zimbabwe and Zambia reported that eliminating tillage and retaining residue can quadruple rainwater infiltration relative to conventional tillage (Thierfelder et al. 2012a, b). Increased soil water content and crop yields have been reported in Kenya, Zimbabwe, Zambia, and Malawi when these two conservation agriculture principles are implemented (Gicheru et al. 2006; Thierfelder et al. 2012c). Mulch cover or crop residue provides benefits in improved water infiltration and reduced soil surface evaporation especially under dry or moisture-limited conditions (Turmel et al. 2014). Rainfall infiltration is improved under no-tillage systems, which increases the amount of soil water available for plants for heavy textured soils (Mrabet 2007). However, a few researchers, showed that infiltration and hydraulic conductivity, was lower under no-tillage than conventional tillage. In brief, CA improves hydraulic conductivity and infiltration and reduces evaporation rate, runoff, and soil erosion by crop residue.

1.10.4 Reducing Soil Runoff and Water Erosion

The runoff and soil loss are major concern of slopy crop land of Himalayan regions, especially those with unstable aggregates in the surface horizon both from the standpoint of sustainability and offsite environmental damage, so it has been, and continues to be, a very important research area. CA with crop residue mulch can provide soil cover to reduce rain impact and provide barriers against runoff (Franzluebbers 2002); this will help to increase moisture infiltration and decrease soil detachment. Research on the western Loess Plateau of China reported that under a wheat–pea rotation system, runoff was alleviated and runoff intensity was reduced with no-tillage with stubble retention; soil loss from erosion was reduced by ~62 % (Zhao et al. 2007). In conclusion, CA could reduce soil detachment and enhance water infiltration that implies a decrease of water runoff; consequently, soil erosion would be reduced (Table 1.4).

Table 1.4 Different tillage systems, runoff, and infiltration

	Five year means as % of rainfall	
Tillage system	Seasonal runoff	Seasonal infiltration
Conventional tillage	20	80
Conservation tillage	1	99

Marongwe et al. (2012)

Fig. 1.12 The effect of organic matter (OM) on the same soil type managed using conventional plow tillage (*left*) or zone tillage for 10 years (*right*). Soil with good tilth is crumbly, well structured, and dark with OM and has no large and hard clods

1.10.5 Increases Soil Organic Carbon (SOC) and Organic Matter

Soil organic matter (SOM) is a vital determinant of soil fertility, productivity, and sustainability and is a key indicator of soil quality (Chivenge et al. 2007). It is also provides essential nutrients for crops and maintains soil aggregation and stability (Fig. 1.12). The CAPs increase SOM content and nutrient availability by utilizing the previous crop residues or growing green manure or cover crops (GMCs) and keeping these residues as a surface mulch rather than burning. There is evidence that elimination of tillage can result in sequestration of carbon (Bessam and Mrabet 2003). Long-term field experiments on CA showed that organic carbon was enhanced by direct drill stubble retained on the soil surface. The contents of OC in soil can also be increased by CA by keeping crop residues containing carbon and nutrients at the soil surface layer (Moreno et al. 2006). NT improves soil quality (soil function), OC, aggregation, conservation of soil, evaporation of water, and soil structure (Araya et al. 2012). Mrabet et al. (2001) observed that increases in SOM of ~14 % with NT, and 3.3 % with conventional tillage over an 11-year period. Generally, there is a trend toward a stratification of SOM at the surface under NT (Franzluebbers 2002). CA improves agricultural productivity, reducing moisture deficits, enhancing nutrient cycling and soil fertility, reducing soil erosion, and increasing carbon sequestration (Kassam et al. 2009; Palm et al. 2013).

1.10.6 Soil Nutrients

Conservation tillage, crop residue management, and crop rotation with pulses or legumes have a significant effect on nutrient distribution, recycling, and transformation in soils (Verhulst et al. 2010). Due to its impact on soil organic carbon and nitrogen mineralization, CA can influence soil nitrogen availability (Bradford and Peterson 2000). In general, soil fertility is built up over time under CA, and fewer fertilizer amendments are required to achieve optimal yields over time. A positive effect of NT has also been recorded on the availability of phosphorus and potassium. According to Mrabet et al. (2001), no-tillage soil had a higher concentration of phosphorus and potassium near the soil surface than tilled soil. Rahman et al. (2008) reported that exchangeable Ca, Mg, and K were significantly higher in the surface soil under NT compared to the plowed soil. Soil nutrient supplies and cycling are enhanced by the biochemical decomposition of organic crop residues at the soil surface that are also important for feeding the soil microbes.

1.10.7 Labor

Soil tillage is among all farming operations the single most energy consuming, and thus, in mechanized agriculture it is an air-polluting operation. In the CA, farmers can save between 30 and 40 % of time, labor and, fossil fuels as compared to conventional cropping.

1.10.8 Soil Biodiversity

Soils under conservation tillage are more favored than conventional tillage in terms of soil fauna and flora activities and biological properties improvement (Busari et al. 2015). Cookson et al. (2008) observed a decreased fungal biomass and increased bacterial biomass with increasing tillage disturbance (Fig. 1.13). They also reported alteration in the composition and substrate utilization of the microbial community with distinct substrate utilization in no-tillage soil. It provides a truly sustainable production system, not only conserving but also enhancing the natural resources and increasing the variety of soil biota, fauna, and flora (including wildlife), and also enhances microbial count in agricultural production systems without sacrificing yields on high production levels. The conservation agriculture systems increase microbial biomass by ~83 % and MBC: total carbon ratio by ~23 % and decrease the metabolic quotient by ~32 %. CA improves agricultural productivity, breaking up pest and disease cycles, minimizing weeds, and enhancing nutrient cycling, soil fertility, and biodiversity of soil biota (Kassam et al. 2009; Palm et al. 2013).

Fig. 1.13 Earthworms: the "unheralded soldiers of mankind" and "farmer's friend"

1.10.9 Reduced Incidence of Weeds

The lower weed infestation is reported in conservation agriculture as there is no soil disturbance and weed seed present in lower soil surface cannot germinate. Several studies tend to indicate reduced incidence of *Phalaris minor*, a major weed in wheat, when zero tillage or NT is adopted, resulting in reduced in use of herbicides (Malik et al. 2005).

1.10.10 Economic Benefits

Several studies observed that the cost of wheat production is reduced by Rs. 2000–3000 ($ 33–50) per hectare as compared to conventional cropping. Cost reduction is attributed to savings on account of diesel, labor, and input costs, particularly herbicides (Sharma et al. 2005). Farmers using CA practices typically reported higher productivity (up to 45–48 % higher) with lesser water, fertilizer, and labor inputs, thereby resulting in higher overall farm profits (Table 1.5). Increased agricultural productivity derived from these ecological services and lower production costs have led to widespread adoption of conservation agriculture throughout North and South America and Australia, where agricultural systems are generally large scale and highly mechanized (Kassam et al. 2009).

Table 1.5 Cost comparison between conventional agriculture and CA

Inputs	Tradition agriculture (USD ha^{-1})	Conservation agriculture (USD ha^{-1})	Cost saving (%)
Fuel	75	25	66.67
Depreciation	115	65	43.47
Maintenance	22	10	54.55
Pesticides	35	45	−28.57
Total costs	247	145	41.30

Source: Farooq and Siddique (2016)

1.11 Disadvantage of Conservation Agriculture

- Initially high investment is required for purchasing of specialized sowing and/or planting implements, and requirement of technical knowledge for better management is short-term drawback of the conservation agriculture.
- Reduction in yield during initial years of adoption, higher cost of inputs, higher labor for weeding, competition for crop residue between animal feed and mulching and its promotion as universal system without considering agroecological and socioeconomic conditions (Baudron et al. 2012; Brouder and Gomez-Macpherson 2013; Corbeels et al. 2014; Giller et al. 2009; Valbuena et al. 2012).
- A particularly important gap is the lack of knowledge on locally adapted cover crops that produce good biomass under the prevailing conditions.
- The success or failure of CA depends greatly on the flexibility and creativity of the farmer and extension and researchers of a particular location.

1.12 Important Barriers for the Adoption of Conservation Agriculture

The CAPs are facing a great challenge between the scientific community and the farmers to change the mind-set and explore the opportunities that offer for natural resource management. The CA is also considered as way to sustainable agriculture (Sangar et al. 2005).A mental change of farmers, technicians, extensionists, and researchers away from conventional method which is soil degrading tillage toward natural resource-conserving systems like no-tillage is the need of the hour (Derpsch 2001). Hobbs and Govaerts (2010), however, reported that probably the most important factor in the adoption of CA is change of mental attitude of farmer to tillage. The following are a few important constraints which restrict wide-scale adoption of CA:

- Lack of appropriate machinery especially for small and medium farmers: Although significant efforts have been made in developing and promoting machinery for seeding wheat in no-tillage systems, successful adoption will require rapid effort in developing, standardizing, and promoting quality implements aimed over a range of crops and cropping sequences (Bhan and Behera 2014).

- The great use of crop residues for animal feed and fuel: Specially under rain-fed or dryland situations, farmers face a scarcity of fodder due to less biomass production of different crops. There is competition between CAPs and animal feeding for crop residue. This is a major problem for adoption of CA (Bhan and Behera 2014).
- Burning of crop residues: For timely sowing of the next crop and without machinery for sowing under CA systems, farmers prefer to sow the crop in time by burning the residue. This has become a common feature in the rice–wheat system in north India. This creates environmental problems for the region (Tripathi et al. 2013).
- Lack of knowledge about the potential of CA to agriculture leaders, extension agents, and farmers: This implies that the whole range of practices in CA, including planting and harvesting, water and nutrient management, diseases and pest control, etc., need to be evolved, evaluated, and matched in the context of new systems (Bhan and Behera 2014).
- Skilled and scientific manpower: Managing CA systems, need for enhanced capacity of researchers to face problems from a systems perspective and to be able to work in close relationship with farmers and other stakeholders. Strengthened knowledge and information sharing mechanisms are urgently needed.

Some area-specific constraints in semiarid areas during the transformation to CA system relate to initial low supply of crop residues and vegetation biomass for soil mulch cover development, to initial short-term competition for crop residue as animal feed, and to weeding during initial phase while soil mulch cover and integrated weed management practice is being established. However, farmers, those are really interested in adoption of CA, are finding solution for these problems locally. Lots of this type of cases have been reported for small and large farms in different parts of the world. It is said that convincing farmers that crop cultivation is also possible with reduced tillage is major problem in promotion and adoption of CA system on a wide scale (Bhan and Behera 2014).

1.13 Policy Issues Required for CA

The CA implies a radical transformation from traditional agriculture. There is need of policy intervention to integrate with existing technologies and how policy issues promote or deter CA. Some of the important policies needed for promotion of CA are given below.

- Scaling up CA practices: More and more support from stakeholders including policy and decision makers at the local, regional, and national levels will facilitate promotion of CA and help farmers to earn more profit. In couple of the last decade, much research work has been done on CA especially in India. However, its expansion to end user is very limited. One of the possible reasons for limited adoption of the technology by the farmers was mind-set of a large group of farmers about tillage (Hobbs and Govaerts 2010). Under such situations, during initial phase farmer's participatory on-farm research to test the technology followed by wide-scale demonstration in subsequent years is required. In India, efforts are being started through a network research project for on-farm evaluation and demonstration of CA technology for its expansion (Bhan and Behera 2014).

- The shift in focus from food security to livelihood security: Myopic "food security" policy based on cereal production must now replace a well-articulated policy goal for livelihood security. This will help the horizontal or vertical diversification of major cropping systems in the IGPs, as continuous use of rice–wheat cropping system has overexploited the natural base by the use of conventional tillage practices. Policy intervention also affects cropping pattern and crop diversification of the region. The policies that affect crop horizontal and vertical diversification are pricing policy, tax and tariff policies, trade policies, and policies on public expenditure and agrarian reforms (Behera et al. 2007).
- Developing, improving, standardizing machinery for sowing, fertilizer application, and harvesting that require least soil disturbance in crop residue management under different soil conditions will be key to success of CA. In hilly areas, small landholders' bullock-drawn equipment will have highly importance. In these situations, the subsidy support from national or local government bodies for manufacturing low-cost implements will help in the expansion of CA practices.
- There is a need to develop and evaluate crop varieties suitable to CA technologies. Farmer's participatory research would appear promising for identifying and developing crop cultivars performing well to a particular region.
- Training on CA are also required for capacity building through policy intervention. Lack of trained technical person at field level is major drawback in adoption and expansion of CA. Efforts to perfectly train all agricultural extension personnel on CA should be made in relevant departments.
- Support for the adaptation and validation of CA technologies in local environments: To promote principles and practices of CA at local conditions, adaptive research is the need of the hour. This should be done in partnership with local people and other stakeholders. The resource-poor and smallholder farmers in India do not have economic access to new seeds, herbicides and seeding machineries etc., this calls for policy frame work to make easily available critical inputs (Sharma et al. 2012).
- Support the manufacturing of CA machinery and ensure its availability: equipments required in CA practices are very complex as well as expensive for a farmer to purchase it. Therefore, a farmer may develop a local hire service industry by providing equipment to hire on custom basis. The new machineries, viz., happy seeder, turbo seeder, laser land leveler, etc., are found useful for CA

practices, but these machines are not suitable for small and marginal farmers groups. These machines need high horse power (>50 hp) tractors for better functioning in field conditions. Small and marginal farmers having small holdings and economic limitations are unable to afford for such heavy machines. They need policy interventions for manufacturing of smaller versions of these machines at the local level (Bhan and Behera 2014).

- Promote payments for environmental services (PES) and fines for faulty practices: by the adoption of CA practices, farmers improve the quality of environment by storing carbon in soil, reducing runoff and soil erosion by mulching and promotion of groundwater recharge. It provides eco-friendly services, so farmers should be rewarded for such services, which improves quality life of all living beings on the planet.

1.14 Conclusions

CA systems are most appropriate way to conserve soil and water resources in slopy area of Indian Himalayan Region (IHR). It is also proven and needs to be promoted as vigorously and widely as possible to compensate the effect of climate change. Factors involved in promoting CA have often not taken into account perceptions of smallholder farmers of climate change and CA as an adaptation strategy. This chapter integrates the smallholder farmers' perceptions of climate change in Himalayan and CA. Smallholder farmers' perceptions related to floods and droughts were significantly associated with adoption of CA.

Acknowledgments We thank the editors and anonymous reviewers for their constructive comments, which helped us to improve the manuscript.

References

Araya T, Cornelis WM, Nyssen J, Govaerts B, Getnet F, Bauer H, Amare K, Raes D, Haile M, Deckers J (2012) Medium-term effects of conservation agriculture based cropping systems for sustainable soil and water management and crop productivity in the Ethiopian highlands. Field Crops Res 132:53–62

Barret DWA, Wiles TL, Barker MR (1972) Spray-seed with bipyridyls in Western Australia. In: Proceedings no-tillage systems symposium, Columbus, 21–22 Feb 1972, pp 83–92

Baudron F, Andersson JA, Corbeels M, Giller KE (2012) Failing to yield? Ploughs, conservation agriculture and the problem of agricultural intensification: an example from the Zambezi valley, Zimbabwe. J Dev Stud 48:393–412

Baumer K (1970) First experience with direct drilling in Germany. Neth J Agric Sci 18:283–292

Behera UK, Sharma AR, Mahapatra IC (2007) Crop diversification for efficient resource management in India: problems, prospects and policy. J Sustain Agric 30(3):97–217

Below T, Artner A, Siebert R, Sieber S (2010) Micro-level practices to adapt to climate change for African small-scale farmers. A review of selected literature. Environ Prod Technol Division, p 953

Bessam F, Mrabet R (2003) Long-term changes in soil organic matter under conventional and no-tillage systems in semiarid Morocco. Soil Use Manag 19:139–143

Bezner-Kerr R, Snapp SS, Chirwa M, Shumba L, Msachi R (2007) Participatory research on legume diversification with Malawian smallholder farmers for improved human nutrition and soil fertility. Exp Agric 43:437–453

Bhan S, Behera UK (2014) Conservation agriculture in India – problems, prospects and policy issues. Int Soil Water Conserv Res 2(4):1–12

Bhatt R, Kukal SS (2015) Direct seeded rice in South Asia. Sustain Agric Rev 18:217–252

Bhattacharyya R, Ved-Prakash RB, Kundu S, Srivastva AK, Gupta HS (2009) Soil aggregation and organic matter in a sandy clay loam soil of the Indian Himalayas under different tillage and crop regimes. Agric Ecosyst Environ 132:126–134

Bhattacharyya R, Tuti MD, Kundu S, Bisht JK, Bhatt JC (2012) Conservation tillage impacts on soil aggregation and carbon pools in a sandy clay loam soil of the Indian Himalayas. Soil Sci Soc Am J 76:617–627

Boulal H, Mateos L, Gómez-Macpherson H (2011) Soil management and traffic effects on infiltration of irrigation water applied using sprinklers. Irrig Sci 29:403–412

Bradford JM, Peterson GA (2000) Conservation tillage. In: Summner ME (ed) Handbook of soil science. CRC Press, Boca Raton, pp G247–G270

Brouder SM, Gomez-Macpherson H (2013) The impact of conservation agriculture on smallholder agricultural yields: a scoping review of the evidence. Agric Ecosyst Environ 187:11–32

Brouder SM, Gomez-Macpherson H (2014) The impact of conservation agriculture on smallholder agricultural yields: a scoping review of the evidence. Agric Ecosyst Environ 187:11–32

Busari MA, Kukal SS, Kaur A, Bhatt R, Dulazi AA (2015) Conservation tillage impacts on soil, crop and the environment. Int Soil Water Conserv Res 3:119–129

Cairns JE, Hellin J, Sonder K, Araus JL, MacRobert JF, Thierfelder C, Prasanna BM (2013) Adapting maize production to climate change in sub-Saharan Africa. Food Secur 5:345–360

Chauhan BS, Singh RG, Mahajan G (2012) Ecology and management of weeds under conservation agriculture: a review. Crop Prot 38:57–65

Chivenge P, Murwira H, Giller K, Mapfumo P, Six J (2007) Long-term impact of reduced tillage and residue management on soil carbon stabilization: implications for conservation agriculture on contrasting soils. Soil Tillage Res 94(2):328–337

Colmenero MR, Bienes R, Eldridge DJ, Marques MJ (2013) Vegetation cover reduces erosion and enhances soil organic carbon in a vineyard in the central Spain. Catena 104:153–160

Cookson WR, Murphy DV, Roper MM (2008) Characterizing the relationships between soil organic matter components and microbial function and composition along a tillage disturbance gradient. Soil Biol Biochem 40:763–777

Corbeels M, Graaff J, Ndah TH, Penot E, Baudron F, Naudin K, Andrieu N, Chirat G, Schuler J, Nyagumbo I, Rusinamhodzi L, Traore K, Mzoba HD, Adolwa IS (2014) Understanding the impact and adoption of conservation agriculture in Africa: a multi-scale analysis. Agric Ecosyst Environ 187:155–170

Das B, Chakraborty D, Singh VK, Aggarwal P, Singh R, Dwivedi BS, Mishra RP (2014) Effect of integrated nutrient management practice on soil aggregate properties, its stability and aggregate-associated carbon content in an intensive rice-wheat system. Soil Tillage Res 136:9–18

Derpsch R (2001) Keynote: frontiers in conservation tillage and advances in conservation practice. In: Stott DE, Mohtar RH, Steinhart GC (eds) Sustaining the global farm. Selected papers from the 10th International Soil Conservation Organisation Meeting held May 24–29, 1999 at Purdue University and the USSA-ARS National Soil Erosion Research Laboratory

Derpsch R (2007) Historical review of no-tillage cultivation of crops. http://www.rolf-derpsch.com/notill.htm#1#1. Accessed 19 May 2014

Derpsch R, Friedrich T (2009) Global overview of conservation agriculture adoption. In: Lead Papers 4th World Congress on Conservation Agriculture. World Congress on Conservation agriculture, New Delhi

Dill GM (2005) Glyphosate-resistant crops: history, status and future. Pest Manag Sci 1:219–224

FAO (2011a) The State of food and agriculture. Women in agriculture. Closing the gender gap for development. Food and Agriculture Organisation of the United Nations, Rome, 2

FAO (2011b) Climatic risk analysis in conservation agriculture in varied biophysical and socio-economic settings of southern Africa. Food and Agriculture Organisation of the United Nations, Rome

FAO (2012) Food and Agriculture Organization of the United Nations, 2012. Available online at http://www.fao.org/ag/ca/6c.html

FAO (2014a) What is conservation agriculture? FAO CA website (http://www.fao.org/ag/ca/1a.html)

FAO (2014b) CA adoption worldwide, FAO-CA website (http://www.fao.org/ag/ca/6c.html)

FAO (2015) Food and Agriculture Organization of the United Nations, conservation agriculture. Available online at http://www.fao.org/ag/ca/6c.html

FAO (2016) Food and Agriculture Organization of the United Nations, Conservation agriculture. Available online at http://www.fao.org/ag/ca/6c.html.

Farooq M, Siddique KHM (2016) Conservation agriculture: concepts, brief history, and impacts on agricultural systems

Faulkner EH (1943) Plowmans folly. Micheal Joseph, London

Franzluebbers AJ (2002) Soil organic matter stratification ratio as an indicator of soil quality. Soil Tillage Res 66:95–106

Friedrich, T., Kassam, A.H. and Shaxson, F. (2009) Conservation Agriculture. In: Agriculture for Developing Countries. Science and Technology Options Assessment(STOA) Project. Karlsruhe, Germany: EuropeanTechnology Assessment Group

Friedrich T, Derpsch R, Kassam AH (2012) Global overview of the spread of conservation agriculture. Field Action Sci Rep 6:1–7

Fukuoka M (1975) One straw revolution. Rodale, Tokyo, p 138

Gathala MK, Ladha JK, Kumar V, Saharawat YS, Kumar V, Sharma PK, Sharma S, Pathak H (2011a) Tillage and crop establishment affects sustainability of South Asian rice–wheat system. Agron J 103:961–971

Gathala MK, Ladha JK, Saharawat YS, Kumar V, Kumar V, Sharma PK (2011b) Effect of tillage and crop establishment methods on physical properties of a medium-textured soil under a seven-year rice–wheat rotation. Soil Sci Soc Am 75:1851–1862

Gathala M, Kumar V, Sharma PC, Saharawat Y, Jat HS, Singh M, Kumar A, Jat ML, Humphreys E, Sharma DK, Sharma S, Ladha JK (2013) Optimizing intensive cereal-based cropping systems addressing current and future drivers of agricultural change in the northwestern Indo-Gangetic Plains of India. Agric Ecosyst Environ 177:85–97

Ghasal PC, Bir D, Yadav A, Prakash V, Verma RK (2014) Productivity and profitability of rice varieties under different methods of establishment. Ann Agric Res 35(3):298–303

Ghosh PK, Das A, Saha R, Kharkrang E, Tripathy AK, Munda GC, Ngachan SV (2010) Conservation agriculture towards achieving food security in north east India. Curr Sci 99(7):915–921

Gicheru PT, Gachene CKK, Mbuvi JP (2006) Effects of soil management practices and tillage systems on soil moisture conservation and maize yield on a sandy loam in semiarid Kenya. J Sustain Agric 27:77–92

Giller KE, Witter E, Corbeels M, Tittonell P (2009) Conservation agriculture and smallholder farming in Africa: the heretic's view. Field Crops Res 114(1):23–34

Govaerts B, Verhulst N, Castellanos-Navarrete A, Sayre KD, Dixon J, Dendooven L (2009) Conservation agriculture and soil carbon sequestration: between myth and farmer reality. Crit Rev Plant Sci 28:97–122

Grace PR, Jain MC, Harrington L, Philip G (2003) Long-term sustainability of the tropical and subtropical rice–wheat system: an environmental perspective. In: Ladha JK, Hill J, Gupta RK, Duxbury J, Buresh RJ (eds) Improving the productivity and sustainability of rice–wheat systems: issues and impact, ASA Special Publication 65. Agronomy Society of America, Madison, pp 27–43

Hanif MA, Bari MS, Abiar Rahman M (2015) Potentiality of carbon sequestration by agroforestry species in Bangladesh. Res Crops 16(3):562–567

Harrington LW (2008) A brief history of conservation agriculture in Latin America, South Asia and Sub-Saharan Africa. PACA, 1st floor, NASC Complex, DPS Marg, Pusa

Hillel D (1998) Environmental soil physics. Academic, San Digeo

Hobbs PR, Govaerts B (2010) How conservation agriculture can contribute to buffering climate change. In: Reynolds MP (ed) Climate change and crop production. CAB International, Cambridge, pp 177–199
Hobbs PR, Sayre K, Gupta R (2008) The role of conservation agriculture in sustainable agriculture. Philos Trans R Soc 363:543–555
Hussain I, Olson KR, Ebelhar SA (1998) Impacts of tillage and no-till on production of maize and soybean on an eroded illinois silt loam soil. Soil Till Res 52:37–49
IPCC (2007) Climate change 2007: impacts, adaptation and vulnerability. Cambridge University Press, Cambridge
Jain N, Bhatia A, Pathak H (2014) Emission of air pollutants from crop residue burning in India. Aerosol Air Qual Res 14:422–430
Jat ML, Malik RK, Saharawat YS, Gupta R, Bhag M, Raj Paroda (2012) Proceedings of Regional Dialogue on Conservation Agricultural in South Asia, New Delhi, India, APAARI, CIMMYT, ICAR, p 32
Kassam AH, Friedrich T (2009) Perspectives on nutrient management in conservation agriculture. Invited paper, IV World Congress on Conservation Agriculture, 4–7 February 2009, New Delhi
Kassam AH, Friedrich T, Shaxson F, Pretty J (2009) The spread of conservation agriculture: justification, sustainability and uptake. Int J Agric Sustain 7(4):1–29
Kassam AH, FriedrichT, Derpsch R, Kienzle J (2015) Overview of theworldwide spread of Conservation Agriculture. Field Actions Science Reports, Special Issue 6, pp 1–7
Kumar V, Saharawat YS, Gathala MK, Jat AS, Singh SK, Choudhary N, Jat ML (2013) Effect of different tillage and seeding methods on energy use efficiency and productivity of wheat in the Indo-Gangetic Plains. Field Crop Res 142:1–8
Kumar A, Biswas AK, Dwivedi SK, Somasundaram J, Srivastava LK, Mishra VN (2015) Conservation agriculture: concept, principles and methodologies. Glob J Multidiscip Stud 4 (9):138–160
Laik R, Sharma S, Idris M, Singh AK, Singh SS, Bhatt BP, Saharawat YS, Humphreys E, Ladha JK (2014) Integration of conservation agriculture with best management practices for improving system performance of the rice– wheat rotation in the Eastern Indo-Gangetic Plains of India. Agric Ecosyst Environ 195:68–82
Lal R (2013) Soils and ecosystem services. In: Lal R, Lorenz K, Hüttl RF, Schneider BU, von Braun J (eds) Ecosystem services and carbon sequestration in the biosphere. Springer, Dordrecht, pp 11–38
Li L, Huang, Zhang R, Bill B, Guangdi L, Kwong YC (2011) Benefits of conservation agriculture on soil and water conservation and its progress in China. Agric Sci China 10(6):850–859
Lindwall CW, Sonntag B (2010) landscape transformed: the history of conservation tillage and direct seeding, knowledge impact in society. University of Saskatchewan, Saskatoon
Lobell DB, Burke MB, Tebaldi C, Mastrandrea MD, Falcon WP, Naylor RL (2008) Prioritizing climate change adaptation needs for food security in 2030. Science 319(5863):607–610
Lybbert TJ, Sumner DA (2012) Agricultural technologies for climate change in developing countries: policy options for innovation and technology diffusion. Food Policy 37(1):114–123
Malik RK, Gupta RK, Singh CM, Yadav A, Brar SS, Thakur TC, Singh SS, Singh AK, Singh R, Sinha RK (2005) Accelerating the adoption of resource conservation technologies in rice wheat system of the Indo-Gangetic Plains, Proceedings of Project Workshop, Directorate of Extension Education, ChaudharyCharan Singh Haryana Agricultural University (CCSHAU), June 1–2, 2005. CCSHAU, Hisar
Marongwe LS, Nyagumbo I, Kwazira K, Kassam A, Friedrich T (2012) Conservation agriculture and sustainable crop intensification: a Zimbabwe case study. Integr Crop Manag 17:2012
Mashingaidze N, Madakadze C, Twomlow S, Nyamangara J, Hove L (2012) Crop yield and weed growth under conservation agriculture in semi-arid Zimbabwe. Soil Till Res 124:102–110
Moreno F, Murillo JM, Pelegrín F, Girón IF (2006) Long-term impact of conservation tillage on stratification ratio of soil organic carbon and loss of total and active CaCO3. Soil Tillage Res 85:86–93

Mrabet R (2002) Stratification of soil aggregation and organic matter under conservation tillage systems in Africa. Soil Till Res 66:119–128

Mrabet, R., 2007. Conservation agriculture in Morocco: a research review. International Workshop on conservation agriculture for sustainable land management to improve the livelihood of people in dry areas. ACSAD, FAO, GTZ, AAAID and UNEP. Damascus, May 7-9,2007. 27p

Mrabet R, El Brahli A (2005) Soil and crop productivity under contrasting tillage management systems in semiarid Morocco. III World Congress on Conservation Agriculture: "Linking Production, Livelihoods and Conservation". Nairobi. October 3–7, 2005, 6p

Mrabet R, Ibno-Namr K, Bessam F, Saber N (2001) Soil chemical quality changes and implications for fertilizer management after 11 years of no-tillage wheat production systems in semiarid Morocco. Land Degrad Dev 12:505–517

National Mission For Sustaining The Himalayan Ecosystem (2016) http://knowledgeportal-nmshe.in/

Niggli U, Fließbach A, Hepperly P, Scialabba N (2009) Low greenhouse gas agriculture: mitigation and adaptation potential of sustainable farming systems. FAO, Rome, April 2009, Rev. 2–2009

Palm C, Blanco-Canqui H, DeClerck F, Gatere L, Grace P (2013) Conservation agriculture and ecosystems services: an overview. Agric Ecosyst Environ 187:87–105

Perszewski R (2005) Ideas leading to no-till's second revolution. http://www.no-tillfarmer.com/pages/Feature-Articles---Ideas-Leading-To-No-Tills-Second-Revolution.Php. Accessed 2 June 2014

Pittelkow CM, Liang X, Linquist BA, van Groenigen KJ, Lee J, Lundy ME, van Gestel N, Six J, Venterea RT, van Kessel C (2014) Productivity limits and potentials of the principles of conservation agriculture. Nature 517:365–368

Pittelkow CM, Liang X, Linquist BA, Groenigen KJ, Lee J, Lundy ME, Gestel N, Six J, Venterea RT, Kessel C (2015) Productivity limits and potentials of the principles of conservation agriculture. Nature 517:365–368

Powles SB, Lorraine-Colwill DF, Dellow JJ, Preston C (1998) Evolved resistance to glyphosate in rigid ryegrass. Weed Sci 46:604–607

Powlson DS, Stirling CM, Thierfelder C, White RP, Jat ML (2016) Does conservation agriculture deliver climate change mitigation through soil carbon sequestration in tropical agro-ecosystems? Agric Ecosyst Environ 220:164–174

Rahman MH, Okubo A, Sugiyama S, Mayland HF (2008) Physical, chemical and microbiological properties of an andisol as related to land use and tillage practice. Soil Tillage Res 101:10–19

Rockström J, KaumbuthoMwalley P, Nzabi J, Temesgen AW, Mawenya M, Barron L, Mutua J, Damsgaard- Larsen JS (2009) Conservation farming strategies in East and Southern Africa: yields and rain water productivity from on-farm action research. Soil Tillage Res 103:23–32

Roper MM, Ward PR, Keulen AF, Hill JR (2013) Under no-tillage and stubble retention, soil water content and crop growth are poorly related to soil water repellency. Soil Tillage Res 126:143–150

Ryan GF (1970) Resistance of common groundsel to simazine and atrazine. Weed Sci 18:614–616

Sangar S, Abrol JP, Gupta RK (2005) Conservation agriculture: conserving resources enhancing productivity. CASA, NASC Complex, New Delhi, 19 p

Sanginga NS, Woomer PL (2009) Integrated soil fertility management in Africa: principles, practices, and developmental process. Free download form CIAT

Sartori L, Peruzzi P (1994) The revolution of no-tillage in Italy: a review of the scientific literature. In: Proceedings of the EC-Workshop-I, Giessen, 27–28 June, 1994, Experience with the applicability of no-tillage crop production in the West-European countries, Wissenschaftlicher Fachverlag, Giessen, 1994, pp 119–129

Sayre K (2000) Effects of tillage, crop residue retention and nitrogen management on the performance of bed-planted, furrow irrigated spring wheat in northwest Mexico. Paper presented at the Conference of the International Soil Tillage Research Organization, 15; Fort Worth; 2–7 July, 2000

Sayre KD, Hobbs PR (2004) The raised-bed system of cultivation for irrigated production conditions. In: Lal R, Hobbs P, Uphoff N, Hansen DO (eds) Sustainable agriculture and the rice-wheat system. Ohio State University, Columbus. Paper 20, pp 337–355

Sharma AR, Singh R, Dhyani SK (2005) Conservation tillage and mulching for optimizing productivity in maize-wheat cropping system in the outer western Himalaya region – a review. Indian J Soil Conserv 33(1):35–41

Sharma AR, Jat ML, Saharawat YS, Singh VP, Singh R (2012) Conservation agriculture for improving productivity and resource-use efficiency: prospects and research needs in Indian context. Indian J Agron 57(IAC Special Issue):131–140

Six J, Ogle SM, Conant RT, Mosier AR, Paustian K (2004) The potential to mitigate global warming with no-tillage management is only realized when practised in the long term. Glob Chang Biol 10(2):155–160

Steiner K (2002) Conservation tillage-gateway to food security and sustainable rural development. African conservation tillage network. Information series no 4

Tanner T, Mitchell T (2008) Entrenchments or enhancement: could climate change adaptation help to reduce chronic poverty? Inst Dev Stud Bull 39:6–15

TerAvest D, Boggs LC, Thierfelder C, Reganold JP (2015) Crop production and soil water management in conservation agriculture, no-till, and conventional tillage systems in Malawi. Agric Ecosyst Environ 212:285–296

Thierfelder C, Wall PC (2010) Rotation in conservation agriculture systems of Zambia: effects on soil quality and water relations. Exp Agric 46:309–325

Thierfelder C, Cheesman S, Rusinamhodzi L (2012a) A comparative analysis of conservation agriculture systems: benefits and challenges of rotations and intercropping in Zimbabwe. Field Crops Res 137:237–250

Thierfelder C, Mulundu M, Rusinamhodzi L (2012b) Conservation agriculture in eastern and southern provinces of Zambia: long-term effects on soil quality and maize productivity. Soil Tillage Res 126:246–258

Thierfelder C, Chisui JL, Gama M, Cheesman S, Jere ZD, Bunderson WT, Eash NS, Rusinamhodzi L (2012c) Maize-based conservation agriculture systems in Malawi: long-term trends in productivity. Field Crops Res 142:47–57

Thierfelder C, Mutenje M, Angeline M, Mupangwa W (2015) Where is the limit? Lessons learned from long-term conservation agriculture research in Zimuto Communal Area, Zimbabwe. Food Secur 7:15–31

Thomas GA, Dalal RC, Standley J (2007) No-till effects on organic matter, pH, cation exchange capacity and nutrient distribution in a luvisol in the semi-arid subtropics. Soil Tillage Res 94:295–304

Tripathi S, Singh RN, Sharma S (2013) Emissions from crop/biomass residue burning risk to atmospheric quality. Int Res J Earth Sci 1(1):24–30

Turmel MS, Speratti A, Baudron F, Verhulst N, Govaerts B (2014) Crop residue management and soil health: a systems analysis. Agric Syst 134:6–16

UNEP (2013) The emissions gap report 2013. United Nations Environment Programme (UNEP), Nairobi. http://www.unep.org/publications/ebooks/Emissionsgapreport2014/portals/50268/pdf/EGR2014_LOWRES.pdf

Valbuena D, Erenstein O, Homann-KeeTui S, Abdoulaye T, Claessens L, Duncan AJ, Gérard B, Rufino MC, Teufel N, van Rooyen A, van Wijk MT (2012) Conservation agriculture in mixed crop–livestock systems: scoping crop residue trade-offs in Sub-Saharan Africa and South Asia. Field Crops Res 132:175–184

Vanlauwe B, Wendt J, Giller KE, Corbeels M, Gerard B, Nolte C (2014) A fourth principle is required to define conservation agriculture in sub-Saharan Africa: the appropriate use of fertilizer to enhance crop productivity. Field Crop Res 155:10–13

Verhulst N, Govaerts B, Verachtert E, Castellanos-Navarrete A, Mezzalama M, Wall P, Deckers J, Sayre KD (2010) Conservation agriculture, improving soil quality for sustainable production systems? In: Lal R, Stewart BA (eds) Advances in soil science: food security and soil quality. CRC Press, Boca Raton

Zhao JF, Huang GB, Xin P, Xie KZ, Xu AX (2007) Conservation tillage effects on run-off and soil erosion. Bull Soil Water Conserv 27:16–19

Conservation Agriculture: A New Paradigm for Improving Input Use Efficiency and Crop Productivity

2

B.P. Meena, A.O. Shirale, M.L. Dotaniya, Pramod Jha, A.L. Meena, A.K. Biswas, and A.K. Patra

Abstract

Conservation agriculture (CA) refers to a set of agricultural practices encompassing minimum mechanical soil disturbance, diversified crop rotation and permanent soil cover with crop residues to mitigate soil erosion and improve soil fertility besides soil functions. The CA aims to conserve, improve and make more efficient use of resources through CA-based technologies. It has many tangible and intangible benefits in terms of reduced cost of production, saving of time, increased yield through timely planting, improved water productivity, adaptation to climate variability, reduced disease and pest incidence through stimulation of biological diversity, reduced environmental footprints and ultimately improvements in soil health. However, weeds are a major biotic interference in CA, posing big defy towards its success unless all the principles are completely followed. Development of post-emergence herbicide and growing herbicide-tolerant crops and also the retention of crop residues as a mulch help in managing weed problems and also improve soil moisture retention. Furthermore, this practice of agriculture improves soil organic carbon content which ultimately leads to an increase in input use efficiency.

Keywords

Conservation agriculture • Crop rotations • Crop residues • Carbon sequestration • Nutrient dynamics • Weed management

B.P. Meena (✉) • A.O. Shirale • M.L. Dotaniya • P. Jha • A.K. Biswas • A.K. Patra
ICAR-Indian Institute of Soil Science, Bhopal 462 038, India
e-mail: bharatmeena24@gmail.com

A.L. Meena
ICAR-Indian Institute of Farming Systems Research, Modipuram 250 110, India

J.K. Bisht et al. (eds.), *Conservation Agriculture*,
DOI 10.1007/978-981-10-2558-7_2

2.1 Introduction

Modern agriculture is facing a serious problem of the decline in crop yield and deterioration in soil quality (Montgomery 2007) despite the use of improved varieties, adequate fertiliser nutrition and plant protection chemicals. Hence, synchronising the food demand of the ever-growing population can only be achieved through an alternative production system which can maintain high yields in consonance with maintenance of ecological equilibrium. Food security is a multidimensional theme, which directly hits the poor and needy and, in turn, the quality of life. The agriculture sector is the starting point for finding sustainable solution to overcome the food crisis. It is imperative to manage critical inputs and resources for a higher food production. The CA is an obvious new paradigm in the twenty-first century to achieve a base for sustainable agricultural production intensification (Friedrich et al. 2012). The CA aims to conserve, improve and make more efficient use of natural resources through integrated management of available soil, water and biological resources combined with external inputs (Meena et al. 2013, 2015a, b; Singh et al. 2014; Kumar et al. 2015). The CA is described by FAO (www.fao.org.ag/ca) as a concept for resource-saving agricultural crop production, which is based on enhancing the natural and biological processes below and above the ground. It contributes to environmental conservation as well as to enhanced and sustained agricultural production. CA involves three linked principles, viz. (i) providing permanent soil cover to mitigate soil erosion and to improve soil fertility and soil functions; (ii) minimum mechanical soil disturbance, i.e. reduced tillage (RT) or zero till/no tillage (NT) and direct seeding; and (iii) diversified, efficient and economically viable crop rotations and such type of rotations that contribute to maintaining soil biodiversity (Kassam et al. 2009). The CA can improve agriculture through improvement in water infiltration, enhanced ground water storage, reduced soil erosion, improved soil aggregates and reduced soil compaction through promotion of biological tillage, enrichment in soil organic carbon (SOC), moderated soil temperatures and enhanced microbial diversity and weed suppression; it also helps to reduce cost of production, saves time, increases yield through timely planting, reduces diseases and pests through stimulation of biological diversity and reduces greenhouse gas (GHGs) emissions.

2.2 Conservation Agriculture: Why?

Input-intensive conventional production system has led to second-generation problems such as deteriorating soil health, declining soil organic matter and increasing multiple deficiencies of N, P, K, S, Zn, Fe and Mn due to their overmining from soils (Ladha et al. 2000; Tiwari 2002), decline in the groundwater table, deterioration of groundwater quality (Humphreys et al. 2010), sodicity and salinisation problems (Tiwari et al. 2009), herbicide resistance, shift in weed flora and pest populations (Hobbs et al. 1997). The problems encountered due to input-intensive chemical-based conventional agriculture are:

(i) Intensive tillage operations for the preparation of fine seedbed for sowing to ensure proper seed germination, improved moisture conservation, weed control and other pests and also application of organic and inorganic fertilisers
(ii) Continuous monocropping system which led to the degradation of soil fertility
(iii) No recycling of crop biomass after harvesting, resulting in mining of nutrients
(iv) Overexploitation of groundwater resources
(v) Burning of crop residues after harvesting
(vi) Indiscriminate use of chemical fertilisers leading to decreasing factor productivity
(vii) Energy-intensive farming (Sharma et al. 2012)

The current production system management is posing a serious threat to food security and livelihood of farmers, especially the poor and marginal farmers. Land and water are the natural resources that are essential for the existence of life. Now, these precious resources are under tremendous pressure due to increasing demographic pressure and other biotic and abiotic stresses. Among the natural resources, land is a limited, finite, inelastic and a highly valuable natural resource and a base for food, feed, fuel and fibre production, besides providing many critical ecosystem services. A decrease in the per capita availability of agricultural land in India from 0.48 ha in 1950 to 0.15 ha in 2000, and likely a further reduction of 0.08 ha by 2020 due to continuous increasing industrialisation and urbanisation (Pande et al. 2012), is a serious cause of concern. A number of persons per ha of the net-cropped area were three in 1951 and six in 2000 and could rise to eight in 2025 in India. The present population of more than 1.2 billion, accounting ~18 % of world population supported by only ~2.4 % land area, is estimated to become 1.4 billion by 2025 and 1.7 billion by 2050 AD, needing annually ~380 million tonnes and 480 million tonnes of food grain, respectively (Yadav and Singh 2000). Hence, there is an urgent need to reorient the present management production system to a system which can improve the input (land, seed, water, labour, fertiliser and energy) use efficiency of advanced crop management technologies, while maintaining the natural resource base (Meena et al. 2015c, d).

2.3 Principles of Conservation Agriculture

The CA is a concept for resource-saving agricultural crop production system that strives to achieve acceptable profits together with high and sustained production levels while at the same time conserving and improving the environmental quality. CA system is based on enhancing natural biological processes above and below the ground. The key elements of CA include adequate retention of crop residues on the soil surface for mulching, diversified crop rotations and cropping systems coupled

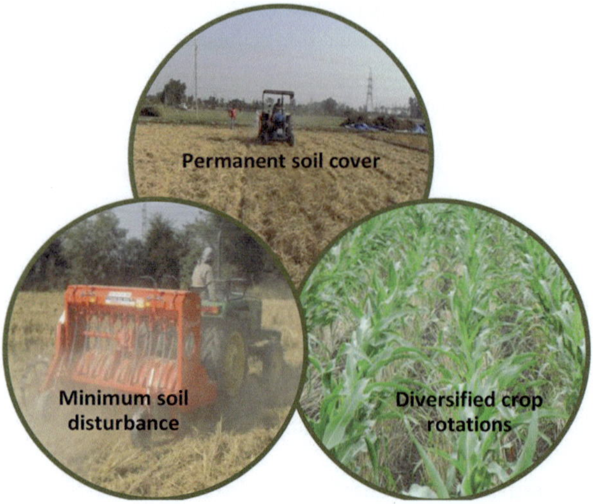

Fig. 2.1 Three linked principles of conservation agriculture

with minimum soil disturbance through controlled traffic (Hobbs et al. 2008), and the use of external inputs such as agrochemicals and nutrients of mineral or organic origin is applied at an optimum level and in a way and quantity that does not interfere with, or disrupt, the biological base. CA is characterised by three core principles (Fig. 2.1) which are linked to each other, viz. minimum soil disturbance, permanent soil cover and diversified crop rotations. These principles are very specific and economically viable in different agroclimatic and socio-economic conditions of the farmers.

2.3.1 Minimum Soil Disturbance

Minimising mechanical soil disturbance is aimed at reducing tillage operations to the minimum necessary for ensuring a fine seedbed, proper germination and satisfactory crop stand. In the past, soil tillage increased the soil fertility due to mineralisation of soil nutrients (Dumanski et al. 2006), but long-term unnecessary tillage operations led to soil erosion and degradation of soil structure (Donovan and Casey 1998). The introduction of ZT or minimum soil disturbance is to curb the negative impact of excessive tillage and also to reduce soil erosion, which ultimately improves soil and water conservation (Li et al. 2007). Minimising soil disturbance also maintains proper aeration in rooting zone, oxidation of organic matter, water movement in soil and also exposing the weed seeds either for germination (Kassam and Friedrich 2009) or as prey to beetles.

2.3.2 Permanent Soil Cover

A permanent soil cover is important to protect the soil against the deleterious effects of exposure to rain and sun and also provide micro- and macroorganisms in the soil with a regular supply of food and alter the soil physical environment for optional growth and development of soil organisms. Crop residue is the principal component of soil cover to protect soil and water erosion from top soil and also playing an important role in building soil organic matter, nutrient recycling and improving soil quality (Chauhan et al. 2012). Crop residues are the principal sources of carbon and it has significant effect on soil physical, chemical and biological soil properties (Kumar and Goh 2000).

2.3.3 Diversified Crop Rotations

Diversified crop rotation is of paramount importance in mitigating the biotic and abiotic problems, arising in monoculture particularly in rice–wheat cropping system. It must be suited to different agroclimatic and farmers' socio-economic conditions. Crop rotation involving legumes helps in minimal rates of build-up of population of pest and diseases through life cycle disruption, biological nitrogen fixation, control of off-site pollution and enhancing soil biodiversity (Kassam and Friedrich 2009; Dumanski et al. 2006). Inclusion of legume crops in rotation can also play an important role in conserving groundwater and soil water. Moreover, the quality of these crops (higher protein content) is better than wheat and other cereals in rice–wheat-grown areas. Particularly in wheat, crop rotation addresses the problems of insect, pest and diseases by integrating crop rotations, which help to break the cycle that perpetuates crop diseases such as wheat rust and pest infestations, resulting in higher yield (Witmer et al. 2003).

Nowadays the controlled traffic is loosely taken as the fourth principle (Sims 2011) of CA to ensure less or no soil compaction due to broad wheel tires of tractors. CA systems are not only about the use of one or two principles, although this sort of practice has been very common in adoption of CA in certain crops and locations in India, such as precision planting using a seed drill or planters without tillage or significantly reduced tillage and zero-till sowing without residue in wheat. Rather, it is a holistic set of management practices that can ensure more production in a sustainable manner. The CA is aimed to enhance the input use efficiency through conserving, improving and making more efficient use of natural resources through integrated management of available soil, water and biological resources combined with external inputs. CA is now considered the principal path to sustainable agriculture. Thus, it is a way to achieve goals of higher productivity, while soil erosion, water losses from runoff and soil physical degradation may be minimised by reducing soil mechanical disturbance and marinating permanent soil cover (Serraj and Siddique 2012). CA also helps to improve soil biodiversity in the natural agroecosystem base (Friedrich et al. 2012). CA system requires a total paradigm shift from conventional agriculture with respect to integrated

Table 2.1 Some distinguishing features of conventional and CA systems

Conventional agriculture	Conservation agriculture
Cultivating land/ploughing soil/tilling the soil using science and technology to dominate nature	Least interference with natural processes in harmony with nature
Excessive mechanical tillage and soil erosion	ZT or drastically reduced tillage (biological tillage)
High wind and soil erosion	Low wind and soil erosion
Residue burning or removing all organic matter (bare surface)	Surface retention of residues (permanently covered soil surface)
Water infiltration is low	Infiltration rate of water is high
Use of ex situ FYM/composts	Use of in situ organics/composts
Green manuring (incorporated)	Brown manuring/cover crops (surface retention)
Kills established weeds but also stimulates more weed seeds to germinate	Weeds are a problem in the early stages of adoption but decrease with time
Freewheeling of farm machinery, increased soil compaction	Controlled traffic, compaction in tramline, no compaction in cropped area
Monocropping/culture, less efficient rotations	Crop diversification and efficient crop rotations
Heavy reliance on manual labour, uncertainty of operations	Mechanised operations, ensure timeliness of operations
Poor adaptation to stresses, yield losses greater under stress conditions	More resilience to stresses, yield losses are less under stress conditions
Productivity gains in the long run are in declining order	Productivity gains in the long run are in incremental order

Source: Sharma et al. (2012)

management of crops, soil, water, nutrients, weeds and farm machinery (Table 2.1). Farmers are interested in sustainable crop production system to adopt and adapt improved crop management practices, and this may be considered as a new paradigm of the twenty-first century for higher crop productivity and sustainability.

2.4 Spread of Conservation Agriculture in India and World

CA is gaining more attention in both rainfed and irrigated ecosystems of the tropics, subtropics and temperate regions all over the world. At present ~157 M ha area (Fig. 2.2) is practised under the CA system, particularly in the USA, Brazil, Argentina, Canada, Australia, China and Russian Federation, which covers more than 11 % of the world's arable land area (Fig. 2.3) and continues to spread at an annual rate of ~10 M ha (Kassam et al. 2014). In India, CA adoption is still at a nascent stage, but CA-based crop management technologies are being practised in ~43 M ha (Jat et al. 2011) including ~1.5 M ha under the rice–wheat system in the Indo-Gangetic Plains (IGP). In South Asia, it covers almost 3 M ha, and ZT in wheat after rice harvest is the most widely adopted CA technology in the IGP of South Asia particularly in India. ZT technology is being adopted particularly in the

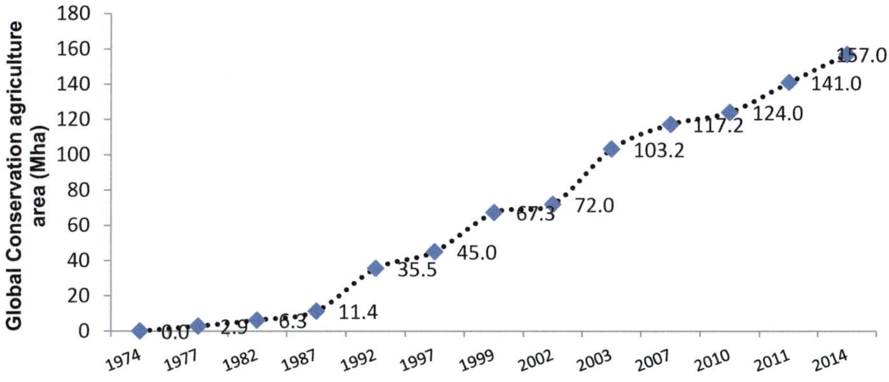

Fig. 2.2 Year-wise adoption of conservation agriculture (Source: FAO 2015)

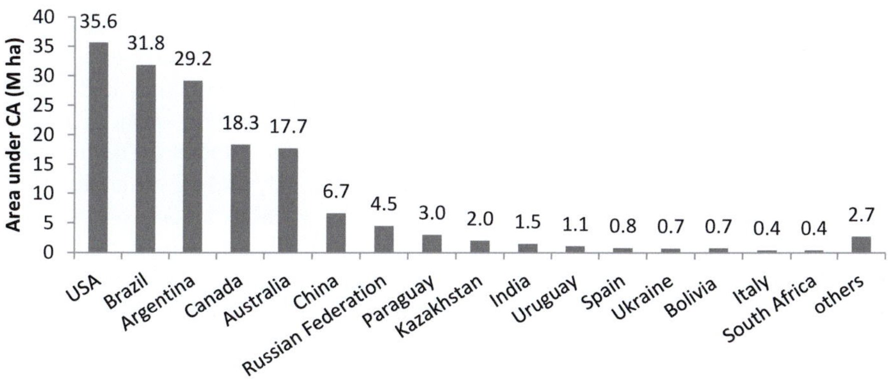

Fig. 2.3 Conservation agriculture system adoption in major countries (Source: FAO 2015)

northwest IGP of India. Although the potential area of this technology is estimated to be 10 M ha, only ~20 % has been achieved till now. CA-based management practices have shown the benefits in enhancing the natural resource base, improving input use efficiency, soil aggregation, soil health, farm productivity and mitigation of climate change.

The ZT has been a success in rice–wheat cropping system due to reduction in cost of production by 2000–3000 per ha, which is the main driver behind its spread (Malik et al. 2005), and improved soil health (Jat et al. 2009; Gathala et al. 2011) too. The potential of C sequestration in C-depleted soils of India is high with the adoption of ZT. Long-term carbon sequestration and build-up of soil organic matter constituted a practical strategy to mitigate GHG emission and impart greater resilience to production systems to climate change (Saharawat et al. 2012) and improved environmental quality (Pathak et al. 2011). Crop residue management provides an opportunity to protect the topsoil, enriched with organic matter, moderate soil temperature, improve soil biological activities (Gathala et al. 2011) and also enhance the water and nutrient use efficiency (Jat et al. 2012; Meena et al. 2015e; Verma et al. 2015a; Ghosh et al. 2016).

Development of non-selective contact and post-emergence herbicide for weed control provided a base for recommendation and allowed for consistent yield and creditable performance, particularly reducing the incidence of the predominant weeds *Phalaris minor* in wheat (Malik et al. 2005).

2.5 Tillage and Crop Establishment: Machinery

Anomalies associated with plough-based farming include deteriorating soil physical environment, declining factor productivity, labour and water shortage, escalating cost of production (Jat et al. 2009; Ladha et al. 2009; Chauhan et al. 2012), emerging new weed biotype and climate change-related challenges. Hence, fundamentally CA requires suitable farm machinery for tillage and crop establishment, fertilisation, weeding, irrigation, crop harvesting and other operations. Crop establishment depends on the external inputs (seed, fertiliser, weedicide, etc.), soil, climatic, machinery and management factors. The first two core principles of CA demand specialised machinery for seeding on unploughed field with crop residue. Direct seeding with ZT machine is one of the best technologies that potentially addresses the issues of labour, energy, water, soil health, etc. (Gathala et al. 2011; Jat et al. 2013) and adaptations to climatic variability (Jat et al. 2009). However, the ZT machinery could not carry out seeding effectively in high quantity of crop stubble, which normally occurs in CA systems. Therefore, the use of the new generation machines like happy seeder, rotary-disc drill, double-disc drill and punch planter may lead to wider adoption of CA (Sidhu et al. 2007). Farmers generally remove the residue or burn prior to wheat sowing, which leads to losses of soil organic matter and nutrients and creates environmental pollution (Singh et al. 2007).

Happy seeder, an improved version of the ZT seed drill machine, developed in Punjab provides an alternate to burning for managing crop residues and allows direct seeding of wheat in high quantity of crop residues (5–9 t/ha of anchored and loose straw). Happy seeder and rotavator are considered to be the efficient equipments for in situ management of rice straw as well as control of weeds (ACIAR 2013; Kang 2013). Qamar et al. (2013) have also noticed a higher growth rate and grain yield in ZT with happy seeder. The advanced version of happy seeder, like turbo seeder, combo happy seeder, post-consumer recycled (PCR) planter and easy seeder, is also being designed for efficient crop establishment and fertiliser placement. Hence, for CA system, the machines need to be more user-friendly to develop the interest of private manufactures for investing in the machine (Akter and Gathala 2014).

2.6 Weed Management in CA Systems

Weed infestations are the major constraints and its control is a major challenge in CA-based systems since weed ecology and management are entirely different in CA systems. Weeds pose serious threat to the companion crop through its competition for nutrients, water, sunlight and space, which cause considerable reduction in yield. Under the tillage system, ploughing and harrowing kill growing weeds mainly by burial, whereas elimination of conventional tillage may cause serious problem regarding weed infestation (Buhler et al. 1994). In CA system, weed seed bank, dispersal mechanism, distribution, growing pattern, weed competition, etc. are more complex than conventional agriculture. Shifts in weed populations from annuals to perennials have been observed in CA systems, and these are known to thrive in CA systems such as Bermuda grass (*Cynodon dactylon*), nutsedge (*Cyperus rotundus*) and Johnson grass (*Sorghum halepense*) which generally reproduce from underground plant storage structure: stolons, tubers or nuts and rhizomes, respectively (Sharma et al. 2012). Similarly, Mukherjee and Debnath (2013) reported higher weed biomass of *Polygonum pensylvanicum*, *Polygonum Persicaria*, *Polygonum orientale*, *Oldenlandia diffusa*, *Cynodon dactylon* and *D. sanguinalis* in Terai region of West Bengal, India. Weed species shifts and losses in crop yield as a result of increased weed density have been cited as major hurdles to the widespread adoption of CA. Hence, it requires special approaches for effective weed management in CA systems such as preventive measures, modified tillage, improved cultural practices (tillage, crop residue, intercropping, cover cropping, competitive crop cultivars, planting geometry, sowing time, nutrient management, etc.), chemical weed control, biological weed control, herbicide-tolerant cultivars and integrated weed management (IWM).

2.6.1 Preventive Measures

Preventive measures are first and the most important steps to be taken to manage weeds in general and especially under a CA system because 'prevention is better than cure'.

Weeds may tolerate compaction and drainage problems better than many crops. As a result, weeds are more competitive and problems are more severe particularly in CA systems. The best way to control weeds is by keeping them out of your fields:

All weed control encompasses all measures taken to prevent the introduction of weeds (Rao 2000), such as the use of clean seed (certified seed), and prevent the dissemination of weed seeds from one area to another or from one crop to another by using clean equipments (Das 2014);

Controlling weeds in ditches and at the edges of fields or around sloughs is an important practice for limiting the spread of weeds, for example, Canada thistle;

Preventing the weed seeds' dissemination and dispersal;

Using well-decomposed compost and manure so that it does not contain any viable weed seeds;

Mechanically removing weeds before they have a chance to set seed is an important form of field sanitation.

2.6.2 Modified Tillage Operations

Modified tillage operations provide an opportunity to suitable weed management tools in CA system. It is observed that crop yields can be similar to conventional and CA systems if weeds are properly controlled and crop stands are uniform. Results of on-farm trials at several locations in Haryana, India, revealed that the population of *Phalaris minor* was considerably lower and grain yield of wheat was comparatively higher under ZT as compared to CT (Gupta and Seth 2007). A review of literature indicated that ZT increased as well as the reduced infestation of certain weed species in different crops (Verma and Srivastava 1989; Singh et al. 1998; Chauhan et al. 2003). The infestation of problematic and troublesome weed, *Phalaris minor*, was lesser in ZT wheat due to minimum disturbance of soil (Chauhan et al. 2001; Agarwal and Goswami 2003), and ZT reduced the weed seed bank as compared to disturbed soils, and there was less multiplication of weed seeds in the succeeding years (Singh et al. 2010). Although ZT significantly increased the population of *V. sativa*, it reduced the population of *C. album* in vertisol in Jabalpur, India (Sharma et al. 2013). Thus, ZT with appropriate crop cultivars and herbicide selection provided advantages over conventional tillage (Tubbs and Gallaher 2005; Verma et al. 2015b; Meena et al. 2016).

2.6.3 Improved Cultural Practices

Cultural practices are aimed to ensure better soil and crop management (Nazir 1994). Successful weed management is not to merely control weeds in a crop field, rather to create a system that reduces weed establishment and minimises weed competition with crops. There are different ways to handle weeds by improved cultural management practices. Effective water management plays a vital role in weed control under the CA System. Soil cover with dead or live mulch and crop cover also one of the pillars of CA. The crop/cover crop residue may also release some toxic substances, which may also suppress weed seed germination process (Ramesh 2015). Crop rotation in CA is a successful approach to reduce the weed pressure. Laser land levelling is an integral part of CA as it provides uniform moisture distribution to the entire field and allows uniform crop stand and growth, leading to lesser weed infestation. Reduction in weed population in wheat after 30 days was recorded under precisely levelled fields in comparison to traditional levelled fields (Jat et al. 2009). Row crop cultivation is also a good approach to accomplish the management of weeds under reduced tillage. It is very effective to combine the chemical approach and row crop cultivation maintaining high residues in the field. Mulch tillage is specially designed to retain more than 30 % crop residues on the surface, suppressing different weeds due to shading or covering effect. Moreover, different types of organic compounds released from mulches through leaching cause inhibition of weed seed emergence (Ball 1992).

2.6.4 Chemical Weed Management

Herbicides are another weed control option, but greater attention must be given to alternative control methods and to ensuring that chemicals are used properly to reduce health risks and environmental damage. Herbicides are less effective, if improperly applied, for instance, at the incorrect time and dose, or without appropriate adjuvants. Most commonly used burn-down herbicides are glyphosate, paraquat, glufosinate, 2,4-D and dicamba. The rate and time of application is very critical in CA systems as the weed control under ZT systems based on vegetation cover is present in the field (Vargas and Wright 2005). Several low-dose, high-potency, selective, post-emergence herbicides and mixtures are available for effectively managing weeds in crops like rice and wheat grown in sequence under CA system. However, the recent development of post-emergence broad-spectrum herbicides provides an opportunity to control weeds in CA system.

In fact, many farmers in India apply isoproturon, a good broad-spectrum herbicide, by broadcasting it with sand or urea. Improper herbicide use has probably contributed to the herbicide resistance in *P. minor* in India and *A. fatua* in Mexico (Malik and Singh 1995). The use of new herbicides or a mixture of herbicides is another alternative and will remain a part of the weed control strategy. In CA systems the presence of residue on the soil surface may influence soil temperature and moisture regimes that affect weed seed germination and emergence patterns over the growing season. This shows that under CA system, farmers have to change the timing of weed control measures in order to ensure their effectiveness. Soil surface residues can interfere with the application of herbicides, so there is a greater likelihood of weed escape if residue is not managed properly or herbicide application timings or rates are not adjusted. Some herbicides intercepted by crop residues in CA systems are prone to volatilisation, photodegradation and other losses. The extent of loss, however, may vary depending upon their chemical properties and formulations. Herbicides with high vapour pressure, e.g. dinitroaniline herbicides, are susceptible to volatilisation loss from the soil surface.

2.6.5 Herbicide-Tolerant Crops

Herbicide-tolerant crops are designed to tolerate specific broad-spectrum herbicides, which kill the surrounding weeds, but leave the cultivated crop intact. Several crops have been genetically modified to be resistant to non-selective herbicides; particularly glyphosate-resistant soybean has been adopted principally because it simplifies weed control to the use of a single herbicide and with a more flexible timing than that required for conventional herbicides. However, it is very difficult to isolate the effects of the adoption of GM crops from other factors which may affect pesticide use (Heimlich et al. 2000). Almost 90 % of all transgenic crops grown worldwide are glufosinate and glyphosate resistant (GR) (Duke and Powles 2008). GR crops offer farmers a vital tool in fighting weeds and are compatible with NT agriculture (Duke and Cerdeira 2005). They give farmers the flexibility to apply herbicides only when needed, to control total input of herbicides and to use herbicides with preferred environmental characteristics. Aulakh et al. (2010) reported that glyphosate-tolerant cotton increased 13–29 % higher yields than the glufosinate-tolerant cotton and conventional cotton.

2.6.6 Integrated Weed Management in CA System

Integrated weed management (IWM) is a holistic approach under CA system. This also includes the biological weed control methods. Biological weed control offers a huge, largely untapped resource for weed control method (Kennedy 1999); it includes a large number of available living entities such as predators, pathogens and other plant competitors of weeds that are exploited to kill or suppress the weeds. Microorganisms that suppress growth of many common agricultural weeds have been identified and commercial development is underway (Stubbs and Kennedy 2012). In short, it is an effective method of weed control in CA systems. IWM is basically a long-term approach which aims to manage weeds rather than controlling weeds. IWM has the potential to restrict weed populations to manageable levels; reduce the economic losses, risk to human health and potential damage to the ecosystem; increase cropping system sustainability; and reduce selection pressure for weed resistance to herbicides. A combination of different weed management strategies such as herbicide rotation, green manures (Kirkegaard et al. 2014), selection of suitable crop cultivars and cropping systems coupled with CA principles may help in weed management (Ramesh 2015).

2.7 Nutrient and Water Management

Tillage practices favourably modify the soil physical and biological environment facilitating root proliferation. These actively growing roots can take up nutrients from a greater soil volume and could improve the nutrient use efficiency. Conservation tillage (CT) and deep tillage increased NPK uptake compared to minimum tillage (MT). Tillage practices along with organic matter (OM) further affected the moisture and nutrient availability to crops. The availability of nutrients at different growth stages of *Sorghum* was increased by deep tillage (Patil and Sheelavantar 2006). In general, crop yields under ZT practices are more stable than under tilled systems with greater efficiency in the use of nutrients (Martin 2006). The retention of crop residues on the soil surface, along with fertilisation with organic manure and involvement of legumes in crop rotation coupled with MT and ZT practices, plays an important role in sustaining soil fertility, improving fertiliser/water use efficiency (WUE) and physical conditions of soils and enhancing crop productivity and agricultural sustainability (Dalal and Chan 2001; Lampurlanes et al. 2001). The chemical, physical and biological fertility of soil is depending on soil organic carbon (Chan et al. 2008).

Crop residues enhanced the soil organic matter and total soil N levels in the long term (Cassman et al. 1996). Mohammad et al. (2012) found that N yield and fertiliser N utilisation by wheat were increased significantly by crop residues under NT compared to the tillage. Higher nutrient use efficiency (i.e. PFP) of applied N under NT than in conventional tillage was probably due to better moisture conservation under NT which might have facilitated plant nutrient uptake. Formation of a layer of crop residues on the soil surface under NT system improved the crop growth rate and nutrient uptake (Sapkota 2012). Sapkota et al. (2014) conducted on-farm trials in seven districts of Haryana, India, for two consecutive years (2010–2011 and 2011–2012) to evaluate three different approaches to SSNM based on recommendations from the Nutrient Expert® (NE) decision support system in NT and CT-based wheat production systems; as a result NT with site-specific approaches for nutrient management can increase yield, nutrient use efficiency and profitability while decreasing GHG in wheat.

Nutrient management in CA is a significant concern of agriculture today. Intensive cropping pattern over the years have mined soil nutrients due to improper replenishment. Increasing the awareness and close monitoring of nutrient budgeting will promote the researchers and farmers to compute the soil nutrient input–output balance sheet in rational ways. In the conventional system of crop cultivation, the tillage operation is higher which promote higher level of soil disturbance and affected the nutrient dynamics in soil. The soil surface, covered with crop residue also modified the soil properties in many ways, especially nutrient availability to crops (Dotaniya 2012). Nutrient management in CA system mainly follows some basis aspects like:

(i) Enhance the soil biological process to protect the soil microorganism population and diversity, so that the SOM is either build-up or maintained.
(ii) Maintain the adequate biomass production and biological N fixation in relation to soil biota activities in terms of soil energy and nutrient stocks.
(iii) Provide adequate access to all plant nutrients by plant root from soil solution, from natural and also from synthetic sources, to fulfil the crop demand.
(iv) Keep soil pH within the acceptable range.

Reduced tillage practices in CA reduced the rate of soil organic C burning, preserved C, enhanced the soil aggregation or reduced the soil bulk density. The soil residue cover directly increased the soil C on the soil surface (Dotaniya et al. 2013). The nutrient management approaches like integrated soil fertility management (ISFM) and integrated natural resources management (INRM) are having more attention on meeting crop nutrient requirement rather than managing soil health and land productivity. In the CA system, the use of both organic and inorganic plant nutrient sources is listed in ISFM and INRM process as per the requirement of crop and soil with respect to temporal and special variability.

The CA is widely affected by the tillage and cropping system, and the soil health improvement, enhancement of crop productivity and wide support to livelihood and the environment are interlinked with them. The old agricultural practices clearly described that healthy agricultural soils constitute biologically active soil system and having wide range of soil microbial diversity and plant nutrient process are in equilibrium with various existing phases and the adequate nutrient supply to plants in combination of ecosystem services. In the present context, CA emerged as a new breakthrough system approach for crop production and soil health. It represents biologically and bio-geo-physically integrated system of nutrient management during crop production and maintaining soil health in the long perspective (Friedrich and Kassam 2009). It could reduce the requirement of external inputs due to generation of high level of internal ecosystem services, which enhanced the crop production factor response in higher magnitude.

The CA practice depends on the climate and resources availability, i.e. in dry tropical and subtropical ecologies with small number of farmers with poor resources, the establishment of CA will take a longer period. The long-term applications of conservation practices improve the soil organic carbon and soil properties, mostly in ~10 cm upper soil layer. Increasing levels of C improved the soil aggregates, water holding capacity, microbial growth and plant nutrient transformation and reduced soil erosion. Application of crop residue on the soil surface conserved the soil moisture and mediated the plant nutrient dynamics. India produces a larger amount of crop residues approximately 500–550 MT annually (IARI 2012); these residues are used through the major practices like burning, incorporation and removal. It affected the N availability in soil solution due to immobilisation with wider C/N ratio of incorporated crop residues and the crop productivity (Bradford and Paterson 2000). Some researchers suggested that temporary immobilisation of N in the ZT system caused the leaching and denitrification losses of mineral N, but in longer way, it reduced the N application rate; and initial N fertiliser requirement during the crop production is high (Dotaniya et al. 2014). In case of P availability, it was more reported in ZT compared to conventional tillage practices. Accumulation of higher P concentration in the surface layer in ZT compared to CT is due to more availability of residue. According to Ismail et al. (1994), after 20 years of ZT practice, availability of P was increased ~42 % in upper 0–5 cm and lower (8–18 %) in 5–30 cm as compared to CT in a silt loam soil.

Increasing the crop residues in CA practice increased the availability of K in surface soil solution and declined with increasing soil depth (Du Preez et al. 2001). According to Govaerts et al. (2007), permanent raised bed enhanced the extractable K concentration 1.65 times in 0–5 cm and 1.43 times higher in 5–20 cm soil under ZT over conventional practice under crop residue retention plots. Micronutrient role in crop production is also important, and the use efficiency of these fertilisers is lower than macronutrient fertilisers, but the role in crop production is vital. Addition of organic residue in CA enhanced the micronutrient concentration, especially cations (Zn, Fe, Cu and Mn), than conventional tillage practices. Zn and Mn concentration are more found under ZT, due to surface placement (0–5 cm) of organic residue compared to conventional tillage (Franzluebbers and Hons 1996). However, CA improved the soil properties, i.e. chemical, biological and physical, and plant nutrient concentration influences the soil biological activities and nutrient transforming process.

An intensive soil tillage and mismanagement of irrigation water and fertilisers under current agricultural practices have accelerated the pace of degradation of irrigated dry lands in India. Increasing water scarcity and concerns of irrigation water quality have further raised serious doubts about the sustainability of current conventional agricultural systems. WUE of crops can be improved by the selection of crops and cropping system based on the availability of irrigation water resources. The latter can be achieved by the selection of irrigation methods, irrigation scheduling, tillage, mulching and fertilisation. Puddling rice paddies reduces percolation of water and leaching of fertilisers, especially N, besides helping in weed control. ZT machines have become particularly relevant in rice–wheat cropping system, where wheat sowing is generally delayed if the conventional methods of pre-sowing irrigation and land preparation are adopted (Mehla et al. 2000). It is now possible to sow wheat soon after rice harvest without primary cultivation, which permits timely sowing (Yadav et al. 2005) and saving irrigation water for field preparation. The advantage of ZT has also been reported for maize after rice in Telangana region of Andhra Pradesh (Reddy and Veeranna 2008). Reducing tillage and optimising N fertilisation are important strategies for soil and water conservation and N use efficiency for sustainable agriculture. Incorporation of crop residues on the soil surface under ZT system minimises water loss through evaporation which enhances higher growth rate of crop leading to higher water productivity (Sapkota 2012).

On a sandy loam (Typic Haplustept) soil in semiarid climate of New Delhi, Saha et al. (2010) evaluated the effect of tillage (ZT and CT) and residue management (incorporation, retention and removal) on soil physical properties after 3 years of continuous maize (*Zea mays*)–Indian mustard (*Brassica juncea*) sequence and reported that ZT optimised water use by ~14 % and 12 % in maize and mustard, respectively, as compared to conventional tillage. Maximum WUE was obtained in conventional tillage with residue incorporation, mainly because of maximum yield in maize (2.93 t/ha) and mustard (1.83 t/ha) obtained under the treatment (Fig. 2.4). Continuous rice–wheat cropping with intensive tillage in the IGP of South Asia has resulted in land degradation and inefficient use of water. Hence replacement of rice with less water-requiring crops such as soybean in RW system and identification of effective strategies for tillage management are prime need to sustain the productivity in the IGP of India. An experiment conducted in the IGP of India by Ram et al. (2013) revealed that soybean and wheat planted on raised beds recorded ~17 % and 23 % higher WUE, respectively, than in flat-bed planting. Similarly, Das et al. (2016) also reported that CA had significantly higher WUE as compared to conventional tillage in second and third year experiments despite the similar water use in pigeon pea (*Cajanus cajan* L.)–wheat (*Triticum aestivum*) system in the IGP region.

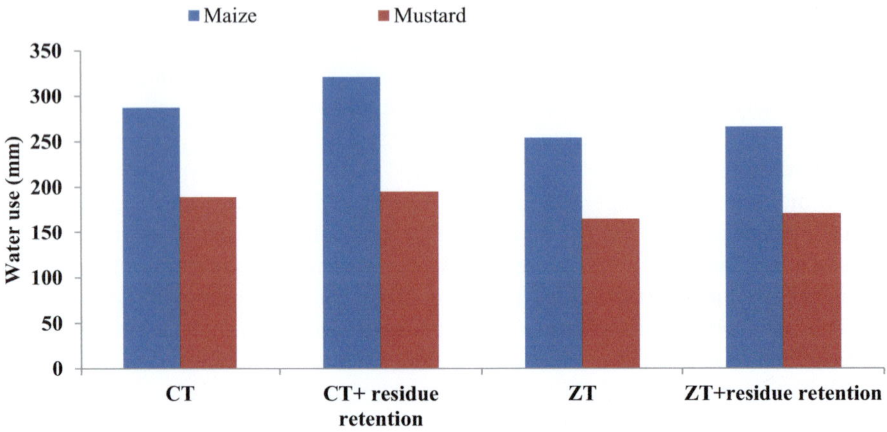

Fig. 2.4 Water use under different CA practices (Source: Saha et al. 2010)

2.8 Crop Productivity Levels in CA

The crop yield can be variable in the CA system (Farooq et al. 2011), for example, a CA may increase crop yield through improving soil fertility by conserving soil, water and sequestering organic carbon in farmland soils (Holland 2004; Goverts et al. 2007; Liu et al. 2010). The real effects of CA on crop yield may depend largely on specific CA practices, regional climate characteristics and cropping systems (Hobbs et al. 2008; Giller et al. 2009). CA systems also include improved efficient cultivars, less inputs and improved production and income and address the emerging problems (Gupta and Seth 2007; Saharawat et al. 2009). The CA technologies involve ZT or minimum-till farming with direct seeding, bed planting and crop residue retention for higher input use efficiency (Abrol and Sanger 2006) , crop productivity and stability (Bhushan et al. 2007), and also CA includes crop diversification and laser land levelling for increasing input use efficiency and higher farm profitability. Sen et al. (2002) recorded significantly higher grain yield of wheat under ZT as compared to conventional system, while Yadav et al. (2005) reported higher yield attributes of wheat, viz. effective tillers, grain per spike and 1000-grain weight. ZT resulted in better aggregate stability, more earthworms, a more open and continuous network of soil pores, more roots in the top 100 mm of soil and the same yield at lower cost. Guo et al. (1995), who studied ZT and minimum tillage in wheat for 25 years in an area where soils are heavy and plant stands are poor, reported that CT helped to preserve surface soil moisture and improved plant stands, soil structure and weed control.

The MT along with crop residue management has been found to be beneficial for improving crop productivity (Sharma et al. 2005; Saharawat et al. 2009; Jat et al. 2012). In similar, CT to wheat with crop residue retention on surface produced grain yield either equivalent to or greater than residue incorporation at sowing in conventional tillage in hills of north-west India (Achrya et al. 1998). Tillage with the incorporation of wheat residue in the cotton–wheat system increases seed cotton yield by 23–39 % (Jalota et al. 2008), and the reduced tillage in cotton and minimum tillage in wheat were also found to be effective with respect to soil disturbance to sustain yield and apparent crop productivity in the cotton–wheat system (Blaise 2014). In China, Zhang et al. (2007) reported relay cropping of cotton in wheat row had land use advantages than monocultures. Zheng et al. (2014) performed a meta-analysis to quantify the actual impacts of different CA practices such as NT/RT only, CT with straw retention (CTSR) and NT with straw retention (NTSR) on crop yields as compared to CT without straw retention and found that each CA practice, CTSR and NTSR significantly increased crop yield by ~4.9 % and 6.3 %, respectively, as compared to CT. Similarly, Ghosh et al. (2015) reported that the mean wheat equivalent yield was 47 % higher with CA as compared to conventional agriculture in maize–wheat crop rotation. The results across IGP in India suggested that double ZT with retention of crop residues produced higher system productivity over conventional and ZT without residues (Jat et al. 2011). A comparison of different CT practices that include ZT drilling, strip till drilling, rotary till drilling, bed planting and conventional sowing and crop residue management practices (retrieval, burning and recycling) in rice–wheat cropping system on crop productivity by Singh and Sharma (2005) over 6 years revealed that the rotary, strip and zero-till drilling and bed planting of rice and wheat provided higher yields (2–8 %) and are cost-effective (9–27 %) and energy efficient (21–32 %).

The in situ recycling of wheat straw provided rice yields of 6.3 t/ha that was 11 % and 7 % higher than residue retrieval and burning, respectively. In another study at New Delhi, direct-seeded rice (DSR) alone gave ~0.5 t/ha lower yield than transplanted rice in rice–wheat cropping system. However, the loss was compensated when brown manuring with *Sesbania* was done or green gram residues were incorporated in the previous summer season, while the highest productivity was recorded under DSR followed by ZT wheat and green gram cropping system (Sharma et al. 2012).

The addition of crop residues in soil before sowing had a positive effect on the restoration of soil fertility, which results in increased yield (Table 2.2). The wheat and rice yield increased with crop residue, left over stubble and fly ash application

Table 2.2 Effect of crop residues on grain yield of rice and wheat

Crop/ treatment	No crop residues	Pre-crop residues at 5 t/ha	Leftover stubbles	Fly ash at 2 t/ha	CD ($P = 0.05$)
Rice					
2001	2.93	3.42	3.12	3.39	NS
2002	3.29	3.68	3.65	3.95	0.31
Wheat					
2001	1.87	2.33	2.25	2.37	0.22
2002	1.96	2.29	2.21	2.42	0.16

Source: Kachroo and Dixit (2005)

(Kachroo and Dixit 2005). Generally, straw retention improved aggregate stability, reduced soil erosion, and increased the infiltration and conservation of soil water, thus enhancing soil productivity (Li et al. 2007; He et al. 2009, 2011). Additionally, straw retention directly increased the input of organic matter and nutrients into soil, in turn improving soil nutrient availability for crop growth (Kaschuk et al. 2010; Qin et al. 2010).

2.9 Energy Conservation and Climate Change

Modern agricultural production system is energy and input intensive and also depends on the external use of fossil fuels for excessive use tillage, overuse of fertilisers, pesticides and other farm operations. Paddy field contributes to 5–20 % of total methane emission throughout the world under the conventional system (Scheehle and Kruger 2006; IPCC2007; Xu et al. 2007). However, CA-based technology may mitigate the risk of methane emission through direct seeding of rice crop into the field by ZT machine without standing water such as upland wheat. Similarly, the extensive tillage operations prior to wheat crop sown can be the source of atmospheric CO_2 when octane-rich fuel is used (H-ur-Rehman et al. 2015). ZT improves the operation field capacity by 81 % and grain yield by ~6 % in the CA system due to reduced tillage, or ZT ensures more timely sowing, precision and quality of field operations; saves labour and irrigation water cost; reduces weather risk in these changing climatic scenarios with improved crop productivity; and generates employment (Barclay 2006; Ladha et al. 2009; Saharawat et al. 2010). Mishra and Singh (2012) reported that the rice–wheat system require maximum energy (38,187 MJ/ha) under the conventional tilled system due to intensive field operations, whereas CA-based ZT system requires less energy and a has high energy output to input ratio as well as higher system productivity. Ram et al. (2010) also reported least cost of production, minimum energy usage, higher water productivity, higher net returns and higher energy use efficiency in CA-based (ZT) maize–wheat cropping system.

ZT reduces the cost of energy by lowering the tractor-operated costs with conventional tillage and lessening the irrigation requirement in ZT wheat than conventional methods, thus reducing the energy costs associated with pumping underground water (Hobbs and Gupta 2003). Crop residue is also an integral part of CA, which can serve as a continuous energy source for soil microorganism and provide ideal conditions for increase microbial abundance (Carter and Male 1992; Salinas-Garcia et al. 2002). The use of CT reduces the energy consumption and emission of carbon oxide emission (Holland 2004). In similar, using the best nutrient management techniques, the negative consequences of extreme climatic situations can be minimised (Subash et al. 2014). Therefore, adoption of CA-based practices can save the environment through a reduction in GHG emissions by decreasing the tillage operations (Erenstein and Laxmi 2008) particularly in IGP regions.

2.10 Carbon Sequestration and GHG Emission

Increasing the GHG emission in the atmosphere enhanced the atmospheric temperature; affected the soil process, crop production and productivity and emergence of new insect pest and caused sudden changes in climatic events. The effects of climate change are assumed to have reached to that level where the irreversible change in the functioning of the earth planet is feared. Today we need to reduce the GHG emission or capture from the atmosphere in a long-lived pool, so that it cannot re-emit to the atmosphere (Kundu et al. 2013). Among the GHGs, CO_2 plays a crucial role, due to its larger concentration and wide sources of emissions. Intensive tillage practices reduced the soil C emitted into the atmosphere. It reduced the soil fertility and productivity adversely. The CA is one of the options in the agricultural system to reduce the GHG emission and also C sequestration through agricultural crops. The reduced burning of crop residue and incorporation of surface cover enhanced the soil C and reduced the rate of C emission into the atmosphere. Minimum tillage practices also improved the plant nutrient efficiency and cut off the volatilisation or denitrification losses during the crop production.

The CA practice can contribute to making the agricultural system more resilient to climate change. It has a powerful mechanism to adopt climate change by increasing resilience to drought, increasing water use efficiency and thermal stress to agricultural crops and also increasing moisture content in soil. The CA aims to increase the annual C rate into the soil through reducing the C losses through erosion and mineralisation. When the crop residues are retained on the soil surface in combination with no tillage, it improved the soil quality and overall resource enhancement. It leads to sustainable improvement in the nutrient use efficiency, nutrient balance and availability and reduced soil moisture loss, which all enhanced the productivity of system in terms of carbon sequestration. In general the adoption of best management practices in CA sequestered the soil C 1.8 ton CO_2 per hectare per year (FAO 2008). Lal (1998) computed the adoption of best management practices of CT on 400 M ha of crop land by the year of 2020 can sequester average total C of 1500–4900 Mg. This figure can be more intensive in agricultural areas. Across the global world, average production of crop residues is ~3.4 billion Mg, its 15 % of C can be converted to passive SOC fractions, and it can promote C sequestration rate 0.2×10^{15} g/year Lal (1987). The carbon sequestration is dependent on C input and output rate; it follows three stages:

i. Steady state: in which *C* input is equal to *C* output of the system.

$$\text{Steady state} \quad \ldots \quad C_{\text{input}} = C_{\text{output}} \ldots \tag{2.1}$$

ii. Soil C depletion: *C* output rate is greater than the input. In this case C sequestration rate is low and the system promotes the emission of GHGs and climate change activities.

$$\text{C depletion} \quad \ldots \quad C_{\text{input}} < C_{\text{output}} \ldots \tag{2.2}$$

iii. Carbon sequestration: this stage is associated with CA; in which system prominently act as sequester rather than C emission. This stage is beneficial for mitigating climate change effect and reduction of GHGs.

$$\text{C sequestration} \quad \ldots \quad C_{\text{input}} > C_{\text{output}} \ldots \tag{2.3}$$

Some soils are poorly managed and the C loss (50–75 %) is higher from the ecosystem. These estimates can be more in high erosion areas in lower input or poor management by the farmers. This loss is more in coarse-textured compared to fine-textured soil. The terrestrial C sequestration C cycle is governed by the climate, soil, crop and management factors and also their complex interlinking effects. The estimates of both GHG emissions and C sequestration are also affected by soil microorganisms; poorly managed soil evolution has greater amount of CO_2 than well-managed soil.

The C sequestration under CA can be more under the situation of maximising the C input factors and reducing the C loss from soils (Table 2.3). The primary cautions are minimising the GHGs from the source points during the crop production, i.e. fertiliser application, volatilisation rate, denitrification in high pH soil, soil erosion, crop residue burning, etc., and in secondary use the best management practices in combination with the C can sequester at an optimum level (Kushwah et al. 2014). The crop diversification is also important for CA, in which N is taken up by various depths with C/N ratios of various root residues.

The SOC scenario mainly varies as per the management and tillage practices. A hypothetical graph is showing the three scenarios and each scenario having its own C sequestration potential: scenario A, SOC realistically increased to 0.25 Mg C/ha/year, whereas in scenario B it is C sequestration which is effective in CA, the same as observed in conventional agriculture, because in conventional agriculture SOC

Table 2.3 Strategies to sequester SOC

(1) Maximising C input	(2) Minimising C loss from soil
Plant selection	Reducing soil disturbance
Species, cultivar, variety	Less intensive tillage
Growth habit (perennial/annual)	Controlling soil erosion
Rotation sequence	Utilising available soil water
Biomass energy crops	Promote optimum plant growth
Tillage	Reduces soil microbial activity
Type	Maintaining surface residues cover
Frequency	Increased plant water use and production
Fertilisation	
Rate, timing, placement	More fungal dominance in soil
Organic amendments	

Source: Franzluebbers (2008)

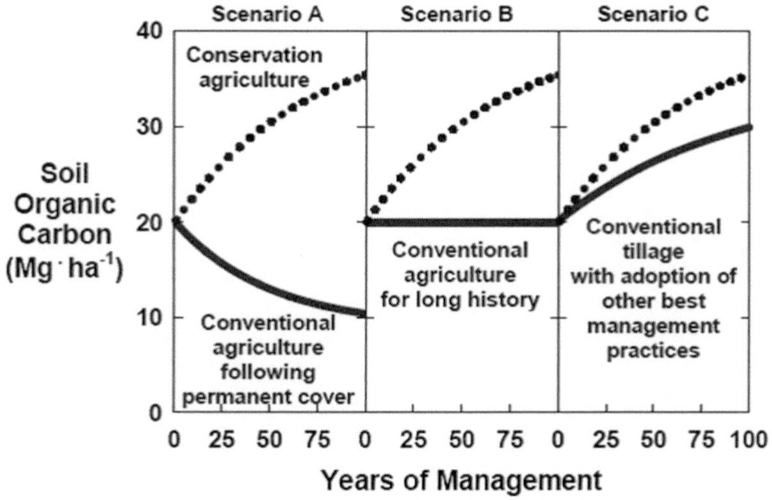

Fig. 2.5 Hypothetical examples of SOC content under CA and three different baseline conditions (Source: Adopted from Franzluebbers (2008)

rate is in a steady-state condition. In scenario C, SOC sequestration in CA would have adjusted to 0.05 Mg C/ha/year, because the conventional agriculture was improved due to CA and the sequestered SOC is at 0.10 Mg C/ha/year (Fig. 2.5).

2.11 Tangible and Intangible Benefits of CA

CA is generally a win–win situation for both farmers and environments, and it has tangible monetary benefits like the reduced labour, irrigation water, fertilisers and more stable yields and improved soil nutrient exchange capacity, thereby resulting in higher overall farm profits. Adoption of ZT wheat reduces the production cost by ₹2000–3000/ha (Malik et al. 2005) due to reduced land preparation costs, early sowing, less seed rate (Piggin et al. 2011) and reduced diesel consumption by 50–60 l/ha (Sharma et al. 2005). Thus it ultimately reduced the cost of production, increased yields and increased net returns (Sidhu et al. 2010). The use of rice residue in ZT wheat may save irrigation water through reducing water losses by evaporation (Chauhan et al. 2012); similarly furrow-irrigated raised-bed system (FIRBS) can help in saving irrigation water from 18 % to 50 % (Jat et al. 2005). In addition to tangible benefits, a large number of intangible benefits can be gained:

- Enhancement of soil quality, i.e. soil's physical, chemical and biological conditions (Jat et al. 2009; Gathala et al. 2011).

- Long-term carbon sequestration, significantly build up in soil organic matter and mitigate GHG emissions and greater resilience to production system to climate change (Saharawat et al. 2012).
- Enhancement of input (water, nutrient and fertilisers) use efficiency (Jat et al. 2012).
- Avoidance of crop residue burning reduces loss of nutrients and environmental pollution, which reduces a serious health hazard (Sidhu et al. 2007).
- Improvement of resource use efficiency through residue decomposition, improvement in soil physical environment and increased recycling and availability of plant nutrients (Jat et al. 2009).
- Residue incorporation or retention as mulch to control weeds, moderate soil temperature, reduce evaporation and improve soil biological activities (Jat et al. 2009; Gathala et al. 2011).

2.12 Constraints in Adoption of CA

Despite several tangible and intangible benefits of CA system, it is very difficult to convince the farmers about the potential benefits of CA beyond its ability to reduce production costs, due to reduction of the mainly tillage operations. The mindset of farmers is a big issue in the adoption of CA technologies particularly about ZT (Hobbs and Govaerts 2010). The major barriers to the adoption and spread of CA practices are as follows:

(i) Lack of trained and skilled human resources.
(ii) Lack of suitable and low-priced machineries (seeding/planting equipment into anchored and loose crop residue conditions) especially for small and marginal land holding farmers.
(iii) Widespread use of crop residues for livestock feed and fuel particularly in rainfed areas.
(iv) Weed management strategies without tillage practices; especially for small and marginal farmers with limited accessibility to purchase costly herbicides, weed infestations are the major obstruction to adoption of CA.
(v) Localised insect pest population densities.
(vi) Biophysical, socio-economic and cultural barriers such as limited access to financial capital, credit opportunities, inability to take risk, short-term priorities and small land farmers.
(vii) Agronomic constraints and lack of CA knowledge among the technicians and extensionist.

2.13 Future Outlook

The CA is not a panacea to arrest all agricultural problems, but it is a new paradigm for raising crops which will offer new opportunity to tackle with diverse agricultural production system and sustain the environmental quality. A lot of research on CA has been conducted, keeping views in the effect of CA on soil health (Baudron et al. 2012), but more research is needed to find the effect of different CA practices on crop yield, weed dynamics and nutrient dynamics, especially in long-term experiments. Future research needs to be identifying best nutrient management strategies with crop residue retention and economically viable crop rotations to boost crop productivity (Vanlauwe et al. 2014). Research should be conducted on weed dynamics to understand weed biotypes development under the CA and weed management research. Development of integrated weed, disease or pest management strategies is paramount for the success of CA systems. Development of low-price CA machineries particularly direct seeders/planters for seed sowing into crop residues (loose and anchored crop residues) especially for small and marginal farmers is also needed. Other research areas include understanding herbicide performance under heavy load of crop residues, nutrient dynamics under residue cover, etc.

2.14 Conclusions and Summary

Continuous shrinking of natural resources, decline in crop yield, deterioration in soil health and rising costs of agricultural inputs in conventional production system pose a threat to food security and livelihood of farmers. In this situation, CA is an obvious new paradigm in achieving higher productivity, improving environmental quality and preserving natural resources. CA systems involve mainly three principles like providing permanent soil cover, minimum mechanical soil disturbance and diversified crop rotations. Several benefits of sustainability have been addressed in ZT wheat in rice–wheat cropping system: reduced costs in fuel and labour, timely planting of crops, higher yield production, reduced weed density, saving of irrigation water and improved input use efficiency because of better crop stands due to good seed and fertiliser nutrient placement. Residue retention or inclusion of cover crops (*Sesbania*, cowpea, mung bean, etc.) on the soil surface is also one of the main CA principles which provide beneficial effects on soil moisture, temperature moderation and weed control. It also minimises water loss through evaporation which enhances higher growth rate of crop leading to higher water productivity. The combination of tillage and crop residue enhanced the SOC in the upper layer of soil. The best crop management practices improved the soil microbial diversity and population, which mediated the nutrient transformation and availability to crop plants. It improved the nutrient use efficiency and reduced the rate of inputs during the crop production. ZT reduced the CO_2 emission and improved the SOC in soil. The environmental hazards contributed by agriculture can be minimised through the CA. Problematic weed infestations are the major constraints in adoption of CA; however modified tillage practices provided an opportunity for effective weed management in CA system. ZT could increase certain weed species, but decrease other weed species depending upon the crops. Development of low-dose, high-potency, selective, post-emergence herbicides and herbicide mixtures are necessary for managing weeds in CA system. Herbicide-tolerant crops also provide opportunity to managing weeds. Therefore, the paradigm shift from tilled field to ZT–CA systems requires a thrust on nutrient management to improve soil and crop productivity and environmental quality and spread rapidly across the globe.

Acknowledgements Authors are highly thankful to Dr. K. Ramesh, Principal Scientist (Agronomy), ICAR-Indian Institute of Soil Science, Bhopal, India, for the needful help during the writing of the book chapter.

References

Abrol IP, Sangar S (2006) Sustaining Indian agriculture-conservation agriculture the way forward. Curr Sci 91(8):1020–2015

Achrya CL, Kapur OC, Dixit SP (1998) Moisture conservation for rainfed wheat production with alternative mulches and conservation tillage in the hills of north-west India. Soil Tillage Res 46:153–163

ACIAR (2013) Fine-tuning the Happy Seeder technology for adoption in northwest India. http://www.aciar.gov.au/project/cse/2006/124dt. Accessed 2 Dec 2013

Agarwal P, Goswami B (2003) Bed planting system for increasing water use efficiency of wheat (Triticum aestivum) grown in Inceptisol (Typic Ustochrept). Indian J Agric Sci 73:422–425

Akter S, Gathala MK (2014) Adoption of conservation agriculture technology in diversified systems and impact on productivity: evidence from three districts in Bangladesh. Contributed paper prepared for presentation at the 88th annual conference of the agricultural economics society, AgroParis Tech, Paris, France, 9–11 April, p 24

Aulakh J, Price AJ, Balkcom KS, Arriaga FJ, Delaney DP (2010) Weed control efficacy and lint yield of herbicide resistant cotton technologies under different tillage systems and row spacing. In: Conservation agriculture impacts – local and global-proceedings of the 32nd southern conservation agricultural systems conference 20–22 July 2010. Tennessee, Jackson

Ball AD (1992) Weed seed bank response to tillage, herbicide and crop rotation sequence. Weed Sci 40:654–659

Barclay A (2006) The direct approach: a return to the ways of their forefathers has seen Indian and Bangladeshi rice farmers reduce their cost for water and address the growing problem of labor shortages. Rice Today 5:12–18

Baudron F, Tittonell P, Corbeels M, Letourmy P, Giller KE (2012) Comparative performance of conservation agriculture and current smallholder farming practices in semi-arid Zimbabwe. Field Crops Res 132:117–128

Bhushan L, Ladha JK, Gupta RK, Singh S, Tirol-Padre A, Saharawat YS, Gathala MK, Pathak H (2007) Saving of water and labour in a rice-wheat system with no-tillage and direct seeding technologies. Agron J 99(5):1288–1296

Blaise Desouza (2014) Conservation agriculture in cotton production. In: Somasundaram J, Chaudhary RS, Subba Rao A, Hati KM, Sinha NK, Vassanda Coumar M (ed) Conservation agriculture for carbon sequestration and sustaining soil health. New India Publishing Agency. ISBN 978-93-93-83305-32–2. pp 193–200

Bradford JM, Peterson GA (2000) Conservation tillage. In: Sumner ME (ed) Handbook of soil science. CRC Press, Boca Raton, pp G247–G269

Buhler DD, Stoltenberg DE, Becker RL, Gunsolus JL (1994) Perennial weed population after 14 years of variable tillage and cropping practices. Weed Sci 34:29–33

Carter MR, Mele PM (1992) Changes in microbial biomass and structural stability at the surface of a duplex soil under direct drilling and stubble retention in North-Eastern Victoria. Aust J Soil Res 30:493–503

Cassman KG, Dobermann A, Sta Cruz PC, Gines GC, Samson MI, Descalsota JP, Alcantara JM, Dizon MA, Olk DC (1996) Soil organic matter and the indigenous nitrogen supply of intensive irrigated rice systems in the tropics. Plant Soil 182:267–278

Chan KY, Cowie A, Kelly G, Singh B, Slavich P (2008) Scoping paper: soil organic carbon sequestration potential for agriculture in NSW. NSW Department of Primary Industries Science Res Tech paper, pp 1–29

Chauhan DS, Sharma RK, Kharub AS, Tripathi SC, Chhokar RS (2001) Effect of crop intensification on crop productivity, profitability, productivity, energetic and soil fertility in rice (*Oryza Sativa* L.) and wheat (*Triticum aestivum* L.) cropping system of North-Western plain. Indian J Agric Sci 71(5):299–302

Chauhan DS, Sharma RK, Chhokar RS (2003) Comparative performance of tillage options in wheat (Triticum aestivum) productivity and weed management. Indian J Agric Sci 73:402–406

Chauhan BS, Singh RG, Mahajan G (2012) Ecology and management of weeds under conservation agriculture: a review. Crop Prot 38:57–65

Dalal RC, Chan KY (2001) Soil organic matter in rainfed cropping systems of the Australian cereal belt. Aust J Soil Res 39:435–464

Das TK (2014) Weeds and their control methods. In: Weed management, pp 9–14. http://www.nsdl.niscair.res.in/bitstream/123456789/568/1/Weeds+and+their+control+methods+Formatted.pdf. Accessed 6 Mar 2014

Das TK, Bandyopadhyay KK, Ranjan Bhattacharyyaa, Sudhishri S, Sharma AR, Behera UK, Saharawata YS, Sahoo PK, Pathak H, Vyas AK, Bhar LM, Gupta HS, Gupta RK, Jat ML (2016) Effects of conservation agriculture on crop productivity and water-use efficiency under an irrigated pigeon pea wheat cropping system in the western Indo-Gangetic Plains. J Agric Sci. doi:http://dx.doi.org/10.1017/S0021859615001264

Donovan G, Casey F (1998) Soil fertility management in Sub–Saharan Africa, World Bank Technical paper no 408. The World Bank, Washington, DC

Dotaniya ML (2012) Crop residue management in rice-wheat cropping system. Lap Lambert Academic Publisher, Saarbrücken, p 116. ISBN 978-3–659–29388–7

Dotaniya ML, Sharma MM, Kumar K, Singh PP (2013) Impact of crop residue management on nutrient balance in rice-wheat cropping system in an Aquic hapludoll. J Rural Agric Res 13 (1):122–123

Dotaniya ML, Datta SC, Biswas DR, Kumar K (2014) Effect of organic sources on phosphorus fractions and available phosphorus in Typic Haplustept. J Indian Soc Soil Sci 62(1):80–83

Du Preez CC, Steyn JT, Kotze E (2001) Long-term effects of wheat residue management on some fertility indicators of a semi-arid Plinthosol. Soil Tillage Res 63:25–33

Duke SO, Cerdeira AL (2005) Potential environmental impacts of herbicide resistant crops. Collect Biosaf Rev 2:66–143

Duke SO, Powles SB (2008) Glyphosate: a once-in-a-century herbicide. Pest Manag Sci 64:319–325

Dumanski J, Peiretti R, Benetis J, McGarry D, Pieri C (2006) The paradigm of conservation tillage. Proc World Assoc Soil Water Conserv 1:58–64

Erenstein O, Laxmi V (2008) Zero tillage impacts in India's rice-wheat systems: a review. Soil Tillage Res 100:1–14

FAO (2008) Investing in sustainable agricultural intensification: the role of conservation agriculture–a framework for action' FAO Rome, August 2008 (Available at www.fao.org/ag/ca/)

FAO (2015) AQUASTAT main database – Food and Agriculture Organization of the United Nations (FAO)

Farooq M, Flower KC, Jabran K, Wahid A, Siddique KHM (2011) Crop yield and weed management in rainfed conservation agriculture. Soil Tillage Res 117:172–183

Franzluebbers AJ (2008) Soil organic carbon sequestration with conservation agriculture in the south eastern USA: potential and limitations. http://www.fao.org/ag/ca/Carbon%20Offset%20Consultation/carbonmeeting/3fullpapersbyconsultationspeakers/paperfranzluebbers.pdf

Franzluebbers AJ, Hons FM (1996) Soil-profile distribution of primary and secondary plant available nutrients under conventional and no tillage. Soil Tillage Res 39:229–239

Friedrich T, Kassam, AH (2009) Adoption of conservation agriculture technologies: constraints and opportunities. Invited paper, IV World Congress on Conservation Agriculture, 4–7 February 2009, New Delhi, India

Friedrich T, Derpsch R, Kassam AH (2012) Global overview of the spread of conservation agriculture. Field Actions Sci Rep 6:1–7

Gathala MK, Ladha JK, Saharawat YS, Kumar V, Kumar V, Sharma PK (2011) Effect of tillage and crop establishment methods on physical properties of a medium-textured soil under a seven-year rice–wheat rotation. Soil Sci Soc Am J 75:1851–1862

Ghosh BN, Dogra P, Sharma NK, Bhattacharyya R, Mishra PK (2015) Conservation agriculture impact for soil conservation in maize–wheat cropping system in the Indian sub-Himalayas. Int Soil Water Conserv Res 3:112–118

Ghosh BN, Meena VS, Alam NM, Dogra P, Bhattacharyya R, Sharma NK, Mishra PK (2016) Impact of conservation practices on soil aggregation and the carbon management index after seven years of maize–wheat cropping system in the Indian Himalayas. Agr Ecosyst Environ 216:247–257

Giller KE, Witter E, Corbeels M, Tittonell P (2009) Conservation agriculture and smallholder farming in Africa: the heretics' view. Field Crop Res 114:23–34

Govaerts B, Fuentes M, Mezzalama M, Nicol JM, Deckers J, Etchevers JD, Figueroa SB, Sayre KD (2007) Infiltration, soil moisture, root rot and nematode populations after 12 years of different tillage, residue and crop rotation managements. Soil Tillage Res 94:209–219

Guo S, Jilin Z, Yuanmin S, ZuHua B (1995) Yield potential and agronomic techniques for wheat after rice. Jiangsu Academy of Agricultural Sciences, Nanjing

Gupta RK, Seth A (2007) A review of resource conserving technologies for sustainable management of the rice-wheat cropping systems of the Indo-Gangetic Plains (IGP). Crop Prot 26:436–447

He J, Kuhn NJ, Zhang XM, Zhang XR, Li HW (2009) Effects of 10 years of conservation tillage on soil properties and productivity in the farming–pastoral ecotone of Inner Mongolia, China. Soil Use Manag 25:201–209

He J, Li HW, Rasaily RG, Wang QJ, Cai GH, Su YB, Qiao XD, Liu LJ (2011) Soil properties and crop yields after 11 years of no tillage farming in wheat–maize cropping system in North China Plain. Soil Tillage Res 113:48–54

Heimlich RE, Fernandez-Cornejo JF, McBride W, Klotz-Ingram C, Jans S, Brooks N (2000) Adoption of genetically engineered seed in U.S. agriculture. In: Fairbairn C, Scoles G, McHughen A (eds) Proceedings of 6th international symposium on the Bio-safety of GMOs, Saskatoon, pp. 56–63

Hobbs PR, Govaerts B (2010) How conservation agriculture can contribute to buffering climate change. In: Reynolds MP (ed) Climate change and crop production. CAB International, Cambridge, MA, pp 177–199

Hobbs PR, Gupta RK (2003) Resource conserving technologies for wheat in the rice–wheat system. In: Ladha JK, Hill J, Gupta RK, Duxbury JM, Buresh RJ (eds) Improving the productivity and sustainability of rice–wheat systems: issues and impact. American Society of Agronomy (ASA), Madison, pp 149–171

Hobbs PR, Giri GS, Grace P (1997) Reduced and zero-tillage options for the establishment of wheat after rice in South Asia. RWCIGP Paper No. 2. Mexico: Rice–Wheat Consortium for the Indo-Gangetic Plains and CIMMYT

Hobbs PR, Sayre K, Gupta R (2008) The role of conservation agriculture in sustainable agriculture. Philos Trans R Soc B 363:543–555

Holland JM (2004) The environmental consequences of adopting conservation tillage in Europe: reviewing the evidence. Agric Ecosyst Environ 103:1–25

Humphreys E, Kukkal SS, Christen EW, Hira GS, Balwinder-Singh, Yadav S, Sharma RK (2010) Halting groundwater decline in north-west India – which crop technologies will be winners. Adv Agron 109:156–199

H-ur-Rehman, Ahmad Nawaz, Abdul Wakeel, Yashpal Singh Saharawat and Muhammad Farooq (2015) Conservation Agriculture in South Asia M. Springer International Publishing Switzerland 2015 Farooq, Siddique KHM (eds) Conservation agriculture. doi:10.1007/978-3-319-11620-4_11

IARI (2012) Crop residues management with conservation agriculture: potential, constraints and policy needs. Indian Agricultural Research Institute, New Delhi, PP vii+32

IPCC (2007) Summary for policymakers. In: Parry ML, Canziani OF, Palutikof JP, Linden PJV, Hanson CE (eds) Climate change 2007: impacts, adaptation and vulnerability. Contribution of working group II to the fourth assessment report of the intergovernmental panel on climate change. Cambridge University Press, Cambridge, pp 7–22

Ismail I, Blevins RL, Frye WW (1994) Long-term no-tillage effects on soil properties and continuous corn yields. Soil Sci Soc Am J 58:193–198

Jalota SK, Buttar GS, Sood A, Chahal GBS, Ray SS, Panigrahy S (2008) Effects of sowing date, tillage and residue management on productivity of cotton (*Gossypium hirsutum* L.)–Wheat (*Triticum aestivum* L.) system in northwest India. Soil Tillage Res 99:76–83

Jat ML, Singh S, Rai HK, Chhokar RS, Sharma SK, Gupta RK (2005) Furrow irrigated raised bed planting technique for diversification of rice–wheat system of Indo-Gangetic plains. J Jpn Assoc Int Coop Agric For 28:25–42

Jat ML, Gathala MK, Ladha JK, Saharawat YS, Jat AS, Vipin K, Sharma SK, Kumar V, Gupta R (2009) Evaluation of precision land leveling and double zero-till systems in rice-wheat rotation: water use, productivity, profitability and soil physical properties. Soil Tillage Res 105:112–121

Jat ML, Saharawat YS, Raj G (2011) Conservation agriculture in cereal system of south Asia: nutrient management perspectives. Karnataka J Agric Sci 24(1):100–105

Jat ML, Malik RK, Saharawat YS, Gupta R, Bhag M, Paroda R (2012) Proceedings of regional dialogue on conservation agricultural in south Asia. APAARI, CIMMYT, ICAR, New Delhi, p 32

Jat ML, Gathala MK, Saharawat YS, Tetarwal JP, Gupta R, Singh Y (2013) Double no-till and permanent raised beds in maize–wheat rotation of north-western Indo-Gangetic Plains of India: effects on crop yields, water productivity, profitability and soil physical properties. Field Crops Res 149:291–299

Kachroo D, Dixit AK (2005) Residue-management practices using fly ash and various crop residues for productivity of rice -wheat cropping system under limited moisture conditions. Indian J Agron 50(4):249–252

Kang MS (2013) Happy Seeder technology is well received by punjab farmers. http://www.bharatsandesh.com/bharat/index.php?option=com_content&task=view&id=16957&Itemid=27dt. Accessed 2 Dec 2013

Kaschuk G, Alberton O, Hungria M (2010) Three decades of soil microbial biomass studies in Brazilian ecosystems: lessons learned about soil quality and indications for improving sustainability. Soil Biol Biochem 42:1–13

Kassam AH, Friedrich T (2009) Perspectives on nutrient management in conservation agriculture. Invited paper, IV World Congress on conservation agriculture, 4–7 February 2009, New Delhi, India

Kassam AH, Friedrich T, Shaxson F, Jules P (2009) The spread of conservation agriculture: justification, sustainability and uptake. Int J Agric Sustain 7(4):292–320

Kassam A, Friedrich T, Derpsch R, Kienzle J (2014) Worldwide adoption of conservation agriculture. 6th World Congress on Conservation Agriculture, 22–27 June 2014, Winnipeg, Canada http://www.ctic.org/WCCA/

Kennedy AC (1999) Soil microorganisms for weed management. J Crop Prod 2:123–138

Kirkegaard JA, Conyers MK, Hunt JR, Kirkby CA, Watt M, Rebetzke GJ (2014) Sense and nonsense in conservation agriculture: principles, pragmatism and productivity in Australian mixed farming systems. Agric Ecosyst Environ 187:133–145

Kumar K, Goh KM (2000) Crop residue management: effects on soil quality, soil nitrogen dynamics, crop yield, and nitrogen recovery. Adv Agron 68:197–319

Kumar A, Bahadur I, Maurya BR, Raghuwanshi R, Meena VS, Singh DK, Dixit J (2015) Does a plant growth promoting rhizobacteria enhance agricultural sustainability? J Pure Appl Microbiol 9(1):715–724

Kundu S, Dotaniya ML, Lenka S (2013) Carbon sequestration in Indian agriculture. Climate change and natural resources management. In: Lenka S, Lenka NK, Kundu S, Subba Rao A (eds) New India Publishing Agency. ISBN 978-93-81450-67-3. 269-289. p 363

Kushwah SK, Dotaniya ML, Upadhyay AK, Rajendiran S, Coumar MV, Kundu S, Rao AS (2014) Assessing carbon and nitrogen partition in kharif crops for their carbon sequestration potential. Natl Acad Sci Lett 37(3):213–217

Ladha JK, Fischer KS, Hossain M, Hobbs PR, Hardy B (2000) Improving the productivity and sustainability of rice-wheat systems of the indo-gangetic plains: a synthesis of NARS-IRRI partnership research, Discussion Paper No. 40. International Rice Research Institute, Los Banos

Ladha JK, Kumar V, Alam MM, Sharma S, Gathala MK, Chandna P, Saharawat YS, Balasubramanian V (2009) Integrating crop and resource management technologies for enhanced productivity, profitability and sustainability of the rice-wheat system in South Asia. In: Ladha JK, Yadvinder-Singh, Erenstein O, Hardy B (eds) Integrated crop and resource management in the rice-wheat system of South Asia. IRRI, Los Banos, pp 69–108

Lal R (1987) Managing the soils of sub-Saharan Africa. Science 236:1069–1076

Lal R (1998) Soil erosion impact on agronomic productivity and environment quality. Crit Rev Plant Sci 17:319–464

Lampurlanes J, Angas P, Cantero-Martinez C (2001) Root growth, soil water content and yield of barley under different tillage systems on two soils in semiarid conditions. Field Crop Res 69:27–40

Li HW, Gao HW, Wu HD, Li WY, Wang XY, He J (2007) Effects of 15 years of conservation tillage on soil structure and productivity of wheat cultivation in northern China. Aust J Soil Res 45:344–350

Liu EK, Zhao BQ, Mei XR, So HB, Li J, Li XY (2010) Effects of no-tillage management on soil biochemical characteristics in northern China. J Agric Sci 148:217–243

Malik RK, Singh S (1995) Little seed canary grass (*Phalaris minor*) resistance to Isoproturon in India. Weed Technol 9:419–425

Malik RK, Gupta RK, Singh CM, Yadav A, Brar SS, Thakur TC, Singh SS, Singh AK, Singh R, Sinha RK (2005) Accelerating the adoption of resource conservation technologies in rice wheat system of the Indo-Gangetic plains. Proceedings of Project Workshop, Directorate of Extension Education, Chaudhary Charan Singh Haryana Agricultural University (CCSHAU), June 1–2, 2005. Hisar, CCSHAU

Martin Diaz-Zorita (2006) Innovations for improving productivity and nutrient use efficiency: no-till grain cropping systems of South America. 18th World Congress of Soil Science July 9–15, 2006 – Philadelphia, Pennsylvania, USA

Meena RS, Yadav RS, Meena VS (2013) Heat unit efficiency of groundnut varieties in scattered planting with various fertility levels. Bioscan 8(4):1189–1192

Meena RS, Meena VS, Meena SK, Verma JP (2015a) The needs of healthy soils for a healthy world. J Clean Prod 102:560–561

Meena RS, Meena VS, Meena SK, Verma JP (2015b) Towards the plant stress mitigate the agricultural productivity: a book review. J Clean Prod 102:552–553

Meena VS, Meena SK, Verma JP, Meena RS, Ghosh BN (2015c) The needs of nutrient use efficiency for sustainable agriculture. J Clean Prod 102:562–563

Meena VS, Meena SK, Verma JP, Meena RS, Jat LK (2015d) Current trend to mitigate climate change and sustainable agriculture: a book review. J Clean Prod 102:548–549
Meena VS, Verma JP, Meena SK (2015e) Towards the current scenario of nutrient use efficiency in crop species. J Clean Prod 102:556–557
Meena RS, Bohra JS, Singh SP, Meena VS, Verma JP, Verma SK, Sihag SK (2016) Towards the prime response of manure to enhance nutrient use efficiency and soil sustainability a current need: a book review. J Clean Prod 112(1):1258–1260
Mehla RS, Verma JK, Gupta, RK, Hobbs, PR (2000) Stagnation in the productivity of wheat in the Indo-Gangetic plains: zero-till-seed-cum-fertilizer drill as an integrated solution. Rice-wheat Consortium Paper Series 8, Rice-Wheat Consortium for the Indo-Gangetic Plains, New Delhi, 12p
Mishra JS, Singh VP (2012) Effect of tillage sequence and weed management on weed dynamics and productivity of dry-seeded rice (*Oryza sativa*)—wheat (*Triticum aestivum*) system. Indian J Agron 57:14–19
Mohammad W, Shah SM, Shehzadi S, Shah SA (2012) Effect of tillage, rotation and crop residues on wheat crop productivity, fertilizer nitrogen and water use efficiency and soil organic carbon status in dry area (rainfed) of north-west Pakistan. J Soil Sci Plant Nutri 12(4):715–727
Montgomery DR (2007) Soil erosion and agricultural sustainability. Proc Natl Acad Sci U S A 104:13268–13272
Mukherjee PK, Debnath P (2013) Weed control practices in maize (Zea mays L.) under and conservation tillage practices. In: Proceedings of 24th Asian-Pacific Weed Science Society Conference, October 22–25, 2013, Bandung, Indonesia p 302
Nazir S (1994) Crop production, 1st edn. National Book Foundation, Islamabad
Pande S, Sharma M, Ghosh R (2012) Role of pulses in sustaining agricultural productivity in the rainfed rice-fallow lands of India in changing climatic scenario. International Crops Research Institute for the Semi-arid Tropics (ICRISAT), Patancheru–502 324 (Andhra Pradesh), India Email: s.pande@cgiar.org. Climate Change and Food Security in India. pp 53–70
Pathak H, Saharawat YS, Gathala M, Ladha JK (2011) Impact of resource-conserving technologies on productivity and greenhouse gas emission in rice-wheat system. Greenh Gases Sci Technol 1:261–277
Patil SL, Sheelavantar MN (2006) Soil water conservation and yield of winter sorghum (*Sorghum bicolor* L. Moench) as influenced by tillage, organic materials and nitrogen fertilizer in semi-arid tropical India. Soil Tillage Res 89:246–257
Piggin C, Haddad A, Khalil Y (2011) Development and promotion of zero tillage in Iraq and Syria. Proceedings of 5th World Congress on Conservation Agriculture, Brisbane, 26–29 Sep 2011, pp 304–305
Qamar R, Ehsanullah RA, Ali A, Ghaffar A, Mahmood A, Javeed HMR, Aziz M (2013) Growth and economic assessment of wheat under tillage and nitrogen levels in rice-wheat system. Am J Plant Sci 4:2083–2091
Qin SP, He XH, Hu CS, Zhang YM, Dong WX (2010) Responses of soil chemical and microbial indicators to conservational tillage versus traditional tillage in the North China Plain. Eur J Soil Biol 46:243–247
Ram H, Kler DS, Singh Y, Kumar K (2010) Productivity of maize (*Zea mays*)–wheat (*Triticumaestivum*) system under different tillage and crop establishment practices. Indian J Agron 55:185–190
Ram H, Yadvinder-Singh Saini, KS, Kler DS, Timsina J (2013) Tillage and planting methods effects on yield, water use efficiency and profitability of soybean–wheat system on a loamy sand soil. Expl Agric pp. 1–9 Cambridge University Press 2013 doi:10.1017/S0014479713000264
Ramesh K (2015) Weed problems, ecology, and management options in conservation agriculture: issues and perspectives. Adv Agron 131:251–303
Rao VS (2000) Principles of weed science, 2nd edn. Science Publishers, Enfield, pp 37–39

Reddy RR, Veeranna G (2008) Rice-zero tillage in maize proved economical in Andhra Pradesh. Indian Fmg 58(5):8–9. Rice-wheat cropping system. In: Abrol IP, Gupta RK, Malik RK (eds) Conservation agriculture- status and prospects, Centre for Advancement of Sustainable Agriculture, New Delhi p. 242

Saha S, Chakraborti D, Sharma AR, Tomar RK, Bhadraray S, Sen U, Behra UK, Purakayastha TJ, Garg RN, Kalra N (2010) Effect of tillage and residue management on soil physical properties and crop productivity in maize (Zea mays)–Indian mustard (Brassica juncea) system. Indian J Agric Sci 80(8):679–685

Saharawat YS, Gathala M, Ladha JK, Malik RK, Singh S, Jat ML, Gupta R, Pathak H, Singh K (2009) Evaluation and promotion of integrated crop and resource management in rice-wheat system in North West India. In: Ladha JK et al (eds) Integrated crop and resource management in rice wheat system of south Asia. IRRI, Los Banos, pp 133–150

Saharawat YS, Singh B, Malik RK, Ladha JK, Gathala MK, Jat ML, Kumar V (2010) Evaluation of alternative tillage and crop establishment methods in a rice-wheat rotation in North Western IGP. Field Crop Res 116:260–267

Saharawat YS, Ladha JK, Pathak H, Gathala M, Chaudhary N, Jat ML (2012) Simulation of resource-conserving technologies on productivity, income and greenhouse gas emission in rice-wheat system. J Soil Sci Environ Manag 3(1):9–22

Salinas-Garcia JR, Velázquez-García JD, Gallardo-Valdez A, Díaz-Mederos P, Caballero-Hernández F, Tapia-Vargas LM, Rosales-Robles E (2002) Tillage effects on microbial biomass and nutrient distribution in soils under rain-fed corn production in central-western Mexico. Soil Tillage Res 66:143–152

Sapkota TB (2012) Conservation tillage impact on soil aggregation, organic matter turnover and biodiversity. In: Lichtfouse E (ed) Organic fertilisation, soil quality and human health. Springer, New York, pp 141–160

Sapkota TB, Majumdar K, Jat ML, Kumara A, Bishnoi DK, McDonald AJ, Pampolino M (2014) Precision nutrient management in conservation agriculture based wheat production of Northwest India: profitability, nutrient use efficiency and environmental foot print. Field Crops Res 155:233–244

Scheehle EA, Kruger D (2006) Global anthropogenic methane and nitrous oxide emissions. Energy J 27:33–44

Sen A, Sharma SN, Singh RK, Pandey MD (2002) Effect of different tillage system on the performance of wheat. Herbicide resistance management and zero tillage in rice-wheat cropping system. CCSHAU, Hissar, pp 115–116

Serraj R, Siddique KHM (2012) Conservation agriculture in dry areas. Field Crops Res 132:1–6

Sharma AR, Singh R, Dhyani SK (2005) Conservation tillage and mulching for optimizing productivity in maize-wheat cropping system in the outer western Himalaya region – a review. Indian J Soil Conserv 33(1):35–41

Sharma AR, Jat ML, Saharawat YS, Singh VP, Singh R (2012) Conservation agriculture for improving productivity and resource-use efficiency: prospects and research needs in Indian context. Indian J Agron 57(3rd IAC Special Issue):131–140

Sharma AR, Singh VP, Singh R (2013) Weed management in conservation agriculture systems – problems and prospects. In: Proceedings of 24th Asian-Pacific Weed Science Society conference, October 22–25, 2013, Bandung, Indonesia, pp 31–35

Sidhu HS, Singh M, Humphreys E, Singh Y, Singh B, Dhillon SS, Blackwell J, Bector VM, Singh S (2007) The happy seeder enables direct drilling of wheat into rice straw. Aust J Exp Agric 47:844–854

Sidhu RS, Kamal V, Dhaliwal HS (2010) Conservation agriculture in Punjab: economic implications of technologies and practices. Indian J Agric Econ 53(3):1413–1427

Sims BG (2011) Some advances in mechanization options for conservation agriculture systems: reflections from the V World Congress of Conservation Agriculture, Brisbane, Australia, September 2011, www.fao.org/ag/ca

Singh KK, Sharma SK (2005) Conservation tillage and crop residue management in Wang QJ, Bai YH, Gao HW, He J, Chen H, Chesney RC, Kuhn NJ, Li HW (2008) Soil chemical properties and microbial biomass after 16 years of no-tillage farming on the Loess Plateau, China, Geoderma 144:502–508

Singh P, Aipe KC, Prasad R, Sharma SN, Singh S (1998) Relative effect of zero and conventional tillage on growth and yield of wheat (Triticum aestivum) and soil fertility. Indian J Agron 43 (2):204–207

Singh S, Ladha JK, Gupta RK, Bhushan L, Rao AN, Sivaprasad B, Singh PP (2007) Evaluation of mulching, intercropping with *Sesbania* and herbicide use for weed management in dry-seeded rice (*Oryza sativa* L.). Crop Prot 26:518–524

Singh VP, Kumar A, Banga A (2010) Current status of zero tillage in weed management. Indian J Weed Sci 42(1&2):1–9

Singh DK, Singh SK, Singh AK, Meena VS (2014) Impact of long term cultivation of lemon grass (Cymbopogon citratus) on post-harvest electro-chemical properties of soil. Ann Agriv Biol Res 19(1):45–48

Stubbs TL, Kennedy AC (2012) Microbial weed control and microbial herbicides. INTECH Open Access Publisher, Rijeka

Subash N, Gangwar B, Singh R, Sikka AK (2014) Identification of climate-resilient integrated nutrient management practices for rice-rice cropping system—an empirical approach to uphold food security. Int J Biometeorol. doi:10.1007/s00484-014-0825-5 (published online)

Tiwari KN (2002) Balanced fertilization for food security. Fertil News 47:113–122

Tiwari VM, Wahr J, Swenson S (2009) Dwindling groundwater resources in northern India, from satellite gravity observations. Geophys Res Lett 36:L18401. doi:10.1029/2009GL039401

Tubbs RS, Gallaher RN (2005) Conservation tillage and herbicide management for two peanut cultivars. Agron J 97:500–504

Vanlauwe B, Wendt J, Giller KE, Corbeels M, Gerard B, Nolte C (2014) A fourth principle is required to define conservation agriculture in sub-Saharan Africa: the appropriate use of fertilizer to enhance crop productivity. Field Crops Res 155:10–13

Vargas RN, Wright SD (2005) UC IPM pest management guidelines: cotton. State wide IPM Program. Division of Agriculture and Natural Resources, University of California Publication 3444, Oakland

Verma UN, Srivastava VC (1989) Weed management in wheat under zero and optimum tillage conditions. Indian J Agron 34:170–179

Verma JP, Jaiswa DK, Meena VS, Meena RS (2015a) Current need of organic farming for enhancing sustainable agriculture. J Clean Prod 102:545–547

Verma JP, Jaiswal DK, Meena VS, Kumar A, Meena RS (2015b) Issues and challenges about sustainable agriculture production for management of natural resources to sustain soil fertility and health. J Clean Prod 107:793–794

Witmer JE, Hough-Goldstein JA, Pesek JD (2003) Ground-dwelling and foliar arthropods in four cropping systems ground-dwelling and foliar arthropods in four cropping systems. Environ Entomol 32:366–376

Xu S, Peter RJ, Denise LM (2007) A process-based model for methane emission from flooded rice paddy systems. Ecol Model 205:475–491

Yadav JPS, Singh GB (2000) Natural resource management for agriculture production. International conference on managing natural resources for sustainable agriculture production, pp 14–18

Yadav DS, Shukla RP, Kumar SB (2005) Effect of zero tillage and nitrogen level on wheat after rice. Indian J Agron 50:52–53

Zhang L, Van Der Werf W, Zhang SP, Spiertz HJ (2007) Growth, yield and quality of wheat and cotton in relay strip intercropping system. Field Crop Res 103(3):178–188

Zheng C, Jiangc Y, Chenc C, Sunc Y, Fenga J, Denga A, Songa Z, Zhang W (2014) The impacts of conservation agriculture on crop yield in China depend on specific practices, crops and cropping regions. Crop J 2:289–296

Resource Conservation Technologies (RCTs) for Climate-Resilient Agriculture in the Foothill of Northwest Himalayas

3

Sanjay Arora and Rajan Bhatt

Abstract

Potential threats to agriculture in the foothills of northwest Himalayas are the climate change. Further, erratic rains, undulating slopes and traditional management practices further strengthen the problem which might be responsible for soil erosion, low NUE, lower grains and finally to the lower livelihoods of the farmers. Climate change phenomenon influences the agriculture over the globe. India especially the northwest Himalayan region is one of the most vulnerable areas. Climate change is projected to reduce timely sown irrigated wheat production by ~6 % by 2020. In the case of late-sown wheat, the projected levels are alarmingly high, to the extent of 18 %. Similarly, a 4 % fall in the yield of irrigated rice crop and a 6 % fall in rain-fed rice are foreseen by 2020 due to climate changes. The warming trend in India over the past 100 years is estimated at 0.60 °C. The projected impacts are likely to further aggravate yield fluctuations of many crops with impact on food security. Thus there is a need for the serious attention on adaptation and mitigation strategies to overcome the problems of climate change more particularly in the northwest Himalayan region. Sustainable food security is further affected by persistent land degradation, land fragmentation, labour problem, overexploitation of natural resources, etc. We need to focus on sustainable production systems by strengthening the ecological foundations and mitigating the adverse effect of the global warming. This requires an integrated approach by considering technological, biophysical,

S. Arora (✉)
Regional Research Station, Central Soil Salinity Research Institute (ICAR), Lucknow, Uttar Pradesh, India
e-mail: aroraicar@gmail.com

R. Bhatt (✉)
Farm Advisory Service Scheme, Tarn Taran, Punjab Agricultural University, 141 004 Ludhiana, Punjab, India
e-mail: rajansoils@pau.edu

J.K. Bisht et al. (eds.), *Conservation Agriculture*,
DOI 10.1007/978-981-10-2558-7_3

socio-economic, political and environmental factors in one basket. Food security and environmental sustainability can be attained by improved land and water management, adopting eco-friendly technologies and initiating good agricultural practices in different agroecosystems. Further, strategic research and technology in agriculture and adoption of sustainable practices are necessary to meet current and future threats to food security.

Keywords
Global warming • Sustainable agriculture • Climate change • NW Himalayan region

3.1 Introduction

The Indian Himalayan Region (IHR), with geographical coverage of over 5.3 lakh km^2, constitutes a large proportion of the hotspot and, therefore, contributes greatly to the richness and representativeness of its biodiversity components at all levels. Out of this 5.3 lakh km^2, 33.13 Mha of the areas are being constituted by the northwest (NW) Himalayas. Further, most of the water used to grow maize crop in the sub-humid foothill region of northwest Himalayas is derived from rainfall. Erratic rains, fragile ecosystems and traditional indigenous management practices are mainly responsible for the current situation (Arora 2006; Arora et al. 2006; Bhatt et al. 2013). The IHR covers 11 states entirely (i.e. Jammu and Kashmir, Himachal Pradesh, Punjab, Uttarakhand, Sikkim, Arunachal Pradesh, Nagaland, Manipur, Mizoram, Tripura, Meghalaya) and two states partially (i.e. hilly districts of Assam and West Bengal). The region represents ~4 % of the total human population of the country and exhibits diversity of ethnic groups which inhabit remote terrains. Further it is reported that the northwest Himalayan region (NWHR) which spreads to an approximate area of 33.13 million hectares, comprising of Jammu and Kashmir, Himachal Pradesh and Uttarakhand which is ~10 % of the country's total geographical area, supports ~2.4 and 4 % of human and cattle population of the country, respectively (Sharma et al. 2013a, b). This region has a diverse climate, topography, vegetation, ecology and land use pattern. The annual average rainfall varies from 80 mm in Ladakh to over 200 cm in some parts of Himachal Pradesh and Uttarakhand in North India. The major natural resources are water, forests, floral and faunal biodiversity. Forests constitute the major share in the land use of the region with only ~15 % of the net sown area and ~162 % cropping intensity (Ashraf 2015). Due to hill and mountainous topography, the region differs from plain in respect to weather and soil parameters, biodiversity, ethnic diversity, land use systems and socio-economic conditions. Growing concerns for deteriorating environment by stakeholders and others seem to have linkage with gigantic cause-and-effect arguments due to deforestation, landslides, large-scale downstream flooding, increasing poverty and malnutrition. Recent estimates indicate that NW Indian Himalayas has considerable area under potential erosion rate which is really alarming (Bhatt et al. 2013). In northwestern Himalayas, on an average, ~17 % of the area falls in very severe category with erosion rates >40 t ha^{-1} $year^{-1}$, while ~25 % area has erosion rate of more than 10 t ha^{-1} $year^{-1}$. The state of Uttarakhand in Western Himalayas has a maximum area of 33 %, under very severe category with erosion rates of >40 t ha^{-1} $year^{-1}$. It calls for serious efforts to employ appropriate conservation measures to check land degradation problems (ICAR and NASC 2010). As per Sharda et al. (2012) ~50 % of the total geographical area of India falls in priority five erosion risk classes. According to Mandal and Sharda (2011), assessment of soil loss tolerance limits (SLTLs) serves as a tool to gauge the potential erosion risk in a given area with regard to long-term sustainability. The analysis has indicated that soil loss tolerance or *T*-value varies from 2.5 to 12.5 Mg ha^{-1} $year^{-1}$ depending upon soil quality governing soil resistibility to erosion and depth at a particular location. Approximately 57 % of the area in the country has permissible soil loss of less than 10.0 Mg ha^{-1} $year^{-1}$, which needs to be treated with appropriate conservation measures. The highest priority needs to be accorded to ~7.5 % of the area where the *T*-value is only 2.5 Mg ha^{-1} $year^{-1}$ due to soil quality constraints. In Punjab ~10 % of the area of the state is under sub-mountainous tract which is also severely affected by the problem of soil erosion in general and gully erosion in particular. Maize-based cropping systems mainly dominated in the region even though water use efficiency (WUE) of maize (a C_4 crop) is approximately double (Nafziger 2014). Therefore, moisture conservation techniques which are suitable for the area will serve the purpose and should be a paramount objective to attain sustainable maize yield in this region (Arora and Bhatt 2006; Arora et al. 2010). Maize along with leguminous crops not only improves the soil fertility status but also improves the livelihood of the farmer of the region (Nafziger 2014). Maize being widely spaced crop provides an opportunity to grow legumes as intercrop without reducing its own yield. The main concept of this legume intercropping is to get increased land productivity and judicious utilization of land resources and farming inputs. Intercropping improved the availability of the resources such as water, light and nutrients. Besides, under rain-fed conditions, rainwater use efficiency can be elevated in maize + black gram intercropping system (Solanki et al. 2011). Mulching, as it interferes the direct hitting of the solar radiations onto the bare soil surface, conserves the moisture, regulates the soil temperature and increases both the water and land productivity in the region (Bhatt and Khera 2006).

Maintenance of nutrient supply is and can be one of the strategies to improve rainwater and nutrient-use efficiency (NUE). The most important management interaction in many drought-stressed maize environments is between soil fertility management and water supply (Meena et al. 2015a, b). Reduced growth rate in nutrient-deficit plants is generally associated with reduced WUE (Bicon 2004). The WUE was increased by application of nitrogen (Ogola et al. 2002). Adequate soil fertility management to remove nutrient constraints on crop production for every drop of water available through rainfall is also considered as one of the promising strategies for rainwater use efficiency. The efficient crop productivity under conditions of limited rainfall and erratic distribution is of great importance in rain-fed agriculture (Rockstrom et al. 2003). Hence, there is a need to identify and assess different RCTs responsible for enhancing the water and NUE even in the era of water-stressed conditions for attaining sustainable yield of crops in any soil and agro-climatic condition (Humphreys et al. 2010). Only then livelihood of the farmers of the region could be improved without adversely affecting the natural resources, viz., soil and water (Maruthi Sankar 1986; Vittal et al. 2003).

The problems of small-sized and fragmented holdings, difficult terrains, limited accessibility and prolonged winters also proved dire necessity for following the conservation approach for sustainable production of food grains. To meet the food demand of increasing population and sustaining the resources for future, conservation approaches like storage of rainwater, conservation tillage, contour bunding, addition of organic matter in the soil, strip cropping, cover cropping, mulching, making vegetative barriers, crop diversification, recycling of residues, sloping land management, adoption of improved varieties according to agro-climatic needs and construction and maintenance of water-harvesting storage are necessary to be adopted (Arora and Gupta 2014; Sharma and Arora 2015) by every stakeholder. In this chapter, we are discussing the different resource conservation technologies which will mitigate the adverse effect of the global warming with a greater stress on the NW parts of the Himalayan region for the judicious use of the natural resources, viz., soil and water, along with improving the livelihood of the farmers of the region. Keeping these points in view, different workers proposed some techniques which they propagate as the resource conservation techniques (RCTs), but their reliability is still under consideration as their success is both site and situation specific. Therefore, one should be very careful in adopting these RCTs for practising climate-smart agriculture (CSA) to move towards sustainable agriculture after mitigating the climate change adverse effects which is the need of the hour.

3.2 Extent of Land Degradation

The Western Himalayan region occupies the major part (~62 %) of the total Himalayan range. About half of the area in the Himalayan region is degraded. Severity of land degradation is in the order of Mizoram > Himachal Pradesh > Uttarakhand > Nagaland > Tripura (Sharda 2011). The major causes of degradation are soil erosion due to water and some adverse soil conditions. The erosion rates are quite high for Shiwalik Hills (>80 t ha^{-1} $year^{-1}$) and shifting cultivation areas (>40 t ha^{-1} $year^{-1}$). According to the Government of India's harmonized database, ~120.7 Mha of the land area is degraded in India (NAAS 2010), 70 % of which is affected by water erosion. The extent of soil erosion due to water erosion varies across the country from <5 t ha^{-1} $year^{-1}$ for dense forest area to more than 80 t ha^{-1} $year^{-1}$ in Shiwalik region (Singh et al. 1992). The extent of water erosion is more severe in Northeastern Himalayas (NEH) (22.27 % of TGA) than Northwest Himalayas (NWH) (12.61 % of TGA). Land degradation due to acid soils is more severe in IHR (13.95 % of TGA) than that of the country (3.72 % of TGA). The extent of acid-affected soils is much more in NEH (29.36 % of TGA) than that in NWH (0.76 % of TGA). Land degradation in forest areas is an acute problem in Indian Himalayas. The area under open forests with canopy less than 40 % is more in the Himalayan region (3.06 % of TGA) than the overall national figure (2.52 % of TGA). It is further revealed that land degradation due to open forests is ~6 times more in NEH (5.54 % of TGA) than NWH (0.95 % of TGA). In NEH, high percentage of area falls under shifting cultivation and indiscriminate forest felling.

3.3 Extent of Soil Erosion in the Himalayan Region

The Himalayas due to steep slopes, fragile geology and intense high storms is intrinsically prone to soil erosion. The problem has been further aggravated by several developmental activities undertaken in the region like road construction, mining and hydropower projects. Recent estimates indicate that ~39 % area of the Indian Himalayas has a potential erosion rate of more than 40 t ha^{-1} $year^{-1}$, which is really alarming. In northwestern Himalayas, on an average, ~17 % of the area falls in very severe category with erosion rates >40 t ha^{-1} $year^{-1}$, while ~25 % area has an erosion rate of more than 10 t ha^{-1} $year^{-1}$. The trend is similar in the northeastern Himalayan states. The states of Uttarakhand in Northwest Himalayas and Nagaland in NEH region have a maximum area of ~34 % and 63.5 %, respectively, under very severe category with erosion rates of >40 t ha^{-1} $year^{-1}$. Overall, ~23 % area in the Himalayan states has potential erosion rates >40 t ha^{-1} $year^{-1}$. It calls for serious efforts to employ appropriate conservation measures to check land degradation problems (Sharda 2011).

Soil loss tolerance limit (T) is defined as the threshold upper limit of soil erosion that can be allowed without degrading long-term productivity of a particular soil. In India, a default soil loss tolerance limit (SLTL) of 4.5–11.2 t ha^{-1} $year^{-1}$ is adopted for planning soil conservation activities. If soil erosion rates are higher than the tolerance, they need to be reduced to maintain sustainable productivity. It is known that a given erosion rate is not equally serious on all soils. On shallow soils with a T-value of 5 t ha^{-1}, erosion at a rate of 12.5 t ha^{-1} could lead to relatively rapid loss of productivity. In contrast, on some deep soils with a T-value of 12.5 t ha^{-1}, erosion at the same rate would not be expected to reduce soil productivity to the same extent. The soil erosion rates in the NE region vary widely from less than 5 t ha^{-1} to more than 40 t ha^{-1} $year^{-1}$. Approximately 29.82 % area in the region falls under the very severe erosion category, whereas very low, moderate and severe erosion classes are 4.47 %, 21.16 %, 16.79 % and 12.9 %, respectively. It was further revealed that due to strong resistance capacity, T-value was more or less consistent at around 7.5 and 12.5 t ha^{-1} $year^{-1}$. As our main emphasis is on the climate change and its consequences on the agriculture, thus there is a need to have an idea regarding climate change (Mandal and Sharda 2011; Sharda et al. 2013).

3.4 Climate Change and Its Consequences

In Asia, more than a billion people could be affected by a decline in the availability of freshwater, particularly in large river basins, by 2050. Glacier melt in the Himalayas, which is projected to increase flooding and rock avalanches, will affect water resources in the next two to three decades. As glaciers recede, river flows will decrease. Coastal areas, especially heavily populated mega-delta regions, will be at greatest risk due to increased flooding from the sea and, in some cases, from river flooding. India is one of the more vulnerable and risk-prone countries in the world. Over the centuries, its population has learned to cope with a wide range of natural and man-made hazards. Rapid population growth, high densities and poverty have led to an increase in vulnerability over the last few decades (Bhatt et al. 2016; Meena et al. 2015c, d).

Climate change is expected to increase the frequency and intensity of current hazards and the probability of extreme events and new vulnerabilities with differential spatial and socio-economic impacts. This is expected to further degrade the resilience of poor, vulnerable communities. It is therefore important to understand a number of processes that are rapidly changing India's landscape, altering livelihood opportunities and wealth distribution, which in turn affect the vulnerability of many communities and stakeholders and their capacity to adapt to long-term risks. India faces a turbulent water future. The country has a highly seasonal pattern of rainfall, with ~50 % of precipitation falling in just 15 days and over 90 % of river flows occurring in just 4 months (Sharma et al. 2013a, b). The Indian mainland is drained by 15 major (drainage basin area >20,000 km^2), 45 medium (2000–20,000 km^2) and over 120 minor (<2000 km^2) rivers, besides numerous ephemeral streams in the western arid region.

The Himalayan glaciers feed India's most important rivers. But rising temperatures mean that many of the Himalayan glaciers are melting fast and could diminish significantly over the coming decades with catastrophic results. In the long run, the water flow in the Ganges could drop by two-thirds, affecting >400 million people who depend on it for drinking water. In the short term, the rapid melting of ice high up in the Himalayas might cause river swelling and floods. The formation of glacial lakes of meltwater creates the threat of outburst floods leading to devastation in lowland valleys. According to a report by the Intergovernmental Panel on Climate Change (IPCC) looking at the threat from climate change to human development and the environment, "only the polar icecaps hold more freshwater than the Himalayan glaciers": "If the current trends of climate change continue, by 2030 the size of the glaciers could be reduced by as much as 80 %," ("Up in Smoke-Asia and the Pacific"). The adverse impact of climate changes includes water crisis and an increased risk of extinction for an estimated 20–30 % of plant and animal species in India if the global average temperature exceeds 1.5–2.5 °C. Climate change will also significantly impact health in India.

The most vulnerable will be the poor, the disabled and the youngest and oldest members of the population as they already face limited access to health facilities and have limited disposable income to cover additional medical costs. The Ganges river basin runs from the central Himalayas to the Bay of Bengal and covers parts of Nepal, India, China and Bangladesh. The basin occupies 30 % of the land area of India and is heavily populated, increasing population density downstream to Bangladesh, the most densely populated country in the world (Rashid and Kabir 1998). Water withdrawal poses a serious threat to the Ganges. In India, barrages control all of the tributaries to the Ganges and divert roughly 60 % of river flow to large-scale irrigation (Adel 2001).

India controls the flow of the Ganges into Bangladesh with over 30 upstream water diversions. The Farraka Barrage, 18 km from the border of Bangladesh, reduced the average monthly discharge of the Ganges from 2213 m^3/s to as low as 316 m^3/s [14 %] (Goree 2004). Climate change will exacerbate the problems caused by water extraction. The Himalayan glaciers are estimated to supply 30–40 % of the water in the Ganges, which is particularly critical in the dry season prior to the monsoon rains. The projected annual renewable water supply for 2025 indicates water scarcity (WWF 2007). Although the Ganges catchment drains virtually all of the Indian Himalayas and water supply per person in the basin ranges from adequate to ample, its dry season outflow (from December to February) to the sea is non-existent. Overall, excessive water diversions threaten to eliminate natural flows and severely damage people's livelihoods in the Ganges. The Indus River basin spans parts of four countries (Afghanistan, Pakistan, India and China) in an area that is more than 30 % arid and much drier than the nearby Ganges river basin (WRI 2003). The Indus River is critical for Pakistan's 160 million people and irrigates ~80 % of its 21.5 Mha of agricultural land (Rizvi 2001). The watershed is also an area of rich biodiversity, particularly where it opens to the Arabian Sea. The Indus River delta is a highly productive area for freshwater fauna and an important region for water birds (Ramsar Convention on Wetlands 2003). The Indus River is extremely sensitive to climate change due to the high portion of its flow derived from glaciers. Temperature controls the rate of glacier melt, which, in turn, provides more water in dry, warm years and less water in cool years. River catchments with a large portion of glacial meltwater experience less variability in water flows.

With global warming, many glaciers will no longer exist to moderate the flow of these rivers (Revenga et al. 2000). Thus communities which depend on glacier water will face more severe water shortages, variability and potentially greater flooding too (Rizvi 2001). The Himalayan glaciers provide the Indus River with 70–80 % of its water (Kiani 2005), the highest proportion of any river in Asia. This is two times the proportion of water that they provide to the Ganges (30–40 %). Himalayan glaciers provide 44.8 % of the water in the upper Indus River in China alone (Yang 1991). The Indus basin is already suffering from severe water scarcity due to over-extraction for agriculture, causing saltwater intrusion in the delta (WRI 2003). In 1995, the Indus River already supplied much less water per person than the minimum recommended by the United Nations (UN) and by 2025 is predicted to suffer even more severe water scarcity (Revenga et al. 2000). Well-managed riparian forests are especially important in minimizing the impacts of climate change on river biota. They provide shade and temperature regulation, can moderate the effect of frequent, short-duration storm events and can support natural water flow regimes. Climate change will exacerbate the impact of deforestation on water regulation. Although the Indus system is currently robust enough to cope with shortages of 10–13 % in river flows, when the river flow drops to 15–20 % below the average, irrigation shortages occur. Climate change will surely exacerbate the problems of irregular and low flow (Sharma and Sharma 2008; Meena et al. 2015e).

3.5 Evidence of Climate Changes

Climate change will affect many sectors, including water resources, agriculture and food security, ecosystems and biodiversity, human health and coastal zones. Many environmental and developmental problems in Asia will be exacerbated by climate change. Under climate change, predicted rainfall increases over most of Asia, particularly during the summer monsoon, and could increase flood-prone areas in East Asia, South Asia and Southeast Asia. In Central and South Asia, crop yields are predicted to fall by up to 30 %, creating a very high risk of hunger in several countries. Global warming causes the melting of glaciers in the Himalayas. This means increased risk of flooding, erosion, mudslides and GLOF in Nepal, Bangladesh, Pakistan, and North India during the wet season. Because the melting of snow coincides with the summer monsoon season, any intensification of the monsoon and/or increase in melting is likely to contribute to flood disasters in Himalayan catchments (Ghosh et al. 2016). In the longer term, global warming could lead to a rise in the snowline and disappearance of many glaciers causing serious impacts on the populations relying on the seven main rivers in Asia fed by meltwater from the Himalayas. Throughout Asia one billion people could face water shortage leading to drought and land degradation by the 2050s (Christensen et al. 2007). Projected sea level rise could flood the residence of millions of people living in the low-lying areas of South Asia such as in Bangladesh and India (Cruz et al. 2007). There is a need to go through some case studies carried out by different workers throughout the NW Himalayan region to mitigate the climate change adverse effects on the environment and finally to improve the livelihood of the farmers.

3.5.1 Case Study 1

Effect of mulching and tillage on soil erosion in submountainous tracts of Punjab.

3.5.1.1 About the Area

The present study was conducted at the Zonal Research Station for Kandi Area (ZRSKA), Ballowal Saunkhri, Nawanshahr district. It is situated at an altitude of 355 m MSL. The area is located in agro-climatic zone-1. The climate of the area is semiarid as per the classification of Thornthwaite (1948). The area is placed in the

NE of the state in the form of the long narrow strip of 10–20 km width, which lies ~30° 41′ to 32° 30′ N latitude and 75° 30′ to 76° 48′ E longitude. The rainfall constitutes the major water resource of the area. Mean annual rainfall of the area is 1000 ± 304 mm, ~80 % of which is received during the kharif season. The summer monsoon rains are received in 20–30 rainstorms, out of which 8–12 storms produce runoff (Hadda and Sur 1987). The runoff and soil loss vary from 35 % and 45 % of the annual rain and 25–225 t ha^{-1} $year^{-1}$, respectively, in the area (Sur and Ghuman 1992). Because of the high temperature and low relative humidity during the summer season, the vegetation cover on the ground is very sparse (Kukal 1987). The problem of high soil erosion is because of the higher slopes and high-intensity short-duration rainstorms. The soils of the area remain dry for 4–5 months in a year and qualify for an ustic soil moisture regime (Sehgal and Sys 1970).

3.5.1.2 Soil Erodibility

Bouyoucos (1935) suggested an index of erodibility:

$$\frac{[\%\,\text{Sand} + \%\,\text{Silt}\,]}{\%\,\text{Clay}}$$

3.5.1.3 Rainfall Characteristics for the Study Period

The meteorological data comprised of daily rainfall, evaporation, maximum and minimum air temperature and top 10 cm soil temperature. A total of 31 rainstorms were received during the study period, out of which 12 produced the significant runoff and soil loss. A total of 547 mm rain was received during the monsoon season which was very much below the normal rainfall of the area. Thus this season was relatively very dry. The distribution of the rainfall was also very erratic. The different characteristics of the received rainstorms during the study period are given in Table 3.1. The methodology used to calculate these characteristics is given below.

3.5.1.4 Rainfall Amount and Duration

Using the recorded rainfall charts, rainfall amounts (ordinate value) are plotted against time (abscissa value). The sum of ordinate values gave the total rainfall amount (mm), whereas the abscissa values gave the total duration of the rainfall (hrs).

3.5.1.5 Average Rainfall Intensity

$$\text{Average intensity}\left(\text{mm h}^{-1}\right) = \frac{\text{Cumulative rainfall(mm)}}{\text{Total duration of rainfall(hrs)}}$$

Table 3.1 Characteristics of the recorded rainstorms at Ballowal Saunkhri during 2002

Date	Rain (mm)	Duration (h)	Average intensity (mm h^{-1})	I_{15} (mm h^{-1})	K.E (m ton ha^{-1})	EI $_{15}$ (m tons ha^{-1}cm h^{-1})
2 August	4.2	0.5	8.4	14	312.3	437.2
3 August	8.7	0.75	11.6	16	317.5	508
4 August	79.5	3.83	20.75	72	375.6	2704.3
6 August	15.2	2	7.6	26	336.2	874.1
10 August	55	2.34	23.5	68	373.4	2519
12 August	14.3	8.25	1.73	6	279.5	167.7
13 August	17.2	8	2.15	11.2	303.7	340.1
2 September	55.2	2.58	21.39	120	395.3	4744.1
3 September	77.2	2.66	29.02	116	394.0	4570.7
7 September	27.8	4.3	6.46	48	360.0	1727.6
8 September	12.2	0.5	24.4	5.6	276.8	155.1
13 September	30.6	14	2.18	56	365.8	2048.9
$\bar{X}$	33.1	4.14	12.5	46.6	341.0	1733.1
SD	25.3	4.0	9.4	40.7	38.9	1638.9
CV (%)	76.4	96.6	71.0	87.3	11.4	94.5

X is mean, *SD* is standard deviation, *CV* is coefficient of variation, I_{15} is maximum 15 min rainfall intensity, *KE* is kinetic energy, E_{15} is rainfall erosion index

3.5.1.6 Maximum 15 Min Rainfall Intensity

The maximum 15 min rainfall intensity was calculated using the fixed time base procedure. Each rainstorm was divided into successive interval of 15 min, and the 15 min having maximum rainfall was selected. The maximum 15 min rainfall intensity (mm h^{-1}) was calculated as follows:

$$\text{Rainfall amount(mm)received in 15 minutes} \times 4$$

3.5.1.7 Kinetic Energy of Rainfall

The kinetic energy of a particular rainfall event was computed using the relationship of Hudson (1984):

$$\text{K.E} = 210.3 + 89\ \text{Log}_{10}\text{I}$$

where

KE is the kinetic energy of the rain (m t ha^{-1})
I is the maximum 15 min rainfall intensity (mm hr^{-1})

Rainfall Erosion Index

Rainfall erosion index of a particular rainfall event was calculated by multiplying the kinetic energy of the rainfall (m t ha^{-1}) and maximum 15 min intensity (cm hr^{-1}).

3.5.1.8 Field Experiment

The field experiment was conducted on erosion plots established on sandy loam soil at ~2 % slope during kharif (summer monsoon season), 2002 at Zonal Research Station for Kandi area, Ballowal Saunkhri, Nawanshahr district. A representative contour was selected. The experiment was laid down in a split plot design with plot size 5 × 1.5 m. The layout is given in Fig. 3.1. The treatments included two tillage treatments, i.e. minimum tillage (T_m) and conventional tillage (T_c), in the main plot and five modes of rice straw mulch applications, i.e. uncovered control plots (M_o), fully covered plots (M_w), lower one-third covered plots ($M_{1/3}$), strip application (M_s) and vertical mulching (M_v), in the sub-plots. In the strip application, rice straw mulch applied in 6″ × 2″ strips was constructed in alternative rows, and in vertical mulching, rice straw applied in the holes (10 cm diameter and 15 cm deep) was made with auger. Thus, there were 6 strips and 17 vertical holes per plot in the M_s and M_v treatments, respectively. Rice straw mulch was applied at 6 t ha^{-1} in all the treatments. The experiment had three replications. Thus in all there were 30 plots. Thirty iron drums (220 Lt. capacity) were embedded in the ground at a fixed interval of 1.65 m.

Proper outlets were provided to remove the excess runoff from the upper area. Care was taken to prevent the runoff from the adjoining plots to enter in a particular plot. The minimum tillage comprised of one discing and cultivations, whereas conventional tillage comprised discing twice followed by three cultivations. Maize (*Zea mays* cv Parkash) was planted on 6 July 2002. All the recommended package of practices was followed. Atrataf was sprayed to check weed growth on 8 July 2002. Half of the nitrogen and recommended dose of phosphorus were drilled at the time of sowing. To save the crop from the prevailing adverse conditions, two life-saving irrigations were applied on 25 July and 1 August 2002. The second dose of nitrogen in the form of the urea was applied about a month after sowing.

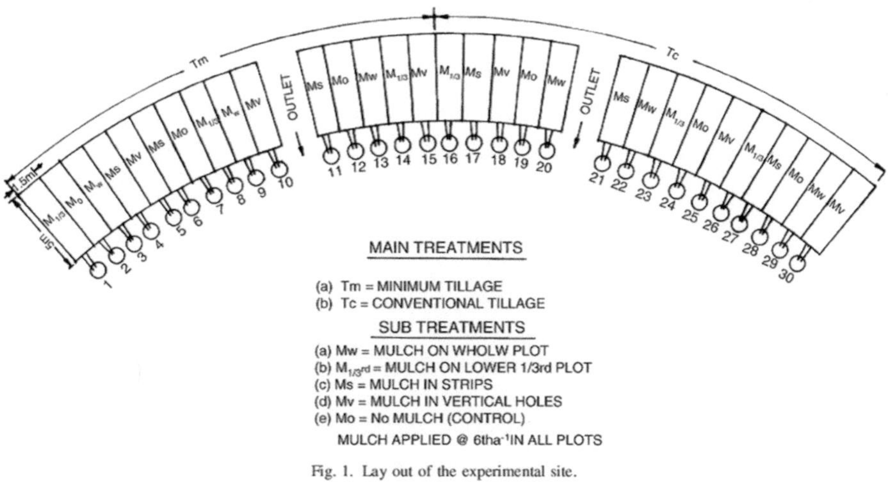

Fig. 3.1 Layout of the carried out experiment (Bhatt 2003)

3.5.1.9 Periodic Soil Loss

After each rainstorm, drums were thoroughly stirred with a wooden stick, and representative samples were taken in the prewashed 1 l plastic bottles. Then further, sub-sampling was done, and a 50 ml representative sample from each bottle was taken in a 50 ml glass beaker, and a pinch of sodium fluoride was added as a dispersing agent. The soil settled down, and clearwater was removed carefully. Then glass beakers containing the soil water suspension were oven-dried at 105 °C for 24 h so that a constant dry weight was obtained (Bhatt 2003).

3.5.1.10 Periodic Runoff

After each rainstorm, the drum depth that was not filled was measured with the help of a metre scale. The drum depth, which was filled with water after a particular rain, was obtained after subtracting it from 86 (standard drum depth). The runoff volume was obtained by multiplying it with the area of the drum.

3.5.1.11 Effect of Tillage and Mode of Mulch Application on Soil Moisture Content

Mulch material was found to be quite effective in conserving soil moisture. However, its role differed to a significant extent under different tillage treatments and with different mode of mulch application. Observations of periodic soil moisture content of surface and sub-surface soil reveal that the minimum tillage (T_m) was more effective in conserving moisture than the conventional tillage (T_c) under all modes of mulch application. As compared to the control, the application of straw mulch at 6 t ha^{-1} in whole plot had 4.0–5.1 % higher soil moisture content in 0–15 cm soil depth and 3.5–7.1 % high soil moisture content in 15–30 cm soil depth under minimum tillage treatment. However, under the conventional tillage treatment, the effect of mulch applied on soil for conserving the moisture was relatively less. The other three modes of mulch application helped in conserving the soil moisture, but the amount of water conserved was relatively small as compared to that under fully covered treatment. The minimum tillage (T_m) was more effective in conserving moisture than conventional tillage (T_c). The greater the surface area covered by that of the mulch material in a particular mode, the greater is its effectiveness in conserving the soil moisture. Thus, among different modes of mulch application, the trend varied as $M_w > M_{1/3} > M_s > M_v > M_o$ as area covered varied as $98 > 33 > 18 > 1.8 > 0$ %.

It has been reported earlier (Hillel 1980) that mulches are most effective in conserving moisture in the wet range when the evaporation is in the constant rate range and controlled by the atmospheric evaporativity. The use of crop residue as surface mulch helps in reducing the soil surface sealing and maintaining high infiltration rate (Wagger and Denton 1992) and reducing vapour losses. Vegetation cover is thus a very important factor, as it absorbs the impact of raindrops and protects the aggregate breakdown which further promotes the greater soil moisture content (Reddy and Reddi 1995).

3.5.1.12 Soil Temperature

Straw mulch provides surface cover on the bare soil and thus acts as a barrier for the sunrays to reach the soil surface. Mulch also reduces the soil temperature by reflecting the sunrays. As compared to the control, the minimum soil temperature in fully covered plots was 1.4–2.0 °C higher in T_m and 1.4–2.4 °C higher in T_c plots. Thus tillage did not significantly affect the minimum soil temperature. However, as compared to the control, the maximum soil temperature in the fully covered mulched plots was 1.9–2.8 °C and 2.5–4.0 °C lower under minimum and conventionally tilled plots, respectively. The straw-mulched plots showed lower temperature than the bare plots, and the trend observed was $M_o > M_{1/3} > M_w$. However, there was no significant effect of the tillage treatment on maximum soil temperature.

3.5.1.13 Runoff

During the study period, 31 rainstorms were received. However, only 12 of them produced significant runoff and soil loss. On the perusal of data in Table 3.2, runoff in conventionally tilled plots (T_c) under all the modes of mulch applications was ~5 % higher than under minimum tilled (T_m) plots. This may be observed due to the fact that minimum tillage promotes aggregation, maintains the soil structure and thus increases the infiltration (Mcgregor et al. 1999) and helps in reducing runoff. As compared to the control, straw mulch was quite effective in decreasing the runoff. The mean runoff decreased from ~50 % under controlled plots (M_0) to 16.8 % under fully covered plots (M_w). The percent runoff was observed to be ~45, 30 and 24 % in vertical mulching (M_v), strip application (M_s) and $M_{1/3rd}$ plots where the lower one-third of plots were applied with mulch. Differentiating behaviour of different modes of mulch application may be due to the different amount of surface cover provided by these treatments. It was observed that M_w, $M_{1/3rd}$, M_s, M_v and M_o had 98, 33, 18, 1.8 and 0 % surface covered. The role of surface cover in decreasing the runoff had been reported earlier by several scientists (Mannering and Meyer 1963; Khera and Singh 1998). Mulch also provided hydraulic resistance to the running water. The greater the surface covered, the greater would be the resistance offered by the mulch material to the runoff (Bhatt and Khera 2006).

However, as far as the interactive effect of tillage and mulch was concerned, it was observed that M_w, $M_{1/3}$, M_s, M_v and M_o had 16.9, 25.9, 31.2, 44.1 and 47.0 % in T_m plots, while these values were increased to 21, 29, 34, 51 and 559 % in T_c plots.

Table 3.2 Temporal variation of gullies in the study catchments (2003–2008)

	Gully texture			Gully density		
Catchment	2003	2008	% increase	2003	2008	% increase
I	758.0	1453.8	91.8	31.7	35.8	12.9
II	439.9	1174.8	167.0	15.5	19.4	25.2
III	722.0	1151.2	59.5	15.8	17.0	7.6
IV	251.2	464.8	85.0	8.6	9.4	9.3

It meant that T_mM_w proved to be the best, while T_cM_o was the worst treatment in checking the runoff. Percent runoff under all the modes of mulch application with T_m was significantly lower than T_c treatment. The application of straw mulch avoided the direct raindrop impact, decreased surface sealing and entrapped the soil particles and enhanced residence time for water to infiltrate into soil (Khera and Singh 1998).

3.5.1.14 Soil Loss

The soil loss was differed significantly affected both by the tillage treatments and by the different modes of straw mulch application. As far as the tillage is concerned, it was observed that on an average T_c (conventionally tilled) plots had ~40 % higher soil loss than the T_m (minimum tilled) plots. The conventional tillage resulted in the breakdown of larger aggregates into the smaller ones which were found to be more erodible (Bhatt 2015). Mulch was quite effective in decreasing the soil loss by providing proper surface cover to the bare soil. The mean soil loss (averaged over all levels of tillage) in M_o, M_v, M_s, $M_{1/3}$ and M_w was observed as 5554, 4691, 2073, 1425 and 548 kg ha^{-1}, respectively. It was due to the fact that different modes of rice straw mulch application provided variable ground cover. The efficiency of this ground cover in decreasing the soil loss was also reported earlier by Khera and Singh (1995). Among different modes of mulching, M_w had highest ground covered area followed by $M_{1/3}$, M_s, M_v and M_o.

However, as far the as interactive effect of tillage and different modes of mulch application was concerned, it was observed that the soil loss was 2063.7, 6258.4, 8058.1, 15,498.9 and 17,152.7 kg ha^{-1} in M_w, $M_{1/3}$, M_s, M_v and M_o under minimum tilled plots (T_m) as compared to the 3118.5, 6769.6, 9869.5, 21,356.4 and 24,947.2 kg ha^{-1} under the T_c (conventional tilled) plots, respectively.

The interaction between modes of mulch applications and tillage was observed to be significant. Except under M_w and $M_{1/3rd}$ treatments, soil loss under the other three modes of mulch application was significantly higher under conventional tillage as compared to minimum tillage. Application of mulch reduces the soil loss as it decreases the erosivity of rain by intercepting the raindrops, increases surface detention and residence time for water to infiltrate and helps in entrapment of soil particles (Mannering and Meyer 1963).

3.5.2 Case Study II

3.5.2.1 Gully Erosion and Its Dynamics in Submountainous Tracts of Punjab

Foothills of lower Shiwalik occupy an area of 2.14 Mha and represent the most fragile ecosystem of the Himalayan mountain range because of its peculiar geological formations. It lies mainly in the states like Jammu and Kashmir (Jammu, Udhampur and Kathua) (0.80 Mha), Punjab (Hoshiarpur, Ropar, Nawanshahr and Gurdaspur) (0.14 Mha), Haryana (Ambala and Yamunanagar) (0.06 Mha) and Himachal Pradesh (Kangra, Una, Bilaspur, Hamirpur, Chamba, Solan and Southern parts of Sirmaur) (1.14 Mha) (Kukal et al. 2006). Prior to the middle of the eighteenth century, the Shiwalik Hills were strictly preserved for hunting and no cultivation, grazing or exploitation of timber were permitted. The increased population of mankind along with the livestock density far exceeding the current carrying capacity of the land, frequent forest fires and mismanagement of land resources resulted in the steady but obvious natural resource degradation, and presently the garden of Punjab has changed to more or less a desert, which is still highly exploited by various anthropogenic activities (Kukal and Bhatt 2006).

Ecological degradation in Shiwalik Hills is the outcome of the overexploitation and mismanagement of soil resources through deforestation, overgrazing and clearance of the vegetation for the agricultural purposes disregarding their slope and topography (Bhatt et al. 2004; Arora and Saygin 2011). The whole belt experiences severe problem of soil erosion because of undulating slopes and highly erodible soils coupled with highly intensive rainstorms and represents the most fragile ecosystem of Himalayan mountain range because of its peculiar geological formations and highly erodible soils (Thakur et al. 2013). It is reported that runoff and soil loss in the region vary from 35 % and 45 % and 25–225 t $ha^{-1}year^{-1}$ (Sur and Ghuman 1994). Minimum tillage (T_m) coupled with rice straw mulching at 6 t ha^{-1} was reported to mitigate the losses of soil erosion to a significant extent (Bhatt and Khera 2006; Arora and Bhatt 2006). Assessing the impacts of climatic and land use changes on rates of soil erosion by water is the objective of many projects (Nearing 2001). Among different types of soil erosion, gully erosion is the most serious one in the region as ~20 % of the area is already under gullies (Kukal and Sur 1992) and spatially advances at an increased rate upslope the catchment as organic carbon and clay content decreased up the slope (Kukal et al. 2005). Thus, the problem of soil erosion is quite serious in the region, and farmers mostly depend on their indigenous technical knowledge (ITKs) to grow their crops sustainably in the region, and many of these techniques are scientifically sound also (Kukal and Bhatt 2006; Bhatt 2013).

Ephemeral gully erosion has been reported to account for ~49 to 79 % of the total soil loss (Zheng et al. 2009). The ~70 to 80 % of the gully erosion control structures have failed in the region (Kukal et al. 2002). The reasons attributed for the failure of gully erosion control structures in the region include lack of information on gully network including distribution and extent of different ordered gullies, gully density, gully texture and behaviour and development of gullies in the region (Kukal et al. 2006). Secondly, the installation of gully erosion control structures is generally done in the highest-order gully. After some time, these structures get silted up along the upstream side, after which the runoff water starts falling down from the crest height of the structure and causes higher erosion losses. The lower-order gullies are seldom tackled in the region while controlling the runoff and soil loss and are generally ignored in all the soil conservation programmes (Kukal et al. 2006). To evaluate the temporal advancement of gullies both in terms of their number and length, two surveys were being conducted in the four catchments of Hoshiarpur district, viz., firstly in 2003 and secondly 5 years thereafter in 2008, this detail survey was conducted in 2003 and then on 2008 in the four catchments of

Hoshiarpur district of Punjab with the objectives (i) to evaluate the spatial variation of vegetation all along the slopes, (ii) to study distribution of gullies on either side of main gully, (iii) to study temporal advancement of gullies from 2003 to 2008 both in terms of gully density and texture and (iv) to formulate strategies for the assessment of gullies and bring out effective techniques for mitigating the adverse effects of gully erosion in particular and soil erosion in general.

3.5.2.2 Study Area and Its Characteristics

The survey was carried out in the four catchments of Hoshiarpur district of Punjab in the Shiwalik region of lower Himalayas in North India. The region lies between 30° 10′ and 33° 37′ N latitude and 73° 37′–77° 39′ E longitude and stretches to ~530 km lengthwise and 25–95 km widthwise. The climate of the region varies between semiarid to subhumid. The maximum temperature (41–42 °C) is recorded in the first fortnight of June, whereas the minimum temperature (5–6 °C) is recorded in the month of January. Majority of the soils range from loamy sand to sandy loam in texture and have low to medium moisture retention capacity and are highly erodible. It is reported that majority of soils (~67 %) are loam followed by loamy skeleton (28 %), sandy skeletal (1.0 %) and sandy (0.7 %) (Sidhu 2002). Soils vary widely in their (shallow to very deep) texture (sand to clay loam), organic carbon (0.1–1.1 %) and pH (5.3–8.4) depending upon the physiography, parent material, vegetation cover and climatic conditions. The area receives an average annual rainfall of ~800–1400 mm with a high coefficient of variation. Approximately 80 % of this rain occurs in the 3 months of monsoon season (July–September) with highly intensive rainstorms. In Shiwalik of Punjab, rainfall aggressiveness (ratio of highest monthly rainfall to total annual rain), an index of rainfall concentration in a period, varies from 55.9 to 502.4 with an average value of 218 ± 21.7 (Singh 2000a). Rainfall constitutes the major water resource and is sufficient to take two crops annually, but because of its ill distribution in time and space, the farmers are unable to utilize it properly for agricultural purposes. Sidhu et al. (2000) reported that in the region, ~40 % land has steep slopes (>15 %), 10 % moderate slopes (8–15 %), 17 % gentle slopes (3–8 %) and 25 % very gentle slopes (10–13 %). Further, Kukal et al. (2006) reported that convex-concave (NE side) and concave slopes (SW sides) dominate in the region with a conclusion that SW sides of a catchment exposed to greater sunshine hours which resulted in greater evaporation, lesser moisture, lesser vegetation, lesser organic matter and greater erosion intensity. In the present study, we surveyed four catchments of Hoshiarpur district of Punjab to study the temporal advancement of gully texture and density from 2003 to 2008. Various factors controlling gully growth are catchment characteristics, viz., area (Burkard and Kostachuk 1997), slope shape (Meyer and Martinez-Casasnovas 1999), gully development parameters, slope steepness (Kukal et al. 1991), surface runoff, precipitation, soil moisture and piping (Stocking 1980).

In the present study, detailed field survey for gully erosion was carried out by dividing catchments into grids of $50 \times 50\ m^2$ each, and then each gully line was sketched on the contour maps (at a scale of 1: 1000) manually (Fig. 3.2) after measuring the distance between wooden pegs laid out in the grids. The gullies up

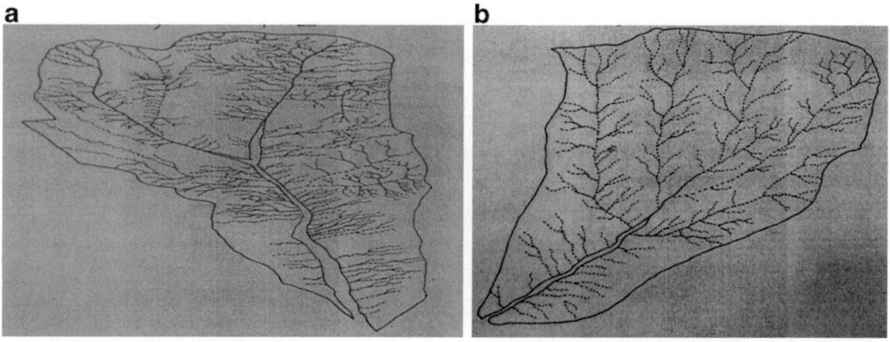

Fig. 3.2 Gully erosion network in catchments I and IV of Saleran watershed, Hoshiarpur, Punjab (Bhatt and Kukal 2015). (**a**) Saleran, Hoshiarpur, catchment IV (**b**) Saleran, Hoshiarpur, catchment I

to the first order were marked on the maps. Slope angles across the slope transacts were measured at certain intervals in the selected catchments with an Abney level and measuring tape to determine the slope shape profiles. Gully heads were also marked in each catchment on the base maps, and their status of activeness was found out. Gullies were classified as first-, second-, third-, fourth- and fifth-order gullies, depending upon extent of their bifurcation. The length of different ordered gullies was measured in each catchment from the gully erosion map. The total length of all the gullies in the catchment was expressed as "gully density" (km/km^2). The number of first-order gullies per unit area was expressed as "gully texture" (number/km^2).

3.5.2.3 Temporal Variation in Gully Erosion

A detailed survey was firstly made in the year 2003 in the Saleran catchments of Hoshiarpur district, and then after 5 years, similar catchments were surveyed again to have an idea regarding the advancement in gullies both from their length and their number. A comparison of "gully density" (km/km^2) and "gully texture" (number/km^2) surveyed in 2003 and 2008 shows that both gully texture and gully density have increased over the years (Table 3.2). Catchment II experienced highest increase in gully texture (167 %) and gully density (25.2 %) followed by catchment I (91.8 % and 12.9 %), catchment IV (85 % and 9.3 %) and catchment III (59.5 % and 7.6 %) (Table 3.2). Interestingly, the percent increase in gully texture was significantly higher than gully density in all the catchments. This indicates that more number of first-order gullies were added every year than the addition in the gully length. In fact the rainfall aggressiveness (ratio of highest monthly rainfall to the average annual rainfall) recorded to be higher (Matharu et al. 2003) in the region leads to the creation of new gullies (Morgan 2005).

It is thus clear that gully networks have been expanding with time in the region. To check this expansion of gully network, gully control strategy as discussed previously needs to be adopted at a wider scale in the region.

First-order gullies were significantly higher both in number and length (Fig. 3.2), and these are the main feeding braches to the higher-order gullies. Thus there is an urgent need to control these gullies rather than to construct the check dams in the highest/main gully which also got silted up after some time.

3.5.2.4 Vegetation Density

The average tree density in different catchments decreased significantly by ~76 % from top-slope to the toe-slope segment (Table 3.2). The decrease was significantly higher in catchment III (21.8 %) compared to catchments I (17.1 %), II (18.4 %) and IV (16.4 %). The decrease in tree density down the slope was also highest in catchment III (83 %) compared to 69–78 % in other catchments. This decrease in tree density down the slope could be due to the reason that the trees growing on or near the toe-slope segment were more prone to be cut by the local population for timber or fuel purposes. Catchment III being smaller in size and more easily accessible to the human beings and animals could have faced more destruction of vegetation.

The average density of bushes increased down the slope by 20 %. However, the increase was 76 % in catchment IV compared to 14.6 % in catchment III and 36.6 % in catchment II. In catchment I, the density of bushes rather deceased down the slope. As in the case of tree density, the density of bushes was highest (28.9 %) in catchment III, followed by catchments I (26.2 %), II (24.3 %) and IV (22.2 %). The overall increase in bush density down the slope could be due to the dominance of *Lantana* spp. which is not liked by animals as fodder. Also this, being thorny and of bad odour is not preferred by the local people for any purpose. Moreover, the higher density of these bushes near the toe-slope segment could be due to higher soil moisture content at the lower slope segments.

Unlike trees and bushes, the density of grasses was higher in catchment II (14.2 %) compared to ~8.6–9.4 % in other catchments. The average grass density also decreased down the slope by ~42 %. However, the extent of decrease in grass density varied in different catchments with 70.6 % in catchment III, followed by 59 % in catchment I and 53 % in catchment IV. However, in catchment II, it did not vary down the slope. The density of vegetation was affected by the slope aspect. In general the slopes facing the Northeast direction had significantly higher tree density than the slopes facing the Southwest direction at top and toe slope. This is due to the fact that slopes facing the Southwest direction experience longer sunshine hours and hence more evaporation losses and greater aridity resulting in lower tree density. However, the average bush and grass density was similar on the two slope aspects, in contrast to the previous studies carried out in the region (Bhatt and Kukal 2015).

The density of trees decreased from top-slope to toe-slope segment on both the slope aspects. However, the decrease was more on slopes facing the Southwest direction (78.8 %) than on slopes facing the Northeast direction (71.8 %). The density of bushes unlike trees increased significantly down the slope. The grass density deceased from top-slope to toe-slope segment by 18.3 % on slopes facing the Northeast in comparison to 60 % on slopes facing the Southwest direction. The highest decrease (95 %) in tree density from top-slope to toe-slope segment was observed on Southwest-facing slopes in catchment III, followed by catchment IV (82 %) and catchment II (78 %), and lowest in catchment I (72 %). The trend of variation in bush density down the slope was opposite to that of tree density. The bushes increased down the slope in almost all the catchments except the Northeast-facing slopes in catchment I, whereas tree density decreased significantly down the slope. The increase in bush density down the slope could be due to the higher prevalence of *Lantana* spp. (Bhatt and Kukal 2015).

The grass density, as in the case of trees, decreased down the slope in all the catchments. As in tree density, the decrease in grass density was more on the slopes facing Southwest than those facing Northeast. The highest decrease (89.3 %) was observed in Southwest-facing slopes of catchment III followed by catchment II (76 %) and catchment IV (70 %) and lowest in catchment I (14 %). The temporal surveys conducted in 2003 and 2008 in the four catchments of the Hoshiarpur district showed that:

(i) First-order gullies are the main culprit which collect the rainwater from each nook and corner of the catchment and supplied it to the higher-order gullies. Governments spend huge amounts to install check dams in the higher-order gullies which prove to be a failure to control the soil erosion at a long time scale. Thus, if we control the first-order gullies in the catchment, then, no water is supplied in the higher-order gullies, and thus extent of soil erosion could be significantly reduced.
(ii) Catchment sides facing SW side reported to have intense gully network as compared to NE because of greater sunshine hours, more evaporation, lesser vegetation, lesser organic matter, poor aggregation and, finally, higher erodibility.
(iii) The severity of gully erosion was observed to be a function of average relief and lamniscate ratio of the catchments as increased steepness results in increased runoff speed which aids in the deepening and widening of gullies. Lower value of lamniscate ratio indicates more erosion in the catchment due to its more compact shape with reduced time of concentration.
(iv) Concave slopes in a particular catchment have been shown to be more prone at the lower slope segments for initiation of gullies as the runoff water concentrates at the lower slope segment where the slope steepness decreases and runoff water starts concentrating. The slope angles were steeper on the upper slope segments, and it decreased in the downslope direction (Bhatt and Arora 2015; Bhatt and Kukal 2015).

3.5.3 Case Study III

3.5.3.1 Soil Erosion in Uttarakhand

Flood damages are generally viewed to occur in plain areas. However, rivers and torrents (seasonal streams with flash flow during monsoon) in the Himalayan region also cause heavy damage to life and property. A recent example is that of flood-induced deluge occurring in Kedarnath valley of Uttarakhand state during mid-June 2013, which took a toll of thousands of lives and destroyed property including agricultural lands which are not retrievable. This has necessitated us for taking adequate flood control measures in the Himalayan rivers/streams. Bioengineering technology has been evolved at CSWCRTI, Dehradun, for treatment of torrents in Shiwalik where mechanical measures like spurs, retaining walls and earthen embankments have been used in conjunction with suitable vegetable species for torrent training. Species like *Arundo donax* (narkul or nada), *Vitex negundo* (shimalu), *Ipomoea* (besharam), bamboo, *Napie*r (hathi ghas), and *Saccharum munja* (munj ghas) have been found suitable for bank protection and vegetative reinforcement of structures.

3.5.3.2 Extreme Rainfall-Induced Disaster in Uttarakhand

A disaster had occurred in Uttarakhand as a result of extreme rainfall occurring during mid-June 2013 resulting in a huge loss of life and property. Under a joint initiative of NARS and Uttarakhand state government, a survey conducted to observe the damage to natural resources inter alia revealed the following:

- The agricultural fields/habitations situated within the high flood level of rivers/streams were washed away, and damage was noticed to adjoining lands also wherever floodwater entered.
- The intensity of damage was more in untreated watersheds compared to the treated ones.
- The maximum mass erosion problem observed was that due landslides/slips, specially along roads. Landslides/slips were more at places where no retaining walls and toe drains were provided and slopes were without vegetation.
- The drainage lines (nalas/gullies) treated with proper bioengineering measures (gabion check dams, earthen gully plugs, etc.) even before 20–30 years back were not much affected by extreme rainfall events (14–17 June 2013).
- The diversion drains constructed by some farmers (at their own initiative) for safe disposal of runoff water saved valuable agricultural land and crops.
- The degraded hill slopes and landslides/slips treated earlier (12 years before) with Geojute technology were found stable and lush green with vegetation.
- Erosion problem was less or absent, and damage was minimized with good agroforestry practices.

Thus there is a need to control this menace of soil erosion by managing the watershed approach which on integrated basis will serve the purpose. The following are some of the case studies from the foothills region of the Shiwalik.

3.5.3.3 Watershed Management Programme

The increasing anthropogenic pressure on the Himalayan resources to meet the ever-increasing demands for material supplies is leading to their widespread degradation. It was established that ecologically relevant destruction took place in rain-fed old croplands within mid-slope and high landscape positions. The continued degradation of the fragile Himalayan region would affect adversely the socio-economic and environmental stability of the region. The major part of the land area of the northwestern hill region is hilly terrain, and a considerable part of this is under forest cover. Thus, a very small area is available for cultivation, and a considerable part of it is under rain-fed having low productivity. Undulating topography, varied climate, scanty cultivated land, overwhelming percentage of small and marginal holding, difficult conditions, high cost and low returns on food grain crops, poor economic condition of the farmers, etc. are the main causes responsible for this situation. The majority of the hill population resides in the ranges of the middle Himalayas having an elevation between 600 and 2000 m and is mostly dependent upon rain-fed agriculture. Farmers are in a habit of using traditional agricultural techniques and methods for crop production and are getting low productivity almost from all the crops being grown in the region. Therefore, more holistic approach to land use and management is needed to cope with increased pressure on soil resources for sustainable food and fibre production while reducing the adverse off-site environmental impacts of agricultural practices. Besides regular watershed programmes launched by the government in the country such as NWDPRA and IWDP, some programmes specially tailored for the hilly regions were implemented as given below (Mishra and Juyal 2015).

3.5.3.4 Integrated Watershed Management Programme (IWMP)

The World Bank-assisted Integrated Watershed Management Programme (IWMP) was implemented in the Shiwalik belt in two phases since 1980–2005 covering the Shiwalik region in the states of Himachal Pradesh, Jammu and Kashmir, Uttarakhand, Punjab and Haryana. The project under IWMP hills I and II covered a total area of 5,84,564 ha at an investment of Rs.108347.1 lakhs.

3.5.3.5 Uttaranchal Decentralized Watershed Development Project (UDWDP)

The World Bank-aided Uttaranchal Decentralized Watershed Development Project (UDWDP) also known as *gamya* was operationalized in Uttarakhand state since 2004 for a period of 8 years (2004–2012) covering a total of 461 gram panchayats (GP) spread over 77 micro-watersheds and covering a total area of 2348 km^2 (part of 11 districts of Uttarakhand state). The project aimed at improving the productive potential of natural resources and to increase income of rural inhabitants in selected watersheds following socially inclusive and sustainable approaches.

3.5.3.6 Watershed Development Project in Shifting Cultivation Areas (WDPSCA)

Shifting or *jhum* cultivation is a serious problem in the northeast hills where an area of ~43.57 lakh ha is affected by this problem which has been further aggravated due to reduction of earlier *jhum* cycle of 20–30 years to 3–6 years now. The Watershed Development Project in Shifting Cultivation Areas (WDPSCA) with 100 % financial assistance was implemented in all NE states with the objectives of protecting the hill slopes of *jhum* areas from excessive soil erosion and to create livelihood support activities and introduce appropriate land uses.

3.5.3.7 Physical and Financial Achievements

The vast area treated/reclaimed in different Himalayan states under various watershed development programmes up to tenth plan (Sharda et al. 2012). It is evident from Table 3.3 that only 13.46 % of TGA of northwest Himalayan states and 15.5 % of TGA northeast Himalayan states have been covered. In comparison to the overall situation at the national level, the treated area in Himalayan states is only ~14 % against 17.20 % of TGA at country level. As the hilly regions are more fragile and ecologically sensitive, they need priority in treatment. The soil conservation measures taken up in watershed development include mechanical measures like bunding, terracing, check dams, agronomic measures, vegetative barriers, alternate land use systems, runoff harvesting and recycling systems, etc. These measures were undertaken on agricultural lands (arable/nonarable) for enhancement of productivity and profitability.

Under IWMP (2009–2010 to 2012–2013), a total number of 862 watershed projects were undertaken covering an area of 3.633 M ha with funding of Rs.680.56 crores. Under Integrated Wasteland Development Programme (IWDP), a total of funds released amounted to Rs.1514.81 crores (2007–2008 to 2012–2013).

Table 3.3 Physical and financial status of treated area in different Himalayan states of India under various watershed programmes (up to 10th plan)

States	TGA (M ha)	Area treated (M ha)	Expenditure (Lakh Rs.)	Percent of TGA treated
Jammu and Kashmir	22.22	1.19	57,609.29	5.34
Himachal Pradesh	5.57	1.45	67,688.94	25.94
Uttarakhand	5.35	1.82	76,316.63	33.93
NW Himalayas	33.14	4.46	201,614.86	13.46
Arunachal Pradesh	8.37	0.5	11,639.43	5.97
Assam	7.84	1.09	27,321.18	13.97
Manipur	2.23	0.44	16,227.47	19.69
Meghalaya	2.24	0.33	11,534.61	14.76
Mizoram	2.11	0.66	27,334	31.37
Nagaland	1.66	0.58	31,677.2	35.19
Sikkim	0.71	0.23	10,167.69	32.44
Tripura	1.05	0.23	11,109.52	21.57
NE Himalayas	26.21	4.06	147,011.10	15.49
Himalayas hill total	59.35	8.52	348,625.96	14.36
All India total	328.66	56.54	1,947,057.24	17.20

Table 3.4 Effect of bioengineering measures on landslide (1964–1994) and minespoil rehabilitation (1984–1996) project

Particulars	Landslide project		Minespoil project	
	Before treatment	After treatment	Before treatment	After treatment
Runoff (mm)	55	38	57	37
Sediment load (t/ha/year)	320	5.5	550	08
Dry weather flow (days)	100	250	60	240
Vegetative cover (%)	<5	>95	10	80

3.5.3.8 Impact of Watershed Development Programme

There is an ample evidence of positive impacts of watershed programmes in terms of reduction in soil and water losses and improved agricultural productivity in normal rainfall years in regions that were bypassed in the conventional green revolution era. Several reviews on the performance of watershed development projects (Palanisami et al. 2002; Joy et al. 2005) in India have diagnosed various limitations of watershed programmes. Participatory Integrated Watershed Management (IWM) approach being adopted in the recent past has shown encouraging results over the previously adopted commodity-based or sectoral approaches. Operational research project on watershed management at Fakot in outer Himalayas which was implemented by the CSWCRTI during 1975–1986 is a successful example of this participatory approach. Similar trends in production increase and environmental benefits have been observed in other watersheds developed under NWDPRA, DPAP, RVP and other bilateral projects implemented during the 1990s.

3.5.3.9 Landslide and Minespoil Rehabilitation

It was observed that as a result of the bioengineering treatment of landslide-affected area (Nalotanala watershed, 60 ha) and minespoil-affected area (Sahastradhara watershed, 64 ha), heavy soil erosion was checked and brought within permissible limit, surface runoff was drastically reduced with attenuation of flood peaks, and the lean period flow increased due to groundwater recharge and biodiversity improved (Table 3.4). Restoration of limestone minespoil area which had excess of calcium resulted in improving the water quality through reduction in calcium content.

3.6 Recommendations

From the foregoing discussion, it can be inferred that treatment of mass erosion-affected areas needs to be brought within the purview of watershed development programmes. Some of the priority areas for treatment are given below:

- *Torrent/river training:* Rivers/torrents/streams which cause flood damages in watersheds should be treated through cost-effective bioengineering technology using mechanical measures (spurs, etc.) and locally adaptable suitable vegetative species which possess requisite hydraulic properties to withstand and dissipate the impact force of flood waters. Technique for *katta crate* structures using locally available riverbed material has been evolved by CSWCRTI, Dehradun. The adjoining lands should be put under permanent tree cover. The area up to 300 m from the stream banks should be banned for habitation and constructional activities.

- *Minespoil rehabilitation*: Areas wherever mining/quarrying has been done must be put back under green cover after mining is over. The use of natural geotextiles (jute/coir nets) has been found effective for quick revegetation.
- *Landslides/slips control*: Landslides/slips particularly occurring along roads, agricultural lands, habitations, etc. should be treated through bioengineering technologies on watershed basis already available. Green technology for road construction should be followed. Treatment of the eroded/degraded/cut slopes along the roads should be mainstreamed into the road construction work. A separate soil conservation wing with the road construction agencies should be created for the purpose. Training/awareness programmes on bioengineering aspects for the highway engineers need to be conducted. However, landslide/slip treatment is possible in areas which are not geologically unstable.
- *Cost estimates:* As mass erosion control requires more intensive measures particularly engineering ones, the cost of treatment per hectare naturally increases. From the past experiences, on an average, the landslide-/slip-affected watersheds may need an investment of Rs.1–1.5 lac per ha. For treatment of torrents, an amount of Rs.15–20 lac per km of torrent length may be required.

In the watershed programmes so far, the focus has been on resource conservation and productivity enhancement on agricultural lands. But in the Himalayan region, due to their geological, topographical and climatic setting, the problem of mass erosion such as landslides/slips, riverbank erosion and high sedimentation is common. In the changing climate scenario, such problems are expected to increase due to forecasting of high-intensity storms and denudation of forest cover. Hence, there is a need to mainstream treatment of such problems areas into the watershed programmes.

3.7 Integrated Approach to Mitigate the Adverse Effect of Climate Change

Delineation of the process or evidence showing control will act as a booster for the farmer and encourage them to come out of this critical condition of poor livelihood. Here we are going to discuss some of the pioneer techniques which proved to be effective in controlling the problem of the soil erosion. Northwest Himalayan region is severely suffering from the problem of soil erosion. Hence soil conservation measures suitable to an area must be adopted to improve the livelihood of the farmers. Soil and water conservation includes conservation farming, storage of rainwater in soil profile, modification of soil profile, addition of organic matter, etc. The various soil and water conservation technologies are land levelling; contour bunding; trenching; contour cultivation; strip cropping; cover crops; mixed cropping; conservation tillage; cultivation of fast-growing and early maturing crops; recycling of crop residues; use of mulches, vegetative grass barriers and agroforestry systems; sloping land management; etc. (Sharma and Singh 2013). Among these mulching was found to be quite effective, but its rate and mode are equally effective (Bhatt and Khera 2006). Recognizing the importance of these techniques, we are going to discuss them one by one for delineating their role in conserving natural resources in hill agriculture.

3.7.1 Contour Farming

In regions with low rainfall, the primary objective of contour farming is to provide the greatest possible conservation of rainwater; in humid regions, on the other hand, the primary purpose is to reduce soil loss by water erosion, although water conservation is also highly important. The purpose of contour farming is to place rows and tillage lines at the right angles to the normal flow of surface water. The resistance developed by the crop rows and by furrows, between the ridges, to the water flow, thus, reduces runoff velocity and gives more time to water to infiltrate in to the soil. Contour tillage operations, thus, hold part of the rainwater and store it in situ, thus reducing runoff and soil erosion and bring about a more uniform distribution of water received through rainfall. Contour farming, in addition to reducing soil loss also increases crop yields. The practice of up and down method of cultivation in many parts of the country is one of the causes of encouraging man-made erosion. On steep slopes, this practice enables rainwater to gain velocity, facilitating runoff water to erode soil as time of concentration decreased to manifolds. Thus, the upper layer of the fertile soil is washed away and silted up in the dams and bunds installed in the highest-order gully and decreased its lifespan. Contour farming is a system of crop husbandry in which all cultivation operations are done on contour, e.g. preparatory tillage, sowing, interculture, etc.; water loss and soil erosion are reduced by contour farming as erosivity of the rainwater decreased to marginal levels. Soil loss decreased to 32.2 % when the site was brought to contour farming as compared to the conventional methods in Dehradun (Ghosh et al. 2015). Grass gave the least runoff and soil loss, and cultivated fallow gave the maximum soil loss (Table 3.5).

3.7.2 The Use of Cover Crops or Mulching

The effect of mulching in improving water productivity has been ascribed to restricted evaporative losses because of decrease in radiant energy reaching the soil to cause phase change from liquid water to gaseous phase, decrease of vapour pressure difference within soil and ambient air and finally because of decrease in

Table 3.5 Soil and water loss on 8 % slope under different cropping systems at Dehradun

Treatment	Rainfall (mm) (June–October)	Water loss (mm)	Water loss as % of rainfall	Soil loss (tonnes/ha)
(Average of 4 years) (8 % slope – Dhulkot silty clay loam)				
Up and down cultivation maize-wheat	1239	670	54.1	28.5
Contour cultivation maize + cowpea-wheat	1239	511	41.2	19.3
Cowpea-wheat	1239	405	32.7	28.3
Giant star grass	1239	31	2.5	1.3
Cultivated fallow	1239	445	35.9	44.0

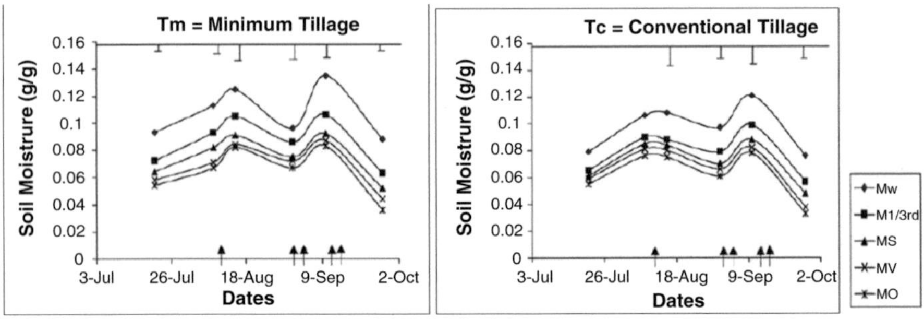

Fig. 3.3 Soil moisture content of surface soil as affected by tillage and different modes of mulch application (Bhatt and Khera 2006)

vapour lifting capacity of the air (Fig. 3.3) and thereby causing yield augmentation (Bhatt et al. 2004; Bhatt and Khera 2006).

Cover crop is a close growing crop raised mainly for protection and maintenance of the soil. Effectiveness of the cover crop depends on close spacing and development of good canopy for interception of raindrops so as to expose minimum soil surface for erosion. Good ground cover canopy gives protection to the land like an umbrella. Erosion from cultivated fields can be reduced if the land has enough crop canopy during the peak season (Sharma et al. 2015); the studies at Dehradun show the canopy development of various legumes crops during rainy season (Fig. 3.4).

3.7.3 Strip Cropping and Hedgerow Cultivation

Strip cropping is a system under which ordinary farm crops are planted in relatively narrow strips, across the slope of the land, and are arranged in such a manner that the strips of erosion permitting crops are always separated by those of close growing or erosion-resisting crops. Seven strip mixture treatments of maize and soybean were tested by Singh and Bhushan (1976) in high rainfall area at Dehradun.

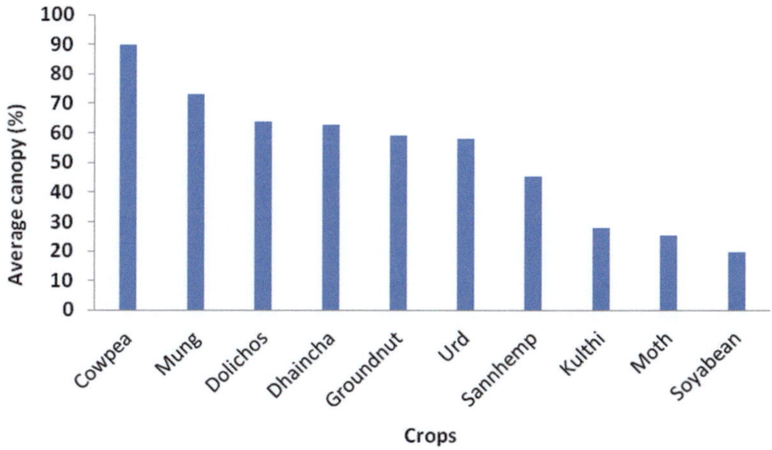

Fig. 3.4 Average crop canopy of different legumes

Crops behave differently in their capacities to produce vegetative cover and root development and consequently affect splash erosion, runoff and soil loss. Legumes, in general, as they produce good cover, are erosion resisting, whereas open-tilled crops like maize and cotton are erosion permitting. Farmers cultivate the lands to satisfy his food requirement and at the same time produce minimum erosion. Strip cropping is the system which meets such requirement. The analysis indicated that with two rows of maize (60 cm apart) and eight rows of soybean (30 cm apart), maximum yield was obtained. Since there was a major contribution of soybean in this production system (3708 kg/ha), maximum economic returns were also obtained under this combination (Table 3.6). Maize and cluster bean in alternate strips of 9 m each at 1.5 % slope in Shiwalik region reduced the runoff and soil loss by 24 % and 43 %, respectively, as compared to maize cultivation along the slope (Nafziger 2014).

3.7.4 Mixed Cropping/Intercropping

Mixed cropping/intercropping is a practice of growing more than one crop in the same field simultaneously. This system of cropping is very extensively adopted by the farmers in India. The most important mixture consists of legume and nonlegume crops. The system gives better cover on the land surface, good protection to soil from beating action of raindrops. A study conducted at different places in the country has shown that mixed cropping/intercropping gives less runoff and soil loss. Studies in maize + cowpea intercropping show that splash erosion reduced from 64 g/57.6 m^2/77 days under pure crop of maize (with 68 % canopy) to 22 g/

Table 3.6 Yield (kg/ha) of maize and soybean under different crop mixtures (4 years average)

	No. of maize rows			No. of soybean rows			
Crop	10:0	8:2	6:4	5:5	4:6	2:8	0:10
Maize	3690	3254	2722	2502	2557	2054	–
Soybean	–	285	358	632	900	1654	2071
Total grain production	3690	3539	3079	3215	3457	3708	2071

57.6 m^2/77 days under intercrop (with 94 % canopy). The results conclusively suggest that intercropping of cowpea (double row) for vegetable purposes is more advantageous both in terms of monetary benefits and soil conservation (Bonde and Mohan 1984). Intercropping of legumes in maize not only is adventitious in reducing runoff and soil loss by 15 and 27 % but also resulted in higher yield of succeeding wheat crop by 61 % (Tomar 1992; Ghosh et al. 2006).

3.7.5 Tillage and Crop Residue Management

Tillage is well-known soil and water conservation technology. It results in improved soil organic matter (SOM), which positively influences physical, chemical and biological properties of soils. Conservation tillage with the maintenance of crop residue cover on 30 % of the soil surface is soundly based within the framework of conservation of natural resources, viz., soil, water, nutrients and SOM and sustained production. The impact of a zero tillage and crop residue management associated with soil health and quality by improving soil properties, minimizing soil erosion and soil water evaporation and conserving soil moisture is well documented. Hence reduced tillage practices have been widely used in the last decades as an attractive alternative over conventional tillage practices because of their potential to reduce production/operating costs and benefit for the environment and can save considerable time with seedbed preparation compared with conventional tillage practices (Al-Kaisi and Yin 2004). Bhardwaj and Sindwal (1998) reported that mulching with weeds @ 4.5–5 t/ha in inter-row of maize spaces reduced runoff from 42 % to 22 % and the soil loss from 12 to 3 t/ha at 4 % slope. Reduction in runoff and soil loss was also reported by Ghosh et al. (2013) by using weed as mulch along with the practice of minimum tillage at 2 % slopy lands. Covering the land surface through crop residues, organic manures, leaf litter and other vegetative materials and its proper management are the most effective measure to reduce evaporation, keep down weeds, improve soil structure and infiltration rate and ultimately increase crop yields. In general, mulches used in *rabi* season crops have always conserved soil moisture and increased considerable yield in many cases. In hilly region, farmers use heavy mulching through organics and FYM for moisture conservation. The results of a study conducted for 3 years to assess the effect of four mulching treatments on yield of wheat crop reported by Tejwani et al. (1975) revealed that maize residues ploughed under and seedbed prepared by sub-surface tillage were beneficial as they gave higher grain yield of wheat as compared to other treatments. In Doon Valley after the harvest of *kharif* crops, *rabi* crops suffer from moisture deficit at sowing and also during crop growth period. Therefore, Sharma et al. (2013a, b) suggested that in a maize-wheat cropping system, maize field should be ploughed immediately after maize harvest followed by planking and covering soil surface by maize stover mulch at 5 t/ha up to wheat sowing to reduce soil moisture losses (up to 33 % at surface layer) and increase germination (43 %), WUE (63 %) and yield of wheat crop as compared to control. The advantage of mulch along with conservation tillage has also been described by Samra et al. (2000) for wheat sowing in maize-based cropping system. In a study in middle Himalayan region, Bhattacharyya et al. (2009) reported that management strategies, such as reduced tillage (ZT and MT) and continuous leguminous cropping sequences (soybean-wheat/lentil/garden pea), played significant roles in total SOC sequestration in soil in the rain-fed hilly agroecosystem. On an equivalent initial soil mass basis, SOC storage of ~15 cm soil depth after 4 years was 26.0 Mg ha^{-1} in continuous NT plots but 23.9 Mg ha^{-1} in continuous CT (Bhattacharyya et al. 2009).

3.7.6 Crop Rotations

One of the basic components of the conservation agriculture is the rotation of the crops as it changed the rhizosphere depths. Rotation of crops is an order in which the chosen cultivated crops follow one another in a set cycle on the same field over a definite period for their growth and maturity with an objective to get maximum profit from least investment without impairing soil fertility. Crop rotations can be an important soil and water conservation practice. In a good crop rotation, the effects of soil-improving crops offset the effects of soil-depleting crops. With the increased availability and use of fertilizers, it is now possible to take more effective care of the fertility aspects of the soil under different rotations. Inclusion of thick growing crops in the rotation protects soil from the impacts of raindrops, besides the interception of runoff (Bhatt and Arora 2015).

3.7.7 Plant Population

The greater the plant population, the greater is the hindrance in running water even at the steeper slopes which finally decreases its erosive potent to cause the soil erosion. Further it is reported that increasing plant population to 1.5 lakh ha^{-1} and using 50 % plants as mulch after 1 month of sowing considerably controlled soil and water losses in maize. But considering increase in yield and reduction in runoff and soil, 1.0 lakh population should be preferred over 55,000 plants/ha (Narain and Singh 2000). Comparing maize plant populations of 40,000, 55,000, 70,000 and 85,000 plants per hectare as a sole crop and in combination with cowpea and black gram as intercrops concluded that maize (70,000 plants) + cowpea gave highest maize equivalent total production over 6 years of 4270 kg/ha (Tyagi et al. 1998). Runoff and soil and nutrient loss followed the trend of canopy cover and plant population, i.e. decreased with increase in plant population of maize or intercropping with legumes. However, Khola et al. (1999) concluded that maize grain yield was higher by 144 kg ha^{-1} under 60 × 20 cm (83,000 plants ha^{-1}) spacing than 60 × 30 cm (55,000 plants ha^{-1}), but succeeding wheat grain yield was slightly lower by 58 kg/ha under the farmer. Thus, having higher plants will certainly improve both the land and water productivity on one side and decrease the magnitude of the soil erosion on the other side.

3.7.8 Green Manuring

Green manuring can be defined as a practice of ploughing or turning into the soil the undecomposed green plant tissues for the purpose of improving physical structure as well as fertility of the soil (Singh et al. 2013). Under in situ green manuring, crops are grown and buried in the same field which is to be green manure. The most common green manure crops grown under this system are sunn hemp (*Crotalaria juncea*), *dhaincha* (*Sesbania aculeata*), etc. Studies conducted at Dehradun indicated that sunn hemp is an excellent green manure crop, which has the inherent ability for improving the fertility status of the soil besides reduction in runoff and soil loss (Sharda et al. 1999a, b). Considerable reduction in runoff amounting to 18.6 % in green manuring (GM)-wheat system has been recorded as compared to the fallow-wheat system. Reduction in soil loss was highly significant (~33 %), mainly due to fast and vigorous growth of sunn hemp after the onset of monsoon which provides dense cover to the soil. A study at Dehradun indicated that growing sunn hemp for green manuring in 60 cm wide contour strips between maize rows and recycling after 1 month as surface mulch reduce erosion and control weed growth. Incorporation of mulch after maize also improved the wheat yield due to moisture conservation and nutritional built up (Table 3.7).

3.7.9 Soil Fertility Approach

Sustenance of soil fertility is important through balanced and integrated plant nutrient management to enhance the productivity and sustainable agriculture in the regions. Preference should be given to locally available organic nutrient sources like FYM, green manuring, etc. Four crop cultivation approaches, namely, (i) maize + cowpea (1:2)-wheat, (ii) maize-wheat + mustard (9:1), (iii) paddy-wheat or paddy (SRI)-wheat and (iv) maize-potato-onion with INM to each crop rotations were demonstrated among 19 farmers' fields under the Farmers Participatory Action Research Programme (FPARP) on more crop per drop of water in the agro-climatic zone of valley land of northwest Himalayas (agro-ecological region 14) in the districts of Dehradun (Uttarakhand) and Sirmaur (Himachal Pradesh) during

Table 3.7 Effect of green mulching with green manuring and land slope on runoff (%) and soil loss (t/ha)

	With green manuring		Without green manuring	
Land slope (%)	Runoff	Soil loss	Runoff	Soil loss
0.5	6.0	3.07	11.2	4.1
2.5	12.2	6.83	15.7	8.6
4.5	18.3	12.80	21.2	17.7
9.5	24.9	23.19	28.2	28.2

2007–2009. Results revealed that irrespective of technology, WUE increases to the tune of ~45 % with yield increase of 41 %, and the corresponding net return increases to 90 %. It is inferred from this study that amongst the crop rotations practiced in this region, maize-potato-onion with INM is an important option for best managing the land, water resource and net return and improving the soil quality under limited irrigation condition, whereas maize-wheat + mustard under rain-fed conditions (Ghosh et al. 2011).

3.7.10 Mulch Farming

The mulch farming is a system in which organic residues or other materials are neither ploughed into the soil nor mixed with it but are left on the surface to serve as mulch. Mulch farming is not only useful for reducing soil and water losses but is also useful for maintaining high soil moisture in the field. Thus, mulching can be used in higher rainfall area for decreasing soil and water losses and in low rainfall region for improving soil moisture. Studies conducted at various places have amply demonstrated that mulching increases both soil moisture and yield of crops. Khola et al. (1998) in their study observed that intercropping of cowpea or soybean with maize either for fodder or in situ mulching reduced maize grain yield by 101–451 kg/ha compared to sole counterpart. However, succeeding wheat grain yield was higher under mulching of cowpea (2744 kg/ha) or sorghum (2683 kg/ha) in maize and lower under sole stands of maize (2291–2349 kg/ha). Tyagi et al. (1998) observed that planting of perennial pigeon pea as a vegetative barrier and incorporation of its non-woody material, cut at 50 cm height in the soil, registered minimum runoff (24 %) and soil loss (5.3 t/ha). Sharma et al. (2010) concluded that legume mulching is a highly beneficial practice for enhanced moisture and nutrient conservation, leading to increased productivity and soil health of maize-wheat cropping system under Doon Valley conditions of north-western India. Beneficial effects of mulching in intercropped legumes with maize on canopy cover and soil and water losses had been reported by several workers (Mohan 1992; Narain and Singh 2000; Singh et al. 2003).

As mulching can be used in higher rainfall period/region for decreasing soil and water loss, similarly, this can also be used in low rainfall period/region for increasing *in-situ* soil moisture. Studies conducted at various places have demonstrated that mulching increases soil moisture and yield of crops. Wheat crop sown in *rabi* season at Dehradun gave significantly higher yield with surface mulch. A study conducted at Dehradun shows that mulches applied immediately after the harvest of maize increased the grain yield of wheat significantly. Straw mulching before sowing was superior giving ~36 % higher yield than control followed by grass mulch (Singh and Bhushan 1978). However, dust mulch was found as the most practical way of moisture conservation, which produced ~22 % higher yield than control. Sharma et al. (1997) conducted a study to find out the optimum quantity of air-dried *Leucaena* leaf mulch and its time of incorporation proceeding to rain-fed wheat sowing.

The equation $Y = a + b\text{M} + c\text{T} + d\text{M}^2 + e\text{T}^2 + \text{FMT}$ was fitted between wheat grain yield (Y) and combination of doses of mulch (M) and timings of its incorporation (T). The results revealed that in addition to recommended dose of chemical fertilizers, air-dried *Leucaena* leaf mulch at 2.25 t/ha may be incorporated into the soil within a period of 30–40 days of maize harvesting. The wheat yield can be obtained ~32 q/ha, which is ~25 % higher than the control (Table 3.8).

3.7.11 Water Harvesting and Recycling

In NWH region receiving high rainfall, the water-harvesting structure can be a source of supplemental irrigation during the time of scarcity for cash crops for high productivity and profitability (Sharda et al. 1986). The harvested water in dugout or embankment-type ponds can be efficiently utilized for supplemental irrigation during lean period to boost crop production. In the middle of Himalayas, where large water-harvesting structures are not feasible owing to topographical limitations, small tanks locally called *tankas* of capacity ranging from 10 to 20 cum have been successfully constructed at farmers' fields. The top length varying from 5 to 6.5 m and width of 3.5–5 m has been found suitable with side slopes of 1:1 and depth of 1.6 m (Juyal and Gupta 1985). They suggested that *tankas* can be suitably lined with 1000 gauge (250 μm) LDPE sheet. Before laying the sheet in excavated tank, protruding stone pieces and roots should be removed. For Kumaon region, two water-harvesting systems have been found feasible, namely, (i) intra-terrace and (ii) micro-watershed. In the former, runoff is collected from a single terrace of ~0.025 ha and recycled back to the same terrace, whereas in the latter, water is collected from a micro-watershed of 0.08 ha and recycled to a command area of ~0.06 ha in the downstream. Both the systems are found to reduce the peak flows (Srivastava 1983).

The harvested water in dugout or embankment-type ponds can be efficiently utilized for supplemental irrigation during lean period to boost crop production. In Doon Valley, there are three options available for utilizing harvested water, viz., (i) pre-sowing irrigation (5 cm), (ii) application at crown root initiation (CRI) stage (5 cm) and (iii) pre-sowing + CRI (total 10 cm). It has been found that between the first two options, increase in crop yield was almost the same. However, owing to

Table 3.8 Effect of different levels and time of incorporation of *Leucaena* mulch on grain and straw yield of wheat

Treatments	Grain yield (kg ha^{-1})	Straw yield (kg ha^{-1})
Leucaena mulch level (t ha^{-1})		
0	2607	4548
2	3207	5287
4	3435	5380
6	3623	5560
LSD ($P = 0.05$)	236	333
Days of incorporation after spread of mulch		
0[a]	3085	5141
15	3292	5176
30	3231	5199
45	3264	5259
LSD ($P = 0.05$)	NS	NS

NS non-significant
[a]Incorporation immediately after maize harvest

Table 3.9 Effect of supplemental irrigation on wheat yields

Particulars	Pre-sowing (5 cm)	CRI (5 cm)	Pre-sowing + CRI (10 cm)	Control
Yield (kg/ha)	3203	3236	3898	2045
Percent increase by supplemental irrigation	80	50	25	–

excessive storage losses, delayed application in the second case results in reduced command area. In the case of the third option, though two irrigations boost the crop yield compared to the first two methods (Table 3.9), corresponding reduction in command area is more (Singh and Bhushan 1980; Singh et al. 1981).

In a study at Dehradun on low-cost water-harvesting and micro-irrigation, this year, three silpaulin-lined (200 GSM) water-harvesting tanks of 7–8 cum capacity were constructed at Ashti integrated with low-cost drip irrigation system operating on low head (3–5 m). The integrated system is more convenient in irrigating the crops for higher productivity and saves labour as compared to hand watering normally practiced in hilly region (Anonymous 2012).

3.7.12 Sub-surface Water Harvesting

In Himalayan foothills, seepage or oozing/trickling of water from the hill slopes as a surface/sub-surface flow is a common phenomenon. Duration and discharge of flow depend on the local geological set-up which forms perched type of aquifers/cracks allowing short duration or sustained release of water from the permeable faces of the hills. Such type of flow can be observed in most of the parts in the Himalayan region. The water can be stored in the harvesting structures which are specially designed allowing the water to flow into the tank from the upper walls (pervious wall) and stored against the lower wall (impervious). Such structure (21 × 8 × 2 m) was constructed in Kalimati village in order to harvest the sub-surface flow. The upper long wall of the tank was constructed with pervious gabion work, and other walls were constructed with masonry work, i.e. impervious, so that the incoming water from the upper pervious wall could be retained in the tank.

The idea was that pore space available in gabion walls between boulders would act as inlet and allow sub-surface water flow to be collected in the tank. In all, 33 families from Kalimati village and 17 from Badasi village participated in developing the water resource for irrigation to 16 ha of land. Stakeholders of Kalimati and Badasi village (50) had the right to share the resource for crop production. The yield of wheat in limited irrigation ranges from 2800 to 3600 kg/ha with an average 3240 kg/ha over the farmer's practice as control. The increase in yield was recorded 44 % which had been achieved because of irrigation and generation of water resource in the village (Samra et al. 2000).

3.7.13 Conservation Bench Terrace

CBT system comprises 3:1 ratio of contributing and receiving area and 20 cm depth of impoundment at the end. A study was initiated with a view to study the efficacy of conservation bench terrace system, under recommended cropping sequence of CSWCRTI, Dehradun. To store the excess runoff from these treatments, a brick masonry tank of size 30 × 15 × 2.5 m was constructed at the experimental site at S.C. Farm, Selakui of CSWCRTI. From the analysis, it has been concluded that the CBT system is most effective in reducing runoff by over 80 % and soil loss by more than 90 % compared to conventional system, i.e. sloping boarder with zero depth of impoundment, crop rotation and maize + cowpea-wheat + mustard. In the levelled bench of CBT system, paddy was taken during *kharif* season, and intercrop of mustard with wheat was taken during *rabi* season. The paddy yield under the CBT system makes the system more remunerative compared to the traditional system of maize-wheat rotation. The CBT system also recorded the highest yield of wheat and mustard crops on an average. The nutrient loss from CBT system was observed to be at minimum. The system has a great scope for harvesting the inevitable runoff particularly in Doon Valley and its subsequent utilization during intervening droughts for sustaining the crop yields (Sharda et al. 1999a, b).

3.7.14 Gravity-Fed Participatory Water Resource Development and Management

Low productivity and cropping intensity and substance level of farming are attributed to poor water resource development in the hilly areas. A major part of rainfall goes unutilized as surface runoff which is not being used for productive purposes. The community is unable to invest huge capital on water resource development due to their poor resource conditions particularly in hilly area. Central Soil and Water Conservation Research and Training Institute (CSWCRTI), Dehradun, initiated water resource development and its management for enhancing livelihood security of hilly farmers in villages, viz., Pasauli, Devthala in Vikasnagar block in the district of Dehradun, Uttarakhand, situated in NW Himalayas.

Water is being tapped from a perennial spring located at 771 m amsl in outer Himalayas. An intake well (IW) was constructed at the source to convey water from the source to the distribution tank (DT) through galvanized iron (GI) pipe (dia: 110 mm) of 2080 m in length, by gravity flow with hydraulic gradient 1.94 %. Distribution tank having a capacity of 50,000 l and inflow at DT is 10 LPS. The tank is located at the head of the command area (25 ha) in order to convey the water through underground PVC pipelines to the farmers' fields, by gravity flow with maximum 26.5 m gravity head. PVC pipeline (1360 m) was laid across the command area in such a way that it can supply the water to the farmers' fields spread in the command area with 10–12 LPS discharge. The command area is divided into 15 unit command areas (UCA), and each UCA covers 1.5 ha. At the head of each UCA, a riser is fitted with the PVC pipeline to uniformly deliver the water. In order to enhance the field application efficiency, collapsible LLDPE pipes are being used by the farmers. Distribution of water for the irrigation to each beneficiary farmer is regulated by the committee constituted by the Water User Association (WUA). Water is being distributed to the farmers' fields on a rotation basis as decided by the WUA, and farmers get their turn in 15 days (irrigation period). In one season, the delivery of water follows the sequence from head to the tail riser, and in the next season, delivery sequence gets reversed in order to negate the clash among the farmers. Productivity of major crops, viz., maize, paddy, wheat and *Toria*, was increased by 48 %. Cropping intensity increased by 29 % due to intervening crop of *Toria* in-between maize-wheat sequence and cultivation of *rabi* wheat in fallow land (Dhyani et al. 2011).

3.8 Concluding Remarks

The northwest Himalayan region of India is suffering severely from the problem of soil erosion, and the farmers of the region are bound to live with a poor living status because of lower water and land productivity. However, this problem could be tackled in a smart way by adopting certain integrated approach as single approach might not be so effective. Proper tillage and mulching in agricultural field help in reducing soil erosion and retain soil moisture. Gully control on sloping lands is effective through vegetation cover. The rainwater harvesting to meet water demands in rain-fed foothill region improves crop yields. Proper residue management not only improves soil physical properties and recycling of nutrients but also protects environment from polluting. With integrated approach, soil health and livelihood of the farmers are certainly improved by mitigating adverse effect of the global warming and the soil erosion.

Acknowledgements Authors are indebted to all the scientists who are currently contributing to the soil conservation research in India presently and did so in the past for improving the livelihood of the farmers in foothill region of NW Himalayas.

References

Adel MM (2001) Effect on water resources from upstream water diversion in the Ganges basin. J Environ Qual 30:356–368

Al-Kaisi MM, Yin X (2004) Stepwise time response of corn yield and economic return to no tillage. Soil Till Res 78:91–101

Anonymous (2012) Annual report 2011–12, Integration of low cost water harvesting and micro-irrigation for resource conservation. In: Patra S, et al. (eds). Central Soil and Water Conservation Research and Training Institute, Dehradun, p 61

Arora S (2006) Preliminary assessment of soil and water conservation status in drought prone foothill region of north-west India. J World Assoc Soil Water Conserv J1–5:55–63

Arora S, Bhatt R (2006) Effect of in-situ soil moisture conservation practices on soil moisture storage and yield of maize (*Zea mays* L.) in a rainfed foothill region of North-west India. Trop Agri 82(3):37–42

Arora S, Saygın L (2011) Soil and water conservation in hilly terrain in north India and central Turkey: issues and challenges. J Soil Water Conserv 10(2):87–93

Arora S, Gupta RD (2014) Soil related constraints in crop production and their management options in North Western Himalayas. J Soil Water Conserv 13(4):305–312

Arora S, Sharma V, Kohli A, Jalali VK (2006) Soil and water conservation for sustainable crop production in Kandi region of Jammu. J Soil Water Conserv 5(2):77–82

Arora S, Hadda MS, Sharma KR, Bhatt R (2010) Soil moisture conservation for improving maize yields through participatory micro-watershed approach in the foothills of the Shivaliks. In: Zaidi PH, Azrai M, Pixley KV (eds) Maize in Asia – emerging trends and technologies. ARMW/CIMMYT, Makassar/Mexico, pp 467–470

Ashraf M (2015) Natural resources of the North Western Himalayas- threats, evaluation and conservation *IN* National seminar on natural resources of the NW Himalayas at Sri Partap College, Srinagar J&K state

Bhardwaj SP, Sindwal NS (1998) Zero tillage and weed mulch for erosion control on sloping farm land in Doon valley. Indian J Soil Conserv 26(2):81–85

Bhatt R (2003) Evaluation of effect of tillage and cover management techniques on the soil erosion in the submontaneous tracts of the Punjab, India. M.Sc Thesis, Punjab Agricultural University, Ludhiana, Punjab, India

Bhatt R (2013) Gully erosion management in foothills of lower Shiwaliks. Soil and Water Conservation Today 8(4): 5–6

Bhatt R (2015) Soil water dynamics and water productivity of rice-wheat system under different establishment methods. Dissertation submitted to the Punjab Agricultural University, Ludhiana

Bhatt R, Arora S (2015) Impact of tillage on soil bio-chemical properties, water storage and environment. J Soil Water Conserv 14(3):278–282

Bhatt R, Khera KL (2006) Effect of tillage and mode of straw mulch application on soil erosion losses in the submontaneous tract of Punjab, India. Soil Till Res 88:107–115

Bhatt R, Kukal SS (2015) Dynamics of gully erosion in lower Shiwaliks. J Soil Water Conserv 14 (1):1–9

Bhatt R, Khera KL, Arora S (2004) Effect of tillage and mulching on yield of corn in the submontaneous rainfed region of Punjab, India. Int J Agric Biol 6:1126–1128

Bhatt R, Kukal SS, Arora S (2013) Resource conservation technologies for improving water productivity. J Soil Water Conserv 12(4):313–320

Bhatt R, Kukal SS, Arora S, Busari MA, Yadav M (2016) Sustainability issues in rice-wheat cropping system. Int J Soil Water Conserv Res (Elsevier) 4:68–83

Bhattacharyya R, Ved-P, Kundu S, Srivastva AK, Gupta HS (2009) Soil aggregation and organic matter in a sandy clay loam soil of the Indian Himalayas under different tillage and crop regimes. Agric Ecosyst Environ 132:126–134

Bicon MA (2004) Water use efficiency in plants biology. CRP Press, Boston

Bonde WC, Mohan SC (1984) Intercropping in maize for effective soil conservation, moisture utilization and maximizing production in Doon valley. Annual report, CSWCRTI, Dehradun, pp 72–73.

Bouyoucos GJ (1935) The clay ratio as a criterion of susceptibility of soils to erosion. J Am Soc Agron 27:738–741

Burkard MB, Kostachuk RA (1997) Patterns and control of gully growth along the shoreline of lake Huron. Earth Surf Proc Land 22:901–911

Christensen JH, Hewitson B, Busuioc A, Chen A, Gao X, Held I, Jones R, Kolli RK, Kwon W-T, Laprise R, Magaña Rueda V, Mearns L, Menéndez CG, Räisänen J, Rinke A, Sarr A, Whetton P (2007) Regional climate projections. In: Solomon S, Qin D, Manning M, Chen Z, Marquis M, Averyt KB, Tignor M, Miller HL (eds) Climate change 2007: the physical science basis, Contribution of Working Group I to the fourth assessment report of the Intergovernmental Panel on Climate Change. Cambridge University Press, Cambridge/New York

Cruz RV, Harasawa H, Lal M, Wu S, Anokhin Y, Punsalmaa B, Honda Y, Jafari M, Li C, Huu Ninh N (2007) Asia. In: Parry ML, Canziani OF, Palutikof JP, van der Linden PJ, Hanson CE (eds) Climate change 2007: impacts, adaptation and vulnerability, Contribution of working group II to the fourth assessment report of the Intergovernmental Panel on Climate Change. Cambridge University Press, Cambridge, pp 469–506

Dhyani BL, Madhu M, Singh C, Kumar A, Bankey B, Mandal D, Muruganandam (2011) Participatory dissemination and assessment of land and water management technologies for livelihood security in rainfed areas of north western Himalayas. Annual report, 2011–2012, CSWCRTI, Dehradun, pp 99–100

Ghosh BN, Khola OPS, Dadhwal KS (2006) Impact of organic matter build up on resource conservation under different crops on 2% slope. Annual report, 2005-2006, CSWCRTI, Dehradun, pp 29–30

Ghosh BN, Sharma NK, Dadhwal KS (2011) Integrated nutrient management and cropping systems impact on yield, water productivity and net return in valley soils of north-west Himalayas. Indian J Soil Conserv 39(3):236–242

Ghosh BN, Sharma NK, Dogra P (2013) Yield maximization and resource conservation through organic input management. Annual Report, 2013, CSWCRTI

Ghosh BN, Dogra P, Sharma NK, Bhattacharya R, Mishra PK (2015) Conservation agriculture impact for soil conservation in maize–wheat cropping system in the Indian sub-Himalayas. Int Soil Water Conserv Res 3(2):112–118

Ghosh BN, Meena VS, Alam NM, Dogra P, Bhattacharyya R, Sharma NK, Mishra PK (2016) Impact of conservation practices on soil aggregation and the carbon management index after seven years of maize–wheat cropping system in the Indian Himalayas. Agric Ecosyst Environ 216:247–257

Goree K (2004) Water-L news. International Institute for sustainable development in collaboration with the third world water forum secretariat and the world water council. Issue 9 (24 Feb–16 Mar) 2004. Bangladesh

Hadda MS, Sur HS (1987) Effect of land modifying measures on erosion, nutrient, water storage and yield of pearl mellet fodder. J Indian Soc Soil Sci 35:480–486

Hillel D (1980) Applications of soil physics. Academic, New York, p 385

Hudson N (1984) The erosivity of rainfall. Soil Conserv 23:62–80

Humphreys E, Kukal SS, Christen EW, Hira GS, Singh B, Sudhir-Yadav SRK (2010) Halting the groundwater decline in north-west India-which crop technologies will be winners? Adv Agron 109:156–199

ICAR, NASC (2010) Degraded and wastelands of India -status and spatial distribution, Directorate of Information and Publications of Agriculture Indian Council of Agricultural Research New Delhi, pp 158

Joy KJ, Parnjpe S, Shah A, Badigar S, Lele S (2005) Scaling up of watershed development projects in India: learning from the first generation projects. Fourth IWMI-Tata Annual Partners Meet, International Water Management Institute, Anand, India. pp.133-134.

Juyal GP, Gupta RK (1985) Construction of LDPE lined 'TANKAS' in hills – a case study. Indian J Soil Conserv 13(1):10–13

Khera KL, Singh G (1995) Effect of paddy straw mulch and rainfall intensity on runoff and soil loss under stimulated rainfall. Ind J Soil Conserv 23(1):20–23

Khera KL, Singh G (1998) Effect of crop cover and field slope on soil erosion in northern plain hot sub humid Punjab. Indian J Soil Conserv 26(1):19–21

Khola OPS, Sharma NK, Khullar AK (1999) Evaluation of different maize growing systems for resource conservation and biomass production under varying fertility conditions. Annual Report, CSWCRTI, 1999–2000, Dehradun, pp 32–33

Kiani K (2005) Water-related crisis feared in 20 years. Dawn: The Dawn Group of Newspapers: Islamabad

Kukal SS (1987) Soil erosion risk assessment survey of a representative area in submontaneous tract of Punjab. M.Sc. thesis, Punjab Agricultural University, Ludhiana

Kukal SS, Bhatt R (2006) An appraisal of the indigenous technology for soil and water conservation in Shiwaliks of Punjab. Department of Soils, Punjab Agricultural University, Ludhiana

Kukal SS, Sur HS (1992) Soil erosion hazards in the foothills of lower Shiwaliks. J Indian Soc Soil Sci 40:162–167

Kukal SS, Sur HS, Gill SS (1991) Factors responsible for soil erosion hazard in submontane Punjab, India. Soil Use Manag 7(1):38–44

Kukal SS, Bawa SS, Khera KL (2002) Evaluation of gully control measures in foothills of lower Shiwaliks-a compendium. Department of Soils, Punjab Agricultural University, Ludhiana

Kukal SS, Singh B, Bhatt R (2005) Runoff and sediment losses in relation to gully erosion in foothills of Shiwaliks. Map India 124:1–8

Kukal SS, Bawa SS, Bhatt R, Kamboj A (2006) Behaviour and patterns of gully erosion in foothills of lower Shiwaliks. Final report submitted to Department of science and Technology (DST), New Delhi

Mandal D, Sharda VN (2011) Assessment of permissible soil loss in India employing a quantitative bio-physical model. Curr Sci 100(3):383–390

Mannering JV, Meyer LD (1963) The effect of various rates of surface mulch on infiltration and erosion. Soil Sci Soc Am Proc 27:84–86

Maruthi SGR (1986) On screening of regression models for selection of optimal variable subsets. J Indian Soc Agric Stat 38:161–168

Matharu GS, Kukal SS, Bawa SS (2003) Rain characteristics in relation to runoff in submontane Punjab. J Indian Soc Soil Sci 51:288–290

McGregor KC, Cullum RF, Mutchler CK (1999) Long term management effects on runoff, erosion and crop production. Trans ASAE 42(1):99–105

Meena RS, Meena VS, Meena SK, Verma JP (2015a) The needs of healthy soils for a healthy world. J Clean Prod 102:560–561

Meena RS, Meena VS, Meena SK, Verma JP (2015b) Towards the plant stress mitigate the agricultural productivity: a book review. J Clean Prod 102:552–553

Meena VS, Meena SK, Verma JP, Meena RS, Ghosh BN (2015c) The needs of nutrient use efficiency for sustainable agriculture. J Clean Prod 102:562–563

Meena VS, Meena SK, Verma JP, Meena RS, Jat LK (2015d) Current trend to mitigate climate change and sustainable agriculture: a book review. J Clean Prod 102:548–549

Meena VS, Verma JP, Meena SK (2015e) Towards the current scenario of nutrient use efficiency in crop species. J Clean Prod 102:556–557
Meyer A, Martinez-Casasnovas JA (1999) Prediction of existing gully erosion in Vineyard parcels of the NE Spain: a logistic modelling approach. Soil Till Res 50:319–331
Mishra PK, Juyal GP (2015) Soil erosion in Indian Himalayan region and mainstreaming of current problems of landslides and floods in the watershed programmes. In: Bhan S, Arora S (eds) Advances in soil and water resource management for food and livelihood security in changing climate. SCSI, New Delhi, pp 1–9
Mohan SC (1992) Effect of green manuring with intercropping on soil fertility and moisture conservation under rainfed conditions, Annual report, 1992–1993, CSWCRTI, Dehradun, p 44
Morgan RPC (2005) Soil erosion and conservation. 3rd edn. Blackwell Publishing ISBN: 1-4051-1781-8 pp 324
NAAS (2010) Degraded and wastelands of India: status and spatial distribution. ICAR, New Delhi
Nafziger (2014) Growth and production of maize: mechanized cultivation. Soils, plant growth and crop production. vol-1, Encyclopedia of life support systems (EOLSS), UNESCO-EOLSS, Miollis, 75732 Paris, pp 331–362
Narain P, Singh RK (2000) Quantification of ground cover effect on runoff and soil loss through artificial and natural rainfall. Annual report, CSWCRTI, Dehradun, p 23–24
Nearing MA (2001) Potential changes in rainfall erosivity in the US with climate change during the 21(st) century. J Soil Water Conserv 56(3):220–232
Ogola JS, Mutuilah WV, Omulo MA (2002) Impacts of gold mining on the environment and human health: a case study in the Migori Gold Belt, Kenya. Environ Geochem Health 24:141–158
Palanisami K, Suresh KD, Chandrasekharan B (eds) (2002) Watershed management: issues and policies for 21st century. Associated Publishing Company, New Delhi, 341 pp
Rashid HE, Kabir B (1998) Water resources and population pressures in the Ganges River Basin, Case Study, Bangladesh
Reddy TY, Reddi GHS (1995) Soil conservation. Princ Agron 35:335–367
Revenga C, Brunner J, Henniger N, Kassem K, Payner R (2000) Pilot analysis of global ecosystems, freshwater systems. World Resources Institute, Washington, DC
Rizvi M (2001) Forecasting water flows in Pakistan Indus river. The International development research centre, Canada
Rockstrom J, Barron J, Fox P (2003) Water productivity in rain-fed agriculture: challenges and opportunities for smallholder farmers in drought-prone tropical agro ecosystems. In: Braker R, Molden D (eds) Comprehensive assessment of water management in agriculture series; water productivity in agriculture: limits and opportunities for improvement; CAB International
Samra JS, Mishra AS, Mohan SC, Dhyani BL, Tomar DS (2000) Participatory generation, assessment refinement and dissemination of rural technology (1996 to 1999). CSWCRTI, Dehradun, p 206
Sehgal JL, Sys C (1970) The soils of Punjab – India. III classification and geographic distribution. Pedologic 20:357–380
Sharda VN (2011) Land degradation and watershed management related issues in the Himalayan Region: Status and strategies. Proceedings of workshop Mountain Agriculture in Himalayan region: status, constraints and potentials. 2–3 Apr 2011. pp 1–22
Sharda VN, Sastry G, Joshi DP (1986) Studies on control of seepage losses. Annual report, pp 34–36, CSWCRTI, Dehradun
Sharda VN, Khybri ML, Sharma NK, Mohan SC, Juyal GP (1999a) Green manuring for conservation and production in western Himalayan Region: (I) effect of runoff, soil loss and fertility improvement. Indian J Soil Conserv 27(1):26–30
Sharda VN, Shrimali SS, Khosla OPS (1999b) Hydrological evaluation of recommended conservation measures on mildly sloping lands. Annual report, 1999–2000, CSWCRTI, Dehradun
Sharda VN, Sikka AK, Juyal GP (2012) Participatory integrated watershed management – a field manual, 2nd edn. CSWCR&TI, Dehradun

Sharda VN, Mandal D, Ojasvi PR (2013) Identification of the soil erosion risk areas for the conservation planning in different states of India. J Environ Biol 34:219–226

Sharma V, Arora S (2015) Land degradation processes and factors affecting crop production in foothills of Jammu Shivaliks. J Soil Water Conserv 14(4):294–299

Sharma BR, Sharma D (2008) Impact of climate change on water resources and glacier melting and potential adoption for Indian agriculture. In Annual convention on climate change and it's effect on agriculture. 4–5 Dec

Sharma NK, Singh RJ (2013) Agronomic practices for erosion control. Pop Kheti 1(3):57–60, Available online at www.popularkheti.info

Sharma NK, Agarwal MC, Tyagi PC, Mohan SC, Singh PN (1997) Effect of *Leucaena* leaves mulch and time of incorporation on wheat yield. Indian J Soil Conserv 25(1):51–54

Sharma BR, Rao KV, Vittal KPR, Ramakrishna YS, Amarasinghe U (2010) Estimating the potential of rainfed agriculture in India: prospects for water productivity improvements. Agric Water Manag 97:23–30

Sharma A, Shanker GRM, Arora S, Gupta V, Singh B, Kumar J, Mishra PK (2013a) Analyzing rainfall effects for sustainable rainfed maize productivity in foothills of Northwest Himalayas. Field Crops Res 145:96–105

Sharma NK, Ghosh BN, Khola OPS, Dubey RK (2013b) Residue and tillage management for soil moisture conservation in post maize harvesting period under rainfed conditions of north-west Himalayas. Indian J Soil Conserv 41(3):287–292

Sharma NK, Chand L, Maurya UK, Singh RJ (2015) Conservation approaches for sustainable agricultural production in North-Western Himalayan Region. In: Bhan S, Arora S (eds) Advances in soil and water resource management for food and livelihood security in changing climate. SCSI, New Delhi, pp 414–429

Sidhu GS, Walia CS, Sachdev CB, Rana KPC, Dhankar RP, Singh SP, Velaythum M (2000) Soil resource of NW Shiwaliks for perspective land use planning. In: Mittal SP, Aggarwal RK, Samra JS (eds) Fifty years of research on sustainable resource management in Shivaliks. CSWCRTI, Chandigarh, pp 23–34

Singh G (2000a) Behaviour and patterns of gully erosion in submontane Punjab. M.Sc. Thesis, Punjab Agricultural University, Ludhiana, India

Singh RV (2000b) Watershed planning and management. Yash Publishing House, Bikaner

Sidhu HS (2002) Crisis in agrarian economy in Punjab: some urgent steps. Econ Pol Wkly 37 (30):3132–3138

Singh G, Bhushan LS (1976) Crop mixture strip planting of maize and soybean. Annual report. CSWCRTI, Dehradun, pp 163–164

Singh G, Bhushan LS (1978) Effect of surface mulches on yield of wheat. Annual report. CSWCRTI, Dehradun, p 42

Singh G, Bhushan LS (1980) Supplemental irrigation to wheat crop from runoff harvested water under dry farming conditions. Ann Arid Zone 19(3):215–220

Singh G, Ram B, Chandra S (1981) Soil loss prediction research in India, Bulletin No. T-12/D-9. CSWCRT, Dehradun, pp 1–70

Singh G, Babu R, Narain P, Bhushan LS, Abrol IP (1992) Soil erosion rates in India. J Soil Water Conserv 47:97–99

Singh R, Dhyani SK, Dubey RK (2003) Conservation tillage and green manure mulching for optimizing productivity in maize-wheat cropping system in the sub-mountainous Himalayan region, Annual Report. CSWCRTI, Dehradun, pp 21–22

Singh S, Singh R, Kumar K, Singh B, Shukla L (2013) Biofertilizers and green manuring for sustainable agriculture. In: Kumar S, Prasad B (eds) Modern technologies for sustainable agriculture, 1st edn. New India Publishing Agency, New Delhi, pp 129–150. ISBN 978-93-81450-61-1

Solanki R, Anju, Poonum, Dhankar R (2011) Zinc and copper induced changes in physiological characteristics of *Vigna mungo* (L). J Environ Biol 32:747–751

Srivastava RC (1983) Film lined tanks for hilly regions. Indian Farming XXXIII(II):27–31

Stocking MA (1980) Examination of the factors controlling gully growth. In: De Boodt M, Gabriels D (eds) Assessment of erosion. Wiley, Chichester, pp 505–520
Sur HS, Ghuman BS (1992) Soil and rainfall water management in medium rainfall alluvial soils. Paper presented at the 57th convention of Indian Soc Soil Sci. Held at Hyderabad from 26–29 Nov
Sur HS, Ghuman BS (1994) Soil management and rain water conservation and use-IV. Alluvial soils under medium rainfall. Indian Soc Soil Sci Bull 16:56–65
Tejwani KG, Bhaskaran AR, Thomus PK, Verma BR, Khybri ML, Bains KS (1975) Research in agronomy for soil conservation. In: Tejwani KG, Gupta SK, Mathur HN (eds) Soil and water conservation research. ICAR, New Delhi, pp 106–145
Thakur KK, Arora S, Pandita SK, Goyal VC (2013) Land use based characterization of soils in relation to geology and slope in micro-watershed of Shiwalik foothills. J Soil Water Conserv 12(4):348–354
Thornthwaite CW (1948) An approach toward a rational classification of climate. Geogr Rev 38:55–94
Tomar DS (1992) Effect of intercropping of maize on soil erosion and yield. Annual report 1992–1993, CSWCRTI, Dehradun, p 38
Tyagi PC, Mohan SC, Joshi BP (1998) Crop cover management in maize legume cropping system for sustainable production. Annual report, 1998–1999 CSWCRTI, Dehradun, p 26–27
Vittal KPR, Singh HP, Rao KV, Sharma KL, Victor US, Ravindra CG, Sankar GRM, Samra JS, Singh G (2003) Guidelines on drought coping plans for rainfed production systems. All India Coordinated Research Project for Dryland Agriculture, Central Research Institute for Dryland Agriculture (CRIDA), Indian Council of Agricultural Research, Hyderabad, p 39
Wagger MG, Denton HP (1992) Crop and tillage rotation. Grain yield, residue cover and soil water. Soil Sci Soc Am J 56:1233–1237
World Resource Institute (2003) Watersheds of the world_CD. The World Conservation Union (IUCN), the International Water Management Institute (IWMI), the Ramsar Convention Bureau, and the World Resources Institute (WRI), Washington, DC
WWF (2007) World's top 10 rivers at risk. WWF International, Gland
Yang Z (1991) Glacial water resources in China. Gansu Science and Technology Press, Lanzhou
Zheng F, Huang C, Romkens M (2009) Ephemeral gully erosion research at the Loess Plateau of China. Paper presented at the annual meeting of the soil and water conservation society, Saddlebrook Resort, Tampa http://www.allacademic.com/meta/p202367_index.html

Role of Biofertilizers in Conservation Agriculture

4

Mahendra Singh, M.L. Dotaniya, Amit Mishra, C.K. Dotaniya, K.L. Regar, and Manju Lata

Abstract

In the present time, chemical fertilizers are more in practice for crop production which affected the soil and environment quality. The higher amount of chemical inputs in agricultural production system affected the sustainability of the agricultural crop production systems, increased cost of cultivation, and caused partial factor productivity decline, and maintaining the global food security and environmental quality became a daunting challenge. Indiscriminate and imbalanced use of fertilizers, mostly urea, and the poor application of organic matter to cropland have led to considerable reduction in soil health. Nowadays our agriculture has shifted to old-age practice like conservation agriculture. It is using old tool and techniques with incorporation of modern science and scientific principles. In general, biofertilizer is organic in nature containing an effective particular microorganism in a concentrated form which originated either from the plant root nodule or from the soil of the rhizosphere. Biofertilizers have emerged as potential environment-friendly inputs that are

M. Singh (✉)
Bihar Agricultural University, Sabour 813 210, India
e-mail: mahendra_saini_soil@yahoo.com

M.L. Dotaniya
ICAR-Indian Institute of Soil Science, Bhopal 462 038, India

A. Mishra
International Rice Research Institute, Bhubaneswar, Odisha 751 003, India

C.K. Dotaniya
College of Agriculture, SKRAU, Bikaner 334 006, India

K.L. Regar
Department of Soil Science and Agricultural Chemistry, Institute of Agricultural Sciences, BHU, Varanasi 221 005, India

M. Lata
University of Rajasthan, Jaipur 302 004, India

J.K. Bisht et al. (eds.), *Conservation Agriculture*,
DOI 10.1007/978-981-10-2558-7_4

benefited for agricultural crop production system. They hold vast prospective in fulfilling the plant nutrient requirements, which are reducing the chemical fertilizer application and minimizing environmental pollution. The bioinoculants are used as a seed treatment or soil treatment, improving plant nutrient availability and finally crop growth and yield. These contain living cells of diverse types of microorganisms and have the potential to solubilize and mobilize plant nutrient elements from insoluble form through biological process and also fix atmospheric nitrogen. The adequate use of biofertilizers helps in maintaining soil quality and thus provides a low-cost approach to manage crop yield along with protecting the environment.

Keywords
Biofertilizer • Climate change • Conservation agriculture • Plant nutrient

4.1 Introduction

By the year of 2020, India needs approximately 280 MT of food grain to fulfill the hungry mouth as against the probable production of 264 nutrient in 2013–2014. The use of chemical fertilizers is practicing in agricultural crop production system. This practice has been more prominent since inception of green revolution in Indian agriculture. The green revolution was a complete package of genetic advancement (high-yielding varieties), nutrient management (organic or inorganic), use of natural resources (water), and contribution of agricultural chemicals (pesticides and fungicides). It increased the crop yield potential of India and made this populated country sufficient during the postindependence period (Dotaniya 2013; Dotaniya and Kushwah 2013; Dotaniya et al. 2014d). The revolution in agricultural field was started during the industrial N fertilizer production with the help of Haber–Bosch process enhanced the fertilizer uses and more dependent on fossil fuel (Rao 2014). Excessive use of chemical fertilizers reduced the soil productivity and adversely affected the soil health. The ideal ratios of NPK application are 4:2:1, which is totally distorted, and increase to 61.7:19.2:1 and 61.4:18.7:1 in Punjab and Haryana, respectively (NAAS 2014). Mostly, the use of nitrogen (N) and phosphorus (P) fertilizers and ignoring other fertilizer applications (Fig. 4.1) caused the multiplant nutrient deficiency and yield decline over a period. The potential and enormous scopes of biofertilizers for promoting sustainable agriculture and conservation agriculture have been known for many years. In this respect conservation agriculture has come into view as an important priority area across the globe for fulfilling the growing demand for safe and healthy food and environmental sustainability associated.

In broad term, conservation agriculture is the integration of ecology with agricultural production system with scientific and modern tools and techniques. The addition of modern technologies improved the worth and ecological integrity of the soil system. The application of these is tempered with traditional knowledge of soil husbandry gained from continuous generations of progressive farmers. This

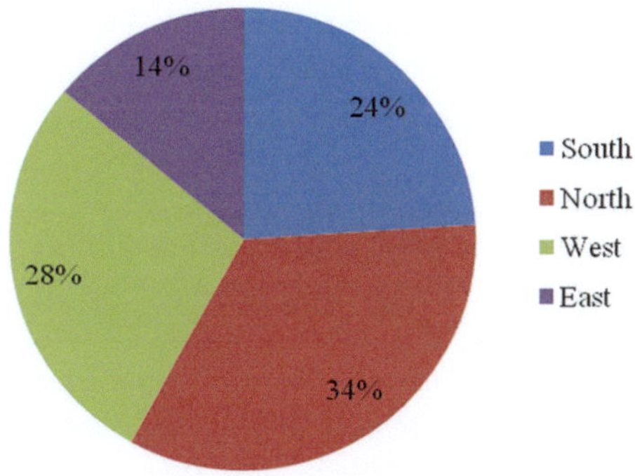

Fig. 4.1 Chemical (NP) fertilizer distributed in regions (NAAS 2014)

holistic approach of knowledge, as well as the potential of farmers to practice this knowledge and innovate and adjust to evolving better conditions, ensures the sustainability of those who practice conservation agriculture. A convergent extension approach therefore is need of the hour (Mukherjee et al. 2012; Mukherjee and Maity 2015). It is the step-like implementation by farmers of complementary, synergistic soil husbandry practices that build into a strong, cost-effective, better productive, and environmentally sound farming system. This system is more sustainable than conventional agriculture due to primary focus of crop production with soil health. For better yield potential adequate plant nutrient supply is much needed (Farooq and Siddique 2015; Rusinamhodzi 2015).

The conservation agriculture promotes least soil disturbance, maximum surface crop residue retention on the crop field, and crop rotation or diversification, and the present need of conservation agriculture is more, due to global climate change (Meena et al. 2013a, b; Meena et al. 2015b, d; Kundu et al. 2013; Kushwah et al. 2014; Singh et al. 2014a, b; Kumar et al. 2015; Ghosh et al. 2016). Across the globe people are promoting organic inputs, to protect the soil health and improve crop production (Meena et al. 2014a, 2015a). Due to global climate change, the modification in plant nutrient pattern of soil (Dotaniya 2015) and also the photosynthesis rate and amount of root exudates affected during the crop growth were observed (Dotaniya et al. 2013b; Dotaniya and Meena 2013). The scientific community gave more attention to combating the adverse effect of climate change and modified the existing agronomic and genetic techniques for better yield potential and environmental safety (Dotaniya et al. 2013c, 2014a). In these issues the use of biofertilizers, addition of farmyard manure (FYM), rational use of chemical fertilizers, and in situ use of crop residues are mainly acknowledged (Dotaniya and Datta 2014; Meena et al. 2015c, e, 2016a, b; Verma et al. 2015a, b).

4.2 Need of Biofertilizers for Crop Production

The blanket use of inorganic chemical fertilizers has led to soil and water pollution and affected the population and diversity of beneficial microorganism in soil. This results to crops more prone to attack of insect pest and drastic decline of the crop yield (Dotaniya et al. 2016a, b). The toxic level of one plant nutrient affected the availability of another nutrient in soil, mainly micronutrients (Mahajan et al. 2003a). Fertilizer demand for agricultural crop production is much higher than the availability in global market. It is expected that by 2020, to achieve the targeted production of 280 MT of food grain, the requirement of plant nutrients will be 28.8 MT, whereas their availability will be only 21.6 MT being deficit of ~7.2 MT. The feedstock/fossil fuels are decreasing, while the cost of fertilizer manufacturing is increasing with faster rate. This is becoming too expensive mainly by small and marginal farmers. Inadequate application of plant nutrient during crop production is also depleting soil health due to widening gap between nutrient removal and supplies. It is one of the major causes of environmental hazards. Apart from the above facts, the continuous use of biofertilizers is cost-effective, environment-friendly, more proficient, productive, and easily accessible to marginal and small farmers over chemical fertilizers (Mahajan et al. 2003b).

4.3 Biofertilizers: Plant Nutrient Potential

Biofertilizers are low-cost and environment-friendly plant nutrient source, which improves the availability of plant nutrients and enhances the soil quality. In India, the history of biofertilizers is more than 50 years old, but its development is slow, which might be due to lack of awareness and slow production of crops. The development of biofertilizers speeds up in the last 20 years, and among the biofertilizers, phosphate-solubilizing bacteria (PSB) placed on top of the list (Rao 2007, 2014). Annual production capacity of biofertilizers in India is ~22,000 and the consumption is 12,000 tonnes. Still, the level of biofertilizer use is quite low and there is a need to enhance the capacity up to 50,000–60,000 tonnes by 2020.

The agricultural growth during the last three decades was accelerated by higher application of chemical fertilizers in combination with other natural resources. This excessive application has ill effect on the soil, which is a serious concern; thus the biofertilizers are considered to be an important alternative source of plant nutrients. They are biologically active, i.e., bacteria, algae, or fungi, with the ability to provide plants with nutrients (Javorekova et al. 2015; Rao et al. 2015). Most biofertilizers belong to nitrogen-fixing, phosphate-solubilizing, and cellulolytic microorganisms. Nitrogen biofertilizers are fixing atmospheric nitrogen into soil in the available forms which are readily usable by plants. These include *Rhizobium*, *Azotobacter*, *Azospirillum*, blue-green algae (BGA), and *Azolla*. The *Rhizobium* group of biofertilizers requires symbiotic association with the root nodules of legumes to fix nitrogen; others can fix nitrogen independently. Phosphate-solubilizing microorganisms secrete various organic acids which enhance the uptake of phosphorus by plants by dissolving rock phosphate and tricalcium phosphates (TCP) in soil (Dotaniya et al. 2013d, 2014b, c, 2015). The Department of Agriculture and Cooperation, Government of India (GOI), started a national project on organic farming from October 2004 onward and established biofertilizer production units with a 25 % capital investment subsidy of the total project cost or

20 lakh per unit, whichever is less. The GOI ensured the quality and production of biofertilizers under Sect. 3 mentioned in essential commodities, Act 1955. Further GOI issues a fertilizer (control) Amendment Order, 2006, with the gazette notification which was S.O. 391 (E) dated on March 24, 2006, for the biofertilizer production. This order was enforced and four biofertilizers came under the FCO, i.e., *Rhizobium*, *Azotobacter*, *Azospirillum*, and phosphate-solubilizing bacteria. It was used as an application to seed, soil, or composting areas with the objective of increasing the numbers of such microorganisms and accelerating certain microbial process to augment the extent of the availability of nutrients in a form which can be assimilated by the plant (Nalawde and Bhalerao 2015). The practical biofertilizer knowledge of applied microbial inoculums is having a long history and passes through generation to generation of farmers mostly in rural areas. It originated with microbial culture of small-scale compost production that has evidently proved the ability of biofertilizer (Abdul Halim 2009; Meena et al. 2015f, g). The mycorrhiza inoculums are such type of biofertilizer that are increasingly being utilized and accepted in the agriculture industry. In Malaysia mycorrhizal biofertilizers are produced in large scale mainly for supplying plant nutrients, reducing the toxic effect of contaminants in soil, controlling the pest and disease of root, and improving soil moisture and soil fertility (Abiala et al. 2013). The availability of substrate material for inoculate is affecting the cost and effectiveness of biofertilizers such as mine sands and agricultural wastes which are well known for cheaper production of biofertilizers. The perception in farming community is the cost of biofertilizers is higher but less effective for crop production. It is often perceived to be more expensive than the chemical fertilizers due to the lack of proper handling and modern technology to manufacture biofertilizer products from available wastes. Besides, the effect on the crop production is slow, but soil properties are more improved, compared to the use of chemical fertilizers. Apart from this, special care, like storage or mixing with inoculant materials, is also needed to handle microbial inoculants so that these remain effective for longer period.

As biofertilizers contain living cells, therefore their effectiveness depends on the surrounding environment. Hence, the potentialities are bound to be inconsistent (Mohammadi and Sohrabi 2012). Rice is one of the major staple food crops grown by most of the farmers with the use of blue-green algae (BGA) and *Azolla* as a plant nutrient supplier. In general, BGA supply N 50–100 kg ha^{-1} through biological N-fixation, and in addition it also supplies plant growth-promoting substances to crop under puddled condition (Mahajan et al. 2003a). *Azolla* is a freshwater fern and can be seen glowing in rice fields and waterbodies and can fix 100–150 kg N ha^{-1} $year^{-1}$ in a rice field together with *Anabaena* (Mahajan et al. 2003b). Some of the major risks associated with biofertilizers include short shelf life, need of suitable carrier materials, vulnerable to high temperature, and need of proper storage that still have to be solved in order to promote the biofertilizers in remote areas.

4.4 Types of Biofertilizers

Biofertilizers are important components of integrated nutrient management in crop production and mostly used in every part of the Indian territory (Fig. 4.2). These fertilizers play a key role in crop productivity and sustainability of soil and also protecting the environment. These are a renewable source of plant nutrients in sustainable agricultural systems. Biofertilizers are products containing living cells of different types of microorganisms, which, when applied to seed, plant surface, or soil, colonize the rhizosphere or the interior of the plant and promote growth by converting nutritionally important elements (nitrogen, phosphorus) from unavailable to available form through a biological process such as nitrogen fixation and solubilization of rock phosphate (Rokhzadi et al. 2008). The biofertilizers accelerate and improve plant growth and protect plants from pest and disease attacks (El-yazeid et al. 2007). The role of soil microorganisms in sustainable development of agriculture has been reviewed by many researchers across the world (Lee and Pankhurst 1992; Wani et al. 1995). Nowadays biofertilizers are more in practice for crop production; various types of biofertilizers are available in the market (Table 4.1).

4.4.1 *Rhizobium*

Particularly *Rhizobium* belongs to family Rhizobiaceae, symbiotic in nature, fixes nitrogen approximately 50–100 kg ha^{-1}. The symbiotic relation between leguminous crops and *Rhizobium* is of vital importance for agricultural crop production system. This process is unique in nature and fixing atmospheric nitrogen into soil with the help of nitrogen fixation bacteria. It reduced the N fertilizer recommended dose during legume crop production. In one way it is saving the crop of cultivation, and in other way, it is protecting the environment by less volatilization loss of N from the crop field. In general, biological nitrogen fixation by terrestrial ecosystem is 135 million metric tons per annum.

Fig. 4.2 Biofertilizers distributed in various regions (NAAS 2014)

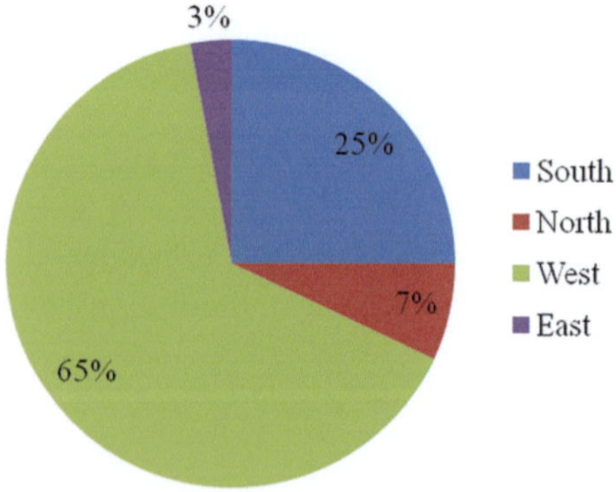

Table 4.1 Groups of biofertilizers based on their nature and function

Groups	Examples
N_2-fixing biofertilizers	
Free living	*Azotobacter*, *Beijerinckia*, *Clostridium*, *Klebsiella*, *Anabaena*, *Nostoc*
Symbiotic	*Rhizobium*, *Frankia*, *Anabaena azollae*
Associative symbiotic	*Azospirillum*
P-solubilizing biofertilizers	
Bacteria	*Bacillus megaterium* var. *phosphaticum*, *Bacillus subtilis*, *Bacillus circulans*, *Pseudomonas striata*
Fungi	*Penicillium* spp., *Aspergillus awamori*
P mobilizing biofertilizers	
Arbuscular mycorrhiza	*Glomus* spp., *Gigaspora* spp., *Acaulospora* spp., *Scutellospora* spp., *Sclerocystis* spp.
Ectomycorrhiza	*Laccaria* spp., *Pisolithus* spp., *Boletus* spp., *Amanita* spp.
Ericoid mycorrhiza	*Pezizella*
Orchid mycorrhiza	*Rhizoctonia solani*
Biofertilizers for micronutrients	
Silicates and Zn solubilizers	*Bacillus* spp.
Plant growth-promoting rhizobacteria	
Pseudomonas	*Pseudomonas fluorescens*

Singh et al. (2014a, b)

In recent years, the global climate change is affecting the crop growth and soil diversity; such events are more focused on the use of biofertilizers mostly in pulse crops. More than 30 M ha agricultural land under pulse cultivation are having wide potential to use the biofertilizers for sustainable crop production. Some of the legume crops, i.e., chickpea, red gram, pea, lentil, black gram, etc., oil seed legumes like soybean and groundnut, and forage legumes like berseem and lucerne are more profitable with biofertilizers being used. The legume roots and nodules are colonized, which are the site for N fixation and commonly known as factories of ammonia production. These nodules reduced the atmospheric N into ammonia, and ammonia is further converted into L-ketoglutaric acid and finally to glutamic acid. *Rhizobium* has the diverse ability to fix atmospheric N in symbiotic association with legumes and certain nonlegumes like *Parasponia* mentioned in Table 4.2 (Saikia and Jain 2007).

4.4.2 The Nitrogen Fixation

The discovery of nitrogen fixation was reported in 1886 by the German researchers Hellriegel and Wilfarth who stated that legume bearing root nodule has a natural capacity to use atmospheric nitrogen. After this, in 1888, Beijerinck isolated the root bacterial strains from root nodules, which was identified as *Rhizobium*

Table 4.2 Rhizobia and the cross-inoculation groups of legumes they nodulate

Names of rhizobia	Legume cross-inoculation groups	Crops
Pea rhizobia (*R. leguminosarum*)	Pea group	Peas (*Pisum* spp.); vetches (*Vicia* spp.); lentils (*Lens culinaris*); faba bean *Vicia faba*)
Bean rhizobia (*R.* phaseoli)	Bean group	Beans (*Phaseolus vulgaris*); scarlet runner bean (*P. coccineus*)
Clover rhizobia (*R.* trifolii)	Clover group	Clovers (*Trifolium* spp.)
Alfalfa rhizobia (*R. meliloti*)	Alfalfa group	Alfalfa (*Medicago* spp.); sweet clovers (*Melilotus* spp.); fenugreek (*Trigonella* spp.)
Chickpea rhizobia (*Rhizobium* sp.)	Chickpea group	Chickpea (*Cicer arietinum*)
Soybean rhizobia (*Bradyrhizobium japonicum*)	Soybean group	Soybeans (*Glycine max*)
Leucaena rhizobia (*Rhizobium* sp.)	Leucaena group	Leucaenas (*Leucaena leucocephala*; *L. shannoni*; *L. lanceolata*; *L. pulverulenta*); *Sesbania grandiflora*; *Calliandra calothyrsus*; *Gliricidia sepium*; *Acacia farnesiana*
Cowpea rhizobia (*Bradyrhizobium* spp.)	Cowpea group	Pigeon pea (*Cajanus cajan*); peanut (*Arachis hypogaea*); cowpea, mung bean, black gram, rice bean (*Vigna* spp.); lima bean (*Phaseolus lunatus*); *Acacia mearnsii*; *A. mangium*; *Albizia* spp.; *Enterolobium* spp., *Desmodium* spp., *Stylosanthes* spp., Kacang bogor (*Voandzeia subterranea*), *Centrosema* sp., winged bean (*Psophocarpus tetragonolobus*), hyacinth bean (*Lablab purpureus*), siratro (*Macroptilium atropurpureum*), guar bean (*Cyamopsis tetragonoloba*), calopo (*Calopogonium mucunoides*), puero (*Pueraria phaseoloides*)

Adopted from Alexander (1978)

leguminosarum. Beijerinck and Lipman discovered the *Azotobacter* spp. in 1901 and 1903, respectively. These discoveries changed the whole scenario of soil biology and motivated the researches to do the biologically oriented research compared to soil fertility. These aspects were further strengthened by Winogradsky in the year 1901 by the isolation of *Clostridium pasteurianum* and function of blue-green algae in N fixation by Stewart in the year 1969, and the new genera were discovered. This story doesn't end with this history, but it reaches many many miles in the present era. Identification of new genera was limited at technical and also less facility of proper tools for taxonomy and phylogeny and also exact measurements of nitrogen fixation capacity (Franche et al. 2008). The nitrogen fixation groups also secreted various types of plant growth-regulating hormones, which affect the soil properties and nutrient uptake pattern from soil to crop.

The molecular nitrogen converted into ammonia is mediated by nitrogenase enzyme, referred to as Mo-nitrogenase and FeMoco. However, *Azotobacter* and other photosynthetic nitrogen vanadiums as a cofactor are called V-nitrogenase or Fe-containing Fe nitrogenase (Rubio and Ludden 2005; Newton 2007). In the N-fixation strains, *nif* gene is responsible for the synthesis of functional group of nitrogenase mostly in 24 kb regions (Arnold et al. 1988). The N-reduction route is a very multifaceted mechanism, but not yet fully elucidated. The reduction of molecular N inside the nodule requires a lot of energy, which is contributed by plant species. The produced photosynthates are transported to nodule; however, dicarboxylic acid further provides the *Bacteroides* carbon and energy substances via symbiosome membrane. The leghemoglobin helped during the respiration and the generation of energy. The biological conversion of the net reduction of molecular N to ammonia (NH_3) generally followed below equation:

$$N_2 + 16\,MgATP + 8\,e^- + 8\,H^+ \rightarrow 2\,NH_3 + H_2 + 16\,MgADP + 16\,Pi$$

Most of the leguminous crop plants have symbiotic relationship with root nodule rhizobacteria called rhizobia; it is gram negative and belongs to proteobacteria division. The rhizobia can be categorized into two groups: (1) alpha genera, *Agrobacterium*, *Allorhizobium*, *Azorhizobium*, *Bradyrhizobium*, *Mesorhizobium*, *Rhizobium*, *Sinorhizobium*, *Devosia*, *Methylobacterium*, *Ochrobactrum*, and *Phyllobacterium*, all harbor nodule-forming bacteria, and (2) beta-proteobacterial *Burkholderia* and *Cupriavidus* (Lindstrom and Martinez-Romero 2007; Lindstrom et al. 2006). Rhizobia produced the nodulation signal with the help of nod, nol, and noe genes. Nod factor is composed of lipo-chitooligosaccharide and helps in N-fixation. The cross-inoculation group of rhizobia and host plant made the compatibility and produce flavonoids during the microbial infection. The flavonoids produced by the host plant induce rhizobial nod genes. This promotes the production of nod factors during the nitrogen fixation. An infection thread passing the root cortex toward a cluster of dividing cells that will become a plant root primordium is shown in Fig. 4.3. Host–rhizobia interaction totally depends on host specificity and the effective strains (responsible for nitrogen fixation) or ineffective strains (non-fixing). Plant root exudates attracting the rhizobia for infection and the genetic code or *nod* factor determine the host specificity in a symbiotic interaction (Abdel-Lateif et al. 2012, 2013).

These interactions determine the biological nitrogen fixation capacity of a healthy strain. Soil fertility status and soil microclimate also affected the rate of interaction, nodulation, as well as amount of N fixation by various cross-inoculation groups.

4.4.3 *Azospirillum*

It belongs to the family Spirilaceae and is found heterotrophic and associative in nature. It is well known for N fixation in plant and its capacity varied 20–40 kg ha^{-1}.

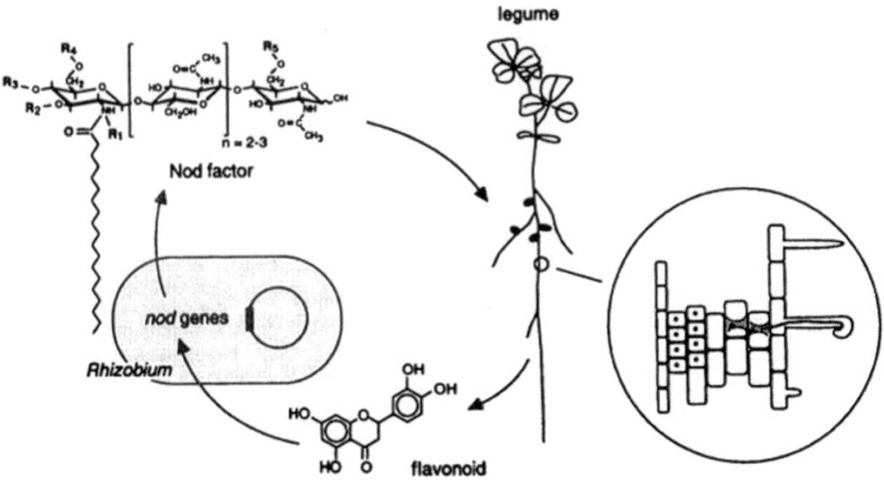

Fig. 4.3 Signal exchange in rhizobium–plant symbiosis (Adopted from Schultze and Kondorosi 1998; Franche et al. 2008)

Apart from this these microorganisms also produce growth-regulating substances which help to protect from soil-borne diseases. A range of species are under this genus, like *A. amazonense*, *A. halopraeferens*, and *A. brasilense*; however, across the globe some of the major species are *A. lipoferum* and *A. brasilense*. *Azospirillum* proved that it improved the leaf area index and ultimately crop yield attributes. The *Azospirillum* form associative symbiosis with many plants, particularly with those having the C_4-dicarboxylic pathway of photosynthesis, because they grow and fix N on salts of organic acids such as malic and aspartic acid. Thus, it is more beneficial for C_4 crops and mostly advising for maize, sugarcane, sorghum (*Sorghum bicolor* L.), pearl millet, etc. It is used as seed treatment; dipping the roots of rice seedlings in ~2 % suspension of *Azospirillum* inoculant enhanced the yield by 100 kg ha^{-1} (Kennedy et al. 2004).

4.4.4 *Azotobacter*

Nitrogen fixation in nature other than symbiotic process is free living in soil. The *Azotobacter* is a free-living and motile genus with thick-walled cysts. It comes under the family *Azotobacteraceae*, aerobic and heterotrophic in nature. *Azotobacters* are present in neutral or alkaline soils and *A. chroococcum* the most commonly occurring species in arable soils. Apart from these *A. vinelandii*, *A. beijerinckii*, *A. insignis*, and *A. macrocytogenes* are other identified species playing an important role in N-nutrient dynamics. Poor organic matter adversely affects the population of *Azotobacter* which hardly ever exceeds about 10^4–10^5 g^{-1} of soil. The antagonistic relationships with other soil microorganisms also limit the population of *Azotobacter* and rate of N fixation in soil. The presence of these free-living N-fixation microorganisms is identified from the rhizospheric soil of most of the crops, i.e., rice, sugarcane, maize, pearl millet, vegetables, and plantation crops (Mazid et al. 2011a). The *Azotobacter* is oval or spherical in shape colonizing the roots on the surface as well as intracellular mode with good agreement with the crop plants. It can fix atmospheric N in the soil that ranged 20–25 kg ha^{-1} under favorable conditions and increased crop yield ~40–50 %. *Azotobacter* improved the seed germination and crop growth due to positive response of B vitamins, NAA, GA, and other chemicals produced during the biochemical process that are showing antagonistic relationship with root pathogens (Mazid et al. 2011b). Mahajan et al. (2008) reported that the application of *Azotobacter chroococcum* enhanced the crop yield in rice–wheat cropping sequence. It improved the soil properties and microbial diversity and population; it acts as a beneficial tool in the area of sustainable agriculture.

4.4.5 Phosphate Solubilizers

Phosphorus is one of the indispensable plant nutrients and structural blocks of genes and chromosomes and actively participated in energy transfer. Crop yield much affected by P applications due to integral role played in optimum growth and reproduction mechanisms. Most of the agricultural crops contain 0.1–0.5 % P. Phosphorus uptake from soil solution mostly in the form of primary orthophosphate ions ($H_2PO_4^-$) and few plants in secondary orthophosphate ions (HPO_4^{2-}) from soil. Most of the Indian soil contains P in the range 100–2,000 mg kg^{-1} (Tandon 1987). Application of inorganic fertilizers during the crop production, more than 80 % applied fertilizers converted into unavailable form with the help of aluminum (Al-P), iron (Fe-P), and calcium (Ca-P) in major soil orders (Dotaniya et al. 2013a, 2014c). In general, the Al-P and Fe-P constitute 1–25 % of total P in soil, mostly in acidic nature of soils. The Ca-P constitutes ~40 % in neutral or calcareous soils. However, Ca-P is much more abundant than Al-P and Fe-P in soil. The labile P in soil solution is closely related to soil pH (Fig. 4.4). The concentration of both $H_2PO_4^-$ and HPO_4^{2-} is equal to 7.2, whereas increasing pH enhanced the availability of PO_4^{3-} .

Applications of phosphatic fertilizers are suggested to apply as a single dose at the time of sowing known as basal dose. The mobility of P ions is slow and plant roots mostly take 2–4 mm distance. However, plant roots can take other nutrients from a larger distance.

Various scientific literatures have examined the potential of different bacterial species to solubilize inorganic phosphate compounds, such as tricalcium phosphate, dicalcium phosphate, hydroxyapatite, and rock phosphate. Some of the bacterial genera are well known for this, i.e., *Pseudomonas*, *Bacillus*, *Rhizobium*, *Burkholderia*, *Achromobacter*, *Agrobacterium*, *Micrococcus*, *Enterobacter*, *Flavobacterium*, and *Erwinia*. In soil and plant, phosphate-solubilizing bacteria are in appropriate number and mediated solubilization. These are aerobic and anaerobic in nature, but under submerged condition, more of the population and

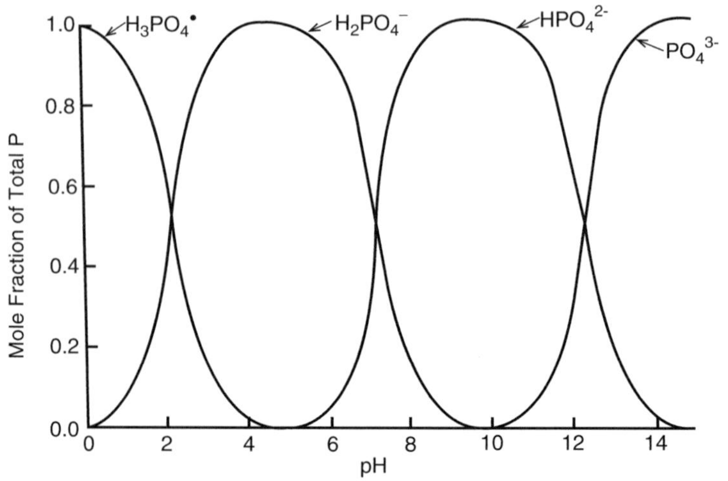

Fig. 4.4 Relationship between solution pH and availability of soluble forms of phosphate

diversity vary greatly. In general, higher amount of organic substrates in rhizospheres is soil; attracting the phosphate-solubilizing bacteria/rhizobacteria; and the population is more in rhizospheric soil compared to non-rhizospheric soil (Ahmad et al. 2016; Bahadur et al. 2016; Das and Pradhan 2016; Dominguez-Nuñez et al. 2016; Dotaniya et al. 2016a; Jaiswal et al. 2016).

Among the P-solubilizing microorganisms, *Pseudomonas* and *Bacillus* and fungi are more dominant in soil. The major microorganisms are producing various types of organic substances and performing various processes like acidification of the medium which solubilizes the insoluble P in soil. During solubilization, diverse organic and inorganic acids alter TCP into di- and monobasic phosphates with higher availability to plant. In India, PSB has the potential to reduce the P fertilizer import by making low-grade rock phosphate utilization in crop production as an enriched phospho-compost (Basak and Biswas 2016).

4.4.6 Potassium Solubilizers

Some of the soil microorganisms are well known for the solubilization of potassium from K minerals. Most of the micas, i.e., orthoclase, mica, illite, and muscovite, are the source of K in soil; with the reaction of microorganisms, produced organic substances solubilize the K and enhanced the soil solution K (Jha and Subramanian 2016; Kumar et al. 2016; Masood and Bano 2016; Meena et al. 2016a, b). These organisms also produce various types of amino acids, growth-promoting compounds (IAA and gibberellic acid), and vitamins, which promote the crop growth and yield (Dotaniya et al. 2016b). *Frateuria aurantia* is well known for the mobilization of K from the mica; it also works in association with other biofertilizers in positive mode and enhances the soil productivity and fertility (Prakash and Verma 2016; Priyadharsini and Muthukumar 2016; Raghavendra et al. 2016; Rawat et al. 2016). Ghosh (2004) listed a range of experiments; those are clearly showing the role of K biofertilizers in crop production, and the K microorganisms are working in the wide range of crops, regions, and climatic conditions. Ranges of K-solubilizing bacteria are present in soil and mobilize the fixed K from clay minerals to soil solutions (Meena et al. 2014b, 2015b, d; Saha et al. 2016a). The application of high K bearing clay mineral with K-solubilizing bacteria can be a good option for mitigating the K requirement in agricultural production system (Saha et al. 2016b; Sharma et al. 2016; Shrivastava et al. 2016; Sindhu et al. 2016; Teotia et al. 2016; Velázquez et al. 2016; Yadav and Sidhu 2016; Yasin et al. 2016; Zahedi 2016).

4.4.7 Zinc Solubilizers

Across the globe, Zn is known as one of the micronutrients and its deficiency limits the growth of crop and adversely affects the crop yield. The Zn fertilizers are more costly and its availability is also limited. In this regard the use of Zn solubilizers can play a vital role in enhancing the yield and quality of crops. The biofertilizers for major plant nutrients like N (*Rhizobium*, *Azospirillum*, *Azotobacter*, BGA), P (phosphate-solubilizing bacteria like *B. megaterium*, *Pseudomonas striata*), and K (*Frateuria aurantia*) are widely accepted. These microorganisms containing biofertilizers supply only major plant nutrients to plant, but the micronutrient deficiency also advocates the micronutrient fertilizers or solubilizes organisms in soil. Some of the microorganism species are isolated, which are responsible for Zn solubilization, i.e., *Bacillus sp*. In the soil system, a few of the common bacterial species are more responsible for the solubilization of micronutrient like Zn, Cu, and Fe from soil. Average earth crust Zn concentration almost 0.008 %, but more than half of the Indian soil showing Zn deficiency symptoms in plants and yield declination was observed in major crops. The external applications of Zn fertilizers, soluble zinc sulfate ($ZnSO_4$), are more common in all crops. The applied soluble Zn fertilizers during crop get into plant available form after reaction with soil constituents. Acid soil cation exchange and high-pH soil chemisorptions ($Zn\text{-}CaCO_3$) or complexation by organic ligands fix the Zn in soil and reduce the concentration in soil solution. The microorganisms that are well known for the solubilization of Zn are *B. subtilis*, *Thiobacillus thiooxidans*, and *Saccharomyces* sp. These strains are used for Zn biofertilizers and get positive response among the farmer communities. Sometimes the application of Zn fertilizers or Zn-containing oxide (ZnO, $ZnCO_3$, ZnS) in combination with Zn biofertilizers (*Bacillus* sp.) is showing better response. It enhanced the soil Zn supply capacity and Zn concentration in soil solution (Samoon et al. 2010).

4.4.8 Phosphorus and Other Nutrient Mobilizers

Arbuscular mycorrhizal fungi (AMF) are obligate biotrophs forming symbiotic relationship with about 90 % of the land plant species (Gadkar et al. 2001). Higher plant phosphorus and other plant nutrients are mediated with mycorrhizal association, in which symbiotic relationship is performed by higher plant and certain fungi (glomus) (Karandashov and Bucher 2005). In this association the intracellular penetration by the fungus to the host roots as well as an outward expansion into the nearby soil is involved. Initially the fungus growth grows between cortical cells; later on it spread the hyphae on the cell wall and penetrates within the cell. It affected the plant nutrient dynamics in soil, especially less mobile nutrients in soil solution with their hyphae, which explore the soil for nutrients, increase surface area for nutrient absorption transport them back to the plant from beyond the root zone. The nutrients like P, Zn, and Cu have been shown to be absorbed and translocated to the host by mycorrhizal fungi. The use of silica biofertilizers improved the crop yield in tropical soils and reduced the insect pest attack in crops (Meena et al. 2013a, b; Garg and Bhandari 2016; Srivastava et al. 2016).

4.5 Importance of Biofertilizers in Conservation Agriculture

The addition of biofertilizers in agriculture plays major role in improving soil fertility, nutrient dynamics, crop residue decomposition, and soil microbial diversity and population and ultimately enhancing the soil health and crop yield. In addition, it helps in reducing the requirement of chemical fertilizers during a particular crop production. On the other hand the AMF produced a heat-stable protein called glomalin, which is a glycoprotein that enhances soil aggregation and helps in soil carbon sequestration. The glomalin and mycorrhizal hyphae together lead to a stable soil structure. The report submitted by various workers that reveals the high potential value of the biofertilizers is summarized in the following points:

1. Application of *Rhizobium* biofertilizers significantly enhanced the agronomic yield attributes in pulse crops under temperate climatic conditions.
2. The *Azospirillum* application in agricultural crops improves the leaf area index, harvest index, and yield attributes.
3. It was observed that seed inoculant green gram by *Rhizobium* under 20 kg N + 45 kg P_2O_5 ha^{-1} fertility level improved the grain and straw yield.
4. The positive effect of *Azotobacter chroococcum* was observed in maize crop; the significant yield improved in biofertilizer-treated plots compared to control. Such type of effect also observed in wheat crop when it was inoculated with *A. chroococcum*.
5. The biofertilizers enhanced the soil enzyme activities in soil; alkaline phosphatase activity was greater in *Azotobacter chroococcum + P fertil*izer compared to control in peach roots.

6. In a greenhouse experiment, *G. fasciculatum* + *A. chrooccocum* + 50 % of the recommended P rate improved the bulb size in onion.
7. Wheat crop application of rock phosphate with biofertilizers (*Azotobacter* + *Rhizobium* + VAM) showed the significant increment in straw and grain yield of wheat.

4.6 Effectiveness of Biofertilizers

Biofertilizers are the living cells, required much attention during transportation and storage and mode of application. Its effect on soil and environment is well known as addressed by scientific communities across the global territory. Regarding its effect on a particular soil and crops, extensive study is needed with good notice of climate and local factors. The conservation agriculture practice is in need of current agriculture situation and a global force of safe ecology promoted the biofertilizer use with fast pace (Saha et al. 2016a, b). In connection to this, little study was conducted under conservation agriculture in different agroclimatic regions or subregion level. At national- and international-level agricultural universities, national research stations are continuously working on various aspects of biofertilizer uses in agricultural production system. This laboratory and field experiment was carried out by the scientific community of the world. Apart from these, in India, National Biofertilizers Development Center actively plays a crucial role in effective use of these microorganisms. These biofertilizers that improved the soil health and crop yield with eco-friendly effect on environment are observed.

4.7 Constraints in Biofertilizer Use

- The main constraints of biofertilizers are poor availability of carrier; it affected the shelf life.
- Most of the chemical fertilizers give the quick response compared to biofertilizers. It affected the marketing of biofertilizers and showing poor awareness among the farmers.
- Poor storage management reduced the viable strains in biofertilizers, and the farmers felt quality assurance prior to application.
- Application and storage need sophisticated instruments and skilled man power to protect the viability of biofertilizer from adverse climatic factors.
- It needs initially high cost of investment.
- One of the major limitations of biofertilizers is mutation during fermentation.
- Poor awareness among the farmers leads to poor resource generation by the industries.

4.8 Biofertilizer Strategies Should Be Followed in Conservation Agriculture

- The study of particular strain effectiveness regarding a particular crop and soil and climatic factors. For this it is needed to strengthen the research and technologies in combination with extension wing.
- The use of biotechnological tools and techniques for the better improvement of N fixation, P solubilizer, and other biofertilizers.
- Standardization of biofertilizer dose in a particular crop and soil.
- Identification of better carrier material for enhancing the shelf life of strains.
- Frequent monitoring of the biofertilizer production units to assure the proper viable count, method of production, storage, etc.
- Wide publicity of biofertilizers for the agricultural crop production use through scientific training, farmer fairs, exhibitions, or media.

4.9 Conclusions

Conservation agriculture is a need of present era; it is a way to fulfilling the need of the food for growing population. Fast overexploitation of nonrenewable natural resources can create food crisis in upcoming years. Agricultural crop production needs to be enhanced in new horizons without deteriorating the natural resources as well as environmental quality. In line with this, low-cost, eco-friendly biofertilizers playing a crucial role for further enhancement of crop yield reduced the use of chemical fertilizers and increased plant nutrient use efficiency vis-a-vis maintaining crop quality as well as environmental aspects. Manufacturing of biofertilizers interlinked and needs more attention for continuous research. Meanwhile, the screen and identification of new efficient microbial strains with respect to crop and climatic and environmental conditions improve the soil and crop sustainability. It is a new concept to the chemical fertilizer applied by farmers to minimize it and promote the biofertilizers in agricultural field, needing extensive market development efforts. With the help of the government, nongovernment organizations (NGOs) and wide publicity through media urge are need present situation. More investment of fund for developing new strains of biofertilizers, maintaining quality, and enhancing shelf life is a major hurdle in its propagation and use.

Acknowledgment Authors are highly thankful to Dr. H. M. Meena, ICAR – Central Arid Zone Research Institute, Jodhpur, India, for the needful help during the writing of the manuscript.

References

Abdel-Lateif K, Bogusz D, Hocher V (2012) The role of flavonoids in the establishment of plant roots endosymbioses with arbuscular mycorrhiza fungi, rhizobia and Frankia bacteria. Plant Signal Behav 7(6):636–641. doi:10.4161/psb.20039

Abdel-Lateif K, Vaissayre V, Gherbi H, Verries C, Meudec E, Perrine-Walker F, Cheynier V, Svistoonoff S, Franche C, Bogusz D, Hocher V (2013) Silencing of the chalcone synthase gene in Casuarina glauca highlights the important role of flavonoids during nodulation. New Phytol 4:1012–1021. doi:10.1111/nph.12326

Abdul Halim NB (2009) Effects of using enhanced biofertilizer containing N-fixer bacteria on patchouli growth. Thesis. Faculty of Chemical and Natural Resources Engineering University Malaysia Pahang. p 145

Abiala MA, Popoola OO, Olawuyi OJ, Oyelude JO, Akanmu AO, Killani AS et al (2013) Harnessing the potentials of vesicular arbuscular mycorrhizal (VAM) fungi to plant growth – a review. Int J Pur Appl Sci Technol 14:61–79

Ahmad M, Nadeem SM, Naveed M, Zahir ZA (2016) Potassium-solubilizing bacteria and their application in agriculture. In: Meena VS, Maurya BR, Verma JP, Meena RS (eds) Potassium solubilizing microorganisms for sustainable agriculture. Springer, New Delhi, pp 293–313

Alexander M (1978) Introduction to soil microbiology. Wiley, New York

Arnold W, Rump A, Klipp W, Priefer UB, Puhler A (1988) Nucleotide sequence of a 24, 206-base-pair DNA fragment carrying the entire nitrogen fixation gene cluster of Klebsiella pneumoniae. J Mol Biol 203:715–738

Bahadur I, Maurya BR, Kumar A, Meena VS, Raghuwanshi R (2016) Towards the soil sustainability and potassium-solubilizing microorganisms. In: Meena VS, Maurya BR, Verma JP, Meena RS (eds) Potassium solubilizing microorganisms for sustainable agriculture. Springer, New Delhi, pp 225–266

Basak BB, Biswas DR (2016) Potentiality of Indian rock phosphate as liming material in acid soil. Geoderma 263:104–109

Das I, Pradhan M (2016) Potassium-solubilizing microorganisms and their role in enhancing soil fertility and health. In: Meena VS, Maurya BR, Verma JP, Meena RS (eds) Potassium solubilizing microorganisms for sustainable agriculture. Springer, New Delhi, pp 281–291

Dominguez-Nuñez JA, Benito B, Berrocal-Lobo M, Albanesi A (2016) Mycorrhizal fungi: role in the solubilization of potassium. In: Meena VS, Maurya BR, Verma JP, Meena RS (eds) Potassium solubilizing microorganisms for sustainable agriculture. Springer, New Delhi, pp 77–98

Dotaniya ML (2013) Impact of various crop residue management practices on nutrient uptake by rice-wheat cropping system. Curr Adv Agric Sci 5(2):269–271

Dotaniya ML (2015) Impact of rising atmospheric CO_2 concentration on plant and soil process. In: Mohanty M, Sinha NK, Hati KM, Chaudhary RS, Patra AK (eds) Crop growth simulation modelling and climate change. Scientific Publisher, Jodhpur, pp 69–86

Dotaniya ML, Datta SC (2014) Impact of bagasse and press mud on availability and fixation capacity of phosphorus in an inceptisol of north India. Sugar Tech 16(1):109–112. doi:10.1007/s12355-013-0264-3

Dotaniya ML, Kushwah SK (2013) Nutrients uptake ability of various rainy season crops grown in a vertisol of Central India. Afr J Agric Res 8(44):5592–5598. doi:10.5897/AJAR2013.7969

Dotaniya ML, Meena VD (2013) Rhizosphere effect on nutrient availability in soil and its uptake by plants -a review. Proc Natl Acad Sci India Sec B: Biol Sci 85(1):1–12. doi:10.1007/s40011-013-0297-0

Dotaniya ML, Pingoliya KK, Meena HM, Prasad D (2013a) Status and rational use of rock phosphate in agricultural crop production- a review. Agric Sustain Dev 1(1):103–108

Dotaniya ML, Prasad D, Meena HM, Jajoria DK, Narolia GP, Pingoliya KK, Meena OP, Kumar K, Meena BP, Ram A, Das H, Chari MS, Pal S (2013b) Influence of phytosiderophore on iron and zinc uptake and rhizospheric microbial activity. Afr J Microbiol Res 7(51):5781–5788. doi:10.5897/AJMR2013.6461

Dotaniya ML, Sharma MM, Kumar K, Singh PP (2013c) Impact of crop residue management on nutrient balance in rice-wheat cropping system in an aquic hapludoll. J Rural Agric Res 13 (1):122–123

Dotaniya ML, Datta SC, Biswas DR, Meena BP (2013d) Effect of solution phosphorus concentration on the exudation of oxalate ions by wheat (*Triticum aestivum* L.). Proc Natl Acad Sci India Sec B: Biol Sci 83(3):305–309. doi:10.1007/s40011-012-0153-7

Dotaniya ML, Datta SC, Biswas DR, Meena HM, Kumar K (2014a) Production of oxalic acid as influenced by the application of organic residue and its effect on phosphorus uptake by wheat (*Triticum aestivum* L.) in an Inceptisol of north India. Natl Acad Sci Lett 37(5):401–405. doi:10.1007/s40009-014-0254-3

Dotaniya ML, Datta SC, Biswas DR, Kumar K (2014b) Effect of organic sources on phosphorus fractions and available phosphorus in Typic Haplustept. J Indian Soc Soil Sci 62(1):80–83

Dotaniya ML, Kushwah SK, Rajendiran S, Coumar MV, Kundu S, Rao AS (2014c) Rhizosphere effect of kharif crops on phosphatases and dehydrogenase activities in a Typic Haplustert. Natl Acad Sci Lett 37(2):103–106. doi:10.1007/s40009-013-0205-4

Dotaniya ML, Pingoliya KK, Lata M, Verma R, Regar KL, Deewan P, Dotaniya CK (2014d) Role of phosphorus in chickpea (*Cicer arietinum* L.) production. Afr J Agric Res 9(51):3736–3743. doi:10.5897/AJAR2013.8427

Dotaniya ML, Datta SC, Biswas DR, Meena HM, Rajendiran S, Meena AL (2015) phosphorus dynamics mediated by bagasse, press mud and rice straw in inceptisol of North India. Agrochimica 59(4):358–369

Dotaniya ML, Datta SC, Biswas DR, Dotaniya CK, Meena BL, Rajendiran S, Regar KL, Lata M (2016a) Use of sugarcane industrial byproducts for improving sugarcane productivity and soil health-a review. Int J Recycl Org Waste Agric. doi:10.1007/s40093-016-0132-8

Dotaniya ML, Meena VD, Basak BB, Meena RS (2016b) Potassium uptake by crops as well as microorganisms. In: Meena VS, Maurya BR, Verma JP, Meena RS (eds) Potassium solubilizing microorganisms for sustainable agriculture. Springer, New Delhi, pp 267–280

El-Yazeid AA, Abou-Aly HA, Mady MA, Moussa SAM (2007) Enhancing growth, productivity and quality of squash plants using phosphate dissolving microorganisms (bio phosphor) combined with boron foliar spray. Res J Agric Biol Sci 3(4):274–286

Farooq M, Siddique K (2015) Conservation agriculture. Springer International Publishing, New York/Dordrecht/London. doi:10.1007/978-3-319-11620-4

Franche C, Lindström K, Elmerich C (2008) Nitrogen-fixing bacteria associated with leguminous and non-leguminous plants. Plant Soil 321:35–59

Gadkar V, David-Schwartz R, Kunik T, Kapulnik Y (2001) Arbuscular mycorrhizal fungal colonization factors involved in host recognition. Plant Physiol 127:1493–1499

Garg N, Bhandari P (2016) Silicon nutrition and mycorrhizal inoculations improve growth, nutrient status, K+/Na+ ratio and yield of *Cicer arietinum* L. genotypes under salinity stress. Plant Growth Regul 78(3):371–387

Ghosh N (2004) Promoting biofertilisers in Indian agriculture. Econ Polit Wkly 5:5617–5625

Ghosh BN, Meena VS, Alam NM, Dogra P, Bhattacharyya R, Sharma NK, Mishra PK (2016) Impact of conservation practices on soil aggregation and the carbon management index after seven years of maize–wheat cropping system in the Indian Himalayas. Agric Ecosyst Environ 216:247–257

Jaiswal DK, Verma JP, Prakash S, Meena VS, Meena RS (2016) Potassium as an important plant nutrient in sustainable agriculture: a state of the art. In: Meena VS, Maurya BR, Verma JP, Meena RS (eds) Potassium solubilizing microorganisms for sustainable agriculture. Springer, New Delhi, pp 21–29

Javorekova S, Jana M, Medo J, Kovacsova S, Charousová I et al (2015) Effect of bio-fertilizers application on microbial diversity and physiological profiling of microorganisms in arable soil. Eur J Soil Sci 4(1):54–61

Jha Y, Subramanian RB (2016) Regulation of plant physiology and antioxidant enzymes for alleviating salinity stress by potassium-mobilizing bacteria. In: Meena VS, Maurya BR, Verma JP, Meena RS (eds) Potassium solubilizing microorganisms for sustainable agriculture. Springer, New Delhi, pp 149–162

Karandashov V, Bucher M (2005) Symbiotic phosphate transport in arbuscular mycorrhizas. Trends Plant Sci 10(1):22–29

Kennedy IR, Choudhury ATMA, Kecskés ML (2004) Non-symbiotic bacterial diazotrophs in crop-farming systems: can their potential for plant growth promotion be better exploited? Soil Biol Biochem 36(8):1229–1244

Kumar A, Bahadur I, Maurya BR, Raghuwanshi R, Meena VS, Singh DK, Dixit J (2015) Does a plant growth promoting rhizobacteria enhance agricultural sustainability? J Pur Appl Microbiol 9(1):715–724

Kumar A, Patel JS, Bahadur I, Meena VS (2016) The molecular mechanisms of KSMs for enhancement of crop production under organic farming. In: Meena VS, Maurya BR, Verma JP, Meena RS (eds) Potassium solubilizing microorganisms for sustainable agriculture. Springer, New Delhi, pp 61–75

Kundu S, Dotaniya ML, Lenka S (2013) Carbon sequestration in Indian agriculture. In: Lenka S, Lenka NK, Kundu S, Rao AS (eds) Climate change and natural resource management. New India publishing Agency, New Delhi, pp 269–289

Kushwah SK, Dotaniya ML, Upadhyay AK, Rajendiran S, Coumar MV, Kundu S, Rao AS (2014) Assessing carbon and nitrogen partition in kharif crops for their carbon sequestration potential. Natl Acad Sci Lett 37(3):213–217. doi:10.1007/s40009-014-0230-y

Lee KE, Pankhurst CE (1992) Soil organisms and sustainable productivity. Aust J Soil Res 30:855–892

Lindstrom K, Kokko-Gonzales P, Terefework Z, Rasanen LA (2006) Differentiation of nitrogen-fixing legume root nodule bacteria. In: Cooper JE, Rao JR (eds) Molecular techniques for soil and rhizosphere microorganisms. CABI Publishing, Wallingford, pp 236–258

Lindstrom K, Martinez-Romero E (2007) International committee on systematics of prokaryotes subcommittee on the taxonomy of Agrobacterium and Rhizobium: Minutes of the meeting, 23–24 July 2006, Århus, Denmark. Int J Syst Evol Microbiol 57:1365–1366

Mahajan A, Choudhary AK, Jaggi RC, Dogra RK (2003a) Importance of bio-fertilizers in sustainable agriculture. Farmers' Forum 3(4):17–19

Mahajan A, Bhagat RM, Trikha A (2003b) Fertilizing bio-fertilizers. Agric Today 6(9):36–37

Mahajan A, Gupta RD, Sharma R (2008) Bio-fertilizers-a way to sustainable agriculture. Agrobios Newsl 6(9):36–37

Masood S, Bano A (2016) Mechanism of potassium solubilization in the agricultural soils by the help of soil microorganisms. In: Meena VS, Maurya BR, Verma JP, Meena RS (eds) Potassium solubilizing microorganisms for sustainable agriculture. Springer, New Delhi, pp 137–147

Mazid M, Khan TA, Zeba HK, Quddusi S, Mohammad F (2011a) Occurrence, biosynthesis and potentialities of ascorbic acid in plants. Inl J Plant Anim Environ Sci 1(2):167–184

Mazid M, Khan TA, Mohammad F (2011b) Cytokinins, a classical multifaceted hormone in plant system. J Stress Physiol Biochem 7(4):347–368

Meena RS, Yadav RS, Meena VS (2013a) Heat unit efficiency of groundnut varieties in scattered planting with various fertility levels. Bioscan 8(4):1189–1192

Meena VD, Dotaniya ML, Rajendiran S, Coumar MV, Kundu S, Rao AS (2013b) A case for silicon fertilization to improve crop yields in tropical soils. Proc Natl Acad Sci India Sec B: Biol Sci 84(3):505–518. doi:10.1007/s40011-013-0270-y

Meena BP, Kumar A, Dotaniya ML, Jat NK, Lal B (2014a) Effect of organic sources of nutrients on tuber bulking rate, grades and specific gravity of potato tubers. Proc Natl Acad Sci, India Sec B: Biol Sci. DOI 10.1007/s40011-014-0398-4

Meena VS, Maurya BR, Verma JP (2014b) Does a rhizospheric microorganism enhance K^+ availability in agricultural soils?. Microbiol Res 169:337–347

Meena BP, Kumar A, Lal B, Sinha NK, Tiwari PK, Dotaniya ML, Jat NK, Meena VD (2015a) Soil microbial, chemical properties and crop productivity as affected by organic manure application in popcorn (*Zea mays* L. var. everta). Afr J Microbiol Res 9(21):1402–1408

Meena RS, Meena VS, Meena SK, Verma JP (2015b) The needs of healthy soils for a healthy world. J of Cleaner Prod 102:560–561

Meena RS, Meena VS, Meena SK, Verma JP (2015c) Towards the plant stress mitigate the agricultural productivity: a book review. J Cleaner Prod 102:552–553

Meena VS, Maurya RB, Verma JP, Aeron A, Kumar A, Kim K, Bajpai VK (2015d) Potassium solubilizing *rhizobacteria* (KSR): Isolation, identification, and K-release dynamics from waste mica. Ecol Eng 81:340–347

Meena VS, Meena SK, Verma JP, Meena RS, Ghosh BN (2015e) The needs of nutrient use efficiency for sustainable agriculture. J Cleaner Prod 102:562–563

Meena VS, Meena SK, Verma JP, Meena RS, Jat LK (2015f) Current trend to mitigate climate change and sustainable agriculture: a book review. J Cleaner Prod 102:548–549

Meena VS, Verma JP, Meena SK (2015g) Towards the current scenario of nutrient use efficiency in crop species. J of Cleaner Prod 102:556–557

Meena RS, Bohra JS, Singh SP, Meena VS, Verma JP, Verma SK, Sihag SK (2016a) Towards the prime response of manure to enhance nutrient use efficiency and soil sustainability a current need: a book review. J of Cleaner Prod 112(1):1258–1260

Meena VS, Bahadur I, Maurya BR, Kumar A, Meena RK, Meena SK, Verma JP (2016b) Potassium-solubilizing microorganism in evergreen agriculture: an overview. In: Meena VS, Maurya BR, Verma JP, Meena RS (eds) Potassium solubilizing microorganisms for sustainable agriculture. Springer, New Delhi, pp 1–20

Mohammadi K, Sohrabi Y (2012) Bacterial biofertilizers for sustainable crop production: a review. ARPN J Agric Biol Sci 7(5):307–316

Mukherjee A, Bahal R, Roy Burman R, Dubey SK (2012) Conceptual convergence of pluralistic extension at Aligarh District of Uttar Pradesh. J Commun Mobil Sustain Dev 7(1&2):85–94

Mukherjee A, Maity A (2015) Public–private partnership for convergence of extension services in Indian agriculture. Curr Sci 109(9):1557–1563

NAAS (2014) Efficient utilization of phosphorus, vol 68, Policy Paper. National Academy of Agricultural Sciences, New Delhi, p 16

Nalawde A, Bhalerao SA (2015) Effect of biofertilizer (*Azotobacter* sp.) on the growth of blackgram *Vigna mungo* (L.) Hepper. Int J Curr Res Biosci Plant Biol 2(5):20–23

Newton WE (2007) Physiology, biochemistry and molecular biology of nitrogen fixation. In: Bothe H, Ferguson SJ, Newton WE (eds) Biology of the nitrogen cycle. Elsevier, Amsterdam, pp 109–130

Prakash S, Verma JP (2016) Global perspective of potash for fertilizer production. In: Meena VS, Maurya BR, Verma JP, Meena RS (eds) Potassium solubilizing microorganisms for sustainable agriculture. Springer, New Delhi, pp 327–331

Priyadharsini P, Muthukumar T (2016) Interactions between arbuscular mycorrhizal fungi and potassium-solubilizing microorganisms on agricultural productivity. In: Meena VS, Maurya BR, Verma JP, Meena RS (eds) Potassium solubilizing microorganisms for sustainable agriculture. Springer, New Delhi, pp 111–125

Raghavendra MP, Nayaka SC, Nuthan BR (2016) Role of rhizosphere microflora in potassium solubilization. In: Meena VS, Maurya BR, Verma JP, Meena RS (eds) Potassium solubilizing microorganisms for sustainable agriculture. Springer, New Delhi, pp 43–59

Rao DLN (2007) Microbial diversity, soil health and sustainability. J Indian Soc Soil Sci 55:392–403

Rao DLN (2014) Recent advances in biological nitrogen fixation in agricultural systems. Proc Indian Natl Sci Acad 80(2):359–378

Rao DLN, Balachandar D, Thakuria D (2015) Soil biotechnology and sustainable agricultural intensification. Indian J Fertilisers 11(10):87–105

Rawat J, Sanwal P, Saxena J (2016) Potassium and its role in sustainable agriculture. In: Meena VS, Maurya BR, Verma JP, Meena RS (eds) Potassium solubilizing microorganisms for sustainable agriculture. Springer, New Delhi, pp 235–253

Rokhzadi A, Asgharzadeh A, Darvish F, Nourmohammadi G, Majidi E (2008) Influence of plant growth-promoting rhizobacteria on dry matter accumulation and yield of chickpea (*Cicer arietinum* L.) under field condition. Am-Euras J Agric Environ Sci 3(2):253–257

Rubio LM, Ludden PW (2005) Maturation of nitrogenase: a biochemical puzzle. J Bacteriol 187:405–414. doi:10.1128/JB.187.2.405-414.2005

Rusinamhodzi L (2015) Tinkering on the periphery: labour burden not crop productivity increased under no-till planting basins on smallholder farms in Murehwa district, Zimbabwe. Field Crop Res 170:66–75

Saha M, Maurya BR, Bahadur I, Kumar A, Meena VS (2016a) Can potassium-solubilising bacteria mitigate the potassium problems in India? In: Meena VS, Maurya BR, Verma JP, Meena RS (eds) Potassium solubilizing microorganisms for sustainable agriculture. Springer, New Delhi, pp 127–136

Saha M, Maurya BR, Meena VS, Bahadur I, Kumar A (2016b) Identification and characterization of potassium solubilizing bacteria (KSB) from Indo-Gangetic Plains of India. Biocatal Agric Biotechnol 7:202–209

Saikia SP, Jain V (2007) Biological nitrogen fixation with non-legumes: an achievable target or a dogma. Curr Sci 92(3):317–322

Samoon HA, Dar SA, Zehra B, Mahdi SS, GI Hassan SA (2010) Bio-fertilizers in organic agriculture. J Phytol 2(10):9–14

Schultze M, Kondorosi A (1998) Regulation of symbiotic root nodule development. Annu Rev Genet 32:33–57

Sharma A, Shankhdhar D, Shankhdhar SC (2016) Potassium-solubilizing microorganisms: mechanism and their role in potassium solubilization and uptake. In: Meena VS, Maurya BR, Verma JP, Meena RS (eds) Potassium solubilizing microorganisms for sustainable agriculture. Springer, New Delhi, pp 203–219

Shrivastava M, Srivastava PC, D'Souza SF (2016) KSM soil diversity and mineral solubilization, in relation to crop production and molecular mechanism. In: Meena VS, Maurya BR, Verma JP, Meena RS (eds) Potassium solubilizing microorganisms for sustainable agriculture. Springer, New Delhi, pp 221–234

Sindhu SS, Parmar P, Phour M, Sehrawat A (2016) Potassium-solubilizing microorganisms (KSMs) and its effect on plant growth improvement. In: Meena VS, Maurya BR, Verma JP, Meena RS (eds) Potassium solubilizing microorganisms for sustainable agriculture. Springer, New Delhi, pp 171–185

Singh DK, Singh SK, Singh AK, Meena VS (2014a) Impact of long term cultivation of lemon grass (Cymbopogon citratus) on post-harvest electro-chemical properties of soil. Ann Agri Bio Res 19(1):45–48

Singh S, Singh BK, Yadav SM, Gupta AK (2014b) Potential of biofertilizers in crop production in Indian agriculture. Am J Plant Nutr Fert Technol 4(2):33–40

Srivastava AK, Pasala R, Minhas PS, Suprasanna P (2016) Plant bioregulators for sustainable agriculture: integrating redox signaling as a possible unifying mechanism. Adv Agron. doi:10.1016/bs.agron.2015.12.002

Tandon HLS (1987) Phosphorus research and agricultural production in India. Fertilizer development and Consultation Organization, New Delhi

Teotia P, Kumar V, Kumar M, Shrivastava N, Varma A (2016) Rhizosphere microbes: potassium solubilization and crop productivity-present and future aspects. In: Meena VS, Maurya BR, Verma JP, Meena RS (eds) Potassium solubilizing microorganisms for sustainable agriculture. Springer, New Delhi, pp 315–325

Velázquez E, Silva LR, Ramírez-Bahena MH, Peix A (2016) Diversity of potassium-solubilizing microorganisms and their interactions with plants. In: Meena VS, Maurya BR, Verma JP, Meena RS (eds) Potassium solubilizing microorganisms for sustainable agriculture. Springer, New Delhi, pp 99–110

Verma JP, Jaiswa DK, Meena VS, Meena RS (2015a) Current need of organic farming for enhancing sustainable agriculture. J Clean Prod 102:545–547

Verma JP, Jaiswal DK, Meena VS, Kumar A, Meena RS (2015b) Issues and challenges about sustainable agriculture production for management of natural resources to sustain soil fertility and health. J Clean Prod 107:793–794

Wani SP, Rego TG, Rajeshwari S, Lee KK (1995) Effect of legume based cropping systems on nitrogen mineralization potential of vertisol. Plant Soil 175(2):265–274

Yadav BK, Sidhu AS (2016) Dynamics of potassium and their bioavailability for plant nutrition. In: Meena VS, Maurya BR, Verma JP, Meena RS (eds) Potassium solubilizing microorganisms for sustainable agriculture. Springer, New Delhi, pp 187–201

Yasin M, Munir I, Faisal M (2016) Can *Bacillus* spp. enhance K+ uptake in crop species. In: Meena VS, Maurya BR, Verma JP, Meena RS (eds) Potassium solubilizing microorganisms for sustainable agriculture. Springer, New Delhi, pp 163–170

Zahedi H (2016) Growth-promoting effect of potassium-solubilizing microorganisms on some crop species. In: Meena VS, Maurya BR, Verma JP, Meena RS (eds) Potassium solubilizing microorganisms for sustainable agriculture. Springer, New Delhi, pp 31–42

Sustainable Agrochemicals for Conservation of Agriculture and Climate Change

5

Shrikaant Kulkarni

Abstract

In the recent past, there has been a paradigm shift in the interest in the form of using natural products as sustainable agrochemicals in preference over synthetic ones with the growing realization in the mind of the public to restrain from using the hazardous and obnoxious, nonbiodegradable, and persistent chemicals. It demands substituting chemical-based conventional agrochemicals with more greener, eco-friendly but compatible alternatives which control pests with a great degree of efficiency, effectiveness, and efficacy, and therefore it is sustainable in action. The chemical-based agrochemicals have already brought about a lot much of ecological degradation by substantially polluting water bodies, reducing fertility of the soil, making their way into the food cycle because of their nonbiodegradability and persistent nature, and thereby posing a threat to biodiversity. Thus, the use of natural products will continue to provide clues to new modes of action, new chemistries for the sake of conservation of agriculture and in enhancing productivity without compromising ecological balance.

Keywords

Toxic • Hazardous • Natural products • Biodiversity • Biodegradability • Persistent

5.1 Introduction

Synthetically derived chemicals have played a pivotal role in enhancing crop productivity by virtue of its great potential in the suppression of pests. However, over the time the cost of indiscriminate use of these xenobiotics to achieve the

S. Kulkarni (✉)
Department of Chemical Engineering, Vishwakarma Institute of Technology, Pune 411037, India
e-mail: shrikaant.kulkarni@vit.edu

J.K. Bisht et al. (eds.), *Conservation Agriculture*,
DOI 10.1007/978-981-10-2558-7_5

targeted productivity has been found to be huge in terms of loss of soil fertility which in turn has led to substantial reduction in crop yield. The consequences of such synthetic chemicals have also resulted in ecological and medical problems which is a great cause of concern. Some of these chemicals have been found to be carcinogenic and have potent health effects, in addition to their nonbiodegradability, and can thus make their way into the food chain of plants and animals (e.g., DDT, BHC, etc.), and consequently such synthetically derived chemicals had to be abandoned or phased out over time (Vetter and Oehme 2000).

To address this problem of evil effects of the use of synthetic agrochemicals on ecology, there is an immense and growing need of exploring other innovative but eco-friendly and sustainable options by designing and developing nonhazardous alternatives which can work efficiently and effectively as substitutes to highly toxic conventional synthetic agrochemicals without compromising the function, efficacy, price, and availability. The purpose of this article is aimed at exploring innovative options for the design and development of the next-generation green, clean, eco-friendly, sustainable, and real-time agrochemicals for safe environment with better health prospects.

5.2 Modern Crop Protection: Necessity and Challenges

There has been a substantial escalation in the costs toward chemicals used for crop protection purpose in the market, and therefore R and D initiatives are being given an impetus for strengthening crop protection industry, e.g., June 2002 merger of Aventis Crop Science by Bayer to form Bayer Crop Science. Revolutionary changes in agriculture sector in particular continue to make substantial contribution for improving upon human health, nutrition, and quality of life around the world. State-of-the-art practices in the areas like plant breeding, agronomy, use of inorganic fertilizers, and synthetic pesticides are the leading examples of these evolutionary changes which have been gaining a lot much of importance in the sustenance of agriculture sector at large (Phipps et al. 2002).

5.3 Concerns of Pesticide Industry

The pesticides have so far played and will play a vital role in public-health initiatives and thereby saving millions of human lives vulnerable to diseases by ridding off insects, rodent vectors, and intermediate hosts (Quistad et al. 2003). However although about ~24 million American workers amounting to ~17 % of workforce are associated with the production, processing, and selling of the nation's food and fiber, high technology agriculture still has led to substantial reduction in workforce being actively engaged in farming. The fallout of this is many consumers are left with not much idea about the intricacies of farming and do not see much advantage in using pesticides in farming (Ratra et al. 2001).

There has been a momentum to movements by various activist groups and media which are coming to the fore with a strong message against either banning or remediation of even the minute quantities of synthetically derived chemicals, like approved pesticides or food additives because of certain issues like:

- Safety
- Toxicity
- Persistency, etc. of these chemicals

All such scientific, but socially uncalled for initiatives for an environment free from pesticide divert consumer attention away from the very real public-health goal of increasing consumption of fruits and vegetables causing distortion of health risks; threaten innovation, jobs, and standard of living; and frequently divert critical resources needed to address health-related risks. The scientific communities and agricultural industry together have an important role to play in these matters so as to better educate consumers and policy makers on the delicate balance between known and hypothetical risks (Rubin 2004). Figure 5.1 shows that a sustainable economy is one which is necessarily economically viable in terms of using natural, financial, and human capital to create value, wealth, and profits; environmentally compatible in respect of using cleaner, more eco-efficient products and processes to remediate pollution, depletion of natural resources, and loss of biodiversity and wildlife habitat and minimize damage to the ecosystem aimed at providing many ecological goods and services to the society; and socially responsible in regard to its behavior in an ethical manner and management of its impacts on production.

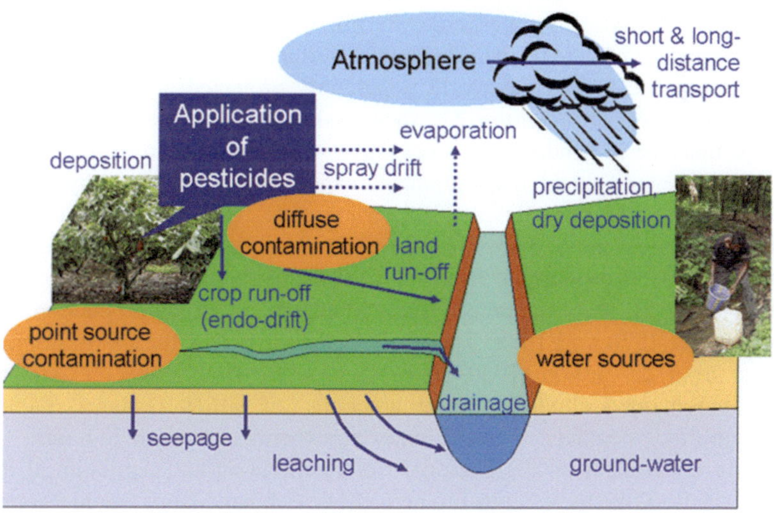

Fig. 5.1 Representative pathways pesticides

5.4 Role of Pesticides in Modern Agriculture

A very vital role is being played by modern crop protection chemicals in today's agriculture in terms of its positive impact on the global economy, preservation of natural resources, and the quality of life across the world. However, it is often either not well conceived or taken cognizance (Ware and Whitacre 2004; Warrior 2000; Wei 1992). Over the past 40 years, there have been substantial advancement in agriculture in areas like modern plant breeding, improved agronomy, and development of inorganic fertilizers and synthetic pesticides which has led to betterment in improving human health and nutrition between 1961 and 2000, and thereafter too food production almost:

- Doubled in developed countries
- Tripled in developing countries, e.g., after the green revolution in India (Tomlin 2009)

5.4.1 Demand for Agrochemicals vis-à-vis Crop Productivity

Advances in modern agricultural practices have successfully supported world population and sustained rise in food production and preservation of world's precious forests reserves and grasslands in most of the developed countries particularly during the last two decades of the twentieth century. This increase in productivity in terms of yield per hectare has led to feeding the world's population which otherwise would have required three to four times more land for production of the same amount of food. However, advances in new agricultural practices and technologies, like innovative pesticides, fertilizers, and biotechnology, have made it possible and will have to be put into practice seriously in the time to come in order to support the ever-increasing world population, projected at ten billion by the middle of the twenty-first century. The increased use of pesticides is responsible for this huge increase in productivity since 1950. Research initiatives in this regard have proved beyond doubt that output of many crops would drop by 50–100 % (Hardstone and Scott 2010; Schuler and Casida 2001). Sans pesticides, e.g., wheat yields in England, climbed from 2 to 6 MT/ha while nearly 1000 years were consumed for yield to increase from 0.5 to 2 MT/ha without pesticides. In Asia, with an increase of ~4 % land area cultivation under cereals, the production rose by twofold during 1970 and 1995. The same is the case with the world population increased from 2.5 billion (1950) to 6 billion (2000); although the land under cultivation remained almost constant at about 1.5 billion hectares, Table 5.1 shows figurative sales data of global major agrochemical markets in $ mn.

A study undertaken by the National Center for Food and Agricultural Policy (NCFAC) proved that mere replacing herbicides with hand labor would not pay off as it still may result in 30–50 % losses in crop productivity (www.ifpri.org). Further to prevent any yield loss vis-à-vis herbicides huge manpower about 70 million workers, ~25 % of the US and 25–75 % population of developing countries would toil daily. US farmers are found to have comparative advantage in terms of productivity in the world, producing over 40 % soybeans and corn, 20 % cotton,

Table 5.1 Sales data of global major agrochemical markets ($ mn)

Country	FY 2013	FY 2012	Change %
Brazil	11,600	9710	+19.5
USA	9200	8503	+8.2
Japan[a,b]	3102	3064	+1.2
Argentina	2506	2381	+5.2
Australia[a,c]	2017	2186	−7.7
Germany[a]	1905	1771	+7.6

[a]At the current rate
[b]Fiscal year 2012 and 2013, ended September 30th
[c]Fiscal year 2012 and 2013, ended September 30th

and 24 % beef of total world's production. America's agricultural production turnover surpasses US $220 billion per annum with over $50 billion worth products exported (2001). This large-scale production has led to US citizens paying less for their food, ~7 % of income, as against 10 % and 75 % of the income spent by their counterparts in other parts of the world (Gianessi and Sankula 2003).

Thus, ~98 % of US citizens today work in sectors other than farming, and in spite of it, they still boast of most abundant and economic food supply across the world. Unfortunately, many consumers have no clue as regards the complexity of farming; do not have knowledge about the multiple diseases, insect, and weed proneness that can affect the entire crops; and see little or no benefit accrued from the use of pesticides. Many consumers in the developed world are still caught unaware of, or ignore, that >99.99 % (by weight) of the pesticides in their diet amount to natural chemicals, which plants are responsible to produce to protect themselves thereby improving upon their defense system Farm Facts (2002).

5.4.2 Watchdog Agencies or Apex Bodies and Pressure Groups

Many activist groups and media sections intentionally blow out of proportion the use of pesticides by raising the hypothetical risks, ignoring the numerous benefits for mitigating their own vested interests. A recent report by Dr. D. Juberg states that activist groups use alleged chemicals and environmental risks as weapons to play with parent's just concerns for the health of their children (Juberg 2003; Koop 2003). Most often these activists surface in the name of promoting legislation, regulation, and litigation that has nothing to do with science, just for the sake of pushing their political agenda even at the cost of technology, free markets, and scientific progress. These groups strongly raise their voice to advocate banning or removing even minute traces of synthetic chemicals, like approved pesticides or food additives (Lomberg 2001; Marrone 1999).

5.4.3 Health Implications of Pesticides

Americans are facing an aggravating problem of obesity. However, to overcome this problem, consumption of fruits and vegetables is a remedy, but in order to divert public attention, scientific and social efforts are underway to create a pesticide-free regime. According to the estimates of Center for Disease Control and Prevention, obesity in America is reaching alarming proportions, and over 25 % are obese (Center for Disease Control and Prevention (CDC) (2000)).

This health epidemic has taken toll of over 300,000 premature deaths annually in the USA, next to tobacco-related deaths. This is a fallout of over consumption of sugars and fats and minimal consumption of fruits and vegetables essential for good health in accordance with the dietary guidelines set by various agencies like of the US Department of Agriculture (USDA), US Department of Health and Human Services (DHHS), National Cancer Institute (NCI), American Dietetic Association (ADA), Produce for Better Health Foundation (PHB), etc. Unfortunately, the activists most often make baseless and biased statements that the foods in the form of fruits like apples are detrimental to children as they are contaminated with toxic pesticides with concentration levels above permissible limits. The science-based apex bodies like the Center for Disease Control (CDC), National Cancer Institute (NCI), American Medical Association (AMA), Food and Dmg Administration (FDA), American Council on Science and Health (ACSH), etc. provide the necessary supportive data but fail miserably in dragging and holding on the attention of general public. A huge chunk of money is poured to for undertaking research and generates reports to assure the public that traces of residues in foods pose minimal or no risk by the watchdog health agencies. No research investigations have so far revealed convincingly that research investigations show that there is no convincing evidence that any food contaminant (including pesticides) is carcinogenic or has established any direct and concrete relationship between contaminant and cancer potential (Carlock and Dotson 2010).

However, according to Dr. Bruce Ames, one of the world's leading cancer experts, lesser is the consumption of fruits and vegetables by the people more is risk carried by them as prospective candidates for cancer and other health-related costs. Reduction in the use of synthetic pesticides will cut down the production of fruits and vegetables which will further reduce its consumption and thereby increase vulnerability to cancer, especially for the disadvantaged ones. Recent studies support the protective value of fruits and vegetables in combating not only various forms of cancer but also other diseases (Benachour and Séralini 2009; Carboni et al. 2004).

The health experts have expressed concern that hypothetical risks are taken care of by spending huge amounts of money over real ones which are neglected in the process unfortunately. According to a Harvard University Center for Risk Analysis survey shows when that when we ignore the cost of our environmental decisions in the form of lesser regulations really more lives are risk prone (Stephenson and Soloman 2007; Thompson et al. 2000; Tomlin 2009).

In the USA, the Center for Disease Control and Prevention (CDC) reports that the life expectancy in early nineteenth century was ~50 years, as against more than 70 today. This increase in life expectancy should be attributed to medical revolution facilitating development lifesaving drugs, accompanied by advancement in public health and improved nutrition. The use of pesticides in food production and public-health programs has checked many human lives by cutting down toxins generated by fungal species as mycotoxins which otherwise are very detrimental to human and health of livestock by way of generating various diseases and disorders (Dhara and Dhara 2002; Penagos 2002).

5.4.4 Testing of Pesticides

It is worth mentioning here that pesticides are thoroughly tested both qualitatively and quantitatively for their chemical composition and risk potential. Environmental Protection Agency (EPA), an apex body, is authorized to pass up health and environmental tests for the registration of the pesticide to take place. These tests take about 9 years for their completion cost of the order of $180 million per compound. Only 1 compound out of 139,000 tested will reach the market ultimately. In spite of this, people are still concerned about the safety of the products and health risks associated with it and disappointed with practices employed by the watchdog agencies like EPA (Johnson and Glynn 2001).

5.5 Natural Products as Green and Sustainable Pesticides

Sustainability refers to living the ways by which we do not deplete the resources of the earth-support systems. Figure 5.2 shows the framework for assessment of sustainability. It further shows that sustainability implies equitable distribution of resources, providing for viable and bearable options from the economical, social, and environmental perspective point of view.

Adequate pest control is meant for increasing crop yield and reduces a land requirement which is not in harmony with the ever-increasing demand of the world's population adopting conventional agricultural practices (Ehrich and Jortner 2010). The costs incurred for the rapid development of modern pesticides include the following:

- Environmental pollution
- Unpredicted human health-risk potentials
- Far-reaching detrimental effects on wildlife and other nontargeted organisms in the food chain

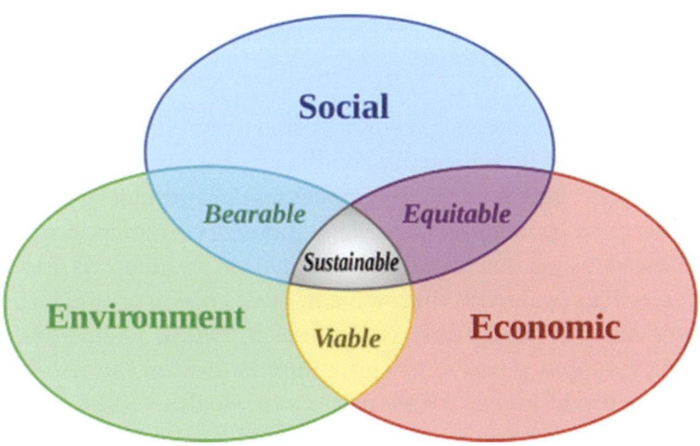

Fig. 5.2 Framework for assessment of sustainability

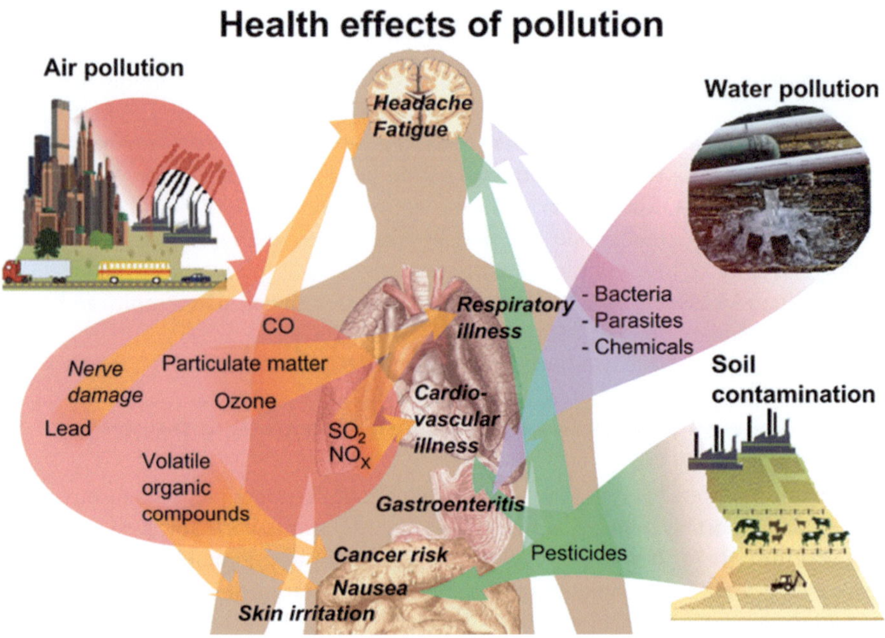

Fig. 5.3 Pesticides are implicated in a range of impacts on human health due to pollution

This has driven the opening up of the new vistas for green chemistry. Green chemistry field in this perspective necessitates the design of new processes and methods that reduce the use and generation of toxic and hazardous substances both to human and ecology. From the Fig. 5.3 shows that pesticides are responsible for inflicting a range of impacts on human health due to pollution.

5.5.1 Natural Products as a Green and Sustainable Substitute to Synthetic Pesticides

Naturally derived products are an excellent option to synthetic counterparts to bring down negative effects on human health and the environment at large. The initiative to adopt and employ green chemistry processes and products and the need for evolving at new crop protection tools and innovative methods of action make discovery followed by commercialization of natural products as green pesticides an attractive as well as profitable alternative in great command and further dragging the attention of all concerned. With the increasing world population, the need to obtain the highest output possible in agriculture is also on the rise tremendously. This has led to the use of higher degree of agrochemicals to control insect pests and plant diseases and to enrich overused soil. Consequently, this has placed an unwarranted heavy burden on the earth's ecosystems and has undergone substantial degradation over time. The striking the balance between the need to bring in sustenance and the need to leave behind enough resources for the generations to come which has further led to agriculture as a sustainable movement such that the future generations need not compromise with the ability to meet their own needs (Ford et al. 2010).

The realization that sustainable industrial processes must be compatible with global environmental concerns has given birth to the evolution of the philosophy and principles of green chemistry and sustainability. These two movements have many goals and ideals in common. Biopesticides are natural products derived from way of congruence between sustainable agriculture and green chemistry. The significant contribution made by natural products to sustainable agriculture and the status of biopesticides in the pesticide market have been recognized by the US EPA as an over the years as an evaluation of how biopesticides can be made to meet the green chemistry goals. Figure 5.4 shows biobased economy wherein agricultural resources are used in stainable manner.

It brings to the notice that biobased economy has its own social benefits as well as social costs too. But benefits supersede costs, e.g., saponins belong to a family of natural products that are structurally designed of aglycones (triterpene or steroidal) and sugars [pentose(s), hexose(s), and/or uronic acid(s)] which possess agricultural, medical, biological, and pharmacological properties as specified in Table 5.1. Saponins are distributed over a wide range in dicotyledonous plants and monocots species. They exist in foods (e.g., beans, peanuts, oats, green peppers, asparagus, garlic, onions, spinach, tomatoes, and potatoes), animal feed (alfalfa and clover), and marine organisms. Apart from this they are found to be present in numerous herbal species as well, e.g., ginseng, quila, horse-chestnut, and beans. They are characterized by their metabolic and regulatory activity responsible for the development of an organism (Cutler and Cutler 1999a, b).

5.5.2 Biopesticides vis-a-vis Tenets of Green and Sustainable Chemistry

Biopesticides play an important role in confirming to the goals of green chemistry in many functional areas.

- An increasing number of measures of safety can be taken care of when it comes to the reduction in potential of fire hazards during the manufacture of crop protection products.
- Human health concerns can be met by reducing the reliance on carcinogens, or endocrine disrupters are key goal of green chemistry.
- Environmental impact is a concern that is prominently found at various stages of industry and government. Sustainable processes are those that will help develop pesticides without compromising the ability of future generations to meet their own needs that will facilitate in reducing the waste to the minimum or no waste in the process.
- Finally, reducing the use of substances with toxic potential benefits in striking the proven balances between management, labor, and regulatory authorities.

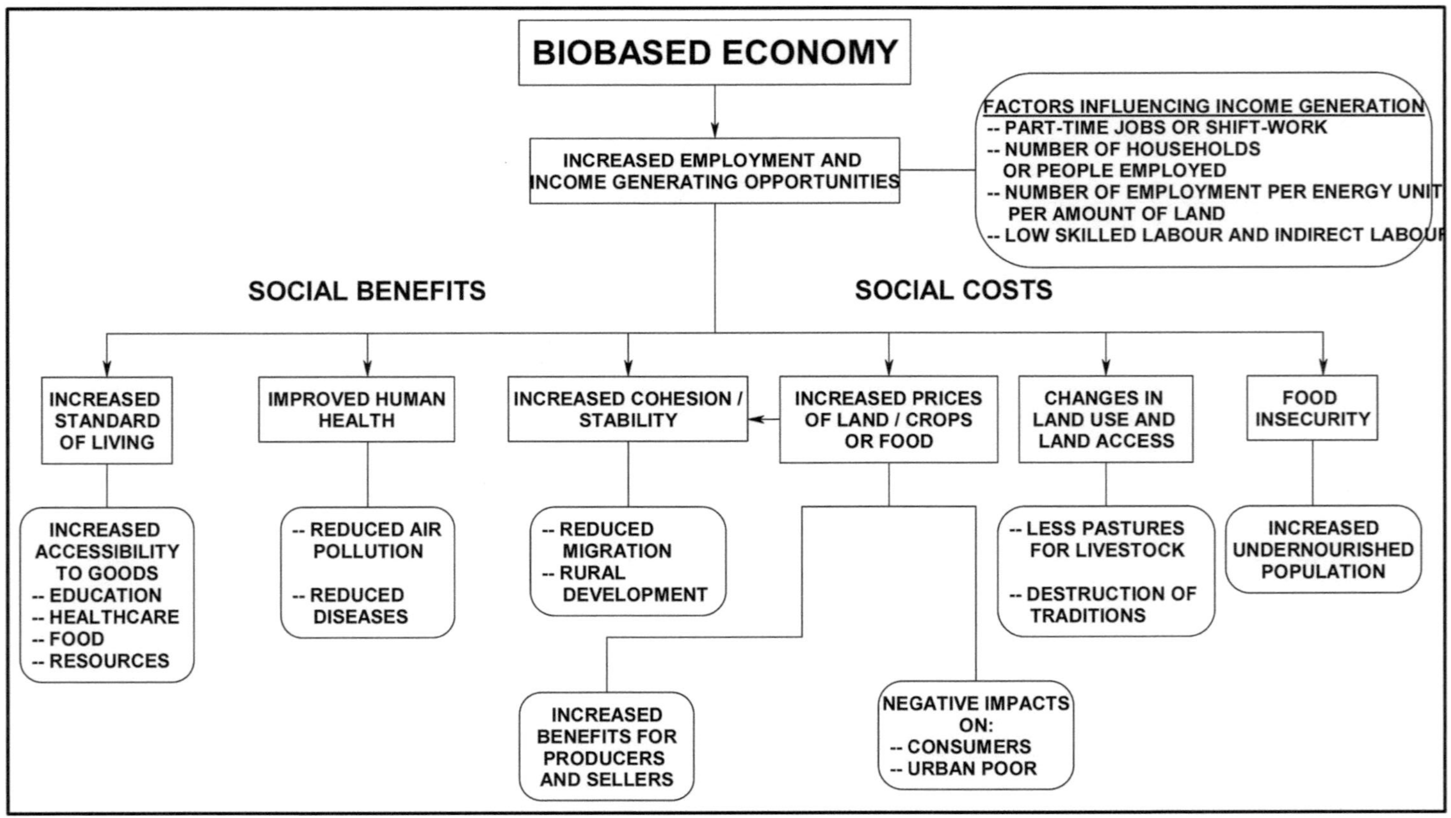

Fig. 5.4 Biobased economy—sustainable use of resources

5.5.3 Safety Issues and Concerns

There has been growing realizations about the safety concerns following some unfortunate incidents while producing agricultural crop protection products.

Some of the industrial events to reckon with in history are as follows:

- Bhopal tragedy, India (1984), at a factory manufacturing the carbamate pesticide, carbaryl. Volatile methyl isocyanine gas that escaped from the plant in one night has taken a toll of over 2500 people. The total number of people who lost their lives in that accident so far ranges between 15,000 and 20,000. Many synthetic processes in chemical industries are vulnerable to fire hazards, explosions, or other unforeseen events such as one which is witnessed in Bhopal (Lappierre and Moro 2001).

Natural products have a comparative advantage over synthetic processes in terms of:

- Lessened safety concerns.
- Biopesticides derived from fermentation used water-based methods which can be undertaken at or near ambient pressures or temperatures.
- Manufacturing using fermentation does not demand continuous monitoring and attendance, thus reducing risk to health substantially.

5.5.4 Human Health

There is a need of the hour for safer alternatives to synthetic pesticides in agriculture. Many existing pesticides are subjected to rigorous examination EPA's Food Quality Protection Act (FQPA). Some pesticides have a shown immediate effect on as contact dermatitis as reported in banana workers in Panama. For example, chlorothalonil, thiabendazole, imazalil, and aluminum hydroxide are some of the pesticides which show immediate effect on contact. Not only are those, the children are more susceptible to long-term effects. A report shows that the children ages 2–5 fed with organic juices, fresh fruits, and vegetables were found to have less organophosphate contents than those fed with fruits and vegetables grown using chemical-based pesticides.

The recognition that in the past the agrochemicals used have had severe and deleterious effects on the human health front has explored new avenues in the form of using natural products as green and sustainable pesticides. Green pesticides are characterized by selectivity and may preferentially attack certain target organisms like insects or plant pathogens, although all natural products are not necessarily nontoxic. Biocides are better placed over synthetic ones in terms of target specificity and high specific activity.

5.5.5 Environmental Impacts

There has been growing awareness among masses after the release of Rachel Carson's Silent Spring in 1962 (JO), about the unintended consequences of pesticides made available by the chemical industry. For example, DDT was used widely as an insecticide to control disease vectors such as mosquitoes but had to be banned in the later stage as it was found to be a persistent chemical and was devastating by making way into the food chain. The egg thinning properties of chlorinated hydrocarbons raised an alarm among public about the hazards chemical pesticides can cause (Vose et al. 2007). There is a greater demand for safer crop protection agents following the detection of pesticides in concentration levels above prescribed limits in runoff and drinking water. Organic halogenated compounds due to their persistent nature and long life are the major culprits responsible for severe environmental damage. Several fungicides that are frequently used in current agricultural practices due to their nonselectivity have evil effects on nontarget organisms (Curl et al. 2003).

The chlorothalonil, a fungicide in itself, is very toxic to fish, shrimp, frogs, beneficial microbes, and earthworms. On degradation its breakdown product is toxic 30 times more than that of the parent product (Coats 2008). The fumigant methyl bromide is one of the most effective ozone-eating substances known. Synthetic pesticides can also dangle into the atmosphere and be transported to areas away from agricultural land with the current of air. Hamers et al. shown that in the Netherlands the rainwater samples collected contained 2000 times more concentration of dichlorvos than the tolerable and prescribed limits for surface waters. Although all natural products are not nontoxic, all are certainly and highly biodegradable. It offers great advantages by reducing the environmental damage while protecting the crops.

In sustainable agriculture, it is of pivotal importance to maintain healthy populations of predatory insects such as ladybirds, laceswings, and predatory wasps. However, synthetic agrochemicals have posed a threat to this population leading to a greater need to use crop protection agents (Iwasa et al. 2004). Biopesticides that are very target specific help conserve and sustain natural predator populations (Maruya et al. 2005).

5.5.6 Waste Minimization

Huge amount of waste generation has been a problem of concern in chemical industry since the advent of synthetic agrochemicals. EPA in this regard has made it mandatory to clean and remediate contaminated soils due to agricultural activities of toxic substances, although it requires huge amount of money and time. Fermentation-derived products have a comparative advantage wherein the by-products are largely biodegradable unlike their chemical counterparts, e.g., in the production of Bt, the by-products can actually be used as fertilizers. Since biopesticide production does not generally consume the toxic and hazardous organic solvents or other substances, the by-products can be disposed off with reduced or concern for impact at the downstream sites. These save cost to production and do away with certain hazardous waste management problems (Heins et al. 2000).

5.5.7 Reducing Toxic and Health Hazardous Substances

Less consumption of hazardous substances has positive impact on strengthening the standard of human life and the environment. Some of the synthetically derived chemicals require toxic intermediates which further require special storage and transportation, expensive facilities able to handle highly dangerous reagents and ultimately at times result in generating toxic by-products. Fermentation-derived biopesticides are bereft of these problems in their production.

5.6 Driving Forces Behind the Development of Green and Sustainable Biopesticides

The sales for conventional chemical pesticides have shown worldwide, a downward trend. There has been a downfall in the global market for pesticides by about \$2 billion as against the projected figure of \$30 billion. As against this backdrop of downturn in the global traditional chemical pesticides, biopesticide market on the other hand has grown up greater than 20 % over the last decade. This rise in biopesticide market by \$12 billion with more than 20 % annual growth is attributed to the commensurate consumer demand for safer foods in America. There has been growing and positive perception among masses about organic foods has offered an excellent opportunity for the consumption of safer, selective, biodegradable, and natural crop protection agents with minimum health risk like biopesticides.

5.6.1 Advantages of Green and Sustainable Pesticides vis-à-vis Synthetic Ones

Biopesticides still only contribute to the extent of 1 % of the global pesticide market; the trends suggest that they are advantageous over their synthetic counterparts in respect of:

- Market share continue to grow as synthetic ones is hazardous and detrimental to productivity.
- Another encouraging factor in the development of biopesticides is it is far more cost-effective and has far less regulatory bottlenecks for new products. For example, new synthetic chemical and biopesticide requires \$185 million and 10 years and \$6 million and 3 years for their development, respectively.
- Companies too are benefited by getting products quickly and efficiently since time taken by the biopesticides for their development is far less which is the need of the day in the current economic climate.

- Biopesticides have become one of the vital tools of the integrated pest management system since their inception following a stiff resistance to conventional pesticides in the wake of perceptional change.
- Many of the new lower-risk synthetic pesticides have a single site mode of action and do not show long-lasting effect due to resistance to it developed by microorganisms and may become redundant or less productive within a few seasons. For example, in 1988, Georghiou and Legumes reported that over 500 species of insects were known to have shown resistance to one or more insecticides (Jones et al. 2009; Salgado 1998).

From these examples it is clear that there are distinct advantages to pursuing the development of natural products as green pesticides over synthetic ones; examples of natural product pesticides can be cited which are available in the market that confirm to the principles of green and sustainable chemistry are given below.

5.7 Representative Examples of Natural Products as Green Pesticides

5.7.1 *Bacillus thuringiensis* (Bt)

Significance—biopesticides are largely looked upon as fresher to the agricultural chemical industry. However, Bt has a much longer background as compared to many synthetic pesticides. Bt is an excellent example of a natural product that can serve as a green pesticide.

Bt has been in use for more than 40 years since its registration by the US EPA in 1961. The Bt-based products contribute heavily to the global biopesticide market. Its use has been on the rise following deregulation of synthetic products due to their toxicity and high incidence of insect resistance. In 1997, among all insecticides Bt was ranked fourth based on fruit crops treated per acre; Bt has shown longevity in terms of its performance and service behavior against newly developed synthetic products which show resistance in less than 5 years (USEPA 2004, 2007, 2011).

5.7.1.1 Properties

- Bt is manufactured by more safer liquid fermentation process, thus avoiding the processes that are prone to fire hazards, or explosion.
- Due to specificity and selectivity in action, it shows overall higher degree of safety to nontarget organisms, including humans.
- Selectively activated in the highly alkaline gut of the insects and is not perturbed by the acidic mammalian gut.

Uses: It is used as fertilizer.

- Can help in addressing the problem of the mass piling of abandoned and obsolete pesticides worldwide, particularly in developing countries. According to

FAO estimates, the figures stand at 100,000 tons of obsolete pesticides like organochlorine compounds, e.g., DDT, dieldrin, and organophosphates including parathion and dichlorvos.
- A number of countries like Mexico, South Korea, Nigeria, Brazil, and India have successfully employed relatively simple Bt fermentation process which involves substitution of simple biomass for chemical feedstock.

5.7.2 Spinosad

Significance: It is an insecticide prepared as a mixture of two naturally occurring compounds derived from the fermentation of *Saccharopolyspora spinosa*.

5.7.2.1 Properties

- The activity of spinosyn A has been found to be much better than compounds such as carbamate and cyclodiene derived from pyrethroids, organophosphate (Nomura and Casida 2011).
- The spinosyns A and D appear to be a novel neurotoxic agents causing hyperexcitation which disrupts central nervous system of the insect.
- The fresh binding sites leading to an additive resistance management benefit for this natural product.
- Photolysis brings about rapid degradation by water and soil with (half-life less than 1 and 9–10 days, respectively).
- Not much prone to leaching or percolation and therefore doesn't pose a threat to groundwater.
- Although less toxic to mammals and birds, slight to moderately toxic toward aquatic species.

Uses: Spinosad's selectivity toward targeted microorganisms and low activity on beneficial insects allow it to be integrated with the pest management programs that depend on natural predators and parasites. US EPA Presidential Green Challenge Award in 1999 was conferred on this fermentation-derived process accomplishing many of the goals of safer chemical products and processes.

5.7.3 Messenger

Significance: Natural defensive response of plants against pathogen attack is used to the advantage for Eden Biosciences unique crop protection product. The messenger represents one of the first and best examples of a natural product based on unraveling the plant defense potential to kill pests and improve yield. Messenger is a new class of nontoxic, naturally occurring proteins, harpins.

5.7.3.1 Properties

When messenger is studied extensively for safety evaluation, it has shown that virtually:

- No adverse effects on mammals, birds, honeybees, plants, fish, aquatic invertebrates, and algae.
- Synthesized without using solvents or toxic reagents but in a water-based fermentation process.
- The process consumes less energy and does not generate hazardous chemical wastes.
- The by-products of the fermentation are biodegradable and don't pose waste management problem.

Uses: Carriers used in the formulation of messenger are food-grade substances providing an end-use product that is environmentally safe and selective.

5.7.4 Serenade

Significance: Serenade biofungicide is derived from a naturally occurring strain of *Bacillus subtilis* QST-713.

Properties:

- A highly effective natural agricultural fungicide derived from plant feed stocks, generating nonhazardous by-products
- Ranges from nontoxic to beneficial and nontarget to organisms such as trout, quail, lady beetles, lacewings, parasitic wasps, earthworms, and honeybees
- Safe to workers and groundwater
- Generates three groups of lipopeptides (iturins, agrastatins/plipastatins, and surfactins) which have synergistic effect to destroy germ tubes and mycelium

Uses:

- Environmentally benign and control over a broad spectrum of microorganisms due its novel, complex mode of action
- Caters to the use in integrated pest management (IPM) programs that utilize spectrum of tools like cultural practices, classical biological control, and other fungicides
- Prevents plant diseases by first covering the leaf surface and physically preventing attachment and access to the pathogens

5.8 Green Solutions for Plant Disease Control

Efficient production of high-quality products worth exporting in plant-based industries will decide its viability. Demand for product in world markets and protection of the market access will be enhanced by the development of natural systems for disease control. Growing public perception and reduced environmental impact are the advantages of biological control as an alternative to synthetic one.

Maintenance and strengthening of natural ecosystems which is often lost in the cropping patterns can be restored by using biological control agents (BCAs). For example, *Trichoderma* is found be a useful BCA, the best strains produce bulk of 6-pentyl-alpha-pyrone (6PAP) which stifles the growth of sapstain fungi, including *Ceratocystis picea*, *Botrytis cinerea* on kiwifruit, and *Armillaria* spp. on trees, kiwifruit vines, and other woody plants, controls pathogens, etc. Soil fumigation in California orchards helped in inhibiting the growth of Armillaria (Dong et al. 1999).

Other examples to name are the following:

- Control of *Botrytis* stem-end rot of kiwifruit with *Trichoderma*
- Control of *Botrytis* in greenhouse tomatoes with *Cladosporium*
- Biological suppression of *Botrytis* in kiwifruit
- Control of silver-leaf disease

Botrytis cinerea cause heavy economic loss by infecting stem wounds of greenhouse tomatoes. A bioassay using stem sections was developed to study wound infection and to screen potential fungal antagonists for activity against *Botrytis*. Strains of *Cladosporium cladosporioides* reduced infection from 80–100 % to 0–10 %. *Trichoderma harzianum* strains gave comparatively less reduction. Similar results were obtained on whole plants. *Penicillium* strains varied widely in activity. The concentration of *Cladosporium* and *Trichoderma* that gave the highest degree of protection was c.108 cfu/mL (Table 5.1).

5.8.1 Disease Suppression by Pine Bark Composts

Significance: The growth of microorganisms such as BCAs can be achieved by the application of composts in a beneficial manner. Research is underway and is aimed at gathering basic biological knowledge on bark-based composts, economic and environmentally viable ways of dealing with voluminous waste bark, and established manipulation of composting process to further disease suppressiveness and wettability in soilless potting mixes (Kömives et al. 2003).

Uses: Such composts:

- Cut down production costs as an alternative to peat, a nonrenewable resource, and add value by using biological methods of disease suppression.
- A cost-effective method of checking soil-borne fungal and bacterial diseases and suppressing weeds may be provided by in situ preparation of bark composts.
- Offering a realistic substitute to methyl bromide, an ozone-depleting biocidal soil fumigant.

5.8.2 Plant Resistance

Mechanism: Plants by themselves by way of using their combination of physical and chemical defense mechanism may exhibit excellent resistance to infection. A failure or a delay in the defenses coming into action can result in affliction to disease. Induced resistance is one of the approaches for disease control; application of compounds called elicitors triggers the plants' defenses. Elicitor-treated plants (sensitized) respond more swiftly and intensely to prevent attack from plant pathogens. For example, yeast cell walls elicitors have shown broad specificity and could induce resistance against major fungal pathogens like powdery mildew on barley, gray mold and stem rot on lettuce, and chocolate spot on beans, stem-down dieback in *Pinus radiata* and *Sclerotinia* in kiwifruit leaves (Umezawa et al. 1998).

The Benefits of Induced Resistance Include:

- Reduction in pesticide use for environmental sustenance
- Broad specificity—an impetus to overall plant immunity and thus has the potential improved pathogen resistance
- Durability—development of sustainable pathogen resistance
- Compatibility—integration with disease control methods

5.9 Effect of Pesticide Use on Climate Change

The overuse and abuse of pesticides have increased over time to destroy insect pests and diseases. However, the major causes of concern are the evil environmental impacts of these agrochemicals on biodiversity, environment, food quality, and human health. Conducive climate change will help support insect conservation and pest status. Climate and weather can substantially affect both the development and distribution of insects. Man-made activities or anthropological activities are responsible for the global warming that has taken place over the last 50 years. The relationship between pesticide use and climate for crops that require bulk of pesticide has been established by an assessment (Tables 5.2 and 5.3).

The British Geological Survey undertook an assessment on the impacts of climate change on the fate and behavior of pesticides on the environment. The approach taken was to examine the effect of the sources of pesticides, their transportation through the subsurface (pathways) and the receptors, particularly the groundwater reservoirs and the rivers. The report concluded that, in the long term, change in land use attributed to changes in climate may have a substantial impact on pesticides in the environment than the direct effect on pesticide fate and transport.

Research shows climate change-induced cropping patterns mean the changes in pesticide consumption will lead to cosmetic shift in the dose of certain classes of already approved pesticide, predominance of incumbent pests, diseases and weed species is predicted under change in climate pattern. It further implies wider and more frequent applications of existing pesticides or the introduction of innovative

Table 5.2 Agricultural, biological, medical, and pharmacological of saponins

Activity	
Adaptogenic	Enzyme
Analgesic	Expectant
Allelochemical	Immunomodulatory
Antiaging	Hypoglycemic
Antifungal	Insecticidal
Antifeeding	Molluscocidal
Anti-inflammatory	Sedative
Antimicrobial	Spermicidal
Antiulcer	Contraceptive
Antiviral	Sweeteners
Antiprotozoal	Hemolytic

Table 5.3 Effect of gel-applied antagonists on *Botrytis* infection of terminal wounds of whole tomato plants

BCA isolate number	Percent *Botrytis* infection	95 % confidence interval
Control	100	54.1–100.0
95-1 (*Trichoderma*)	33	4.3–77.7
806 (*Trichoderma*)	50	11.8–88.2
677 (*Cladosporium*)	0	0–45.9
712b (*Cladosporium*)	0	0–45.9
724 (*Cladosporium*)	0	0–45.9

products, the defense of crops to pesticides may reduce at higher ozone concentrations, and rise in pesticide use may reduce crop yield which may induce the manufacturing of more effective products (Kagabu 2003). Rise in temperature due to global warming would mean the pesticides to be in the gas phase and consequently can transport over longer atmospheric distances from the sources. However, degradation processes will also take place at a faster pace, so the overall effect of rise in temperatures is yet to be unraveled, the rate of atmospheric degradation of vapor-phase pesticides may increase by virtue of rise troposphere ozone concentrations, and changed pattern in seasonal rainfall may bring about changes in the spatial and temporal distribution of wet deposition of airborne pesticides and consequently their degradation products (Hamers et al. 2002).

Higher temperatures will lead to rapid vaporization and degradation of pesticide residues in both soil and surface waters. Increased winter rainfall will wet soils and the runoff rich in pesticide may move fast from the surface through the soil matrix, to drains, and below soil layer; surface runoff may become more prominent causing further erosion of pesticide-rich soil from fields making their way to drains and surface waters; shorter recharge periods could allow less residence time for pesticide transport and more for degradation; and higher soil water contents will promote degradation rates due to temperature effects, while summers may showcase cracking of shrink-swell clay soils. The net effect may be increased in the pesticides movement by transport routes in the winter; dry soils have a lower biodegradation tendency than wet soils. Periodic high groundwater levels can intercept pesticides and other diffuse agricultural contaminants in the unsaturated and soil, reducing the residence time for pesticide degradation and giving way to seasonal rise in the pesticide levels in groundwater; more frequent localized groundwater flooding of agricultural land may accumulate and contaminate soils to a greater degree.

Receptors mean significant reduction in mean river flows which may bring about a reduction in the dilution potential of surface water reservoirs and consequently increasing pesticide concentrations, substantial reduction in annual minimum groundwater levels in the future is very much expected and projected but the repercussions on source yields and pesticide exposure at receptors are not certain, the impact of climate change on base flow to groundwater-dominated rivers has not yet been systematically and scientifically studied, and the implications for changes in pesticide exposure are still very uncertain as well.

5.10 Concluding Remarks and Future Perspectives

It is of pivotal importance for the scientific community and the agricultural industry to actively explore ways and means to build a partnership with cultivator groups, food chain members, medical and public-health experts to better educate people from different walks of life like consumers, and politicians and policy makers on known and real vis-a-vis hypothetical risks and to further appreciate the tremendous benefits that can be accrued using modern agricultural practices. But beyond all a sound convergent extension approach is needed of the hour (Mukherjee et al. 2012; Mukherjee and Maity 2015).

Agrochemicals have an important role to play in the sustenance of the agricultural products without compromising soil fertility, toxicity both to human and ecology, biodegradability, cost, availability, etc. The eco-friendly control measures that promote plant health by making it resilient to host of plant diseases and by promoting the use of biocontrol agents, natural products, and elicitors are gaining a lot much of importance as substantial ecological degradation has already been brought about after using chemical-based agrochemicals over the years. However, certain conventional control methods too can offer an economic and environmentally safe crop protection strategy and can pay off.

Disease suppression using composts and mulches has long been used successfully and efficiently by "organic" gardeners and growers for the said cause. The preferential and intentional use of biological control agents for disease control is a recent as against biological control of insects and weeds. Many of the soils that are conducive for naturally suppressing plant diseases are rich in organic matter that supports the growth of beneficial microorganisms, e.g., *Trichoderma* and *Pseudomonas* which are known to suppress the activity of soil-borne plant pathogens (Sparks et al. 1998).

Acknowledgments The author is indebted to Vishwakarma Institute of Technology Management Trustee, Mr. Bharat Aggarwal for his constant support and encouragement. The author extends his thanks to Director and HOD of the Chemical Engineering Department for their timely guidance.

References

Benachour N, Séralini GE (2009) Glyphosate formulations induce apoptosis and necrosis in human umbilical, embryonic and placental cells. Chem Res Toxicol 22:97–105

Carboni P, Sherer TB, Zhang N, Taylor G, Na HM, Greenamyre JT et al (2004) Rotenone, deguelin, their metabolites, and the rat model of Parkinson's disease. Chem Res Toxicol 17:1540–1548

Carlock LL, Dotson TA (2010) Metam-sodium. In: Krieger RI (ed) Hayes' handbook of pesticide toxicology, vol 2, 3rd edn. Academic, Boston, pp 2293–2306

Center for Disease Control and Prevention (CDC), U. S. Department of Health and Human Services (2000) Physical activity and good nutrition: essential elements for good health, at a-glance. CDC, Atlanta

Coats JR (2008) Toxicology of synthetic pyrethroid insecticides in fish: a case study. In: Di Giulio RT, Hinton DE (eds) The toxicology of fishes. CRC Press, Boca Raton, pp 805–817, Council on Science and Health : 5–6

Curl CL, Fenske RA, Elgethun K (2003) Environ Health Perspect 111:377

Cutler HG, Cutler SJ (eds) (1999a) Biologically active natural products: agrochemicals. CRC Press, Boca Raton/Delhi

Cutler HG, Cutler SJ (1999b) Biologically active natural products agrochemicals. CRC press, Boca Raton

Dhara VR, Dhara R (2002) Arch Environ Health 57:391

Dong G, Delaney TP, Bauer D, Beer SV (1999) Plant J 20:207

Ehrich M, Jortner BS (2010) Organophosphorus-induced delayed neuropathy. In: Krieger R (ed) Handbook of pesticide toxicology. Elsevier, Amsterdam, pp 1479–1504

Farm Facts (2002) American Farm Bureau Federation. www.fb.org

Ford KA, Casida JE, Chandran D, Gulevich AG, Okrent RA, Durkin KA et al (2010) Neonicotinoid insecticides induce salicylate-associated plant defense responses. Proc Natl Acad Sci U S A 107:17527–17532

Gianessi LP, Sankula S (2003) The value of herbicides in U.S. crop production NCFAP

Hamers R, Smit MGD, Murk AJ, Koeman JH (2002) Chemosphere 45:609

Hardstone MC, Scott JG (2010) Apis mellifera more sensitive to insecticides than other insects? Pest Manag Sci 66:1171–1180

Heins SD, Manker DC, Jimenez DR, McCoy RJ, Marrone PG, Orjala JE (2000) US Patent 6,103,228

International Food Policy Research Institute (2002) www.ifpri.org

Iwasa T, Motoyama N, Ambrose JT, Roe RM (2004) Mechanism for the differential toxicity of neonicotinoid insecticides in the honey bee Apis mellifera. Crop Prot 23:371–378

Johnson MK, Glynn P (2001) Neuropathy target esterase. In: Krieger R (ed) Handbook of pesticide toxicology, vol 2. Academic, San Diego, pp 953–965

Jones DK, Hammond JJ, Relyea RA (2009) Very highly toxic effects of endosulfan across nine species of tadpoles: lag effects and family-level sensitivity. Environ Toxicol Chem 28:1939–1945

Juberg DR (2003) Are children more vulnerable to environmental chemicals? American

Kagabu S (2003) Molecular design of neonicotinoids: past, present and future. In: Voss A, Ramos G (eds) Chemistry of crop protection. Wiley-VCH, Weinheim, pp 193–212

Kömives T, Gullner G, Rennenberg H, Casida JE (2003) Ability of poplar (*Populus* spp.) to detoxify chloroacetanilide herbicides. Water Air Soil Pollut Focus 3:277–283

Koop CE (2003) Foreword in are children more vulnerable to environmental chemicals? American Council on Science and Health, p 236

Lappierre D, Moro J (2001) It was five past midnight in Bhopal. Full Circle Publishing, New Delhi
Lomberg DR (2001) The skeptical environmentalist. Cambridge University Press, Cambridge, pp 341–342
Marrone P (1999) Outlook Agric 28:149
Maruya KA, Smalling KL, Mora MA (2005) Residues of toxaphene in insectivorous birds (*Petrochelidon* spp.) from the Rio Grande, Texas. Arch Environ Contam Toxicol 48:567–574
Mukherjee A, Bahal R, Roy Burman R, Dubey SK (2012) Conceptual convergence of pluralistic extension at Aligarh District of Uttar Pradesh. J Commun Mobil Sustain Dev 7(1&2):85–94
Mukherjee A, Maity A (2015) Public–private partnership for convergence of extension services in Indian agriculture. Curr Sci 109(9):1557–1563
Nomura DK, Casida JE (2011) Activity-based protein profiling of organophosphorus and thiocarbamate pesticides reveals multiple serine hydrolase targets in mouse brain. J Agric Food Chem 59:4860–4867
Penagos HG (2002) Int J Occup Environ Health 8:14
Phipps RH, Park JR (2002) Environmental benefits of genetically modified crops: global and European perspectives on their ability to reduce pesticide use. J Anim Feed Sci 11:1–18
Quistad GB, Barlow C, Winrow CJ, Sparks SE, Casida JE (2003) Evidence that mouse brain neuropathy target esterase is a lysophospholipase. Proc Natl Acad Sci U S A 100:7983–7987
Ratra GS, Kamita SG, Casida JE (2001) Role of human $GABA_A$ receptor β3 subunit in insecticide toxicity. Toxicol Appl Pharmacol 172:233–240
Rubin AL (2004) Metam sodium (sodium *N*-methyldithiocarbamate). Risk characterization document. California Environmental Protection Agency, Medical Toxicology Branch, Department of Pesticide Regulation, Sacramento
Salgado VL (1998) Pestic Biochem Physiol 60:91
Schuler F, Casida JE (2001) The insecticide target in the PSST subunit of complex I. Pest Manag Sci 57:932–940
Sparks TC, Thompson GD, Kirst HA, Hertlein MB, Larson LL, Worden TV, Thibault ST (1998) J Econ Entomol 91:1277
Stephenson GR, Soloman KR (2007) Pesticides and the environment. Canadian Network of Toxicology Centres Press, Guelph
Thompson GD, Dutton R, Sparks TC (2000) Pest Manag Sci 56:696
Tomlin CDS (2009) The pesticide manual: a world compendium, 15th edn. British Crop Production Council, Alton
Umezawa H, Aoyagi T, Nishikiori T, Okuyama A, Yamagishi Y, Hamada M, Takeuchi T (1986) J Antibiot 39:737
USEPA (U.S. Environmental Protection Agency) (2004) Interim reregistration eligibility decision (IRED); diazinon. EPA 738-R-04-006. U.S. Environmental Protection Agency, Washington, DC
USEPA (U.S. Environmental Protection Agency) (2007) White paper on the potential for atrazine to affect amphibian gonadal development. U.S.EPA, Washington, DC, Available: http://www.epa.gov
USEPA (U.S. Environmental Protection Agency) (2011) Carbofuran cancellation process. www.epa.gov
Vetter W, Oehme M (2000) Toxaphene. Analysis and environmental fate of congeners. In: Paasivirta J (ed) The handbook of environmental chemistry. New types of persistent halogenated compounds. Springer, Heidelberg, pp 237–287
Vose SC, Holland NT, Eskenazi B, Casida JE (2007) Lysophosphatidylcholine hydrolases of human erythrocytes, lymphocytes and brain: sensitive targets of conserved specificity for organophosphorus delayed neurotoxicants. Toxicol Appl Pharmacol 224:98–104
Ware GW, Whitacre DM (2004) The pesticide book, 6th edn. MeisterPro Information Resources, Willoughby
Warrior P (2000) Pest Manag Sci 56:681
Wei ZM, Laby RJ, Zumoff CH, Bauer DW, He SY, Collmer A, Beer SV (1992) Science 257:85

Wijeyesakere SJ, Richardson RJ (2010) Neuropathy target esterase. In: Krieger R (ed) Hayes' handbook of pesticide toxicology, vol 2, 3rd edn. Elsevier, Amsterdam, pp 1435–1455

Winrow CJ, Hemming ML, Allen DM, Quistad GB, Casida JE, Barlow C (2003) Loss of neuropathy target esterase in mice links organophosphate exposure to hyperactivity. Nat Genet 33:477–485

Xu H, Zhang N, Casida JE (2003) Insecticides in Chinese medicinal plants: survey leading to jacaranone, a neurotoxicant and glutathione-reactive quinol. J Agric Food Chem 51:2544–2547

Yamada S, Takayama Y, Yamanaka M, Ko K, Yamaguchi I (1990) J Pest Sci 15:95

Role of Genetic Resources of Forages in the Present Changing Climatic Scenario

6

Gitanjali Sahay, Pradeep Saxena, Shahid Ahmed, Tejveer Singh, and Seva Nayak

Abstract

Agricultural biodiversity is vital to an ecosystem approach; it enhances soil fertility, increases productivity and safeguards the crops and livestock. Genetic resources of forages have an important role to prevent disaster caused by climate change, following the measures of crop diversification, use of drought-tolerant crops, use of resistant cultivars and effective management of pest or disease outbreak by promoting regional/localised planning and initiative. The prevailing climate change scenario creates the need for a broader vision of risk management and responding to climate change. Plant genetic resources provide a basis for adaptation, to cope with the impact of climate change on agriculture and livestock production. There is a need to act at global, regional, national and local levels. The real challenge, however, will have to be handled in India by quick and focussed adaptation strategies for reducing susceptibilities, strengthening resilience and consolidating the overcoming ability of the farming communities. Building resilience in agro-ecological system and farming communities and improving adaptive capacity by building strong genetic resource base of different forage crops, i.e. contingent crops like fodder cereals, legumes and grasses, germplasm introduction in search for potential forage/fodder crop across the globe and sustainable uses of forage genetic resources are the way to cope. The chapter deals with current status of forage germplasm and the recent fodder varieties. The rationale of this chapter is to consider the status of forage genetic resources in India and their needs for enhancing agriculture entangled with

G. Sahay (✉) • P. Saxena • S. Ahmed • T. Singh • S. Nayak
Crop Improvement Division, ICAR-Indian Grassland and Fodder Research Institute, Jhansi 284003, India
e-mail: sahayg1@rediffmail.com

J.K. Bisht et al. (eds.), *Conservation Agriculture*,
DOI 10.1007/978-981-10-2558-7_6

livestock farming; emphasis has been to create alertness among the workers involved in forage crop improvement and works to coordinate to meet out the national food security with reference to changing climate.

Keywords
Genetic resources • Forages • Climate change • Coping • Adaptation

6.1 Introduction

Crop and livestock are the two main components of mixed farming system, which influence our agricultural economy and provide sustenance. The value of livestock in India's economy can be deciphered from the reality that ~90 million farming families, cultivating ~140 M ha area, are also keeping ~90 million milch animals. In fact, the crop-livestock farming system approach is considered as one of the most effective means for making agriculture resilient to extreme weather events (Hoffmann 2013). Presently the human race is facing many global challenges which include accomplishing food security for a swiftly expanding population, reducing the risk of climate change by reducing the net release of greenhouse gases into the environment due to various human activities and meeting the escalating demands for energy when the reserve fossil energy is reducing, and there are uncertainties about future supply (Lobell et al. 2008). The model of monsoon rainfall, which is the lifeline of country's agriculture, appears to have changed visibly, which necessitates frequent last-minute changes in cropping strategy. The aim of the present article is to highlight the importance of forage genetic resources and sensitise the scientists, workers and breeders engaged in forage crop breeding/improvement programmes to focus on meeting national food and fodder nutritional security for its sustainable management in view of changing climatic scenario (Fig. 6.1).

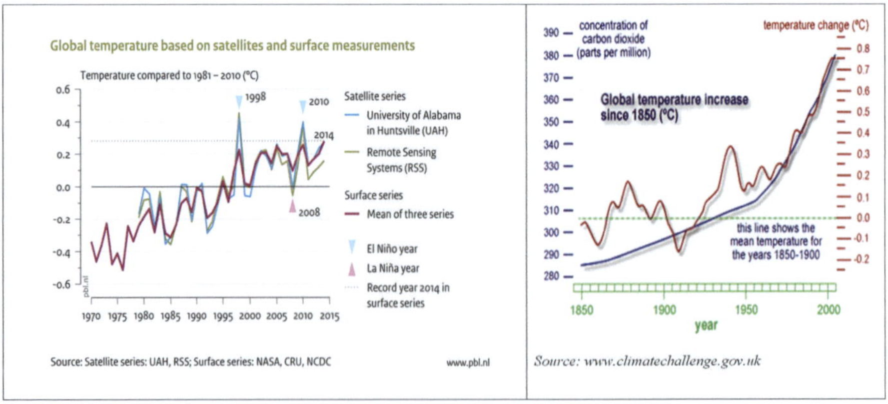

Fig. 6.1 Observed globally averaged temperature and CO_2 concentration anomaly

6.2 Climate Change Scenario

Climate is generally defined as average weather or scrupulously as the statistical description in terms of the mean and variability of relevant quantities over a period of time, ranging from months, thousands or millions of years. The term climate change refers to a statistically significant variation in mean stage of climate or in its variability existing for an extensive period, characteristically a decade or longer. Natural internal phenomena, external forces or persisting anthropogenic changes in the composition of atmosphere or inland use may lead to climate change. The fourth and fifth assessment report of the Intergovernmental Panel on Climate Change (IPCC 2007, 2014) authenticates that over the last 100 years, average global temperature has increased by 0.74 °C (Fig. 6.1), and the anticipated increase in temperature by 2100 is about 1.8–4.0 °C.

The severity and speed of climate change is presenting an unprecedented challenge, and insufficient technical and financial means are making adaptation to a changing climate very difficult. The major consequences of climate change are drought, floods, extreme temperatures both heat and cold, seasonal cycle imbalance or shifting of seasonal cycle. High temperature is drearily widespread over the country; however, there are spatial variances in the anticipated rainfall changes. In future the limits of maximum and minimum temperature are also expected to increase; however, night temperatures show escalating pattern. Prevailing changes in climate are forecasted to lead to changes in the areas suitable for growing certain crops and reduction in yield in a pronounced manner, mainly in rainfed crops. Under inadequate water supply situations, there is no coping mechanism for rainfall inconsistency. Heavy rains often resulting in flooding are damaging to crops and soil structure. As the roots have to breathe, most crops do not withstand continuous waterlogged conditions. Irreversible habitat damage occurs due to the erosion of top soil from the major growing areas by flooding (Burney et al. 2010; Meena et al. 2013, 2015a, b; Singh et al. 2014; Kumar et al. 2015).

The climate change shows direct effect on crop growth by affecting physiology, phenology and morphology. It shows indirect effect by affecting soil fertility, irrigation availability, pests, flood, droughts and rise in sea level. Climate change finally has socio-economic effects by affecting food demand, costs and benefits, policy, trade and farmer response. Climate change creates the need for a broader vision of risk management, and responding to climate change, plant genetic resources provide a basis for adaptation (Datta 2013).

In a case study in Uttarakhand region (Shiva and Bhatt 2013), it was noticed that during winter months (November to March) in years 2006–2009, there was no considerable rainfall; however, during 2006–2008, scanty rainfall occurred in the months of February and March only at high-altitude areas. Heavy loss to rabi crops was noticeable due to other places remaining dry even in the month of March. Sporadic rainfall occurred in March and April 2009 at high altitudes; the whole winter season was dry from October 2008 to June 2009 in mid and low altitudes. Thirty percent loss to rabi crops due to drought ensued at high altitudes (above 1,800 m). In the years 2006, 2007 and 2008, the monsoon showers started few days earlier than normal. In year 2008 huge amount of harm to crops like potato, amaranth, pulses and vegetables resulted due to heavy rains lashing the entire Uttarakhand Himalaya, from 20 May to mid September consequently to the westerly disturbances and early monsoon. Outbreak of damping off disease was reported causing great loss to agriculture in 2008. Such unanticipated situations are clear indications of major shifting of climate cycle in the Uttarakhand region of Indian Himalayas.

It has been observed that seasonal rainfall pattern has been disturbed greatly in the past few years. Winters are going almost dry, without significant rainfall in lower altitudes and without snow in higher altitudes. Winter season seems to be reduced, while spring season appears to be shifted by a month earlier. Overall pattern of rainfall in both seasons has changed considerably in the last decade with monsoon rains becoming extremely localised in both space and time being clear indicator of climate change. Intense showers have replaced evenly well-spread rainfall over time. Duration of rains reduced to few minutes with increased intensity in place of 1–2 h. This trend was more prevalent in 2008–2009 with cloud burst which has also increased considerably. Continuous rains for 7 days in monsoon season, popularly called as 'Sagain', have become obsolete which was crucial for groundwater recharge. According to the local people, most springs have dried up in 2001, 2006, 2007 and 2009. Restoration of biodiversity of the region, organic farming and training farmers was taken up by Navdanya team in the area (Shiva and Bhatt 2013). Most impressive achievement has been the understanding of situation, participatory rural appraisals bringing out solutions by the locals. Climate tours or yatras were carried out by group of villagers and scientists to assess the situation and suggest remedial measures. Next phase would include more training to adapt, mitigate the impact and compel the government to make investments in agriculture, afforestation, irrigation and preservation of genetic resources (Fig. 6.2).

Maximum farmers in north-eastern hill region are smallholders and practise slash-burn cultivation for subsistence with low production. With harsh environmental conditions, mountainous terrain, regular natural catastrophes and irregular rainfall, farmers depend on rainfall patterns. Deforestation leads to landslides. Also farmers lack access to agricultural information such as pest and disease management or risk reduction strategies. Many people have access to a radio or fixed phone but none has computer or Internet. The e-Arik (e-agriculture) project (Saravanan 2011) was started to spread climate-smart agriculture practices for improving food security. A convergent extension approach is therefore required (Mukherjee et al. 2012; Mukherjee and Maity 2015).

Fig. 6.2 Effective management of pest or disease outbreak by promoting regional/localised planning and initiative/use of resistant cultivars

6.3 Adverse Effects of Climate Change

6.3.1 Plant Response to Drought

Due to increasing evaporation, transpiration and atmospheric holding capacity for water vapour as a result of elevated temperature, the moisture stress on plants is directly enhanced. The breakdown of organic matter in soil is speeded by higher temperatures leading to decrease of organic matter level, terminating into lower soil moisture retention and additional moisture stress being an indirect effect. Indication of drought varies with plant species and intensity of the water deficit. Both short-term and long-term effects of drought occur. Initial symptoms of drought are articulated as wilt, curl, roll, yellow, marginal scorch, interveinal necrosis, chlorosis and premature leaf drop, dieback, poor growth or stunting. During the initial stages of development, the plants are particularly susceptible to extreme weather events, strongly and detrimentally affecting crop yields (Victor 2011; Christine and Feller 2014).

6.3.2 Effect on Morphology and Physiology

High temperature, low relative humidity and scanty water affect physical and physiological conditions of plants. The bulk of roots and root system of crops are in the top soil layer, i.e. 12–15 in. of soil; hence the lack of water initially damages root system of plants. In the absence of optimum moisture and temperature, the feeder roots dry and the water-absorbing capability of plants declines drastically, which in turn effects transpiration, nutrient uptake by plants and gaseous exchange. Most plants lower their osmotic potential during drought by accumulating inorganic ions, amino acids, sugars, sugar alcohols as well as organic acids (Kramer 1983; Mattson and Haack 1987). The roots become corklike under water stress and high-temperature conditions; thus the uptake of mineral nutrient gets altered during drought (Kramer 1983). Several minerals, i.e. calcium, chlorine, potassium, magnesium, nitrogen, sodium, can accumulate to higher than normal levels in the above-ground tissues of drought-stressed plants (Bruce et al. 2002; Bhutta 2007).

Damage becomes more intense if drought occurs during periods of active plant growth when cell elongation, expansion of cell membrane, metabolic activities in cell protoplasm and thickening of cell wall demand high water availability. In water deficit condition, all the growth processes get arrested also stomata openings lower in frequency; thus CO_2 intake gets reduced; hence the rate of photosynthesis gets decreased. Low photosynthesis is responsible for reduced carbohydrate and secondary metabolite production, all these affect growth and maintenance of crops. Due to lesser or no production of secondary metabolites, viz. alkaloids, tannins, etc., by plants which are mainly responsible for plant self-defense mechanism against pests and pathogens, the crops become vulnerable for pest attacks (Dixon 2001).

6.3.3 Effect on Pathogen Scenario

The elevated temperature, changed nutrient status of drought-stressed plants may change the pathogen scenario. Under changing climate conditions, the physiological changes in host plants may lead to higher disease resistance, but it may be overcome more rapidly by faster disease cycles, with greater chances of pathogen developing to overcome the host plant resistance. Based on this knowledge, generalised postulations are drawn to develop drought strategies in order to manage the pests of forage crops, grasses and trees. Also some management strategies are also suggested to counteract the drought impacts on forage crops. Insect pest outbreaks may be affected sometimes by the drought which is well known to change the physiology of host plants which lead to changes in pests feeding on them (Sharma et al. 2010). Unusually cool, wet conditions can also lead to intense insect infestation even though excessive moisture drenches the soil-residing insects. Climatic factors affecting growth, spread and survival of crop diseases are temperature, precipitation, humidity, dew, radiation, wind speed, movement patterns and occurrence of extreme events. Humidity, elevated temperature and more precipitation favour spread of plant diseases, since moist vegetation favours germination of spores, and the abundant propagation of microbes such as fungi and bacteria influences the life cycle of soil nematodes (Garrett et al. 2006). It is known that in the arid regions, the disease infestation goes down; however, some diseases (e.g. powdery mildews) flourish in hot, dry conditions, as long as there is dew formation at night.

6.3.4 On Gene Expression of Plant

Under drought-like conditions, changes that occur at the level of molecules also (Jacobsen et al. 1986; Heikkila et al. 1984) have been recorded in case of barley and maize, the quantitative and qualitative changes in protein synthesis. It was found that in plants during drought condition, the ensuing high temperature induces production of heat-shock proteins (Walbot and Cullis 1985; Way et al. 2005) and found that genes up-regulated in coding for aldehyde dehydrogenase (associated with osmoregulation) and delta pyrroline-5 carboxylate synthetase (responsible for biosynthesis of proline and protects plant cell from dehydration). Also fatty acid alpha oxidases (for repair, permeability and fluidity of cell membrane) get affected by short-term and long-term drought stress. Up-regulated genes, concerned with cellular metabolism, cellular transport, signal transduction and transcriptional regulation as well as hydrophilic and heat-soluble proteins and downregulated genes (involved in cell wall, cellulose and germin-like protein synthesis, responsible for leaf growth, changes in leaf structures, stomatal closures and changes in root-shoot ratio), get affected by water stress (Bray 2004).

The climate change poses alarming challenge to the development of livestock sector in India. The predicted rise in temperature due to climate change is expected to intensify the heat stress in dairy animals; the predicted negative impact of climate change on Indian agriculture would also unfavourably affect cattle production by increasing the feed and fodder shortages. The livestock sector which would be major sufferer of climate change is itself a large source of methane emissions and greenhouse effect. Compared to developed countries in India, the emission rate per animal is much lower due to huge cattle population; the total annual methane emissions are about 9–10 Tg from enteric fermentation and animal wastes. It is known that the livestock production system is vulnerable to climate changes and at the same time donor to the phenomenon; climate change is capable of posing increasingly alarming challenge to the progress of livestock sector in India. In the economic advancement of the country, livestock sector has an important role since it contributes over one-fourth (~26 %) to the agricultural GDP and provides employment to 18 million people in principal or subsidiary status. Therefore, formulation of adequate adaptation and mitigation options for this sector is required in response to shift in climate cycle. Additionally it is also human intervention that is required to physically modify the environment and nutritional management practices to combat the negative effects of climate change on cattle production and health (Beede and Collier 1986; West 1999).

6.3.5 Coping with Climate Change

Several alleviation options are available for methane emissions from livestock (CAST 2004). In India, the likelihood of capturing or preventing emissions from animal manure storage is limited as it is widely used for fuel in the form of dung cakes. Thus improving rumen fermentation efficiency may lead to mitigate methane emissions. Nutritional technologies like diet manipulation, direct inhibitors, feed additives, propionate enhancers, methane oxidisers and probiotics may address (Moss 1994). Encouraging results with reduction potential of range 6–32 % have been observed in field experiments in India involving some of these options. Diet management by increasing green fodder reduced methane production by ~6 % (Singhal and Mohini 2002). Depending on the ratio of concentrate in diet, increase in concentrate reduced the methane by 15–32 % (Singh and Mohini 1999). The methane lessens ~9 % from molasses-urea supplementation (Srivastava and Garg 2002) and also ~21 % by using feed additive monensin (De and Singh 2000). The livestock development strategy should definitely focus on minimising potential production losses on the one hand and on the other increase the effort for methane mitigation from livestock sector since this would be crucial in escalating milk production by reducing energy loss from animals through methane emissions.

Since the developing countries of tropics and subtropics probably are going to suffer major damage, therefore, it is necessary not to waste time in developing and executing climate-resilient systems (Meena et al. 2015e, 2016; Verma et al. 2015a; b; Ghosh et al. 2016). Disruption in rainfall cycle and temperature regimes would affect the local water balance and upset the optimal cultivation period available for particular crops, hence throwing food and agriculture production out of gear. The farmers of dry land regions with rainfed agriculture, marginal condition and only one crop per year would face worst impact of climate change. Loss of multiple cropping zones is expected to be major blow to food production in South Asia. The maximum affected areas are anticipated to be the double and triple cropping zones (Meena et al. 2015c, d). To compensate most of this loss, an effort must be made to convert today's single cropping zones into double cropping zones. Efficient water harvesting and equitable management can address this at the foremost.

Careful management of resources like soil, water and biodiversity would be required to overcome the impact of climate change on agriculture. Agriculture should be made sustainable by production systems capable of making most of the environmental goods and services without damaging them. Susceptibility can be reduced by incorporating the climate change impacts directly into the design and execution of the development programmes. It is necessary to prepare the farmers who are today disoriented by the rapid fluctuations in weather conditions, affecting their agriculture, by large-scale literacy programmes. Farmers are unable to overcome these recent anthropogenic changes by their own conventional know-how (Fig. 6.2). The developing nations are facing considerable decrease in cereal production potential. If temperature shoots by 2 °C, India's rice production would definitely go down by almost a t/ha. By the year 2050, about half of India's major wheat production area could shrink due to heat stress, affecting productivity. Wheat yield losses in India, for every 1 °C rise, are probable to be ~7 MT year 1, or ~$1.5 billion at current prices (Sahai 2012).

6.4 Need for Initiatives at Different Levels

To cope with the impact of climate change on agriculture and livestock production, India will need to act at global, regional, national and local levels.

6.4.1 Global

Given that agriculture is the lifeline and bears the worst consequences of climate change, India must be firm and have serious discussions to make sure emission reduction. On the other hand, the raised temperatures would be favourable for the agriculture in temperate regions as the warmer environment would permit their single crop zones to become two even three crop zones along with paying for the adaptation simultaneously.

6.4.2 Regional

To protect the Himalayan ecosystems and minimise the melting of glaciers, there is need for regional cooperation at SAARC level. Urgent talks are necessary to maintain the flow in our major rivers like Ganga and Brahmaputra (Tibetan plateau) for supporting agriculture. Regional plans for tackling climate change across similar agroecologies would be beneficial for protection agriculture and ruminants of the region.

6.4.3 National

There are needs to start now as adaptation approaches have long lead times. Relevant policies and monetary support are mandatory. Various food, fodder and livelihood strategies are obligatory in rural areas to abate the risk.

6.4.4 Local

At the local level, the real action for mitigation and adaptation is required. The chase of sustainable agricultural development at the local level is integral for facing climate change effects. Localised technologies for less loss to food and nutrition have to be developed at the agroecological unit. Reducing emissions from agriculture would decrease the input costs for the farmers, and production systems would become more sustainable. Rapid and targeted coping strategies would have to meet out the real challenge to agriculture future of the country.

6.5 Adaptation Measures

Adaptation demands devising of methods to lessen the susceptibilities, strengthen resilience and build the adaptive capacities of farmers. Bringing sustainability in agricultural production systems in place of looking for maximising crop and cattle outputs will help farmers to handle and overcome the vagaries of climate change. The more diversified the agrosystems, the more efficient the network of insects and microbes that control pests and disease. Developing resilience in agrosystems and farming communities, improving adaptive capacity and decreasing GHG emissions are the methods to cope.

To prevent disaster caused by drought, the following measures are suggested (Fig. 6.2):

- Crop diversification
- Use of drought-tolerant crops
- Use of resistant cultivars
- Effective management of pest or disease outbreak by promoting regional/ localised planning and initiative

6.6 Importance of Forages

Indian agriculture is the single largest component of India's GDP (~26 %), largest employer (two-thirds of workforce); substantial share (~15 %) in total exports, ~ 60 % of net sown area (~142 M ha), is rainfed (FAO 2015). The whole rural economy revolves around cattle production. Since ages the ruminant livestock farming has been major part of Indian agricultural production system. The major cattle production systems are peri-urban, urban, rural area and transhumant where cattle, buffalo, sheep and goat are the chosen animals. Livestock production contributes ~7 % to GDP, and ~25 % of agricultural GDP is contributed by this sector.

Cattles are a source of employment and crucial livelihood for ~70 % in rural areas. Population dynamics and demand, predicted population by 2025, is 1400 million and urbanisation from 27.8 % to 58 %, and urbanisation brought a marked shift in lifestyle of people and food habits towards more milk products, with ensuing increase in demand for cattle products. However, at present there is 35 % and 64 % deficit in the fodder and feed, respectively. In animal feed and fodder supply, coarse grains and four major cereal crops, viz. maize barley, *Sorghum* and pearl millet, have a major role to play along with other forage genetic resources. The data/information on availability of forages varies widely in the country. Forage production as well as its utilisation is governed by various factors such as cropping pattern, climatic and socio-economic conditions, type of livestock, etc. More studies are needed to understand the vulnerability at local scale that helps in making appropriate policies to enhance the adaptation capacity of farmers not only in the interest of crop husbandry but also animal husbandry. Thus there is need for building strong genetic resource base of different forage crops.

The diversity at genetic level makes the species capable to adjust to changing environments and overcome the biotic and abiotic pressures. Agricultural biodiversity is essential for agroecosystem approach that favours and improves soil fertility, fosters high production and protects crops, cattle, fish and soil. Indian agrobiodiversity is dispersed in five agroclimatic zones, each with distinct agroecosystem, having unique gene pools and consists of landraces, primitive forms and wild relatives of different crops including forage species. For the development of Indian dairy and allied sectors, the forages which are considered to be orphan crops do play an important role. More efforts to explore, collect and introduce new crops from exotic sources, their evaluation, conservation and restoring indigenous resources would be useful in enriching forage gene pool of India. Prioritised utilisation in national crop improvement programmes of forages and potential wild and weedy relatives have been chalked out (Sirohi and Michaelowa 2007).

6.7 Germplasm Introduction in Forage/Range Plants: Search for Potential Forage/Fodder Crop Across the Globe

India is a country having large amount of biodiversity in forage crops, thanks to its geographical position and its agroclimatic conditions. Its cultural diversity also plays a significant role in enriching its diversity by introducing new crops in India. New species which were introduced in India and performing well in specific agroclimatic conditions having potential value have resulted in identification of several promising types which can adapt themselves to the harsh environmental and degraded soil conditions, give economic yield in different agroclimatic regions and thereby ensure feed and nutritional quality and provide additional income to the resource poor farmers of remote, backwards, tribal, hilly and other difficult areas of the country (Singh et al. 2009). National Bureau of Plant Genetic Resources (NBPGR), New Delhi, is an ideal organisation engaged in the various activities related to enrichment of plant genetic resources in the country. Introducing new cultivated species and better improved varieties of forages and range is required to strengthen the forage and range improvement programme (Arora et al. 1975). Wild and weedy relatives of these plants play a vital role in improvement because they represent a part of crop gene pool particularly resistant to biotic and biotic stresses and have been the donors of many other useful traits. Acquisition of more exotic germplasm is the priority; the flow of plant genetic resources from everywhere to India is slowing down gradually. Secondly, we are not getting the trait-specific material as in the past under this present circumstance; the exiting genetic resources have to be judiciously evaluated to plan and conduct crop improvement programme for current and future requirements.

6.7.1 Sustainable Uses of Forage Genetic Resources

The systematic work on the collection, evaluation, documentation and conservation of germplasm of forage species, together with wild and weedy taxa, was paid grave attention in the last few decades. Initially, the activities related to germplasm resources in forage plants started with the collection and evaluation of the local ecotypes of selected species by state departments of agriculture/agricultural colleges of the State Agricultural Universities in the states. We need to adapt modern technology to give us efficient evaluation techniques. The NBPGR with the help of other ICAR research organisations, State and Central Agricultural Universities, State Department of Agriculture, other autonomous body and non-governmental organisation (NGO), etc. is dedicated to save and conserve the biodiversity of forages (Table 6.1).

The pioneer institute of grassland and fodder in India, i.e. Indian Grassland and Fodder Research Institute, has the mandate of collection, evaluation, characterisation, documentation and conservation of forage genetic resources. The major activities related to genetic resources of forages at IGFRI, Jhansi include the Conservation in Mid Term Storage Module of the Institute at Jhansi. The

Table 6.1 Current status of forage germplasm at National Genebank NBPGR, New Delhi

	Crop group	Source	Current status		Remarks
			Species	Accessions	
Germplasm holding	All crops	–	1584	3,96,189	–
	Millets and forages	–	178	56,472	–
Germplasm exchange division introductions	Forages	From Australia, Italy, USA, UK, Brazil, Germany, Egypt, Ethiopia, Bulgaria, Philippines, Singapore, Costa Rica, Zimbabwe, Japan, Russia and New Zealand	–	13,181	The introduced germplasm was supplied to IGFRI and AICRP on forage crops located at Jhansi, UP
Long-term storage	Forage crops	–	206	5594 IC No.s	National identity number allotted at NBPGR, most of these are from IGFRI, Jhansi

Source: PGR Portal NBPGR, New Delhi

rationale of germplasm conservation is maintaining integrity over prolonged time period. IGFRI being a National Active Germplasm Site (NAGS) for forage crops presently holds >9,232 accessions representing 53 forage crops (Table 6.2) in Medium Term Storage (MTS) module.

These germplasm accessions have been characterised for morphological and agronomic traits, screening against biotic as well as abiotic stresses, and the observations for fodder yield and quality have been carried out. With publication of fifteen (15) germplasm catalogues and two descriptors of forage crops, i.e. deenanath grass, berseem, teosinte, siratro, cowpea, Guinea grass, *Cenchrus*, forage maize, oat, cluster bean, pearl millet, Napier, white clover, *Stylosanthes*, forage sorghum and descriptors, were developed in Egyptian clover (*Trifolium alexandrium*) and *Dichanthium-Bothriochloa* complex. In addition there have been registrations of novel genetic stocks (22), and there has been development of core collections in *Sorghum* and *Cenchrus ciliaris*.

The vast diversity of forage crops provides support to cope up the climate change. However, in spite of the fact that these resources are endangered, the efforts to conserve them are still inadequate. Recently at the meeting of National Advisory Board for Management of Genetic resources NBPGR, New Delhi, there were detailed discussions with regard to characterisation of the available germplasm along with protecting the collections and modernisation of MTS; modified MTA was discussed for clarity for public and private sectors.

Table 6.2 Present germplasm status at MTS module IGFRI, Jhansi

Crop	Scientific names
Cereal fodder crop: 2978 accessions	
Jowar, maize, bajra, oat, barley	*Sorghum bicolor, Zea mays, Pennisetum glaucum, Avena sativa, Hordeum vulgare*
Cultivated legumes: 2545 accessions	
Guar, *Lablab*, lucerne, berseem, cowpea	*Cyamopsis tetragonoloba, Lablab purpureus, Medicago* spp, *Trifolium* species, *Vigna* spp
Range legumes: 713 Accessions	
Stylosanthes, Subabul, Desmanthus, Clitoria, Lathyrus, wild *Vigna,Vigna*, horse gram, sword bean, jack bean, jointvetch, wild groundnut, gambhari, *Centrosema*, wild pea, *Siratro, Neotonia, Rhynchosia*, red clover, sainfoin	*Stylosanthes* spp, *Leucaena leucocephala, Desmanthus virgatus, Clitoria ternatea, Lathyrus sativus, Vigna vexillata,Vigna* spp, *Macrotyloma uniflorum, Canavalia gladiata, Canavalia virosa, Aeschynomene* spp, *Arachis* spp,*Calotropis ensiformis, Centrosema* spp, *Pisum* spp, *Macroptilium* spp, *Neotonia* spp, *Rhynchosia* spp, *Zornia* spp, *Desmodium* spp, *Trifolium repens,Trifolium pratense, Onobrychis viciifolia*
Range grasses: 2996 accessions	
Buffel grass, *Heteropogon*, marvel grass, *Chrysopogon*, Rhodes grass, Guinea, *Sehima, Pennisetum*, sewan grass, tall fescue, orchard grass, prairie grass,Timothy, Makhanmalai grass, *Themeda, Bracharia*, Johnson grass, *Setaria, Iselema, Vetiveria*	*Cenchrus* spp, *Heteropogon* spp, *Dichanthium annulatum,Chrysopogon fulvus, Chloris gayana, Panicum maximum, Sehima nervosum, Pennisetum* spp, *Lasiurus sindicus, Festuca arundinacea, Dactylis glomerata, Bromus unioloides, Phleum pratense, Lolium perenne, Themeda arundinacea, Brachiaria* spp, *Sorghum halepense, Setaria* spp, *Iselema* spp, *Vetiveria* spp

Source: IGFRI Jhansi website

With the CRP (Consortium Research Programme) on agrobiodiversity proposed in XIIth plan approved in principle, the major objectives are regeneration of all accessions for sufficient seed for LTS and maintenance of active collection at National Active Germplasm Site (NAGS).

Mutual give and take of germplasm or PGR can be described as germplasm exchange; PGR may be a seed, a plant or plant part that is useful in crop breeding, research or conservation because of its genetic attributes (Dhillon and Agrawal 2004; Hamilton et al. 2005). The word genetic resources was used for the first time in 1967 in the second technological meeting on plant exploration and conservation co-hosted by FAO and International Biological Programme to gauge the threat of genetic loss in diversity of crop plants and to delineate a global strategy for PGR conservation. PGR exchange offers immense prospects for better economic growth and potential stability. There is a continuous need of genetic resources for developing varieties resistant to various abiotic and biotic stresses and to improve the quality and quantity traits. Acquisition of diverse and superior germplasm and their conservation is an important concern (Paroda et al. 1986; Swaminathan and Jana

1992; Chandel 1996; Singh et al. 2003, 2004, 2008; Tyagi et al. 2008). Due to advent of IPR regime and Convention on Biological Diversity (CBD), a paradigm shift has been noticed in free flow of genetic resources to a restricted exchange and slow down in sharing of germplasm; there is need for judicious evaluation of existing genetic resources for crop improvement for our current and future requirements as well (Brahmi et al. 2005; Tyagi et al. 2008).

6.7.2 Fodder Crop Varieties

Varietal development of fodder crops, research work on forages by Pathak and Roy 1995; Pandey and Roy 2011, got the required momentum after establishment of Indian Grassland and Fodder Research Institute (IGFRI) Jhansi, in 1962 by ICAR. Later, in 1970, All-India Co-ordinated Research Project on Forage Crops was started, which presently has 21 coordinated centres and cooperating testing locations all over the country for multilocational research and testing, and several high-yielding good-quality forage varieties have been developed and released. Subsequently relevant programmes on grasses and legumes were created at IGFRI, Jhansi; CAZRI, Jodhpur; HAU, Hisar; GBPUAT, Pantnagar; TNAU, Coimbatore; PAU, Ludhiana; and GAU, Anand. In addition to this research, work on forage aspects was intensified in coordinated projects on *Sorghum*, maize and pearl millet. Impact of germplasm utilisation has been that the variability by different exploration trips and correspondence with national and international institutes has been utilised to develop superior varieties through direct selection or through utilisation in hybridisation programme. A list of some promising forage varieties is given (Tables 6.3 and 6.4).

Table 6.3 Fodder crop varieties developed at IGFRI, Jhansi

Crop	Varieties	GFY (t/ha)	Areas for cultivation
Oats	Bundel Jai 822	44–50	Central zone
	Bundel Jai 851	40–50	Whole country
	Bundel Jai 2001–2003	40–50	South and Northwest India
	Bundel Jai 2004	40–50	All India except central zone
	Bundel Jai 991	35–40	Hill zone
Cenchrus ciliaris	Bundel Anjan 1	30–35	Whole country
	Bundel Anjan 3	30–35	Whole country

(continued)

Table 6.3 (continued)

Crop	Varieties	GFY (t/ha)	Areas for cultivation
Dinanath grass	Bundel Dinanath 1	55–60	Whole country
	Bundel Dinanath 2	60–65	Whole country
NBHybrid	Swetika	120–160	Central, NE and North India
	DHN-6	100	North Karnataka
Bajra	AVKB-19	50–60	Whole country
	JHPM-05-2	70–80	Whole country except south zone
Guinea grass	Bundel Guinea 1	40–50	Punjab, Himachal Pradesh, Central Uttar Pradesh, Maharashtra, Tamil Nadu
	Bundel Guinea 2	50–55	Rainfed semiarid, tropical, subtropical and humid tropics
	Bundel Guinea 4	75–81	All Guinea grass-growing areas
Sehima nervosum	Bundel Sen Ghas-1	18–20	Semiarid tropical and subtropical areas across the country
Chrysopogon fulvus	Bundel Dhawalu Ghas-1	26–30	Rangelands under rainfed condition across the country
Heteropogon contortus	IGHC-03-4	25–30	Rangelands under rainfed condition across the country
Berseem	Wardan	65–70	Whole country
	Bundel berseem 2	65–80	Central, NW zone
	Bundel berseem 3	65–80	NE Zone
Lucerne	Chetak	45–50	Lucerne growing areas
Cowpea	Bundel Lobia-1	25–30	Whole country
	Bundel Lobia-2	25–30	North zone
	Bundel Lobia-4	23–26	Northeastern zone
Guar	Bundel Guar 1	25–35	Whole country
	Bundel Guar 2	30–40	Whole country
	Bundel Guar 3	30–40	Whole country
Field bean	Bundel Sem 1	25–35	Whole country

Source: IGFRI Jhansi website

Table 6.4 Recent and other fodder crop varieties developed

Crop	Variety	Centres
Cereal fodder crops		
Sorghum	Gujarat Forage Sorghum-5 (GFS-5)	Surat
	Gujarat Forage Sorghum (AS-16)	Anand
	Gujarat Forage Sorghum Hybrid-1 (GFSH-1)	Gujarat
	CSV 30F	Rahuri
Oats	OS-377	Hisar
	SKO-90	Srinagar
	SKO-96	Srinagar
	JO-2000-61	Jabalpur
	Narendra Jayee-10 (NDO-10)	Faizabad
	JHO-2009-1	Jhansi
	JHO-2010-1	
Pearl millet	AFB-3	Anand
	RBB-1	Bikaner
	JHPM-05-2	Jhansi
	Avika Bajra Chari-19	Jhansi
	NDFB-1	Faizabad
	NDFB-2	Faizabad
	NDFB-3	Faizabad
	NDFB-5	Faizabad
	NDFB-11	Faizabad
	BAIF-Bajra-1	Urulikanchan
	PAC-981 (Nutrifeed)	Advanta Private Ltd.
	APFB-09-1	Hyderabad
Grasses		
Bajra Napier hybrid	CO (BN)-5	Coimbatore
	CO (CN)-4	Coimbatore
	NB-21	Ludhiana
	RBN-13	Maharashtra
	BNH-10	Urulikanchan
Guinea grass	JHGG-04-1	Jhansi
	JHGG-08-1	Jhansi
	DGG-1	Jhansi
Anjan grass/buffel grass	GAAG-1	Anand
	RCCB-2	Bikaner
Marvel grass	Marvel-93	Rahuri
	Phule Marvel-06-40	Rahuri
	Phule Marvel Phule	Rahuri
Setaria grass	S-18	Palampur
Sewan grass	CAZRI-30-5	Jodhpur
	RLSB-11-50	Bikaner

(continued)

Table 6.4 (continued)

Crop	Variety	Centres
Tall fescue grass	Hima-14	Palampur
	Hima-4	Palampur
	OS-377	Hisar
	OS-403	
	SKO-90	Srinagar
	SKO-96	Srinagar
	JO-2000-61	Jabalpur
	Narendra Jayee-10 (NDO-10)	Faizabad
Leguminous fodder crops		
Cowpea	UPC-628	Pantnagar
	UPC-621	Pantnagar
	UPO-06-2	Pantnagar
	MFC-08-14	Mandya
	MFC-09-1	Mandya
	HCP-46	Hisar
	Cowpea-74	Ludhiana
	Fodder cowpea-CO (FC)-8	Coimbatore
Guar	HFG-156	Hisar
	HG-884	Hisar
	HG-2-20	Hisar
Rice bean	KRB-19	Kalyani
	RBL-1	Ludhiana
Stylosanthes	*Stylosanthes scabra*	Rahuri
Berseem	HB-1	Hisar
	HB-2	Hisar
Lucerne	L-9	Ludhiana
	Anand Lucerne-4	Anand
	Anand-23	
	Sirsa Type-9	Punjab, Haryana
	Sirsa-8	Sirsa
	Type-9	Sirsa
	NDRI Selection No.1	Karnal
	RRB-07-1	Bikaner
	Lucerne CO-2	Coimbatore
Senji/sweet clover	FOS-1	Hisar
	PC-5	Ludhiana
Metha/Fenugreek	T-8	Anand
Sem (*Lablab* bean)	Bundel Sem-1 (JLP-4)	Jhansi
Gobhi Sarson	Him Sarson-1	Palampur
	Neelam	

Source: information based on recommendations of 68th to 72nd meetings of Central subcommittee on crop standards, notification and release of varieties for agricultural crops, http://seednet.gov.in and www.iasri.res.in/agridata

6.8 Contingent Crops: Fodder Cereals, Legumes and Grasses

With reference to the changing climate, drought-tolerant forages such as *Brachiaria* spp. can provide year-round fodder supply when established in addition to Napier grass (*Pennisetum purpureum*) 10–25 t/ha/year. These can be easily established in the farm and field bunds for continuous supply of fodder. Maize stover is a source of feed during the dry season, but its crude protein is too low ($\leq$4.0 %) to sustain milk production. Intercropping maize with *Lablab purpureus,* cowpea improves maize stover yield (4,375 kg/ha), protein content (7.6 %) and grain yield compared to the maize monocrop. Legumes provide a soil cover, conserve soil moisture, control soil erosion and suppress weeds in addition to providing a source of protein. Forage legume supplements increase dry matter intake and improve rumen function and laxative influence on the alimentary system. Supplementing dairy cows feed with fodder trees and green forage improves milk yield.

Fodder trees contain 20–25 % crude protein compared to grasses (<10 %) CP. Lactating cows consuming average daily amount of 3 kg of fresh tree fodder over a period of 10 months had an average daily weight gain of 0.55 kg while exhibiting no apparent health problems. *Setaria* grass variety S-18 of CSKHPKVV, Palampur, is resistant to cold, drought and frost and recommended for Himachal Pradesh and Uttarakhand. Hybrid-3 Napier (variety Swetika-1) is suitable for low pH conditions in Andhra Pradesh, Karnataka, Kerala, Himachal Pradesh and Assam under irrigated conditions. Tall fescue variety EC-178182 of CSKHPKVV, Palampur, is tolerant to drought, acidic and alkaline soils and recommended for subtemperate and temperate grasslands and pastures of hill zone of the country. Tall fescue new variety Hima-14 of CSKHPKVV, Palampur, has good quality and no problem of diseases and pest, broad and soft leaves, high-tillering and quick regeneration and is recommended for hill zone of Himanchal Pradesh, Uttarakhand and Jammu and Kashmir.

Many potential legumes for establishment in pastures, viz. siratro, Stylo, *Dolichos*, etc., have also been introduced along with the germplasm of other forage species. Researchers (Hand book of Agriculture) have also enabled to work out a range of cropping systems which can provide green fodder throughout the year with high yield (about 150–250 t/ha) and better quality in diverse agroclimatic conditions. Forage production technology will help significantly in improving animal health as well as milk production, making agriculture resilient to extreme weather events.

6.9 Way Forward

6.9.1 Strategic Research for the Future

Some priority areas for technical and financial investments for climate change are:

- For the use in breeding of new varieties, there should be evaluation of traditional varieties for valuable traits like tolerance to higher temperatures, salinity and drought, feed conversion efficiency as well as disease resistance.
- Official and participatory plant breeding to generate climate-resilient crop varieties that can withstand high temperatures, drought and salinity.
- Evolving short-duration varieties that can mature before peak heat phase arrives.
- In order to combat the yield losses due to heat during growth period, the selection of genotypes of crops with more per day yield potential be encouraged.
- For increasing milk yielding capacity of indigenous cattle with lessened methane emissions, development of balanced feed and fodder ration.

6.9.2 Some Specific Recommendations

Grassland is a vital ecological system type which has agricultural importance as well as can cover other less useful sites like very steep locations and above timberlines. In addition to evident focus required on water management and conservation and pest management, agriculture and livestock production as such requires to become sustainable and ecologically sound to adjust to the climate change turmoil.

- A special package for adaptation needs to be developed including crops, cattle, poultry and agroforestry; farmhouses and homestead gardens supported by nurseries should be encouraged to make up deficits from climate-related yield losses. To reduce the risk, production model should be diversified.
- Conservation of genetic diversity of various crops along with accompanying knowledge, as enterprise with locals, must get the highest priority.
- The crisis of seed availability must be attended by decentralised seed production programmes incorporating local communities. There should be production and stocking of seeds of main crops and contingency crops (in case of delayed, failed monsoon or floods) including fodder and green manure plants which are specific to the agro-ecological unit.
- In order to keep track of changes in pest and disease profile and to anticipate the new pest and disease outbreaks, there should be early warning systems.
- There should be promotion of balanced feed mixtures capable of enhancing milk production and alleviating methane emission.

Farmers should be given special insurance to cover climate change. The information regarding rainfall and weather in real time should reach to farmers through mobile telephones from Gyan Chaupals and village resource centres with satellite connectivity with data from the government's agrometerological services.

6.10 Conclusions

Climate change demands fodder production for harnessing the full potential of both cultivated and noncultivated lands. The achievement of any crop improvement programme is related with the availability of enough genetic variability available in traditional usable form. The present genetic variability in any crop germplasm conserved in genebanks for present and future use can be categorised into three groups: cultivated type, cross-compatible wild type and cross-incompatible wild type. In cultivated type, the genetic variability exists in poor agronomic background or in genetic background not adjusted to breeding or target climate for the straight away use in conventional breeding programmes. The usage of genetic variability in wild species for cultivar development is hampered primarily by linkage drag and various incompatibility barriers between cultivated and wild species. In such situations, pre-breeding proves to be a unique tool to increase the use of genetic variability present in cultivated as well as wild-type germplasm.

Pre-breeding encompasses most of the activities related with recognition of desirable traits and/or genes from unadjusted germplasm (donor) which cannot be directly used in breeding populations (exotic/wild species) and to transfer these traits into well-adapted genetic backgrounds (recipients) resulting in generation of an intermediate set of material which can be used readily by the plant breeders in specific breeding programmes to create new varieties with a broad genetic base; there is need for identification of trait-specific germplasm accessions. Breeding varieties of dual-purpose fodder and cereal crops of shorter duration will help in addressing the demand of food as well as fodder, and in the present-changing climatic situations, the grasslands and pastures have a vital role to play in maintaining the progress of animal husbandry which is an essential part of Indian agriculture. To achieve this goal, extensive efforts have been made at national level to advance the herbage quantitatively and qualitatively by growing potential and improved grass/legume species on marginal, submarginal and degraded habitats. In response to prevailing challenge of climate change, more access to range of forage varieties which can help farmers tackle drought or flood will be required. Thus there is need to develop an action plan for the management of forage genetic resources in accordance with recent vision document for the benefit of our client, i.e. farmers. The focus of research and development should be for studying to increase usage of available genetic resources through subset approaches, pre-breeding to boost the utilisation of genetic resources in various forage crop improvement programmes, further enrichment of forage crop genetic diversity by exploration and correspondence and maintenance of existing genetic diversity by ex situ rejuvenation and midterm storage. There are opportunities to incorporate forage crops in the existing cropping systems: more government focus on rehabilitation of degraded grasslands on environmental concerns and probabilities of including range grasses and legumes in degraded forest, forest management projects, watershed and other plantations. Therefore, efficient conservation and sustainable use of forage genetic diversity plays a significant role, being the basis of food and nutrition security against the background of deteriorating natural resources along with altering climate scenario.

Acknowledgements Authors are thankful to the Head Crop Improvement Division, Director IGFRI, Jhansi, for providing all the necessary help and facilities. Authors are also thankful to the editors, ICAR-Vivekananda Institute of Hill Agriculture Almora, for providing the opportunity to compile this book chapter.

References

Arora RK, Mehra KL, Hardas MW (1975) The Indian gene centre: prospects for exploration and collection of herbage grasses. Forage Res 1:11–12

Beede DK, Collier RJ (1986) Potential nutritional strategies for intensively managed cattle during thermal stress. J Anim Sci 62:543–554

Bhutta WM (2007) The effect of cultivar on the variation of spring wheat grain quality under drought conditions. Cereal Res Commun 35:1609–1619

Brahmi P, Khetarpal RK, Dhillon BS (2005) Access to plant genetic resources: the changing scenario In: Biodiversity: status and prospects. Narosa Publication, New Delhi, pp 124–133

Bray EA (2004) Genes commonly regulated by water deficit stress in *Arabidopsis thaliana*. J Exp Biol 55:2331–2341

Bruce WB, Edmeades GO, Barker TC (2002) Molecular and physiological approaches to maize improvement for drought tolerance. J Exp Bot 53:13–25

Burney JA, Davis SJ, Lobell DB (2010) Greenhouse gas mitigation by agricultural intensification. Proc Natl Acad Sci 107:12052–12057

CAST (2004) Climate change and greenhouse gas mitigation: challenges and opportunities for agriculture. Task Force Report 141, Council for Agricultural Sciences and Technology, USA. p 120

Chandel KPS (1996) Hindustani centre provides India and the world a trove of genetic diversity. Diversity 12:21

Christensen JH, Kumar KK, Aldrian E, An SI, Cavalcanti IFA, Castro M, Dong W, Goswami P, Hall A Kanyanga JK, Kitoh A, Kossin J, Lau NC, Renwick J, Stephenson DB, Xie SP, Zhou T, (2013) Climate phenomena and their relevance for future regional climate change. The physical science basis. contribution of working group I to the fifth assessment report of the Intergovernmental Panel on Climate Change. Cambridge, UK/New York,pp 1217–1308

Christine B, Feller URS (2014) Impacts of drought stress on water relations and carbon assimilation in grassland species at different altitudes. Agr Ecosyst Environ 188(15):212–220

Datta D (2013) Indian fodder management towards 2030: a case of vision or myopia. IJMSSR 2:2, ISSN: 2319-4421

De D, Singh GP (2000) Monensin enriches UMMP supplementation on in vitro methane production in crossbred calves. In: Proceedings of the Xth animal nutritional conference, Karnal, India, Animal Nutrition Society of India, NDRI. p 161

Dhillon BS, Agrawal A (2004) Plant genetic resources: ownership, access and intellectual property rights. In: Plant genetic resources: oilseed and cash crops. Narosa Publication, New Delhi. pp 1–20

Dixon RA (2001) Natural products and plant disease resistance. Nature 411:843–847
FAO (2015) Statistical pocketbook. World Food and Agriculture, Rome, pp 44–122
Garrett KA, Dendy SP, Frank EE, Rouse MN, Travers SE (2006) Climate change effects on plant disease: genomes to ecosystems. Ann Rev Phytopathol 44:489–509
Ghosh BN, Meena VS, Alam NM, Dogra P, Bhattacharyya R, Sharma NK, Mishra PK (2016) Impact of conservation practices on soil aggregation and the carbon management index after seven years of maize–wheat cropping system in the Indian Himalayas. Agr Ecosyst Environ 216:247–257
Hamilton S, Rudraidh, Javier E (2005) Intellectual property rights and germplasm exchange: the new rules. In: Planning rice breeding programs for IMPACT
Heikkila JJ, Papp JET, Schultz GA, Bewley JD (1984) Induction of heat shock protein messenger RNA in maize mesocotyls by water stress, abscisic acid, and wounding. Plant Physiol 76:270–274
Hoffmann U (2013) Agriculture section B: – a key driver and a major victim of global warming. pp 3–5
IPCC (2007) Fourth assessment peport. Intergovernmental Panel on Climate Change Secretariat. Geneva, Switzerland
IPCC (2014) Climate change 2014: impacts, adaptation, and vulnerability: working group II contribution to the fifth assessment report. Cambridge University Press, Cambridge, UK/New York
Jacobsen JV, Hanson AO, Chanyores PC (1986) Water stress enhances expression of amylase gene in barley leaves. Plant Physiol 80:350–359
Kramer PJ (1983) Water relations of plants. Academic, Orlando, p 489
Kumar A, Bahadur I, Maurya BR, Raghuwanshi R, Meena VS, Singh DK, Dixit J (2015) Does a plant growth promoting rhizobacteria enhance agricultural sustainability? J Pure Appl Microbiol 9(1):715–724
Lobell DB, Burke MB, Tebaldi C, Mastrandrea MD, Falcon WP, Naylor RL (2008) Prioritizing climate change adaptation needs for food security in 2030. Science 319(5863):607–610
Mattson WJ, Haack RA (1987) The role of drought in outbreaks of plant-eating insects. Bioscience 37:110–118
Meena RS, Yadav RS, Meena VS (2013) Heat unit efficiency of groundnut varieties in scattered planting with various fertility levels. Bioscan 8(4):1189–1192
Meena RS, Meena VS, Meena SK, Verma JP (2015a) The needs of healthy soils for a healthy world. J Clean Prod 102:560–561
Meena RS, Meena VS, Meena SK, Verma JP (2015b) Towards the plant stress mitigate the agricultural productivity: a book review. J Clean Prod 102:552–553
Meena VS, Meena SK, Verma JP, Meena RS, Ghosh BN (2015c) The needs of nutrient use efficiency for sustainable agriculture. J Clean Prod 102:562–563
Meena VS, Meena SK, Verma JP, Meena RS, Jat LK (2015d) Current trend to mitigate climate change and sustainable agriculture: a book review. J Clean Prod 102:548–549
Meena VS, Verma JP, Meena SK (2015e) Towards the current scenario of nutrient use efficiency in crop species. J Clean Prod 102:556–557
Meena RS, Bohra JS, Singh SP, Meena VS, Verma JP, Verma SK, Sihag SK (2016) Towards the prime response of manure to enhance nutrient use efficiency and soil sustainability a current need: a book review. J Clean Prod 112(1):1258–1260
Moss AR (1994) Methane production by ruminants – literature review of I. Dietary manipulation to reduce methane production and II. Laboratory procedures for estimating methane potential of diets. Nutr Abstr Rev Ser B 64:786–806
Mukherjee A, Maity A (2015) Public–private partnership for convergence of extension services in Indian agriculture. Curr Sci 109(9):1557–1563

Mukherjee A, Bahal R, Roy Burman R, Dubey SK (2012) Conceptual convergence of pluralistic extension at Aligarh District of Uttar Pradesh. J Community Mobilization Sustain Dev 7 (1&2):85–94
Pandey KC, Roy AK (2011) Forage crops varieties. IGFRI, Jhansi, pp 1–88
Paroda RS, Arora RK, Chandel KPS (1986) Plant genetic resources: Indian perspective. NBPGR, New Delhi
Pathak PS, Roy MM (1995) Agrosilvipastoral farming systems for optimizing forage and energy resources in rainfed areas. In: Singh RP (ed) Forage production and utilization. IGFRI, Jhansi, pp 154–178
Suman S (2012) Coping with climate change. LEISA, India, p 14, 2
Saravanan R (2011) Case study: e-Arik: using ICTs to facilitate "climate-smart agriculture" among tribal farmers of North-East India
Sharma HC, Srivastava CP, Durairaj C, Gowda CLL (2010) Pest management in grain legumes and climate change. In: Climate change and management of cool season grain legume crops. Business Media, Springer Science, Dordrecht. pp 115–140
Shiva V, Bhatt VK (2013) Climate change at the third pole – the impact of climate instability on Himalayan ecosystems and Himalayan communities. Ann Neurosci 17:223
Singh GP, Mohini M (1999) Effect of different levels of monensin in diet on rumen fermentation, nutrient digestibility and methane production in cattle. Asian Aust J Anim Sci 12 (8):1215–1221
Singh RV, Chand D, Tyagi V, Singh AK, Singh SP, Singh S (2003) Important crop germplasm introduced into India. Indian J Plant Genet Resour 16:214–221
Singh RV, Singh RV, Chand D, Tyagi V, Singh AK, Singh SP, Singh S (2004) Important crop germplasm introduced into India. Indian J Plant Genet Resour 17:196–205
Singh RV, Chand D, Tyagi V, Singh AK, Singh SP, Singh S (2008) Important crop germplasm introduced into India during 2006. Indian J Plant Genet Resour 20:64–69
Singh AK, Varma N, Yadav SK, Mohanty A, Singh S, Singh S (2009) Indian forage genetic resources: perspectives and strategies. Progress Agric 9(2):250–256
Singh DK, Singh SK, Singh AK, Meena VS (2014) Impact of long term cultivation of lemon grass (Cymbopogon citratus) on post-harvest electro-chemical properties of soil. Annals Agri Bio Res 19(1):45–48
Singhal KK, Mohini M (2002) Uncertainty reduction in methane and nitrous oxide gases emission from livestock in India. Project report, Dairy Cattle Nutrition Division, NDRI, Karnal, India. p 62
Sirohi S, Michaelowa A (2007) Sufferer and cause: Indian livestock and climate change. Published online: 17 March 2007 # Springer Science + Business Media Climatic Change. 85: 285–298
Srivastava AK, Garg MR (2002) Use of sulfur hexafluoride tracer technique for measurement of methane emission from ruminants. Ind J Dairy Sci 55(1):36–39
Swaminathan MS, Jana S (1992) Biodiversity: implications for global food security. Macmillan, Madras
Tyagi V, Singh AK, Chand D, Singh RV, Dhillon BS (2008) Plant introduction in India during pre and post- CBD periods. An analysis. Indian J Plant Genet Resour 19:436–441
Verma JP, Jaiswa DK, Meena VS, Meena RS (2015a) Current need of organic farming for enhancing sustainable agriculture. J Clean Prod 102:545–547
Verma JP, Jaiswal DK, Meena VS, Kumar A, Meena RS (2015b) Issues and challenges about sustainable agriculture production for management of natural resources to sustain soil fertility and health. J Clean Prod 107:793–794
Victor DG (2011) Global warming gridlock: creating more effective strategies for protecting the planet. Cambridge University Press, Cambridge, UK
Walbot V, Cullis CA (1985) Rapid genomic changes in higher plants. Annu Rev Plant Physiol 36:367–396

Way H, Chapman S, McIntyre L, Casu R, Xue GP (2005) Identification of differentially expressed genes in wheat undergoing gradual water deficit stress using a subtractive hybridization approach. Plant Sci 68:661–670

West JW (1999) Nutritional strategies for managing the heat stressed dairy cow. J Anim Sci 77 (Suppl 2/ J):21–35

Toward Climate Change and Community-Based Adaptation-Mitigation Strategies in Hill Agriculture

7

Renu Jethi, Kushagra Joshi, and Nirmal Chandra

Abstract

Climate change is being experienced by the hill community in the form of rising temperatures, extreme climatic events, changing rainfall pattern, and frequently occurring drought-like situations. In hills of Uttarakhand, agriculture and animal husbandry are the main occupations, and therefore, food and nutritional security much more depends on favorable climatic conditions. Crop productivity in the hilly areas is getting reduced because of low soil fertility and higher pest and disease infestations. The persistent changes in weather conditions have resulted in overall decrease in the water level in almost all the water sources of the area. The area under irrigation is decreasing gradually due to drying of most of the water bodies for irrigation resulting in low productivity of crops. Rising temperature has led to the shift in forest biodiversity. Pine trees have replaced broad leaf tree species which is posing a major threat to fodder availability and have resulted in increased forest fire incidents. People in hill region have learned to live and survive with risks for thousands of years, but the present rate of climate change is very rapid which demands much attention for the socioeconomic concerns in the area. Women in hills are involved in maintaining and promoting agricultural genetic diversity by selecting, conserving, and propagating seeds. Participatory adaptation strategies with women involved at every stage must be framed as they have rich traditional knowledge and experience with respect to crop adaptation to enhance food security. Drought-resistant crop varieties should be promoted to address drought-like situations and instances of less rain during cropping seasons. Dual-purpose crop varieties which can be grown for grain as well as for fodder could save a lot of time and drudgery of women involved in carrying back-breaking load up and down in hilly terrains. Developmental policies and strategies should be supportive to enhance access and control of

R. Jethi (✉) • K. Joshi • N. Chandra
ICAR-Vivekananda Institute of Hill Agriculture, Almora 263 601, Uttarakhand, India
e-mail: renujethi2009@gmail.com

J.K. Bisht et al. (eds.), *Conservation Agriculture*,
DOI 10.1007/978-981-10-2558-7_7

men and women over natural resources in order to conserve natural resources, alleviate poverty, and ensure that vulnerable groups including women are enabled to cope with the climate change.

Keywords
Climate change • Adaptation • Mitigation • Himalayas

7.1 Introduction

The Indian Himalayas have a critical influence over the climate. It plays an important role in moderating the monsoon and region's rainfall pattern. The perennial snows and vast glaciers are the water towers feeding innumerable rivers, streams, rivulets, and springs throughout the entirety of northern India. The Himalayas are among the youngest mountains in the world and are acutely fragile (Sorkhabi 2007), most threatened ecosystems in the world and susceptible to disaster, including flash floods, cloud bursts, melting snow in the mountains, and droughts and flood in the Tarai (Aryal 2015). Hills and mountains in the northwestern Himalayan (NWH) region are experiencing the consequences of climate change more than any other region (Anthwal et al. 2006); it acts as an indicator of climate change. Here, climate change impacts people ecologically, socially, economically, and also culturally. The climate change is a real threat as it affects water resources including freshwater bodies, agriculture, vegetation and forest, snow cover, and ecological processes such as landslides and floods and has long-term effect on food security and human health (Schmidubher and Tubeillio 2007). Impact of climate change is more pronounced in ecologically fragile mountain areas such as the Himalayas where rapid altitudinal change results in high degree of variation in relief, temperature, rainfall, natural vegetation, water regimes, and other associated phenomena. The climate change is a major and growing challenge in mountain areas where "even small shifts in temperature can jeopardize the fragile balance of natural environments, which are defined by extreme climatic conditions, steep topography, and a wide variety of ecological zones and associated microhabitats with distinct biodiversity" (Kotru et al. 2014).

In Uttarakhand, elevation of the Himalayas ranges from 300 to 3,600 meter (m) spanning the Great, Middle, and sub-Himalayan region. The Great Himalayan region is largely snow-covered peaks and few temporary/seasonal habitations. The Middle Himalayan region has mainly fertile valleys and dense forest cover. The sub-Himalayan regions are Tarai and plain areas which are heavily farmed. Many important reports suggest that due to its unique topography, the Himalayan belt of the country including Uttarakhand is more susceptible to climate change. In Uttarakhand, ~70 % population resides in rural areas and depends mainly on agriculture for livelihood. Out of the total reported area of 53.48 lakh ha, only 7.66 lakh ha (14 %) is under cultivation. In northwestern Himalayan ecosystem, agriculture is mostly rainfed with only 18–20 % area under irrigation. The cropping intensity of Uttarakhand is ~158 % which is higher than the national average (~134 %). Out of the total geographical area of 53,485 km^2, most of the area is under forests (34,651 km^2) and wastelands, thus leaving only 7,488 km^2 of land for cultivation. About half of the total cultivated land is submarginal and 21 % of the landholding is between 0.5 and 1 ha (GOU 2014). In the past few years, Uttarakhand has witnessed a series of extreme climate events, particularly so in 2009, 2010, 2012, and 2013. The flash floods of 2013 inflicted immense loss of human lives, infrastructure, and property, and Rudraprayag, Chamoli, Uttarkashi, Bageshwar, and Pithoragarh districts were worst affected.

The glacial ecosystem has been tremendously affected by climate change and variability. Some of the reported climate-related consequences in the hills of Uttarakhand include receding glaciers and ascending snow line, depleting natural resources, erratic rainfall (leading to flash floods as seen in June 2013 disaster), irregular and less winter rains, advancing cropping seasons, changes in the flowering behavior of plants, shifting of cultivation zones of fruits (the zone has moved by 1,000–2,000 m), less intensity of snow in winters, rise in temperature, drying up of perennial streams, etc.

7.2 Changes in Temperature, Rainfall Pattern, and Snow Cover

The rising of temperature in hills can cause rapid melting of glaciers, consequently impacting the freshwater supply and its quality. The increase in water temperatures leads to changes in physiochemical property of water with increased microbial population and adversely impacts human health. The health of a water body, such as a river, depends upon its ability to effectively self-purify through biodegradation, which is hindered when there is a reduced amount of dissolved oxygen. Consequently, when precipitation events do occur, the contaminants are flushed into water bodies and drinking reservoirs leading to significant health implications (IPCC 2007).

According to a report (GoU 2012), a net increase in temperature in the Himalayan region in the 2030s is forecasted to increase between 1.7 and 2.2 °C (0.6 ± 0.7 °C) with respect to the 1970s. Various studies have reported rapid changes in amount, intensity, frequency, and form of precipitation. The increase in atmospheric aerosol might have begun to affect the monsoon in the Himalayas, offsetting increasing trend in monsoon precipitation that would have been caused by increase in atmospheric greenhouse gases alone (Shrestha et al. 2000). In the hilly region of India, people are increasingly getting overwhelmed with liquid precipitation rather than solid precipitation that were earlier received in the form of snow (Rautela and Karki 2015). This is perceived to be responsible for reduced snowfall duration in the region. As a result the entire region is suffering from scarcity of water. This trend of increasing temperature has affected rainfall pattern and its distribution across the world particularly over the Indian monsoon by modulating atmospheric moisture content (Singh and Kumar 1997; Goswami et al. 2006; Rajeevan et al. 2006, 2008; Singh et al. 2008; Pattanaik and Rajeevan 2010).

Rainfall has become more erratic and intense for the last 10–15 years. It has been observed that intense rainfall has instigated various disastrous events, most commonly landslides, flash floods, etc. Climate change-related hazards such as drought and flood create stress on rural livelihoods by reducing existing livelihood options and by creating volatility and unpredictability in streams of livelihood benefits (Agarwal and Perrin 2008; Conway and Schipper 2011).

In the Himalayan region, snow cover is highly sensitive to climate change. In many studies snowfall is found to have decreased in amount, as well as changed in timing over the past 20–25 years in NWH region. Himalayan glaciers are melting faster due to rising temperature. A significant rise of 1.6 °C from 1901 to 2002 has been reported in the northwestern Himalayas (Bhutiyani et al. 2007). The seasonal mean and maximum and minimum winter temperatures from 1985 to 2008 also have increased over the Himalayas (Shekhar et al. 2010). The study also reported changes in the seasonal snowfall pattern in the western Himalayas from 1989 to 2007 witnessing a decreasing trend in the winter snowfall. Despite the complexity of observations and the lack of on-site measurements, an overall pattern of warming has been apparent, with evidence of receding glacier and snow cover decrease recorded in the Himalayan region (Armstrong 2010; Kang et al. 2010; Bamber 2012; Bolch et al. 2012).

More specifically, snowfall events are thought to oscillate in two important ways, firstly reduction in the intensity and quantity of snowfall and secondly change in the timing of snowfall. Reduced snowfall results in less snow in glaciers and reduced stream flow. The shorter period of snowfall prevents the snow from turning into hard ice crystals which are liable to melt when summer arrives (Shiva and Bhatt 2009). As reported by Dobhal et al. (2004), Gangotri glacier – the source of the Ganga – is receding at 20–23 miles/year. Milam glacier is receding at 30 m/year and Dokriani is retreating at 15–20 m/year. The receding of glaciers has accelerated with global warming. Some of the most overwhelming effects of glacial meltdown occur in the form of sudden overflow of glacial lakes and glacial lake outburst floods (GLOFs). According to the Intergovernmental Panel on Climate Change (IPCC), "glaciers in the Himalayas are receding faster than in any other part of the world and if the present rate continues, the likelihood of their disappearing by the year 2035 and perhaps sooner is very high if the earth keeps getting warmer at the current rate." According to the IPCC report, the total area of glaciers in the Himalayas will shrink from 1,930,051 to 38,000 mile^2 by 2035. There is a sharp decrease in overall precipitation especially during winters. Less snow results in less moisture for growing crops. This has led a huge impact on agriculture and horticulture.

7.3 Impact of Climate Change on Mountain Agriculture

Agriculture is dependent on appropriate amalgamation of weather and associated factors and is thus highly susceptible to climate variability. A slight change in these factors can have a severe impact on the crop yield. Eighty-five percent of the population of Uttarakhand is directly or indirectly dependent on agriculture for its sustenance and income. Any reduction in production or productivity of crops adversely impacts the food security and the ultimate source of income of these farm families. The quality and productivity of land have been declining over the years. Due to highly erratic rainfall pattern and lowering of level of water streams, farmers are gradually leaving cultivation of some crops like paddy cultivation. Frequent floods and severe rains have led to washing away of fertile land (Satendra 2003; Meena et al. 2013, Meena et al. 2015a; Singh et al. 2014; Kumar et al. 2015; Ghosh et al. 2016). Based on different studies and consultations held by Uttarakhand Centre on Climate Change, Uttarakhand State Action Plan for Climate Change (2012) has reported the main trends of climate change as overall less and more erratic rainfall, increasing atmospheric temperatures, increased frequency of intense rainfall events, less or absent winter rains, overall decreased water availability, warmer and shorter winter, etc.

Temperature, precipitation, and their associated seasonal patterns are determining components of agricultural production. Rising atmospheric temperature as a result of climate change adversely affects crop production. The area under irrigation has gradually decreased due to lack of sufficient water for irrigation resulting in low crop production. Erratic, unpredictable, insufficient rainfall when needed has led to drastic reduction in cultivation of rice (both irrigated and nonirrigated) and other crops such as wheat, cauliflower, potatoes, etc. In addition, untimely hailstorms also damage crops. Winter rains are crucial for soil moisture but this too has become highly erratic (Meena et al. 2015c, d, e; Verma et al. 2015a).

In hills, snow is critical for the maintenance of proper level of soil moisture. This has positive ramification on crops such as wheat, barley, etc. Underground seepage has suffered hugely. Drizzle has become now very uncommon, which was particularly effective as it resulted in water seepage and improved soil moisture. Fog is also more frequent in comparison to earlier years adversely affecting wheat, barley, lentils, and even vegetable production. In some areas of hills, farmers have significantly deviated from their traditional crop calendar. Overall seasonal precipitation determines the crop yield over large areas, but stress and dry spells threaten productivity, even a few hours at critical growth stages (Huntingford et al. 2005). The excessive destruction by wild animals especially wild boars and even monkeys is also one of the reasons why people have switched over to other crops and fruit cultivation and left their land barren. Some environmentalists relate the animal menace as one of the ill effects of climate change. Forest structure is changing at a fast rate where broad-leaved oak forests are rapidly converting into chir pine forests or scrubs. There are many examples where an oak forest has completely converted into pine forest within ~20 years. This rapidly changing forest structure has led to habitat destruction and threat to wildlife. Due to scarcity of food in the forest areas, many wild animals encroach into nearby residential areas and agricultural land for survival. As a result wild animal menace has become a burning issue. A preliminary survey reveals that 30–70 % damage to the agricultural crops is done by displaced wild animals. Due to monkey menace, farmers of many villages have stopped cultivating vegetables and other crops. Monkeys, boars, and leopards have become commonly sighted animals in the villages of Kumaon and Garhwal region of Uttarakhand.

Climate change has also affected insect population in the region. Being poikilotherms, the behavior, distribution, development, survival, and reproduction of insects get influenced by temperature (Yamamura and Kiritani 1998; Petzoldt and Seaman 2010). The major factors of climate change like elevated CO_2 level, rising temperature, and deficient soil moisture therefore affect population dynamics of insect pests and thus significantly enhance the extent of crop yield losses (Reddy 2013).

Reduced insect population has affected seed dispersal, thereby affecting production of cash crops like potato, rye, sesame, amaranth, and kidney bean. Orchards are also affected adversely by climate change. In most of the areas in Uttarakhand due to low and erratic rains, trees could not bear fruits in time. Those villagers with traditional knowledge and agricultural scientists identify changes in crop productions and productivity with rapid changes in climatic conditions that have been witnessed in the hilly regions of Uttarakhand. The phenological responses of plants, particularly the early flowering in plants and crops, are considered to be prominent biological indicators of climate change (Parmesan and Yohe 2003). Dormancy of plants is broken early due to reduced winter periods which are reflected in the form of early flowering. As a result most of the plants flower at the time when weather conditions are unfavorable for their growth and survival. Hailstorms which are common during the period of flowering also result in major crop loss. Blooming which takes place even before local pollinators are active results in less fruiting and thus low productivity. Early flowering in almost all agricultural, horticultural, and forest tree species is reported to be a common observation, particularly in *Rubus* spp., *Malus domestica*, and *Rhododendron arboreum* (Rautela and Karki 2015) (Table 7.1).

The agriculture and climate change work in a vicious circle, from the agricultural practices reducing the sustainability of forest as an ecosystem, stabilizing climate, and contributing to the climate change, which in turn increases the vulnerability of agricultural practices and production. Increase in temperature and erratic rainfall directly affects the agriculture production and food supply. Insufficient rain and increased temperature cause drought-like situations, whereas intense and erratic rain for short period reduces groundwater recharge by accelerating runoff and causes flash floods. Such situations induce negative effects in the agriculture (Malla 2008) (Figs. 7.1 and 7.2).

Table 7.1 Impacts of climate change on agriculture

Temperature-related changes
Extended frost-free periods
High average winter temperature
Fewer extreme cold temperatures in winter
Higher night temperature both in summer and winter
Increased temperature variability
Rainfall-related changes
More variability of summer precipitation
More intense rainfall results in more runoff
Higher absolute humidity
Other climate-related changes
Low wind speeds
Enlarged tropospheric ozone
Excessive loss of soil carbon
Faster plant growth and maturity
More growth of weeds and vines
Under elevated atmospheric CO_2
Weeds migrate northward and are less sensitive to herbicides/weedicides

Source: Sharma and Dobriyal (2014)

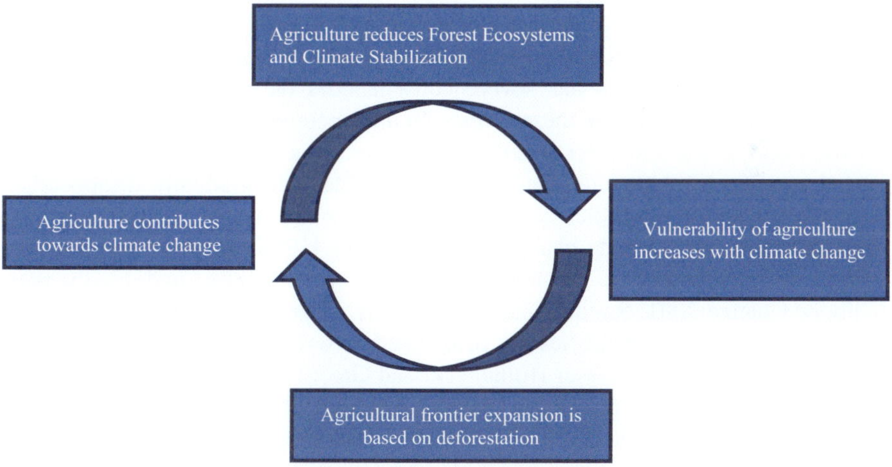

Fig. 7.1 Vicious circle of agriculture and climate change

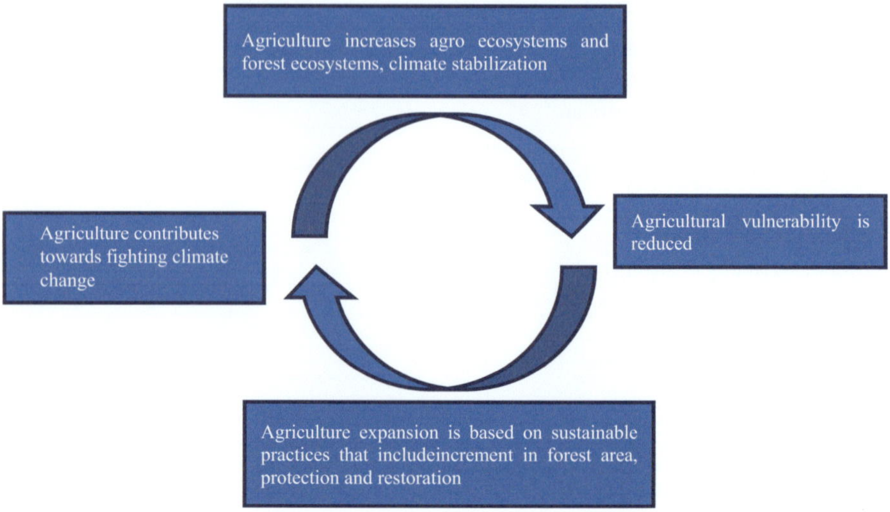

Fig. 7.2 Virtuous circle of agriculture and climate change (Adopted and modified from Rodas 2015)

7.4 Impact of Climate Change on Forest Resources

In the Himalayan region, majority of population is directly dependent on agriculture and forest resources for their livelihood. Any change in structure, composition, and distribution of forests has severe consequences on communities and their livelihood which depends on these forests. In hills, people are dependent on forests for fuelwood, grass, and other biomass. Due to changes in weather pattern, broad-leaved species, a highly nutritious source of fodder, are gradually disappearing from the forest. Some studies suggest that the forest ecosystem can be seriously impacted by further climate change. Even a rise of 1–2 °C in global temperature will result in changes in species compositions, productivity, and biodiversity (Climate Change 2001; Ravindranath et al. 2006). Uttarakhand contains rich floral diversity. Important tree species are distributed across different latitudinal zones, i.e., the sub-Himalayan, Lesser Himalayan, and Great Himalayan. Prominent trees of sub-Himalayan zone include sal (*Shorea robusta*), khair (*Acacia catechu*), and sheesham (*Dalbergia sissoo*). Most important tree species in Lesser Himalayan zone are chir or pine (*Pinus roxburghii*), oak (*Quercus leucotrichophora*), rhododendron (*Rhododendron arboreum*), and alders (*Alnus nepalensis*). Similarly, firs (*Abies alba*), bhojpatra or birch (*Betula utilis*), stunted rhododendron (*Rhododendron campanulatum*), and junipers (*Juniperus* spp.) are the prominent species at the Great Himalayan zone from where alpine meadows start. Alpine meadows, which start from ~3,000 m, are homes of valuable herbs and flower species. Many studies reported that floral wealth of alpine meadows is also at decline.

Rawat (2013) reported that the hill community perceives that climate change affects the phonological events like early flowering and less fruiting in trees. The maximum effects can be seen on *Quercus leucotrichophora*, *Rhododendron arboreum*, *Rubus ellipticus*, *Pyrus communis*, etc. Phenological changes occur due to irregular and erratic rainfall, less snowfall, increasing temperature, and decreasing moisture content in soils. As a result of these climatic conditions, plants have changed their new leaf formation time and flowering and fruiting time, which is now 15–20 days ahead as compared to the timings 10–15 years ago. These studies have reported that in the Himalayan region forests around the villages have depleted and degraded due to road construction activities, cutting of trees for fuelwood, loss of vegetation due to glacial action and other natural disasters, frequent forest fires, overgrazing by animals, and decreased regeneration rates of trees in the forest area.

Deforestation has also increased the physical and social vulnerability in hilly areas; ~65 % of the total land in the state comes under the forest. However, satellite imageries point toward a huge reduction in the actual forest cover. The communities traditionally preserved forests in the Himalayan region as it provides safety against many hazards, e.g., flash floods, slides, avalanche, and rock falls (Gupta and Nair 2012; Verma et al. 2015b; Meena et al. 2016). The degradation of forests in the mid-slopes of the Himalayan region has reduced the availability of fodder and leaf litter, with serious implication on livestock. Fuelwood is still the primary source of energy for cooking in most of the villages in hills. Fuelwood collection which is primarily being done by women has become more burdensome. In hills, dependence on forest resources generally increases during drought-like conditions or other natural calamities (Rautela and Karki 2015; Meena et al. 2015b). Changes in oak forest cover and climate change have adverse impact on the availability of wild species of mushroom, i.e., *Morchella esculenta*.

Forest resources which also include fruits, roots, leaves, mushrooms, and medicinal plants are traditionally collected from forests by people of higher hills. In the last few years, collection of high-value *Ophiocordyceps sinensis*, colloquially known as caterpillar fungus (kira jari or yarsagumba), has become a major source of income inducing people of high hills to visit high altitudes during its harvest season. It appears in the high altitude in the month of May or June with the melting of the snow (Rautela and Karki 2015). From the many studies reported, income from this activity is quite enough for many families of the area to sustain all through the year. According to the people of high hills, due to decrease in growth of the caterpillar fungus, its collection is decreasing every year. It is feared that in years to come it may become an extinct fungus.

7.5 Impact of Climate Change on Food Security

The climate and weather condition of hills determines food security of the hill communities, who depend primarily on agriculture and animal husbandry for their subsistence. Traditional millets have been replaced with rice and wheat (Rawat

2013). The practice of consuming vegetables and other food items which are wild in nature is also disappearing because of declining mixed forests. It is generally difficult to find wild varieties in pine or mono-species forest. This has reduced nutrition, variety, and taste in food, particularly of women and poor as they are frequent visitors to the forests. Traditional grains and vegetables have almost disappeared from the diet of new generation. The inclination to grow and consume at least a bit of healthy traditional food every season is still evident among women, particularly older women, even in areas where cash crops are popularly practiced. In hills, people are now becoming more and more dependent on the market to buy food such as vegetables, lentils, and even grains. Consumption of milk and milk products has gone down due to the decline in animal population primarily by reduced availability of fodder and grass from the forest.

Bees and butterflies are considered to be indicators of climate change and are used for predicting various environmental variations. Their specific survival-related ecological requirement that includes temperature, humidity, food plants, and egg-laying habitats makes them most susceptible to climate change. Reduced nectar availability due to dry spells, drought-like conditions, and phenological changes is perceived to be responsible for their reduced population (Forister and Shapiro 2003; Rakosy and Schmitt 2011). These changes have adversely affected the food security of people with reduced availability of varied, safe, tasty, culturally acceptable, and nutritious food for all, particularly women and other vulnerable sectors. This has had a negative impact on the health of both humans and livestock.

7.6 Impact of Climate Change on Livestock

The population of domesticated animals has also become less dense. As a result of forest degradation, availability of grass and fodder has gone down. Replacement of traditional crops with cash crops has resulted in reduction in the availability of fodder. With the reduced number of livestock, the availability of cow dung for farm yard manure has also reduced considerably and has negatively impacted the quality of soil and production. According to Houghton et al. changes in atmospheric temperature, humidity, wind speed, and other climate factors influence livestock performance like growth, milk production, wool production, and reproduction. Other findings of Singh et al. (2012) examined perceived climate change impacts and adaptation strategies to sustain livestock production adopted by livestock in NWH. Majority of the respondents of NW Himalayas perceived that climate change has adversely affected productive and reproductive performance of livestock. Other changes include increased incidence of livestock diseases and parasitic infestation and decreased availability of feed, fodder resources, and water.

7.7 Climate Change and Gender Issues

In hills, women play a major and crucial role in agricultural economy. They have rich and diversified knowledge as well as skill acquired through managing natural resources, livestock care, and agriculture-based livelihood and practice indigenous ways of maintaining good health. Women have contributed a great deal in maintaining and promoting agricultural genetic diversity (Shiva 2015). Due to natural calamities in the past 5 years, women experienced extended workload and physical and mental exertion. The unpredictability in seasons and rainfall is affecting water availability, soil moisture, forest regeneration, and eventually food production. With limited access to resources and almost negligible participation in decision-making processes at household and community levels, women will be most seriously affected by reduced food and nutritional security (PANAP 2011).

Women collect fuel and fodder from forest which along with other household responsibility put them under mental and physical exertions. Unusual events and extreme conditions enhanced pressure on them along with increasing risks and vulnerability. As compared to other parts of the country, women in hills are more vulnerable to nutritional problems. Difficult living conditions in hills such as fragmented fields, steep slopes, and women's continuous toil in the fields for 14–15 h consistently, coupled with household chores, make their workload overwhelming. Various studies show that women perceive that work involved in cash crops is much greater compared to traditional crops, and this has increased women's workload many times over. Most of the women in hills suffer from lower back pain due to carrying heavy loads over long distances; they also suffer from various skin problems due to long exposure to sun. Due to the use of agrochemicals, women are exposed to several health hazards and gynecological infection. In case of rice transplantation, arthritis and intestinal and parasitic infections may take place due to long hours of work in mud and water (Pandey 2001).

Increased temperature due to climate change is leading to an increase in weed infestation, and as weeding is essentially a women's task, it has added workload of women farmers. Since women's role and responsibilities are closely associated with natural resource management and farming, women will be the worst sufferer of climatic variabilities. Women's limited access and control over these resources increase their vulnerability to climate change and hinder their adaptation capacity. Already marginalized women bear the brunt of increasing food insecurity.

7.8 Adaptation and Mitigation Strategies Toward Climate Change

Adaptation generally refers to the actions that help in better coping with the changing conditions. Adaptation strategies are constantly renewed through learning by doing, experimenting with new ideas, and knowledge-making processes that allow hill communities to adjust and modify their actions with changing climatic conditions. Those living in the rural and hill areas are well aware about the changes taking place in their surrounding due to climate change. The most common phenomena as a result of climate change are varying temperatures, frequent and longer drought periods, erratic rainfall, untimely flowering in some plant species, slowly disappearing natural resources, and occurrence of new diseases. Hill communities are somehow adapting to these changes with the resources available to them. However, presently they lack information, services, technologies, assets, mobility, and the ability to make choices and decisions (Joshi and Bhardwaj 2015; Chulu 2015).

In some areas of hills, farmers have already started taking adaptation measures which include replacing grain crops with vegetables (Gum et al. 2009). These crops give higher production than grain crops even in smaller areas with less resources and labor. This also provides them with much needed cash and nutrition. Production of off-season vegetables together with cash crops can be a strategy toward ensuring sustainable livelihood. With increasing temperature and shortened and less harsh winters, farmers of hill areas can start planting crops early. According to a study, genetic and species diversity in field has been maintained in Johar, Byans, and Niti valley areas through mixed cropping system of Barahanja, meaning 12 grains. In this system, 12 different crops are grown together which ensure favorable conditions for each other and thus improve crop productivity. This practice provides them protection against the entire crop failure and is an effective instrument of food security (Rautela and Karki 2015). The traditional varieties or traditional knowledge system (TKS) of millet when grown along with other modern crops provide for contingency when conditions are unfavorable. Buck wheat, for example, is one of the shortest duration crops that takes only 4–5 weeks from sowing to flowering and thus suppresses weed infestation and prevents soil erosion. The crop at the same time has multiple usages as it is used as vegetable and is also an important source of cattle feed (Rautela and Karki 2015).

In hills, women manage the conservation of seeds. Their traditional knowledge on this aspect is very strong. This knowledge is crucial for adapting to climatic variability and environmental changes. In the Himalayan mountain region, 80–90 % seed requirements of crops are met through traditional seed management and exchange systems, where the role of women is very important. Women are custodians of traditional knowledge related to seed conservation and maintain a diverse genetic pool of this valuable resource through in-situ conservation (Dhakal and Leduc 2010).

In order to meet fodder and fuelwood requirement, planting fodder trees/plants along the boundaries of the terraced fields can be promoted. Studies conducted for about a decade at ICAR-VPKAS, Almora, have revealed that introduction of improved grasses and legumes like pangola (*Digitaria eriantha*), giant star (*Cynodon plectostachyus*), *Panicum coloratum*, *Panicum repens*, and *Setaria kazungula* is a promising technology for the improvement of grassland and other forage production systems (Bisht et al. 1999). The climatic conditions in hills are also favorable for growing medicinal plants that can fetch high prices and have a large and ready market. People of Niti area in Uttarakhand are cultivating jambu faran (*Allium stracheyi Baker*), kala jeera (*Carum persicum*), and sea buckthorn

(*Hippophae rhamnoides*) at many places and earning their livelihood (Rautela and Karki 2015).

This is an irony of the hills that despite of plenty of water sources and adequate rainfall, there exists acute shortage of water not only for raising the crops but also for drinking. This is mainly due to rains getting confined within 3–4 months and several dry spells. In order to cope with this situation, efforts must be made for developing rainwater conservation mechanism along with sprinklers and drip irrigation systems (especially for horticultural crops) for the efficient use of available water. There are several more options available which should be taken up vigorously in order to address the water crisis. These include the repair of *naula* (traditional wells), construction of infiltration tanks and recharging of natural water springs, and plantation of broad-leaved tree species that enhance water retention property of the soil and strengthen slope stability (Acharya 2012). Community water tanks should also be constructed to reduce drudgery of walking long distances to collect water. Forests have significant emission removal capability which can further be enhanced by operational major afforestation and reforestation initiatives like the National Mission for a Green India besides continued strengthening of the present protection regime of forests (Kishwan et al. 2009; Yadav 2013).

Women constitute the largest percentage of the hill population and are invariably affected the most by these changes. But their experiences, perceptions, and viewpoints rarely are heard in various decision-making platforms, both locally and globally, in relation to mitigation and adaptation strategies. The different challenges faced by women and men need to be recognized, and it is crucial that gender perspectives are introduced in adaptation and mitigation strategies. This perspective will prove to be very effective in reducing the impact of climate-induced changes and in enhancing food security (Nelson 2011).

Studies on climate change need to be *multidimensional* as people's lives are affected by various factors such as access to natural resources, globalization, poverty, climate change, etc. Since these factors are closely interwoven, an *integrated approach* is imperative when designing adaptation and mitigation strategies in agriculture, forestry, health, and livelihoods. Assessment of the effects of global climate change on agriculture might help to deal with the farming system and to enhance agricultural production (Table 7.2).

7.9 Conclusions and Future Prospective

Hill regions are more susceptible to climate change events. Natural resources are depleting with the change in seasonal pattern and climate variables. Climate change-induced changes in the region include receding glaciers, erratic rainfall, less and irregular winter rainfall, reduction in snowfall in winter, raised temperature, drying up of perennial streams, etc. The major impact of climate change is presently being perceived on the agricultural sector that accommodates highest proportion of the workforce of the hilly areas. These changes have resulted in fluctuations in the flowering behavior of plants, shifting of cultivation zones of

Table 7.2 Impact of climate change on livelihood and proposed mitigation mechanism

Climatic changes	Impact on livelihood systems	Coping and adaptation mechanism	Potential future risks
Decrease in rainfall and change in rainfall timing	Reduced agricultural productivity and agro-biodiversity	Growing cash crops; change in timing of sowing of crops	Increasing livelihood and food security
Longer dry spells	Drying up of water sources; less flow in springs and streams	Traditional system of water sharing, change in time of sowing of crops	Scarcity of water for drinking and agriculture, adverse effect on workload of women
Warmer winters, lee snow, and more precipitation in the form of rain instead of snow	Increased incidences of crop infestation, weed crop failure	Increased use of insecticides, pesticides, and weedicides	Increasing livelihood and food security
Better survival of wild animals due to deceased snowfall and less harsh winters	Increased incidences of animal attack on agriculture	No coping mechanism	Huge crop loss, low agricultural productivity, and more migration

various agricultural crops, and advancing cropping seasons, and many species of flora and fauna are pushed toward the verge of extinction. These changes have brought disruption in social and economic life of hill communities and enhanced their vulnerability against the climate change.

Women in hill regions are particularly vulnerable to climatic variability due to their limited adaptive capacities that arise from already prevailing social inequalities and ascribed social and economic roles. Women play a significant role in maintaining household chores and in managing natural resources. Women are repositories of indigenous knowledge which contributes toward survival of the community and toward their own adaptation in extreme climatic conditions. In climatic stress conditions, they are forced to turn toward creating indigenous innovations which are cost-effective and less time consuming. However, their role is hardly ever recognized while framing developmental and environmental policies and strategies. Women and men have different ascribed roles in the society, face different challenges, and demonstrate different reactions and methods for coping. These issues should be addressed in research, development, disaster preparedness, and adaptation and mitigation strategies. Climate change affects men and women differently according to their respective vulnerability and adaptation capacities. Therefore, all aspects related to climate change (i.e., mitigation, adaptation, policy development, decision-making) must consider gender perspective and ensure that men and women are able to adapt these climatic variabilities. Although women have low capacity to adapt, the share of the adaptation burden falls disproportionately on them. This makes the consideration of the impact of climate change on gender most imperative. Gender-blind adaptation strategies are very harmful as they may increase already existing inequalities.

The major concerns of the hill communities are their dependence on subsistence farming, forests, and rainfall. The strategies need to be developed to decrease their reliance on these factors which will increase their resilience to climate change impact. It is high time for policy formulators, political leaders, and administrators to put climate change at the center stage of development strategies as it has become a crosscutting issue. Multidisciplinary groups are required to be formed for developing climate change mitigation and adaptation strategies.

Acknowledgements We thank the editors and anonymous reviewers for their constructive comments, which helped us to improve the manuscript.

References

Acharya A (2012) Managing 'Water Traditions' in Uttarakhand, India: lessons learned and steps towards the future. In: Johnston NBR et al (eds) Water, cultural diversity, and global environmental change: emerging trends, sustainable futures? doi:10.1007/978-94-007-1774-9

Agarwal A, Perrin N (2008) Climate adaptation, local institutions and rural livelihoods. IFPRI working paper W081–6. pp 350–3676

Anthwal A, Joshi V, Sharma A, Anthwal S (2006) Retreat of Himalayan glaciers – indicator of climate change. Nat Sci 4:53–59

Armstrong RL (2010) The glaciers of the Hindu Kush-Himalayan region. International Centre for Integrated Mountain Development, Kathmandu

Aryal P (2015) Climate change and food security: Nepal perspective. Climate Change 1:105–109

Bamber J (2012) Climate change: shrinking glaciers under scrutiny. Nature 482:482–483

Bhutiyani MR, Kale VS, Pawar NJ (2007) Long-term trends in maximum, minimum and mean annual air temperatures across the North-western Himalaya during the twentieth century. Clim Chang 85:159–177

Bisht JK, Chandra S, Mani VP, Singh RD (1999) Fodder production and management strategies for hills, Tech. Bull.. VPKAS, Almora, pp 13–36

Bolch T, Kulkarni A, Kaab A, Huggel C, Paul F, Cogley JG, Frey H, Kargel JS, Fujita K, Scheel M, Bajracharya S, Stoffel M (2012) The state and fate of Himalaya glaciers. Science 336:310–314. doi:10.1126/science.1215828

Chulu J (2015) A feminist perspective that poverty is gendered: do women have lesser access to resources in comparison with men? Available online at http://ssrn.com/abstract=2663381

Climate Change (2001) The scientific basis, summary for policy makers and technical summary of the working group 1. Intergovernmental Panel on Climate Change, Switzerland

Conway, Schipper ELF (2011) Adaptation to climate change in Africa: challenges and opportunities identified from Ethiopia. Glob Environ Chang 21:227–237

Dhakal TD, Leduc B (2010) Women's role in biodiversity management in Himalayas. Sustain Mt Dev 57:16–17, ICIMOD

Dobhal DP, Gergan JT, Theyyan RJ (2004) Recession and morphogeometrical changes of Dokriani glacier Garhwal Himalaya. Curr Sci 86:692–696

Forister ML, Shapiro AM (2003) Climatic trends and advancing spring flight of butterflies in lowland California. Glob Chang Biol 9:1130–1135

Ghosh BN, Meena VS, Alam NM, Dogra P, Bhattacharyya R, Sharma NK, Mishra PK (2016) Impact of conservation practices on soil aggregation and the carbon management index after

seven years of maize–wheat cropping system in the Indian Himalayas. Agric Ecosyst Environ 216:247–257
Goswami B, Venugopal V, Sengupta D, Madhusoodanan M, Xavier P (2006) Increasing trend of extreme rain events over India in a warming environment. Science 314:1442–1445
GoU (2012) State action plan on climate change: transforming crisis into opportunity, revised version, June 2012. Government of Uttarakhand, pp 22–23
Govt of Uttarakhand (2014) Uttarakhand action plan for climate change. Government of Uttarakhand. http://www.moef.gov.in/sites/default/files/Uttarakhand%20SAPCC.pdf
Gum W, Singh PM, Emmett B (2009) Even the Himalayas have stopped smiling: climate change, poverty and adaptation in Nepal. Oxfam International, Lalitpur
Gupta AK, Nair SS (2012) Ecosystem approach to disaster risk reduction. National Institute of Disaster Management, New Delhi, Pages 202
Huntingford C, Lambert FH, Gash JHC, Taylor CM, Challinor AJ (2005) Aspects of climate change prediction relevant to crop productivity. Phil Trans R Soc B 360:1999–2009
IPCC (2007) Assessment of adaptation strategies-options, constraints and capacity. Climate change 2007-impacts, adaptation and vulnerability. Contribution of working group II to the fourth assessment report. Cambridge University Press, Cambridge
Joshi K, Bhardwaj N (2015) Women and natural resource management: a study of 'Communities of Practice' prevailing in women farmers' community for management of water and forests of lesser Himalayan region in India. Int J Adv Res 3:363–374
Kang S, Xu Y, You Q, Flügel WA, Pepin N, Yao (2010) Review of climate and cryospheric change in the Tibetan Plateau. Environ Res Lett 5:1–8
Kishwan J, Pandey R, Dadhwal VK (2009) India's forest and tree cover: contribution as a carbon sink, technical paper. Indian Council of Forestry research and Education, Dehradun
Kotru R, Choudhary D, Fleiner R, Khadka M, Pradhan N, Dhakal M (2014) Adapting to climate change for sustainable agribusiness in high mountain watersheds: a case study from Nepal. ICIMOD Working Paper 2014/1, 15 International Centre for Integrated Mountain Development, Kathmandu, p 1
Kumar A, Bahadur I, Maurya BR, Raghuwanshi R, Meena VS, Singh DK, Dixit J (2015) Does a plant growth promoting rhizobacteria enhance agricultural sustainability? J Pure Appl Microbiol 9(1):715–724
Malla G (2008) Climate change and its impact on Nepalese agriculture. J Agric Environ 9:62–71
Meena RS, Yadav RS, Meena VS (2013) Heat unit efficiency of groundnut varieties in scattered planting with various fertility levels. Bioscan 8(4):1189–1192
Meena RS, Meena VS, Meena SK, Verma JP (2015a) The needs of healthy soils for a healthy world. J Clean Prod 102:560–561
Meena RS, Meena VS, Meena SK, Verma JP (2015b) Towards the plant stress mitigate the agricultural productivity: a book review. J Clean Prod 102:552–553
Meena VS, Meena SK, Verma JP, Meena RS, Ghosh BN (2015c) The needs of nutrient use efficiency for sustainable agriculture. J Clean Prod 102:562–563
Meena VS, Meena SK, Verma JP, Meena RS, Jat LK (2015d) Current trend to mitigate climate change and sustainable agriculture: a book review. J Clean Prod 102:548–549
Meena VS, Verma JP, Meena SK (2015e) Towards the current scenario of nutrient use efficiency in crop species. J Clean Prod 102:556–557
Meena RS, Bohra JS, Singh SP, Meena VS, Verma JP, Verma SK, Sihag SK (2016) Towards the prime response of manure to enhance nutrient use efficiency and soil sustainability a current need: a book review. J Clean Prod 112(1):1258–1260
Nelson V (2011) Gender, generations, social protection and climate change: a thematic review. Overseas Development Institute, London
PANAP (2011) Documentation of climate change perceptions and adaptation practices in Uttarakhand, Northern India. By Beej Bachao Andolan-Save Seed Campaign in collaboration with Pesticide Action Network Asia and the Pacific

Pandey H (2001) Understanding farm women. National Research Centre for Women in Agriculture, Bhubneshwar, p 15
Parmesan C, Yohe G (2003) A globally coherent fingerprint of climate change impacts across natural systems. Nature 421:37–42
Pattanaik D, Rajeevan M (2010) Variability of extreme rainfall events over India during southwest monsoon season. Meteorol Appl 17:88–104
Petzoldt C, Seaman A (2010) Climate change effects on insects and pathogens. Climate change and agriculture: promoting practical and profitable responses. Available online at http://www.climateandfarming.org/pdfs/FactSheets/III.2Insects.Pathogens.pdf
Rajeevan M, Bhate J, Kale KD, Lal B (2006) High resolution daily gridded rainfall data for the Indian region: analysis of break and active monsoon spells. Curr Sci 91:296–306
Rajeevan M, Bhate J, Jaswal A (2008) Analysis of variability and trends of extreme rainfall events over India using 104 years of gridded daily rainfall data. Geophys Res Lett L18707:1–6
Rakosy L, Schmitt T (2011) Are butterflies and moths suitable ecological indicator systems for restoration measures of semi-natural calcareous grassland habitats? Ecol Indic 11:1040–1045
Rautela P, Karki B (2015) Impact of climate change on life and livelihood of indigenous people of higher Himalaya in Uttarakhand. Am J Environ Prot 3:112–124
Ravindranath NH, Joshi NV, Skumar R, Saxena A (2006) Impact of climate change on forests in India. Curr Sci 90:354–361
Rawat VS (2013) People perception on climate change and their influence on various aspects around tones valley of Garhwal Himalaya. Environ Ecol Res 1(3):150–154
Reddy PP (2013) Impact of climate change on insect pests, pathogens and nematodes. Pest Manag Hortic Ecosyst 19:225–233
Rodas O (n.d) The relation between agriculture and climate change. Retrieved 12 Oct 2015, from ISID: www.iisd.org/pdf/2011/redd_hanoi_2011_ag_climate.ppt
Satendra (2003) Uttaranchal and its vulnerability. In: Disaster management in the hills. Concept Publishing House. p 57
Schmidubher J, Tubeillio FN (2007) Global food security under climate change. PNAS 104:19703–19708
Sharma P, Dobriyal P (2014) Climate change and agricultural sector in Uttarakhand. J Stud Dyn Chang 1(1):6–14
Shekhar MS, Chand H, Kumar S, Srinivasan K, Ganju A (2010) Climate-change studies in the western. Ann Glaciol 51:105–112
Shiva V (2015) Women's indigenous knowledge and biodiversity conservation. In: Pojman LP, Pojman P, Mcshane K (eds) Environmental ethics: readings in theory and application. Cengage Learning, Boston, pp 383–388
Shiva V, Bhatt VK (2009) Climate change at the third pole – the impact of climate instability on Himalayan ecosystems and Himalayan communities. Available online at http://www.wfo-oma.com/climate-change/case-studies/climate-change-study-in-uttarakhand-india.html
Shrestha AB, Wake CP, Dibb JE, Wayewski PA, Whitlow SI, Carmichael GR, Ferm M (2000) Seasonal variations in aerosol concentrations and compositions in the Nepal Himalaya. Atmos Environ 34:3349–3363
Singh P, Kumar N (1997) Impact assessment of climate change on the hydrological response of a snow and glacier melt runoff dominated Himalayan river. J Hydrol 193:316–350
Singh P, Kumar V, Thomas T, Arora M (2008) Changes in rainfall and relative humidity in different river basins in the northwest and central India. Hydrol Process 22:2982–2992
Singh SK, Meena HR, Kolekar DV, Singh YP (2012) Climate change impacts on livestock and adaptation strategies. J Vet Adv 2(7):407–412
Singh DK, Singh SK, Singh AK, Meena VS (2014) Impact of long term cultivation of lemon grass (Cymbopogon citratus) on post-harvest electro-chemical properties of soil. Ann Agric Biol Res 19(1):45–48
Sorkhabi R (2007) An inquiry into 'Fragile' Himalaya. Himal J 63. Available online https://www.himalayanclub.org/hj/63/4/an-inquiry-into-fragile-himalaya/

Verma JP, Jaiswa DK, Meena VS, Meena RS (2015a) Current need of organic farming for enhancing sustainable agriculture. J Clean Prod 102:545–547
Verma JP, Jaiswal DK, Meena VS, Kumar A, Meena RS (2015b) Issues and challenges about sustainable agriculture production for management of natural resources to sustain soil fertility and health. J Clean Prod 107:793–794
Yadav SK (2013) GIS based approach for atmospheric carbon absorption strategies through forests development in Indian situations. In: Proceedings of environ, info environmental informatics and renewable energies 2013 Pt. II. Available online at https://inis.iaea.org/search/searchsinglerecord.aspx?recordsFor=SingleRecord&RN=45021615
Yamamura K, Kiritani K (1998) A simple method to estimate the potential increase in the number of generations under global warming in temperate zones. Appl Entomol Zool 33:289–298

Agroforestry for Natural Resource Conservation, Livelihood Security and Climate Change Mitigation in Himalayan Agroecosystems

8

D.R. Palsaniya and P.K. Ghosh

Abstract

Agroforestry is widely known to increase resilience and reduce vulnerability of agricultural production systems to climate change effects. Agroforestry may potentially improve livelihood through simultaneous production of food, fodder, timber and firewood as well as mitigation and adaptation to climate change. This chapter critically examines the role of agroforestry in natural resource (soil, water, nutrients, biodiversity) conservation, livelihood security (goods and services to society), providing social and economic well-being to people and carbon sequestration. Moreover, the role of agroforestry is more pronounced and relevant in degraded and fragile agroecosystems, such as the Himalayas. Agroforestry in the Himalayas contributes significantly to ecological, social and economic functions and serves as complementary to natural forests. Therefore, a winning strategy for natural resource conservation and human welfare in the fragile Himalayas can be achieved by promoting scientific agroforestry practices to reconcile food production with conservation in climate change scenario. Agroforestry needs to be strengthened by developing appropriate innovative technologies, community participation, domestication, strong policy support and establishing strong market linkages. Similarly, efforts are needed to promote science-based decision-making, and future thrust is required to remove the remaining uncertainties and to also carefully test the agroforestry practices against other land uses in Himalayan ecosystems in order to know their suitability and to what extent agroforestry can serve mankind and ecosystems.

Keywords

Agroforestry • Natural resource • Climate change • Mitigation • Ecosystems

---

D.R. Palsaniya (✉) • P.K. Ghosh
ICAR-Indian Grassland and Fodder Research Institute, Jhansi 284003, Uttar Pradesh, India
e-mail: drpalsaniya@gmail.com

J.K. Bisht et al. (eds.), *Conservation Agriculture*,
DOI 10.1007/978-981-10-2558-7_8

8.1 Introduction

Agroforestry is an age-old land use practice defined as land use systems in which woody perennials (trees, shrubs, palms, bamboo, etc.) are grown on the same piece of land with herbaceous plants and/or animals, either in spatial arrangement or in time sequence and in which there are both ecological and economic interactions between the trees and non-tree components (Beets 1989). A more recent definition defines agroforestry as a dynamic, ecologically based, natural resource management system that, through integration of trees on farms and in the agricultural landscape, diversifies and sustains production for increased social, economic and environmental benefits for land users at all levels (Leakey 2011). Researchers and planners are increasingly recommending agroforestry systems as a sustainable form of land use for augmenting biomass production from agricultural systems (Nair 1984; Kang and Wilson 1987; Singh et al. 1998; Anonymous 2000; Dhyani et al. 2007, 2011), natural resource conservation (Palsaniya et al. 2011b, 2012c) problems soil management (Singh et al. 1989a, b, 1997; Singh 1995; Palsaniya et al. 2011a) and recently for mitigation of climate change by sequestering carbon in both standing biomass and soil (Dhyani et al. 2013; Ajit et al. 2013; Palsaniya et al. 2011a; Gupta et al. 2015; Meena et al. 2015d, e; Verma et al. 2015a, b, 2016; Ghosh et al. 2016). The integrated approach associated with agroforestry makes it the most self-sustaining and ecologically sound land management system (Singh et al. 1997). In agroforestry trees, crops and animals are integrated in a long-term conservative, sustainable, productive and eco-friendly system where greater emphasis is placed on perennial MPTs (multipurpose trees) that are planted once but yield benefits over a long period of time (Dhyani et al. 2007, 2011; Palsaniya et al. 2010b, 2013). It reduces the risks associated with agriculture and increases sustainability of production system besides conserving and even improving the soil.

The present area under agroforestry in India is estimated to be 25.32 M ha which is 8.2 % of the total geographical area (Dhyani et al. 2013). This area can easily be expanded by adding another 28 M ha from fallows, cultivable fallows, pastures, groves and wastelands. In spite of widely recognised social, ethnic and religious significance of trees and the prevalence of various traditional forms of agroforestry in different parts of India (Tejwani 1994), recognition of trees as components of farming systems has been rather limited (Nair 2001). This may be due to the traditional divide between agriculture and forestry in government policies and programmes and other issues associated with trees. But even then, the significance and scope of agroforestry in degraded and fragile agroecosystems, such as the Himalayas, are immense and, therefore, need to be promoted on large scale.

8.2 Agroforestry in the Himalayan Agroecosystem

The hill agroecosystem in the Himalayas is facing diverse challenges due to demographic pressure, natural resource degradation and climate change. Agroforestry can address some of these challenges, and as a result agroforestry has received greater attention especially in the Himalayas from researchers and planners for its ability to contribute significantly to economic growth, poverty alleviation, natural resource conservation and environmental quality. Hill farming systems are the gradually evolved typical agroforestry system modified by the farmers over the time in response to change in social, cultural, economical and technological needs. Mountain farmers usually retain few trees on their fields to get fuelwood, fruits, fodder and timber and to maintain microclimatic by reducing soil and air temperature, irradiance and wind speed.

8.2.1 Prominent Agroforestry Practices in Himalayas

Agroforestry is practised in the Himalayas as way of life since time immemorial. However, organised agroforestry research is four decades old in India. Today, agroforestry is recognised as an important part of the 'hill agriculture' in Himalayas. Gill and Deb Roy (1992) identified six common agroforestry systems based on the nature of components as agrisilviculture (crop + trees), silvipastoral (tree + pasture/animals), agrihorticultural (crop + fruit trees), agri-silvipastoral (fruit trees + trees + pasture/animals), horti-silvipastoral (fruit trees + trees + pasture/animals) and agri-horti-silvipastoral (crop + fruit trees + pasture/animals). Similarly, Nair (1985) described 18 agroforestry practices followed in different parts of India, viz. improved shifting cultivation, tree fallows planting, taungya, alley cropping, live fences, multipurpose trees on croplands, woodlots, protein bank, trees as shelter for crop and animals, trees for soil conservation on bunds and terraces, trees for water management, home gardens, plantation crop combinations, livestock under trees, dune fixation, aquaforestry, apiculture with forestry and irrigated forestry. Major agroforestry practices followed in the Himalayas are listed in Table 8.1.

Livestock is an integral part of hill agriculture and contributes greatly to the hill economy. It is estimated that about 16 % of wood, 61 % of tree fodder and 25 % of

Table 8.1 Common agroforestry practices followed in the Himalayas

Agroforestry practice	Part of Himalaya
Fodder/timber trees	Throughout the Himalayas
Crops/grasses with fruit trees	Orchards in J and K and Himachal Pradesh Himalayas
Seasonal grazing	High-altitude Western Himalayas
Taungya	Eastern Himalayas
Woodlots for soil conservation	Sloppy and landslide-prone Himalayas
Live hedges and boundary plantation	Terrace farming areas of Himalayas
Farm woodlots	Throughout the Himalayas
Shelterbelts and windbreaks	In wind-prone areas, especially alpine regions
Shaded perennial systems with plantation crops	Eastern Himalayas

Table 8.2 Agroforestry systems priority in Himalaya

Agroclimatic zone	Agroforestry system					
	Agri-silvi culture	Agrihorti silviculture	Agrihorti culture	Hortipasture	Silvipasture	Specialised
Western Himalayas	4	4	5	4	4	Orchards
Eastern Himalayas	5	3	5	4	3	Jhum

No. 1–5 indicate the priority practised model (1 for least and 5 for highest practised)

Table 8.3 Suitable crops and MPTs for Himalayan region

Region	Promising MPTs	Crops
Western Himalayan region	*Ulmus wallichiana*, *Ailanthus excelsa*, *Morus alba*, *Robina pseudoacacia*, *Cedrus deodara*, *Grewia optiva*, *Celtis australis*, *Pinus* spp., *Bauhinia variegata*, *Albizia chinensis*, *Toona ciliata*, *Ulmus laevigata*, *Malus domestica*, *Prunus domestica*, *P. armeniaca*, *P. persica*, *P. deueis* and *Pyrus communis*	Paddy, mustard, maize, wheat, rajma, berseem, peas, potato, cauliflower, cabbage, tomato, chillies, etc.
Eastern Himalayan region	*Acacia auriculiformis*, *Alnus nepalensis*, *Bamboos*, *Paraserianthes falcataria*, *Gmelina arborea*, *Artocarpus heterophyllus*, *Parkia roxburghii*, banana, papaya, mango, pineapple, litchi, etc.	Paddy, maize, mustard, *Colocasia*, pea, potato, cabbage, turmeric

grass requirement of local people are met from the government forest in Himachal Pradesh (Khurana and Negi 1993). On the basis of the diagnosis and design (D&D) exercises (Pathak et al. 2000), the prominent agroforestry system of Himalayas was enumerated on priority scale (Table 8.2).

8.2.2 Suitability of Components for Agroforestry

The selection of components (perennials, annual crops and livestock) for agroforestry systems has a great significance, and their selection requires a judgement based on the knowledge of the crops, adaptations, productivity, utility, family needs and marketing opportunities (Singh et al. 1997; Palsaniya et al. 2008, 2009a, b, 2013). The local agro-climate, soil type, human and animal preference, nutritional value, digestibility, specific equipment needed for cultivation and species compatibility should also be given due consideration while selecting components for agroforestry (Singh et al. 1997; Palsaniya et al. 2010b, Palsaniya et al. 2011a). The components of agroforestry should be less competitive for space, light, water and nutrients and have strong synergy. Suitable MPTs and crops for Himalayas are listed in Table 8.3 (Palsaniya et al. 2012b, 2013).

Select and use trees, crops and planting/harvesting patterns that are suitable for the site, are compatible with planned practices and provide desired economic and environmental returns (Palsaniya et al. 2012a). Further, the recycling and efficient use of residues, by-products and wastes from different components need to be given priority while deciding components for agroforestry (Yadav et al. 2009; Palsaniya et al. 2011a; Meena et al. 2013, 2015a; Singh et al. 2014; Kumar et al. 2015). The selection of species should be from the homologous zones with similar eco-climatologically areas or bioclimatic as the plant life is very selective for their habitat.

8.2.3 *Ulmus–Grewia–Bauhinia–Alnus–Themeda–Arundinella* Natural Silvipastoral Cover

The natural silvipastoral resources in the Indian Himalayan region is categorised as *Ulmus–Grewia–Bauhinia–Alnus–Themeda–Arundinella* cover (Palsaniya and Dhyani 2012; Palsaniya et al. 2011a; Dabadghao and Shankarnarayan 1973). This silvipastoral cover comprises over 23.04 M ha of area and is located in the Western and Eastern Himalayas and *Terai* belt. The main fodder trees in this belt are *Ulmus wallichiana* in J and K, *Grewia optiva* in Himachal Himalayas, *Bauhinia purpurea* in the Uttarakhand and *B. variegata* and *Alnus nepalensis* in the northeastern Himalayan belt. The other forage trees of significance are *Salix alba*, *Morus alba*, *Populus ciliata* and *Celtis australis* in Western Himalayas while *Litsea polyantha*, *Artocarpus heterophyllus*, *Streblus asper*, *Albizia*, *Gmelina arborea* and *Symingtonia* in the eastern part of the Himalayas. *Ficus*, conifers, *Ailanthus excelsa*, *Prunus armeniaca*, *Cedrela toona*, *Terminalia belerica*, *Bamboo* and *Quercus* spp. are also used as important source of fodder in the Himalayas. The mean tree density (no. of trees/ha) under rainfed condition was 46 in J and K, 21–87 in Himachal Pradesh, 5–25 in Uttarakhand, 50 in Manipur, 100 in Assam and 150–200 in Meghalaya.

As understory component, there are 37 major perennial grasses, 32 annual grasses and 34 dicots in the subtropical belt, while the temperate–alpine grasslands found beyond 2,100 m altitude in the temperate and cold arid areas have 47 perennial grasses, 5 annual grasses, 68 dicots and few temperate shrubs and are devoid of trees. The major grass species are *Arundinella bengalensis*, *Themeda anathera*, *Bothriochloa pertusa*, *B. bladhii*, *Heteropogon contortus*, *Ischaemum barbatum*, *Chrysopogon* and *Cymbopogon*. As we move towards higher altitudes (above 2,100 m) in the subtemperate and temperate belt, perennial grass species like *Agropyron*, *Agrotis*, *Dactylis*, *Poa*, etc., dominates. This is one of the most productive silvipastoral covers of India due to favourable climate and vegetation composition. In general, the subtemperate and temperate grasses are finer, luxuriant, nutritious and easily digestible. These grazing lands are extensively being used both by sedentary and migratory herders. Several nomadic tribes such as the Gujjars, Bakarwals, Gaddis and Changpas of the Himalayas, who rear sheep and goat, migrate to subalpine and alpine pastures during summers and come back to lower altitudes during winters. The average dry biomass productivity of natural grasses in this zone is 1.96 t/ha/year with a carrying capacity of 0.70 adult cattle unit (ACU)/ha/year (Singh and Misri 1993). The region-specific dry biomass productivity (t/ha/year) reported is 3.1–8.8 from Himachal Pradesh (Sharma and Sood 1994), 1.0–5.8 from Jammu and Kashmir (Misri and Singh 1995), 2.86–10.85 from Sikkim and 0.1–1.4 from Uttarakhand (Sah and Saxena 1990).

8.3 Natural Resource Conservation Through Agroforestry

Agroforestry practices help in soil, water, nutrients and biodiversity conservation (Grewal 1993; Dhyani et al. 2005; Pandey 2007; Palsaniya et al. 2011a). Reduction in run-off and soil loss and improvement in soil nutrients were observed in agroforestry as compared to agriculture or cultivated fallow land use systems under different agroclimatic conditions. However, the natural resource conservation is more pronounced under fragile agroecosystems of the Himalayas. Some regional examples of multifunctional agroforestry systems in the Himalayas are enumerated in Table 8.4.

Table 8.4 Successful examples of role of agroforestry in Himalayas

Part of Himalaya	Challenges	Changes observed due to agroforestry	Reference
Western Himalayas	Reducing soil and water loss in agroecosystems in steep slopes	Contour tree rows (hedge rows) reduced run-off and soil loss by 40 % and 48 %, respectively	Narain et al. (1997)
Sikkim Himalayas	Enhancing litter production and soil nutrient dynamics	N-fixing trees increase N and P cycling through increased production of litter and caused greater release of N and P, maintained soil organic matter, higher N mineralisation	Sharma et al. (1996a, b)
Meghalaya Himalayas	Enhancing tree survival and crop yield	Crop yield did not decrease in proximity to *Albizia* trees	Dhyani and Tripathi (1998)
Himalayas	Restoration of abandoned agricultural sites	Biomass accumulation (3.9 t/ha in agroforests compared to 1.1 t/ha in degraded forests), improvement in soil physico-chemical characteristics, C sequestration	Maikhuri et al. (2000)
Kurukshetra Himalayas	Improvement of sodic soils	Increase in microbial biomass, tree biomass, soil carbon and available N	Kaur et al. (2002)

8.3.1 Soil Conservation and Wasteland Rehabilitation

Research conducted in Northeastern Himalayas at different slopes (less 50 %, 50–100 % and top slopes above 100 %) revealed 98 % reduction in soil loss under agri-horti-silvipastoral system (Singh 1988). Silvipastoral system comprising of *Alnus nepalensis* + pineapple + *Panicum maximum* or *Setaria sphacelata* + *Stylosanthes guianensis* in 1:1 ratio was found to be a sustainable practice in soils having 30–60 % slope. In Shiwaliks (annual rainfall, 800–1,500 mm), the average soil losses with run-off water were very less in different agroforestry systems as compared to agricultural systems (Grewal 1993). A three-tier silvipastoral system comprising *Leucaena leucocephala* as top canopy, *D. cinerea* as middle canopy and *C. ciliaris* + *S. hamata* as lower canopy reduced the soil loss by 11 times against a bare land (Yadava and Varshney 1997). Srivastava et al. (2003) reported that silvipasture along with staggered contour trenching in Kangra watersheds reduced the soil erosion and improved productivity, survival of plants and output from the silvipastoral system. Treatment with staggered contour trenching at 1 m vertical interval trapped 37 t/ha/year silt in the watershed.

Wastelands, landslide, mine spoils and torrents may also be effectively checked and rehabilitated through silvipastoral system in integration with other measures (Sastry et al. 1981; Singh 1995; Dadhwal and Katiyar 1996; Singh et al. 1997). At Nalota Nala, 4 ha of landslide area on Dehradun-Mussoorie road was successfully stabilised and rehabilitated using bioengineering practices including plantation of *Ipomoea carnea*, *Vitex negundo*, napier grass, *Erythrina suberosa*, *D. sissoo* and *A. catechu* within 10 years. The sediment load reduced to 5.5 t/ha/year from 320 t/ha/year with the improvement of vegetation cover from < 5 % to 95 % (Sastry et al. 1981) and dry weather flow increased to 250 days from just 100 days after the cessation of monsoon. Similarly, Dadhwal and Katiyar (1996) observed that the vegetal cover increased from 10 % to 80 % over a period of 14 years and debris flow decreased from 550 t/ha/year to about 8 t/ha/year due to different biological measures (trees, shrubs and grasses) adopted for rehabilitation of mine spoil area in Himalayan region. Hazra and Singh (1994) reported that soil and water conservation treatments along with agroforestry helped in reducing soil loss from 41 to 9.5 t/ha from barren hillocks and from 20.5 to 5.5 t/ha from wastelands.

8.3.2 Nutrient Recycling and Soil Health

Agroforestry is widely recognised for soil health management. Agroforestry is supposed to promote more efficient nutrient cycling than agriculture (Singh and Singh 1996; Singh et al. 1996; Kumar 2007; Palsaniya et al. 2009a, 2013; Yadav et al. 2009). The nutrient cycles in forest ecosystems are closed where nutrient inputs from atmospheric deposition, biological nitrogen fixation, litter fall, plant residue decay and weathering of primary soil minerals are in balance with nutrient losses due to leaching, denitrification, run-off, erosion and plant removal. On the other hand, there is a net output of nutrients from the soil via crop harvest removals in agricultural systems resulting in net negative balances if nutrients are not replaced through external inputs (fertilisers, manures, etc.). Trees can provide nitrogen and other nutrient inputs in agroforestry systems through biological nitrogen fixation (BNF), deep nutrient capture and litter fall. The presence of active nodules in roots of leguminous species indicates that BNF can supply considerable nitrogen inputs to crops via litter in soils. Sometimes, the non-fixing trees, such as *Cassia*, accumulate more nitrogen in their leaves than nitrogen-fixing legumes, probably because of their greater root volume and ability to capture nutrients (Garrity and Mercado 1994; Meena et al. 2015b, c) which can be added to the soil as green leaf manuring. *Gliricidia* and *Sesbania* are also known for their nitrogen fixation and green-manuring potential. Annual above-ground litter N content was high (256 kg/ha) with an annual release of 208 kg/ha, i.e. 81 % of the total litter fall in *Leucaena* (Sandhu et al. 1990). Deep nutrient capture by tree roots at depths where crop roots are not present is considered as an additional nutrient input in agroforestry systems because such nutrients are otherwise leached as far as the crop is concerned. They become an input on being transferred to the soil via tree litter decomposition (Yadav et al. 2009).

About 20 % of the N released from tree pruning or litter is taken by the current crop (Palm 1995; Giller and Cadisch 1995). Much of the remaining 40–80 % of the applied organic N is incorporated into soil organic matter (SOM) (Haggar et al. 1993). Because of the slow release of N and P from SOM, organic inputs have a greater residual effect on soil fertility than inorganic fertilisers. The rate of decomposition is mainly regulated by edaphic, climatic variables and resource quality (Singh et al. 1989, 1996; Yadav et al. 2008). High-quality organic inputs (high N but low lignin and polyphenol contents) will release nutrients rapidly. In contrast, low-quality organic inputs (poor in N and high in polyphenol and lignin) will release nutrients slowly or even immobilise (Palm 1995). This has practical implications for attaining synchrony between nutrient release and crop nutrient demand and, hence, for efficiency of nutrient use. Maharudrappa et al. (1999) reported that incubation of litter of different MPTs enhanced nutrient availability. The available N decreased slightly at 30 days of incubation compared to control. However, available N increased at the end of 60 days of incubation. This indicated temporary immobilisation of nutrients by microorganisms. Unlike N, the release of P to the available pool from litter was recorded at 30 days of incubation. Mineralisation of organic matter played a major role in P release for plant growth. The decomposition of litter material with narrower C:P ratio is likely to increase the available P as compared to those with higher C:P ratio. Incubation of litter also increased available K in all the treatments. The release of K to soil was more dependent on the quantity and quality of the litter. The increase in available K may be attributed to the fact that K is not strongly bound in organic structures, unlike that of N and P.

The perspectives and potentials for nutrient management need to be studied and quantified in agroforestry systems by measuring processes such as mineralisation, immobilisation, denitrification, volatilisation and leaching, along with changes in the soil organic matter pools. Legumes inclusion in the agroforestry systems plays a significant role in the nitrogen economy of the production system. Pulses constitute the cheapest source of protein to man and animals, enhance soil quality by adding organic matter and improve soil structure and water infiltration (Palsaniya and Ahlawat 2007). They act as nurse crop and supply nitrogen to the associated crops and thereby contribute to the conservation of energy by reducing the need for N fertilisation (Palsaniya and Ahlawat 2009). Therefore, introduction of suitable legumes in rangelands, pastures, silvipastures and agroforestry has great significance.

8.3.3 Nutrient Conservation

Agroforestry conserves nutrients in two ways. Agroforestry reduces run-off and thereby minimises the dissolved nutrient losses. Secondly, trees capture nutrients present in leaching water and also from deeper layers and accumulate them in biomass and finally add in the soil through recycling processes. Studies conducted in Shiwaliks revealed that average nutrients losses with run-off water were very less in different agroforestry systems as compared to agricultural systems (Grewal 1993). Similarly, in a three-tier silvipastoral system comprising *Leucaena leucocephala* as top canopy, *D. cinerea* as middle canopy and *C. ciliaris* + *S. hamata* as lower canopy, the loss of total soluble salts, dissolved nitrogen and potassium was reduced at the tune of 69 %, 67 % and 43 %, respectively, as compared to bare land (Yadava and Varshney 1997). The organic carbon, available nitrogen and phosphorus in silvipastoral system were also improved by 53 %, 23 % and 8 %, respectively, against their initial status.

8.3.4 Water Conservation

Trees in agroforestry conserve water by reducing run-off and increasing infiltration and percolation and, thus, recharge groundwater aquifer. The decaying roots also facilitate infiltration by acting as channels for water to percolate. Trees also improve water holding capacity through favourable effect on soil properties. Research conducted in Northeastern Himalayas on water conservation at different slopes revealed 99 % reduction in run-off under agri-horti-silvipastoral system (Singh 1988). In Shiwaliks, different agroforestry systems recorded reduced run-off as compared to agricultural systems (Grewal 1993). A silvipastoral system comprising *Leucaena leucocephala* as top canopy, *D. cinerea* as middle canopy and *C. ciliaris* + *S. hamata* as lower canopy reduced the run-off loss by six times compared to a bare land (Yadava and Varshney 1997). A 4 ha of landslide area on Dehradun-Mussoorie road was successfully stabilised by plantation of *Ipomoea carnea*, *Vitex negundo* and napier grass with *Erythrina suberosa*, *D. sissoo* and *A. catechu*. The significance of all these measures was reflected on the dry weather flow, which used to last hardly for 100 days after the cessation of monsoon, increased to 250 days after the treatments (Sastry et al. 1981). Dadhwal and Katiyar

(1996) observed that the monsoon run-off decreased from 57 % to 37 % due to agroforestry-based rehabilitation in mine-spoiled area of the Himalayan region. Moreover, the lean period flow days increased from 60 to 240 days. Hazra and Singh (1994) reported that agroforestry reduced run-off from 70 % to 30 % with rise in water table from 1 to 4 m. Srivastava et al. (2003) reported that silvipasture along with staggered contour trenching at 1 m vertical interval retained 227–424 m^3/ha run-off in the Kangra watershed of the Himalayas.

Inclusion of trees and shrubs on the agricultural landscape influences adjoining microclimate (temperature, RH, light, wind), changes the surface energy balance and alters the water requirement, actual water use, water use efficiency and productivity of adjacent crops or intercrops. Trees impact the water relations of the crop by affecting the loss of water through evaporation, transpiration and mining of water from deep soil layers. In erosion-prone areas wind breaks, shelterbelts or agroforestry buffers provide significant reductions in the amount of wind-blown soil and subsequent abrasion of plant parts and cuticular damage. Loss of cuticular integrity or direct tearing of the leaves reduces the ability of the plant to control water loss. Moreover, wind breaks, shelterbelts and agroforestry reduce the air turbulence in the vicinity and conserve soil moisture by reducing ET from the soil and crops. This enhances crop growth and increases its water use efficiency. However, multipurpose tree species in agroforestry themselves consume some of the available water, and, therefore, in the area immediately adjacent to the trees, competition for water between crops and trees has a negative impact on yield (Palsaniya et al. 2012a).

Relatively little attention has been given to water management practices (Palsaniya et al. 2013) as far as agroforestry research is concerned. Under irrigated condition, irrigation water applied to intercrop also partially fulfils the water requirement of the tree component. Under rainfed condition, agroforestry or specifically silvipastures are raised without any provision of irrigation. However, irrigation is essentially required during initial establishment along with life-saving irrigations. Sprinkler and drip system of irrigation is a better option for higher crop growth, yield, profitability and water and input use efficiency (Tewari et al. 2012). Similarly, there is scope of adopting various in situ and ex situ rainwater harvesting techniques based on soil type, slope and agro-climate. Moreover, adoption of watershed-based agroforestry has been found highly effective in enhancing productivity, profitability, cropping intensity, employment opportunities and water availability besides promoting ecosystem services especially in degraded agroecosystems (Palsaniya et al. 2010a, 2011b, 2012c, d).

Agroforestry has also been advocated as a means of managing excess water during floods or that has accumulated in the fields in canal command areas (Ram et al. 2011). *Eucalyptus* as a biodrainage species has widely been recognised for managing rising saline water tables. Inclusion of multipurpose tree species in existing farming system also provides satisfactory drainage control and amelioration of existing salinity (Dunin 2002). The sewage-contaminated waste water from big cities can also be used safely by adopting appropriate agroforestry systems (Singh et al. 2010; Bradford et al. 2003). Turner and Ward (2002) emphasised that agroforestry systems have potential for improving water use efficiency by reducing the unproductive components of the water balance (run-off, soil evaporation and drainage). Agroforestry systems could double rainwater utilisation compared to annual cropping systems mainly due to temporal complementarity and use of run-off in monsoon regions like India (Lovenstein et al. 1991; Droppelmann and Berliner 2003). An ideal combination of annual crop and perennial trees in the agroforestry system uses the soil moisture more efficiently than the sole cropped trees or crops because water uptake by the trees is mainly from deeper layers, whereas the crop could better utilise topsoil water.

8.3.5 Biodiversity Conservation

The Himalayas is one of the most important hot spots of world biodiversity. Out of 59 M ha of land comprising the total geographical area of the Indian Himalayas, 22 M ha are suffering from one or other kinds of land degradation. The effects of ecosystem degradation in uphills are felt locally as landslides; as shortage of food, fodder, fuelwood and water; and also far away in the Indo-Gangetic lowlands as floods resulting in huge loss to agriculture, property and human life. Moreover, land degradation is also threatening Himalayan biodiversity (Myers 1988) which in turn can decrease ecosystem functioning and services. Pandey (2007) and Jose (2009) suggested five major roles of agroforestry in conserving biodiversity:

(a) Agroforestry provides habitat for species that can withstand a certain level of disturbance in agroecosystems.
(b) Agroforestry helps preserve germplasm of socially useful and associated species.
(c) Agroforestry helps reduce the rates of conversion of natural habitat by providing goods and services alternative to traditional agricultural systems that may involve clearing natural habitats.
(d) Agroforestry provides connectivity and acts as a stepping stone by creating corridors between habitat remnants and thereby conservation of area-sensitive plant and animal species.
(e) Agroforestry helps conserve biological diversity by providing other ecosystem services such as erosion control and water recharge, thereby preventing the degradation and loss of surrounding habitat.

Although agroforestry may not entirely reduce deforestation (Angelsen and Kaimowit 2004), but it may act as effective buffer to deforestation. Trees in agroecosystems in Uttarakhand Himalayas have been found to support threatened cavity-nesting birds and offer food and habitat to many species of birds (Pandey and Mohan 1993). Agroforestry systems, in some cases, do support as high as 50–80 % of biodiversity of comparable natural systems (Noble and Dirzo 1997). It also acts as buffer to parks and protected areas (Bhagwat et al. 2008). Agroforestry allows noncrop–crop spillover of a diversity of functionally important organisms (Anand et al. 2010). Areca nut agroforestry systems of south Meghalayan Himalayas conserve 160 species of plants (83 tree species, 22 shrub species, 41 herb species and 14 climber species) in addition to providing cash income, medicine, timber, fuelwood and edibles for household consumption and sale (Tynsong and Tiwari 2010).

8.4 Agroforestry for Climate Resilience and Mitigation

The Intergovernmental Panel on Climate Change in its fourth assessment report concluded that the increasing greenhouse gases have resulted in global warming by 0.74 °C over the last 100 years (IPCC 2007). Moreover, the IPCC have predicted an increase of 1.8–4.0 °C temperature by the end of this century. Such changes are expected to adversely affect agriculture including crops, trees, soils, livestock and fisheries. The indirect effects of climate change on water resources, soil organic carbon and other basic soil processes are likely to be very critical. The adverse effects of climatic change are expected to be relatively high in the Himalayas due to dependence of large population on relatively less agricultural land, degraded natural resource base, fragile landscape, low availability of appropriate technologies and changing land use patterns posing a serious threat to the food, fuel, timber and fodder security in this part of the world.

Agroforestry has to play an important role in reducing vulnerability, increasing resilience of farming systems and buffering households against climate-related risks in the Himalayas. Agroforestry systems are promising land use practices to increase above-ground and soil C stocks to mitigate greenhouse gas emissions. Trees outside forests in India store about 934 Tg C at the rate of 4 Mg C/ha (Kaul et al. 2011). The net annual carbon sequestration rates for fast-growing but short rotation agroforestry crops such as poplar and *Eucalyptus* have been reported to be 8 Mg C/ha/year and 6 Mg C/ha/year, respectively (Kaul et al. 2010). Makundi and Sathaye (2004), Thornton and Herrero (2010) and Nair et al. (2010) found agroforestry as an attractive land use system for carbon sequestration because of the following reasons:

(a) It sequesters carbon in vegetation and in soils depending on the pre-conversion soil C.
(b) The more intensive use of the land for agricultural production reduces the need for slash-and-burn or shifting cultivation.
(c) The wood products produced under agroforestry serve as substitute for similar products unsustainably harvested from the natural forest.
(d) To the extent that agroforestry increases the income of farmers, it reduces the incentive for further extraction from the natural forest for income augmentation.
(e) Agroforestry practices may have dual mitigation benefits as fodder species with high nutritive value can help to intensify diets of methane-producing ruminants while they can also sequester carbon.

The C sequestration potential of tropical agroforestry systems in recent studies is estimated between 12 and 228 Mg/ha with a median value of 95 Mg/ha. In India, average sequestration potential in agroforestry has been estimated to be 25 Mg C/ha with a substantial variation in different regions depending upon the biomass production. The above-ground biomass accumulation in a central Himalayan agroforestry system has been found to be 4 t/ha/year compared with 1.1 t/ha/year at the degraded forestland (Maikhuri et al. 2000). Carbon storage of the central Himalayan forests ranges from an average of about 175 t C/ha for chir pine forests to 400 t C/ha for oak- and sal-dominant forests (Singh 2009). The studies on various Van Panchayat-managed forests in the Himalayas revealed that the mean carbon sequestration rate for the Indian Himalayas was 3.7 t/ha/year and that of Nepal was 1.88 t/ha/year. These rates were more or less similar to 2.79 t/ha/year under normal management conditions where local people extract various forest products to meet their sustenance needs (Rawat 2012). Different agroforestry land use systems existing on arable and nonarable lands in the Kwalkhad watershed of middle Himalayan region of Himachal Pradesh, India, were evaluated for C sequestration and C credits (Goswami et al. 2014). Out of eight land use systems existed, agrisilvihorticulture (ASH) system (14.78 Mg/ha) and agrihortisilviculture (AHS) system (14.45 Mg/ha) sequestered a high amount of C than silvipasture (SP), pure agriculture or grassland and abandoned land, though not significantly more than agrisilviculture (AS) or agrihorticulture (AH). C stocks in soil (0–40 cm) exceeded C stocks in plants by a factor of 15.81 for AHS system. SP, ASH and AS systems, with their higher C mitigation potential of 1.71, 1.52 and 1.43, respectively, were more suitable land use systems for C mitigation in the region. The ASH system produced the most (21.49) C credits on a per-hectare basis. In small-holder bamboo farming system in Barak Valley, Assam, a traditional home garden system, C estimate in above-ground vegetation ranged from 6.51 (2004) to 8.95 (2007) Mg/ha with 87 %, 9 % and 4 % of the total C stored in culm, branch and leaf, respectively, with a mean rate of C sequestration of 1.32 Mg/ha/year. Alternate land use systems, viz. agroforestry, agrihorticulture and agrosilviculture doubled the soil organic carbon content compared to sole cropping system in a period of 6 years in the northeastern Himalayan region of India (Das and Itnal 1994). In another study at same location, it was observed that intercropping of coconut (*Cocos nucifera*) with guava (*Psidium guajava*) increased SOC from 3.4 to 7.8 g/kg or sequestered 877 kg C/ha/year after 38 years (Singh et al. 2012). Gupta et al. (2015) reported that the above-ground and below-ground herbage carbon density in grasslands of Himachal Himalayas was 1.62 and 1.81 t/ha, respectively, and it was higher than herbage carbon density in different plantations (Table 8.5). Among the plantations, herbage carbon density was higher in *P. roxburghii* and mixed forest, while the soil carbon density was higher under *U. villosa* followed by mixed plantation.

However, a major issue for future research is that the C sequestration estimates available globally are mostly derived through biomass productivity and often do not take into account the carbon sequestration in soils. The literature reviewed here indicates that agroforestry systems have the potential to sequester large amounts of above- and below-ground carbon compared to treeless farming systems. Therefore,

Table 8.5 Carbon density in different vegetation systems in Himachal Himalayas

	Carbon density (t/ha)							
	Herbs		Shrubs		Trees			
System/plantation	AG	BG	AG	BG	AG	BG	Soil	Total
Grassland	1.62	1.81	–	–	–	–	42.18	45.61
Quercus leucotricophora	0.56	1.02	0.71	0.73	34.12	7.67	41.91	86.72
Mixed forest	0.67	1.20	–	–	42.97	9.66	43.13	97.63
Ulmus villosa	0.46	0.89	1.07	1.13	67.48	15.18	49.13	135.34
Acacia mollissima	0.52	0.95	–	–	37.2	8.37	41.33	88.37
Eucalyptus tereticornis	0.64	1.10	–	–	46.38	10.44	37.95	96.51
Pinus roxburghii	1.08	1.04	–	–	46.66	8.40	28.76	85.94

innovative policies and rigorous research results are required to exploit the mostly unrealised potential of carbon sequestration through agroforestry in both subsistence and commercial agroforestry.

8.5 Agroforestry for Livelihood Security in the Himalayas

Agroforestry is contributing to livelihoods of Himalayan people where people have a long history of tree growing in landscape. Himalayan fruit-based orchards are well known. Large numbers of non-timber tree products are produced from agroforestry for consumption, subsistence and income, and thus, support livelihoods locally. In the Eastern Himalayas, guava (Meghalaya) and lemon (Assam) based agrihorticultural systems gave 2.96- and 1.98-fold higher net return, respectively, in comparison to farmlands without trees. Average net monetary benefits to guava- and lemon-based agroforestry systems were Rs. 20,610/ha and 13,787.60/ha, respectively. Such systems are the most useful livelihood improvement strategies in the rainfed agriculture of the Himalayas (Bhatt and Misra 2003) through a convergent approach from all the stakeholders (Mukherjee et al. 2012; Mukherjee and Maity 2015).

Domestication of minor forest fruit trees and their integration in existing agroforestry systems offer a significant opportunity for livelihoods and nutritional and economic security of poor people in the Himalayas. Already, such wild edible fruits and other plants form an important constituent of traditional diets in the Sikkim Himalayas where about 190 species are eaten and 47 species are traded in local market. Wild edible fruit species have high carbohydrate content ranging between 32 % and 88 % (Sundriyal and Sundriyal 2001). Such fruit trees can be taken up for domestication in agroecosystems on priority action. The trees in agroforestry produce valuable fodder for animal production especially during lean period. Leaf fodder yield of 1.11 t/ha is reported from 2-year-old plants of *Grewia optiva* in Kashmir. This yield is likely to be double from full-grown trees (Joshi and Talapatra 1980). The green fodder yield from the same species in Kangra district of Himachal Pradesh was reported to be 12–15 kg/adult tree (Negi et al. 1979). In

Bauhinia variegata, average fodder yield of 15–20 kg/tree is reported (Negi et al. 1979), while *Bauhinia purpurea* based system yielded green leaf fodder of 1,425 kg/ha from tree alone on gravelly lands of Doon Valley (Vishwanathan et al. 1998).

Trees in agroecosystems are particularly valued as host to insects that yield marketable products such as silk, lac and honey. The woodcarving industry is also emerging as an important source of income to local artisans in the Himalayas like elsewhere in the world. Suitable community plantations of non-timber forest products in tribal areas through joint forest management (JFM) can potentially serve dual purpose of conserving the useful species as well as livelihood improvement of local people. Moreover, wood-based value chains and small-scale industries should be promoted to further strengthen the livelihood of local people.

8.6 Conclusions and Future Focus

The role of agroforestry in sustaining the Himalayan agroecosystems is now amply clear. However, it will require appropriate research interventions, adequate investment, suitable extension strategies, postharvest processing and value addition, market infrastructure, convergence and strong policy support. A major role for agroforestry in the near future will emerge in the domain of ecosystem services. Prevalence of a variety of traditional and commercial agroforestry practices in the Himalayas offers great opportunity for carbon sequestration, livelihood improvement, biodiversity conservation, soil fertility enhancement and poverty reduction. In order to promote agroforestry systems as an important option for livelihood improvement, natural resource conservation, climate change mitigation and adaptation and sustainable development in the Himalayas, research, policy and practice will have to progress towards effective community participation, maintenance of the traditional agroforestry systems, creation of new systems, designing context-specific silvicultural and farming systems, biodiversity conservation and maintaining a continuous cycle of regeneration–harvest–regeneration as well as locking the wood in non-emitting uses such as woodcarving and durable furniture. Domestication of useful fruit tree species; strengthening the markets for non-timber forest products; addressing the research needs and policy for linking knowledge to action; genetic improvement of identified potential MPTs; in-depth studies on tree–crop–environment interactions; developing contract and organic farming in relation to agroforestry; developing integrated approaches like INM, IPM, IWM, etc., under agroforestry systems; and developing decision support systems for replication of successful agroforestry practices and effective strategies for agroforestry technology transfer and capacity building should also be focussed in our agroforestry plans, policies and research.

Acknowledgements The authors are highly thankful to anonymous reviewers for their valuable comments that helped us in improving this manuscript.

References

Ajit, Dhyani SK, Ramnewaj, Handa AK, Prasad R, Alam B, Rizvi RH, Gupta G, Pandey KK, Jain A, Uma (2013) Modelling analysis of potential carbon sequestration under existing agroforestry systems in three districts of Indo-gangetic plains in India. Agroforest Syst. doi:10.1007/s10457-013-9625-x

Anand MO, Krishnaswamy J, Kumar A, Bali A (2010) Sustaining biodiversity conservation in human-modified landscapes in the Western Ghats: remnant forests matter. Biolog Cons 143 (10):2363–2374

Angelsen A, Kaimowit D (2004) Is agroforestry likely to reduce deforestation? In: Schroth G, Fonseca GAB, Harvey CA, Gascon C, Vasconcelos HL, Izac AMN (eds) Agroforestry and biodiversity conservation in tropical landscapes. Island Press, Washington, DC, pp 87–106

Anonymous (2000) National Agricultural Policy, Department of Agriculture and Cooperation, Ministry of Agriculture, Government of India. p 4

Beets WC (1989) The potential role of agroforestry in ACP countries. CTA/EEC, Wageningen

Bhagwat SA, Willis KJ, Birks HJB, Whittaker RJ (2008) Agroforestry: a refuge for tropical biodiversity? Trends Ecol Evol 23(5):261–267

Bhatt BP, Misra LK (2003) Production potential and cost-benefit analysis of agrihorticulture agroforestry systems in Northeast India. J Sust Agric 22:99–108

Bradford A, Brook R, Hunshal CS (2003) Wastewater irrigation in Hubli-Dharwad, India: Implications for health and livelihoods. Environ Urbaniz 15:157–170

Dabadghao PM, Shankarnarayan KA (1973) The grass cover of India. Indian Council of Agriculture Research, New Delhi

Dadhwal KS, Katiyar VS (1996) Land degradation and environmental pollution by mining and landslides. In: Biswas TD, Narayanasamy G (eds) Soil management in relation to land degradation and environment. Indian Society of Soil Science, New Delhi, pp 114–126

Das SK, Itnal CJ (1994) Capability based land use systems: role in diversifying dryland agriculture in dry land area. Bull Indian Soc Soil sci 16:92–100

Dhyani SK, Tripathi RS (1998) Tree growth and crop yield under agrisilvicultural practices in north-east India. Agrofor Syst 44:1–12

Dhyani SK, Sharda VN, Samra JS (2005) Agroforestry for sustainable management of soil, water and environmental quality: looking back to think ahead. Range Managet Agrofor 26(1):71–83

Dhyani SK, Samra JS, Ajit, Handa AK, Um (2007) Forestry to support increased agricultural production: focus on employment generation and rural development. Agric Econ Res Rev 26 (2):179–202

Dhyani SK, Tewari RK, Palsaniya DR, Singh R, Yadav RS, Venkatesh A (2011) Adoption of agroforestry land use in semi arid Bundelkhand region – a case study from Garhkundar-Dabar watershed. Indian J Agrofor 13(1):10–12

Dhyani SK, Handa AK, Uma (2013) Area under agroforestry in India: an assessment for present status and future perspective. Indian J Agrofor 15(1):164–187

Droppelmann K, Berliner P (2003) Runoff agroforestry – a technique to secure the livelihood of pastoralists in the Middle East. J Arid Environ 54:571–577

Dunin FX (2002) Integrating agroforestry and perennial pastures to mitigate water logging and secondary salinity. Agric Water Managet 53:259–270

Garrity DP, Mercado JAR (1994) Reforestation through agroforestry: smallholder market-driven timber production on the frontier. In: Raintree J, Fernandez H (eds) Marketing multipurpose tree species in Asia. Winrock International, Bangkok

Ghosh BN, Meena VS, Alam NM, Dogra P, Bhattacharyya R, Sharma NK, Mishra PK (2016) Impact of conservation practices on soil aggregation and the carbon management index after seven years of maize–wheat cropping system in the Indian Himalayas. Agric Ecosyst Environ 216:247–257

Gill AS, Deb Roy R (1992) Agroforestry research situation in India. Agriculture Situation in India 47:345–354

Giller KF, Cadisch G (1995) Future benefits for biological nitrogen fixation. Plant Soil 174:255–277

Goswami S, Verma KS, Kaushal R (2014) Biomass and carbon sequestration in different agroforestry systems of a Western Himalayan watershed. Biolog Agric Hort 3(2):88–96

Grewal SS (1993) Agroforestry systems for soil and water conservation in Shiwaliks. In: Gill AS, Deb Roy R, Bisaria AK (eds) Agroforestry in 2000 AD for the semi arid and arid tropics. NRCAF, Jhansi, pp 82–85

Gupta B, Sarvade S, Mahmous A (2015) Effect of selective tree species on phytosociology and production of understory vegetation in mid-Himalayan region of Himachal Pradesh. Range Managet Agrofor 36(2):156–163

Haggar JP, Tanner EU, Beer JW, Kass DCL (1993) N dynamics of tropical agroforestry and annual cropping systems. Soil Biolog Biochem 25:1363–1378

Hazra CR, Singh DP (1994) Soil and water conservation and silvipasture for wasteland improvement at Gaharawa watershed. Range Managet Agrofor 15(1):35–42

IPCC (2007) Climate change 2007: impact, adaptation and vulnerability. In: Pary et al (eds) Contribution of working group II to the fourth assessment report of the intergovernmental panel on climate change. Cambridge University Press, Cambridge

Jose S (2009) Agroforestry for ecosystem services and environmental benefits: an overview. Agrofor Syst 76(1):1–10

Joshi HB, Talapatra (1980) The chemical composition and nutritive value of the Himalayan tree fodder Bhimal (*Grewia oppositifolia* Roxb.). Indian J Dairy Sci 13:68–76

Kang BT, Wilson GF (1987) The development of alley cropping as a promising agroforestry technology. In: Steppler HA, Nair PKR (eds) Agroforestry: a decade of development. ICRAF, Nairobi, pp 227–243

Kaul M, Mohren G, Dadhwal V (2010) Carbon storage and sequestration potential of selected tree species in India. Mitig Adapt Strateg Glob Chang 15(5):489–510

Kaul M, Mohren G, Dadhwal V (2011) Phytomass carbon pool of trees and forests in India. Clim Chan 108(1):243–259

Kaur B, Gupta SR, Singh G (2002) Carbon storage and N cycling in silvipastoral systems on sodic soils in north west India. Agrofor Syst 54:21–29

Khurana DK, Negi YS (1993) Economic potential of agroforestry in hill farming system- a case study of Himachal Pradesh. In: Songkram L et al (eds) Forestry economic research in Asia. Faculty of Forestry, Kasetsort University, Thailand, pp 93–101

Kumar BM (2007) Litter dynamics in plantation and agroforestry systems of the tropics-a review of observations and methods. In: Batish DR, Kohli RK, Jose S, Singh HP (eds) Ecological basis of agroforestry. CRC Press, Boca Raton

Kumar A, Bahadur I, Maurya BR, Raghuwanshi R, Meena VS, Singh DK, Dixit J (2015) Does a plant growth promoting rhizobacteria enhance agricultural sustainability? J Pure Appl Microbiol 9(1):715–724

Leakey R (2011) Non-wood forest products: cultivating the key to better agriculture and improved livelihoods. Non Wood News 22:3–4

Lovenstein HM, Berliner PR, Van Keulen H (1991) Runoff agroforestry in arid lands. Forest Ecol Managet 45:59–70

Maharudrappa A, Srinivasmurthy CA, Nagaraja MS, Siddaramappa R, Anand HS (1999) Decomposition rates of litter and nutrient release pattern in a tropical soil. J Indian Soc Soil Sci 48:92–97

Maikhuri RK, Semwal RL, Rao KS, Singh K, Saxena KG (2000) Growth and ecological impacts of traditional agroforestry tree species in Central Himalaya, India. Agrofor Syst 48:257–271

Makundi WR, Sathaye JA (2004) GHG mitigation potential and cost in tropical forestry – relative role for agroforestry. Environ Develop Sustain 6:235–260

Meena RS, Yadav RS, Meena VS (2013) Heat unit efficiency of groundnut varieties in scattered planting with various fertility levels. The Bioscan 8(4):1189–1192

Meena RS, Meena VS, Meena SK, Verma JP (2015a) The needs of healthy soils for a healthy world. J Clean Prod 102:560–561
Meena RS, Meena VS, Meena SK, Verma JP (2015b) Towards the plant stress mitigate the agricultural productivity: a book review. J Clean Prod 102:552–553
Meena VS, Meena SK, Verma JP, Meena RS, Ghosh BN (2015c) The needs of nutrient use efficiency for sustainable agriculture. J Clean Prod 102:562–563
Meena VS, Meena SK, Verma JP, Meena RS, Jat LK (2015d) Current trend to mitigate climate change and sustainable agriculture: a book review. J Clean Prod 102:548–549
Meena VS, Verma JP, Meena SK (2015e) Towards the current scenario of nutrient use efficiency in crop species. J Clean Prod 102:556–557
Meena RS, Bohra JS, Singh SP, Meena VS, Verma JP, Verma SK, Sihag SK (2016) Towards the prime response of manure to enhance nutrient use efficiency and soil sustainability a current need: a book review. J Clean Prod 112(1):1258–1260
Misri B, Singh JP (1995) Grass covers of India and their production potential. In: Singh RP (ed) Forage production and utilization. IGFRI, Jhansi, pp 60–75
Mukherjee A, Maity A (2015) Public–private partnership for convergence of extension services in Indian agriculture. Curr Sci 109(9):1557–1563
Mukherjee A, Bahal R, Roy Burman R, Dubey SK (2012) Conceptual convergence of pluralistic extension at Aligarh district of Uttar Pradesh. J Community Mob Sustain Dev 7(1&2):85–94
Myers N (1988) Threatened biotas: 'hot spots' in tropical forests. The Environ 8:187–208
Nair PKR (1984) Soil productivity aspects of agroforestry. ICRAF, Nairobi, p 164
Nair PKR (1985) Classification of agroforestry systems. Agrofor Syst 3:97–128
Nair PKR (2001) Agroforestry in our fragile world: challenges and opportunities for sustainable development. Forerunner to the encyclopedia of life support systems, Chapter 1.25, pp.375–393, vol. I. UNESCO/EOLSS, Paris/Oxford.
Nair PKR, Nair VD, Kumar BM, Showalter JM, (2010) Carbon sequestration in agroforestry systems. In: Donald LS (eds) Advances in agronomy, Academic Press 108:237–307
Narain P, Singh RK, Sindhwal NS, Joshie P (1997) Agroforestry for soil and water conservation in the western Himalayan valley region of India 1. Runoff, soil and nutrient losses. Agrofor Syst 39:175–189
Negi SS, Pal RN, Enrich C (1979) Tree fodders in Himachal Pradesh – an introduction to six most common fodder trees in H.P. and their feeding value for cattle. GTZ, Eshborn, p 68
Noble IR, Dirzo R (1997) Forests as human dominated ecosystems. Science 277:522–525
Palm CA (1995) Contribution of agroforestry trees to nutrient requirements in intercropped plants. Agrofor Syst 30:105–124
Palsaniya DR, Ahlawat IPS (2007) Crop productivity, quality and nutrient uptake of pigeonpea-wheat cropping system as influenced by sulphur management. Indian J Agric Sci 77 (10):660–663
Palsaniya DR, Ahlawat IPS (2009) Sulphur management in pigeonpea (*Cajanus cajan*)-wheat (*Triticum aestivum*) cropping system. Indian J Agron 54(3):272–277
Palsaniya DR, Dhyani SK (2012) Natural silvipastoral covers of India. Indian J Forest 35 (2):157–166
Palsaniya DR, Singh R, Tewari RK, Yadav RS, Dwivedi RP, Kumar RV, Venkatesh A, Kareemulla K, Bajpai CK, Singh R, Yadav SPS, Chaturvedi OP, Dhyani SK (2008) Socio economic and livelihood analysis of people in Garhkundar-Dabar watershed of central India. Indian J Agrofor 10(1):65–72
Palsaniya DR, Dhyani SK, Tewari RK, Ramesh S, Yadav RS (2009a) Marketing issues and constraints in agroforestry. In: Chaturvedi OP, Venkatesh AR, Yadav S, Badre A, Dwivedi RP, Singh R, Dhyani SK (eds) Agroforestry: natural resource sustainability, livelihood and climate moderation. Satish Serial Publishing House, Delhi, pp 563–578
Palsaniya DR, Ram N, Yadav RS (2009b) Tree-crop interactions and their management in agroforestry. In: Chaturvedi OP, Venkatesh AR, Yadav S, Badre A, Dwivedi RP, Singh R,

Dhyani SK (eds) Agroforestry: natural resource sustainability, livelihood and climate moderation. Satish Serial Publishing House, Delhi, pp 241–262

Palsaniya DR, Singh R, Venkatesh A, Tewari RK, Dhyani SK (2010a) Grass productivity and livestock dynamics as influenced by integrated watershed management interventions in drought prone semi arid Bundelkhand, India. Range Manag Agrofor A 4–6

Palsaniya DR, Tewari RK, Singh R, Yadav RS, Dhyani SK (2010b). Farmer – agroforestry land use adoption interface in degraded agro-ecosystem of Bundelkhand region, India. Range Managet Agrofor 31(1):11–19

Palsaniya DR, Dhyani SK, Rai P, (2011a) Silvipasture in India: Present Perspectives and Challenges Ahead. Scientific Publishers, Jodhpur, India: p. 207

Palsaniya DR, Singh R, Yadav RS, Tewari RK, Dhyani SK (2011b). Now it is water all the way in Garhkundar – Dabar watershed of drought prone semi arid Bundelkhand, India. Curr Sci 100 (9): 1287–1288

Palsaniya DR, Khan MA, Tewari RK, Bajpai CK (2012a) Tree-crop interactions in *Psidium guajava* based agrihorticulture system. Range Managet Agrofor 33(1):32–36

Palsaniya DR, Singh HK, Ram M, Rai P, Singh UP, Yadav RS (2012b). Effect of pruning on growth and forage production in *Albizia Procera* based silvipastoral system in semi arid Bundelkhand, India. Range Managet Agrofor 33(1):96–98

Palsaniya DR, Singh R, Tewari RK, Yadav RS, Dhyani SK (2012c) Integrated watershed management for natural resource conservation and livelihood security in Semi Arid Tropics of India. Indian J Agric Sci 82 (3):241–247

Palsaniya DR, Singh R, Tewari RK, Yadav RS, Kumar RV, Dhyani SK (2012d) Integrated watershed management for sustainable agricultural production in Semi-Arid Tropics of India. Indian J Agron 57(4):310–318

Palsaniya DR, Kumar S, Chaudhary M et al (2013) Crop planning and management in agroforestry. In: Antony JR, Lal SB (eds) Agroforestry: theory and practices. Scientific Publishers, India, pp 206–220

Pandey DN (2007) Multifunctional agroforestry systems in India. Curr Sci 92(4):455–463

Pandey DN, Mohan D (1993) Nest site selection by cavity nesting birds on *Melia azedarach L.* and management of multiple use forests. J Bombay Nat Hist Soc 90:58–61

Pathak PS, Pateria HM, Solanki KR (2000) Agroforestry systems in India: a diagnosis and design approach. Indian Council of Agriculture Research, New Delhi

Ram J, Dagar JC, Lal K, Singh G, Toky OP, Tanwar VS, Dar SR, Chauhan MK (2011) Bio-drainage to combat water logging, increase farm productivity and sequester carbon in canal command areas of northwest India. Curr Sci 100(11):1673–1680

Rawat VS (2012) Van panchayat forest management as option in conserving biodiversity and carbon storage. Indian J Sci 1(1):32–35

Sah VK, Saxena AK (1990) Variation in structure, biomass and species diversity of grazing lands in Himalaya. Range Managet Agrofor 11:115–121

Sandhu J, Singh M, Ambasht RS (1990) N release from decomposing litter of *Leucaena leucocephala* in the dry tropics. Soil Biol Biochem 22:859–863

Sastry G, Mathur HN, Tejwani KG (1981) Landslide control in North-Western outer Himalayas. Bulletin No. R8/D6, CSWCR&TI, Dehradun

Sharma VK, Sood BR (1994) Effect of introduction and cutting management on the productivity of a natural grassland. Range Manag Agrofor 15:11–14

Sharma R, Sharma E, Purohit AN (1996a) Cardamom, mandarin and nitrogen fixing trees in agroforestry systems in Indias Himalayan region. I. Litterfall and decomposition. Agrofor Syst 35:239–253

Sharma R, Sharma E, Purohit AN (1996b) Cardamom, mandarin and nitrogen fixing trees in agroforestry systems in Indias Himalayan region. II. Soil nutrient dynamics. Agrofor Syst 35:255–268

Singh A (1988) Hydrological behaviour of experimental watersheds. Annual Report ICARRC for NEH Region, Barapani, pp 169–171

Singh G (1995) An agroforestry practice for the development of salt lands using *Prosopis juliflora* and *Leptochloa fusca*. Agrofor Syst 27:61–75

Singh V (2009) Biomass stock and carbon sequestration rates in banj oak (*Quercus leucotrichophora, a. Camus.*) forest under different disturbance regimes in central himalaya. Ph.D. thesis, Kumaun University, Nainital

Singh P, Misri B (1993). Rangeland resources-utilization and management in India. Paper presented at international symposium on grassland resources held at Huehot, Inner Mangolia, August, 16–20

Singh G, Singh NT (1996) Effects of land use practices on soil organic carbon dynamics of sodic soils. In: Lal R et al (eds) Carbon sequestration in soils. Vol. III proceedings of international symposium on carbon sequestration in soil. The Ohio State University, Columbus, 22–26 July

Singh G, Abrol IP, Cheema SS (1989a) Effects of spacing and lopping on mesquite-Karnal grass agroforestry system in a sodic soil. Exp Agric 25:401–408

Singh K, Chauhan HS, Rajput DK, Singh DV (1989b) Report of a 60 month study on litter production, changes in soil chemical properties and productivity under poplar (*P. deltoides*) and Eucalyptus (*E. hybrid*) interplanted with aromatic grasses. Agrofor Syst 9:37–45

Singh G, Singh H, Singh J (1996) Effects of Populus deltoids litter on its saplings, associated agricultural crops and on the properties of an alkali soil. J Trop Forest Sci 9:536–545

Singh G, Singh H, Dagar JC, Singh NT, Sharma VP (1997) Evaluation of agriculture forestry and agroforestry practices in moderately alkali soil in north western India. Agrofor Syst 37:279–295

Singh G, Singh H, Bhojvaid PP (1998) Amelioration of sodic soils by trees for wheat and oat production. Land Degrad Develop 9:536–545

Singh VS, Pandey DN, Chaudhry P (2010) Urban forests and open green spaces: lessons for Jaipur, Rajasthan, India, RSPCB Occasional Paper No. 1/2010, Rajasthan State Pollution Control Board, Jaipur, p 23

Singh AK, Ngachan SV, Munda GC, Mohapatra KP, Choudhary BU, Das A, Rao CS, Patel DP, Rajkhowa DJ, Ramkrushna GI, Panwar AS (2012) Carbon management in agriculture for mitigating green house effect. ICAR Research Complex for NEH Region, Umiam, 377p

Singh DK, Singh SK, Singh AK, Meena VS (2014) Impact of long term cultivation of lemon grass (Cymbopogon citratus) on post-harvest electro-chemical properties of soil. Annals of Agri Bio Res 19(1):45–48

Srivastava AK, Misri B, Singh B (2003) Resource management in silvipastures in Shivaliks. Indian Forest 129(4):446–456

Sundriyal M, Sundriyal RC (2001) Wild edible plants of the Sikkim Himalaya: nutritive values of selected species. Econ Bot 55:377–390

Tejwani KG (1994) Agroforestry in India. Oxford & IBH Publishing Co. Pvt. Ltd, New Delhi, p 181

Tewari RK, Dhyani SK, Palsaniya DR, Singh R, Yadav RS (2012) Farmer's participatory action research programme, phase-II. Technical Bulletin 3/2012, NRCAF, Jhansi

Thornton PK, Herrero M (2010) Potential for reduced methane and carbon dioxide emissions from livestock and pasture management in the tropics. Proc Natl Acad Sci U S A 107 (46):19667–19672

Turner NC, Ward PR (2002) The role of agroforestry and perennial pasture in mitigating water logging and secondary salinity: summary. Agric Water Managet 53:271–275

Tynsong H, Tiwari BK (2010) Diversity of plant species in arecanut agroforests of south Meghalaya, north-east India. J Forest Res 21(3):281–286

Verma JP, Jaiswa DK, Meena VS, Meena RS (2015a) Current need of organic farming for enhancing sustainable agriculture. J Clean Prod 102:545–547

Verma JP, Jaiswal DK, Meena VS, Kumar A, Meena RS (2015b) Issues and challenges about sustainable agriculture production for management of natural resources to sustain soil fertility and health. J Clean Prod 107:793–794

Vishwanathan MK, Samra JS, Sharma AR (1998) Biomass production of silvipasture systems of gravelly lands of Doon Valley. Bulletin no. T-39/D-26, CSWCRTI, Dehradun, p 126

Yadav RS, Yadav BL, Chhipa BR (2008) Litter dynamics and soil properties under different tree species in a semi-arid region of Rajasthan, India. Agrofor Syst 73:1–12

Yadav RS, Palsaniya DR, Tewari RK, Gupta G (2009) Litterfall, decomposition and nutrient release and its implications in agroforestry interventions. In: Chaturvedi OP, Venkatesh A, Yadav RS, Bdre A, Dwivedi RP, Ramesh S, Dhyani SK (eds) Agroforestry: natural resource sustainability, livelihood and climate moderation. Satish Serial Publishing House, Delhi, pp 317–334

Yadava RB, Varshney RB (1997) Multistorey silvipastoral approach for sustaining soil fertility in degraded rangelands of Bundelkhand. Range Managet Agrofor 18(2):195–200

Elevated Carbon Dioxide (CO_2) and Temperature vis-a-vis Carbon Sequestration Potential of Global Terrestrial Ecosystem

9

M.L. Dotaniya, S. Rajendiran, B.P. Meena, A.L. Meena, C.K. Dotaniya, B.L. Meena, R.L. Jat, and J.K. Saha

Abstract

Industrialization across the world during the twentieth century enhanced global economic growth rate in quantum speed, but the global greenhouse gases (GHGs) level in the atmosphere also augmented in the manifold along with the economic growth. The elevated level of GHGs particularly CO_2 due to burning of fossil fuels and agricultural and industrial process increased the atmospheric temperature. The global mean surface temperature has increased by about 0.74 °C over the last 100 years. The Intergovernmental Panel on Climate Change (IPCC) projected that it will rise further to 1.4–6.4 °C units by the end of the twenty-first century. The consequences of global warming is felt everywhere, viz., poor and uncertainty of rainfall, prolonged dry spell, more incidence of insect and pest, reduction in soil carbon, more water stress to crop plants, reduction in crop productivity, etc. The rising atmospheric CO_2 affected the plant physiological processes and soil microorganism activities and biochemical pathway of plant nutrients as well as carbon (C) source-sink relationships of terrestrial ecosystems. Hence mitigation and sequestration of C as stable pool in soil is the need of hour to counter the increasing CO_2 in the

M.L. Dotaniya (✉) • S. Rajendiran • B.P. Meena • J.K. Saha
ICAR-Indian Institute of Soil Science, Bhopal 462 038, India
e-mail: mohan30682@gmail.com

A.L. Meena
ICAR-Indian Institute of Farming Systems Research, Modipuram 250 110, India

C.K. Dotaniya
College of Agriculture, SKRAU, Bikaner 334 006, India

B.L. Meena
ICAR-Central Soil Salinity Research Institute, Karnal 132 001, India

R.L. Jat
ICAR-Indian Institute of Pulses Research, Kanpur 208 017, India

J.K. Bisht et al. (eds.), *Conservation Agriculture*,
DOI 10.1007/978-981-10-2558-7_9

atmosphere. Soil is a major store of C, containing three times as much as C in the atmosphere and five times as much as forests. But a minor variation in such big C pool in terrestrial ecosystems may have a significant effect on the C flux. Due to environmental complexities, it is difficult to quantify soil C changes. However, researchers across the world are trying to estimate the soil carbon dynamics under different land-use and management systems to bridge the gap in the carbon sequestration potential. Also use of available agronomic and management options can be effectively utilized to combat the adverse effect of elevated CO_2 and temperature; thereby, crop productivity and sustainability of the terrestrial ecosystem are maintained for a long run.

Keywords
Carbon sequestration potential • Greenhouse gases • Plant growth • Root exudates • Soil enzymatic activities

9.1 Introduction

The climate is one of the important parts of the global ecosystem and significantly affected the various cycles and processes related to human, animal, and plant system. In the twentieth century, scientists anticipated the need of the future and tried to develop a database for the development of agricultural production and close monitoring of ecosystem changes. This is very helpful for predicting the consequences of various parameters of agricultural crop productivity. The increasing industrialization and burning of fossil fuels enhanced the greenhouse gases (GHGs) into the atmosphere. The most imminent of this is increased concentration of GHGs, namely, carbon dioxide (CO_2), methane (CH_4), and nitrous oxide (N_2O), in the atmosphere. The concentration of CO_2, CH_4, and N_2O have increased markedly by 30 %, 145 %, and 15 %, respectively, as a result of human activity since the industrial revolutions (1750) (IPCC 2007). Among the GHGs, CO_2 plays an important role in global climate change phenomena. The atmospheric concentration of CO_2 has increased from 280 ppm in the pre-industrial era to 385 ppm in 2008 (+37.5 %) and is presently increasing at the rate of 2 ppm $year^{-1}$ or 3.5 Pg $year^{-1}$ (1 Pg or pentagram = 1 Gt = 1 gigaton = 1 billion metric ton). The increase in CO_2 emission by human activities is attributed to fossil fuel combustion, deforestation and biomass burning, soil cultivation, and drainage of wetlands or peat soils. The world population of 7.3 billion in 2015 is increasing at the rate of 1.3 % $year^{-1}$ and is projected to be 9.5 billion by 2050 before stabilizing at 10 billion toward the end of the twenty-first century. There exists a strong positive correlation between population growths and CO_2 emission or the energy demand (PRB 2015). The People's Republic of China accounted for ~13 % of the world's C emissions in 2002, and that share is projected to rise to 18 % by 2025 (Lal 2004a).

Increase in concentration of the GHGs particularly CO_2 in the atmosphere is leading to global warming. Over the last 100 years (1906–2005), global mean surface temperature has increased in the order by 0.74 ± 0.18 °C. The rate of increase in temperature is highest in recent decades. Fourteen of the last 17 years (1995–2011) rank the warmest years since 1850. The IPCC has projected that by the end of the twenty-first century, the temperature increase is likely to be in the range of 2–4.5 °C. The increasing temperature modifies the plant nutrient factors and various nutrient cycles in nature. It also affected the crop productivity and quality of produce and extreme weather conditions creating unusual dry spell and heavy rainfall. Worldwide, the net effect of climate change will be to decrease stocks of organic carbon (C) in soils, thus releasing additional CO_2 into the atmosphere and acting as a positive feedback, further accelerating climate change. It is a global concern to decrease the atmospheric global warming gas contribution and enhance the soil organic carbon into the soil. Soil organic carbon is an important index of soil quality because of its relationship to crop productivity (Lal 2014, 2015a, b). Agricultural activities have profound influence on changes in soil organic carbon both in the short and the long terms. The estimated potential of agricultural intensification on soil organic carbon (SOC) sequestration in soils of India ranges between 12.7 and 16.5 Tg C year^{-1} (Lal 2003b). The overall aim should be to increase SOC density and distribution of SOC in the subsoil and improve aggregation.

Elevated atmospheric CO_2-induced temperature and precipitation changes directly or indirectly influence the major key soil processes. A proper understanding is required, in order to better predict the likely impact of elevated atmospheric CO_2 on soil. The objective of this chapter is to assess the impact of elevated atmospheric CO_2, temperature on various key soil processes, and carbon sequestration potential of the global terrestrial ecosystem.

9.2 Trend of Changes in Greenhouse Gases and Global Warming

The climate change phenomena are complex in nature, and every nation is worried about its consequences on ecosystems. A lot of researchers are working on various climate change-related issues and trying to find the exact data related to diverse entities. The climate change models are predicting the global GHG emission and temperature and its associated consequences. The role of soil microorganisms is changing, and reducing biodiversity is also closely related to elevated temperature and GHG level. The changing pattern of global warming is a matter of concern for the sustainable stay of human on the Earth. Increasing temperature promotes vast shifting on soil biota with respect to abundance, composition, mode of function, and structure (Courty et al. 2010; Meena et al. 2013a, b, 2015a, b; Singh et al. 2014; Kumar et al. 2015; Ghosh et al. 2016). The changing trends in GHG concentration and temperature are measured by various governmental and private organizations at various levels in every country. The Intergovernmental Panel on Climate Change (IPCC), established by the United Nations in 1988, is not doing original research but assessing the published literature in peer-reviewed and non-peer-reviewed sources and giving the predictions related to climate change. It provides the internationally accepted data regarding climate change. IPCC produced a report which was supported by the United Nations Framework Convention on Climate Change (UNFCCC). It is a landmark international treaty in which the central goal is to "stabilize greenhouse gas concentrations in the atmosphere at a level that would prevent dangerous anthropogenic interference with the climate system." The IPCC is closely working on the climate-related scientific, technical, social, and economic aspects as well as options for adaptation and mitigation. Some of the main parameters are discussed below.

9.2.1 Greenhouse Gases

The main distribution of GHGs in the atmosphere is water vapor (36–72 %) > CO_2 (9–26 %) > CH_4 (4–9 %) and ozone (3–7 %). The CO_2 concentration increased 40 % from 280 ppm in 1750 to ~400 ppm in 2015. The elevated concentration of GHGs is mainly due to anthropogenic activities like burning of fossil fuel, industrial emissions, and combustion of wood, coal, oil, and natural gas (Table 9.1). The large portion of emitted CO_2 is taken up by the plant during the photosynthesis, but the decreasing green vegetation reduced the total amount in the lower figure.

The increasing concentration of GHGs in the atmosphere increases the global temperature. In general terms, it is defined that those gases that absorb and emit radiation within the thermal infrared range in the atmosphere are called GHGs. It is the main cause of greenhouse effect contributed by the primary gases, i.e., water vapor, carbon dioxide, methane, nitrous oxide, and ozone, in the Earth's atmosphere. The science behind the increasing global temperature due to GHGs is based on physics principles. The sun's energy is predominantly short-wave radiation and arrives to the Earth's atmosphere. Approximately 30 % of this energy is reflected back into space with the help of the clouds, aerosols, gases, and Earth's surface. The remaining incoming energy, 20 %, is absorbed by the atmosphere and 80 % by the Earth's surface. The Earth's surface gets warm and emits the energy in long-wave infrared, and this is absorbed by the GHGs and increases the temperature of the atmosphere. This whole process is called the greenhouse effect. This effect is mainly governed by the characteristics of gas and its abundance in the atmosphere. If we image the Earth temperature in the absence of GHGs, then it would be 32 centigrade cooler. Since this temperature is well below the freezing point of

Table 9.1 Elevated level of GHGs

Gas	Pre-industrial level	Current level since 1750
CO_2	280 ppm	388 ppm
CH_4	700 ppb	1745 ppb
N_2O	270 ppb	314 ppb
CFC-12	0	533 ppt

(Source: Adapted from Lenka et al. 2013a)

water, it is apparent that the planet would be much less hospitable to life in the absence of the greenhouse effect (Anderson 2013).

The global warming potential (GWP) of gases plays a crucial role in mounting global atmospheric temperature. The IPCC provides the generally accepted values for GWP, which changed slightly between 1996 and 2001. An exact definition of how GWP is calculated is to be found in the IPCC's 2001 Third Assessment Report. The GWP is defined as the ratio of the time-integrated radiative forcing from the instantaneous release of 1 kg of a trace substance relative to that of 1 kg of a reference gas:

$$GWP(x) = \frac{\int_0^{TH} a_x \cdot [x(t)]dt}{\int_0^{TH} a_x \cdot [r(t)]dt}$$

where *TH* is the time horizon over which the calculation is considered; a_x and a_r are the radiative efficiency of any substances and reference gas due to their unit increase in the atmosphere (i.e., $Wm^{-2}\ kg^{-1}$), and $[x(t)]$ is the time-dependent decay in abundance of the substance following an instantaneous release of it at time $t = 0$. The denominator contains the corresponding quantities of the reference gas (i.e., CO_2). The radiative efficiencies a_x and a_r are not necessarily constant over time. While the absorption of infrared radiation by many greenhouse gases varies linearly with their abundance, a few important ones display nonlinear behavior for current and likely future abundances (e.g., CO_2, CH_4, and N_2O). For those gases, the relative radiative forcing will depend upon abundance and hence upon the future scenario adopted. Carbon dioxide has a GWP of exactly 1 (since it is the baseline unit to which all other greenhouse gases are compared). Some of the GHGs are listed below with GWP (Table 9.2).

Table 9.2 Global warming potential of GHGs

Gas	GWP
CO_2	1
CH_4	21
N_2O	310
HFC-23	11,700
HFC-32	650
HFC-125	2800
HFC-134a	1300
CF_4	6500
C_2F_6	9200
C_4F_{10}	7000
C_6F_{14}	7400
SF_6	23,900

(Source: Adapted from Mohanty et al. 2015)

9.2.2 Temperature

As per the data analyzed by the NASA's Goddard Institute for Space Studies (GISS), researchers found that the mean global temperature of Earth has increased 0.8 °C since 1880. Over the twentieth century, the average temperature of the Earth has increased by 0.4–0.8 °C and expected 1.4–5.8 °C by the twenty-first century. The increasing temperature enhanced the more chances of weather extremes across the globe. The extremes are called in the form of economical (insurance costs), social (loss of life), and environmental (destruction of habitat) losses. The elevated temperature increased the rate of glaciers melting and increasing the sea levels, for example, Montana's Glacier National Park has 150 glaciers in the year 1910 but now only 27. The major loss of coral biodiversity was observed in 1998, which was highly susceptible to minute changes in aquatic temperature. The global energy balance can be described with the help of radiative forcing (RF). The change in concentration of atmospheric GHGs and aerosols, solar radiation, and land cover modified the energy balance of the system and the drivers of climate change (Myhre et al. 2013). In simple words RF is a concept in which quantitative comparisons of the strength of different natural and human agents are causing climate change. It may be positive or negative changes and used for the comparison of warming and cooling influences on the climate change. If we pooled the data of GHGs emitted from the 1750 baseline, then the RF increased 1.46, 0.48, 0.15, and 0.17 regarding CO_2, CH_4, N_2O, and CFC-12, respectively. The RF can be related through a linear relationship to the global mean equilibrium temperature change at the surface as follows (Ramaswamy et al. 2001):

$$\Delta Ts = \lambda RF$$

where,

ΔTs = global mean equilibrium temperature change at the surface
λ = climate sensitive parameters

The equation predicts the linear view of global mean climate change between two equilibrium states. This is simple to identify the ranking as well as quantify the effect of climate change, but it is limited to measuring the overall influence of climate change. The accumulative RF of CO_2, CH_4, and N_2O is positive 2.3 (+2.1–2.5) W m^{-2} and increased due to increasing concentration of gases in the atmosphere since the industrial era. According to the IPCC, AR3, the CO_2 RF increased ~20 % in the duration of 1995–2005, which was the highest change for any decade in the last 200 years. The increasing level of GHGs influences the ozone level of the stratosphere, which is acting as a life-saving layer in the atmosphere regarding regulating the transmittance of ultraviolet light to the Earth's surface. This light causes various types of harmful effects on the human, animal, and plant ecosystem. O_3 is highly phytotoxic, and its exposure caused a symptomatic and chronic effect (Ghude et al. 2008). Across the globe, tropospheric O_3 concentration increased from the pre-industrial level of 10–20 ppb to the present level of 40 ppb; and it is increasing by 1–2 % per year in Northern Hemisphere (Fiore et al. 2008).

It is believed that the amount of atmospheric warming due to the increase of tropospheric O_3 is comparable to, or possibly even greater than, the amount of warming due to CO_2. Meehl et al. (2007) mentioned that the data interpretation of various climate change models showing tropospheric O_3 level will increase 20–25 % between 2015 and 2050 and further increase by 40–60 % by 2100, if the present emission trends continued. Increasing the Earth temperature with minute fraction affected the Earth's ecosystems, soil biodiversity, crop production activities, and melting of glacier caps (Grover et al. 2015). The increase in rate of ice melting leads to raise in sea levels is a major concern mainly for water-located countries and cities situated on the bank of the ocean. On the side of food security, elevated temperature affected the yield potential of agriculture crops (Chen et al. 2014). The elevated temperature may have some beneficial effects in the temperature-limiting areas, whereas in the sufficiency areas, the photosynthate transfer of economic part is reducing due to a more respiration rate. Plants transfer ~40 % of total photosynthates in soil C as a rhizodeposition (Jones et al. 2009; Dotaniya et al. 2013c, d). This value might be higher for the temperature-sensitive crops. These root exudates act as a source of food for soil microorganisms and mediate the plant nutrient transformations (Dotaniya and Meena 2013). The elevated temperature modified the crop plant physiology and C level in plant and soil (Drigo et al. 2013). The crop management factors also need to modify with respect to elevated level of GHGs and temperature. Therefore, it is an urgent need to carefully monitor the global climate change effect on natural resources and different ecosystems and compute the single and multifactor effect for understanding the clear picture.

9.3 Impact of Climate Change on Soil System

Elevated atmospheric CO_2 has direct as well as indirect impacts on soil processes and properties through imposed changes in soil temperature, soil water, and nutrient competition. Soil processes in general have been broadly classified into structural (physical) and hydrological processes and biogeochemical processes. These are often working simultaneously in the soil and involved in influencing the various soil quality parameters in any soil ecosystem. The structural and hydrological processes include aggregation, dispersion, removal of topsoil (erosion) and runoff, soil water availability, and distribution and its movement. These processes mainly affect the soil structure, soil water and aeration, and other related properties. Important chemical and biochemical processes that occur in the soils are mineralization, immobilization, dissolution, precipitation, sorption, desorption and oxidation, and reduction (Sanyal and Majumdar 2009). They are involved in influencing the dynamics of several nutrients, soil carbon dynamics, and ultimately soil biodiversity by creating nutrition competition (Dotaniya et al. 2013a, b). The key soil processes like soil-forming processes, biogeochemical processes, and hydrological processes are likely to be influenced by global changes. All these are temperature- and moisture-dependent and microbial-mediated processes. Thus, elevated atmospheric CO_2-induced temperature and precipitation change results in changes of these key soil processes which will subject the soils to physical and chemical degradation, soil erosion, salinization, decreased water availability and changes in C and N dynamics, decreased nutrient storage in soil, and depletion of soil biodiversity (Meena et al. 2015c, 2016; Verma et al. 2015a, b). These unfavorable changes pose a big threat to soil productivity, soil and water quality, and sustainability of the production system (Benbi and Kaur 2009; Emmett et al. 2004). The potential impacts of elevated atmospheric CO_2 on soil processes and resultant unfavorable changes in soil quality parameters are briefed on the following discussions.

9.3.1 Soil Genesis

Climate, vegetation, parent material, topography, and time are the factors which affect the soil formation. Among these, climate and vegetation are considered as active factors. Changing global atmospheric CO_2 may affect the climate particularly temperature, rainfall, and vegetation in an ecosystem. These changes accelerate the soil-forming process by modifying the processes of chemical, biochemical, or mineralogical changes in soil parent material over a period of time at prevailing topographic conditions.

Changes in the clay mineral surfaces or the bulk composition of the clay fraction of soils are brought about by a small number of transformation processes, listed below (Brinkman 1982). Each of these processes can be accelerated or inhibited by changes in external conditions due to global atmospheric CO_2 change.

- Cheluviation, which dissolves and removes especially aluminum and iron by chelating organic acids
- Hydrolysis by water containing carbon dioxide, which removes silica and basic cations
- Dissolution of clay minerals by strong mineral acids, producing acid aluminum salts and amorphous silica
- Ferrolysis, a cyclic process of clay transformation and dissolution mediated by alternating iron reduction and oxidation, which decreases the cation exchange capacity by aluminum interlayering in swelling clay minerals
- Reverse weathering, i.e., clay formation and transformation under neutral to strongly alkaline conditions, which may create, e.g., montmorillonite, palygorskite, or analcime

These changes in the long run may alter soil-formation processes in different degrees as per the local environmental conditions prevailing in a particular region result in changes in world soil distribution scenario.

9.3.2 Soil Water

Climate change can affect soil hydrological cycle through multiple pathways because many climatic variables such as precipitation, temperature, and CO_2 concentration as well as their interactions are often complex, dynamic, and nonlinear. The actual impacts of individual variables and/or their interactions, which may differ seasonally and geographically, can be adequately assessed. The rise in temperature increases the rate of evapotranspiration and causes soil drying and lowering of groundwater table (Abtew and Melesse 2012). Rapid melting of snow and large glaciers of the Himalayas may increase the stream flow of some north Indian rivers and increase the availability of water for crop growth for short periods, but in the long run the scenario may change (Gosain et al. 2006). The decrease in atmospheric precipitation in some parts will result in a decrease in water infiltration and water storage in the soil. The extreme variability in precipitation pattern, spatially and temporally, may lead to higher surface runoff and erosion, which radically modify the field water balance and its component.

9.3.3 Soil Carbon and N Cycle

The changes in climate are likely to influence the rate of accumulation and decomposition of soil organic matter (SOM), both directly through changes in temperature and moisture and indirectly through changes in plant growth and rhizodepositions (Das and Hati 2010). Increased temperature and frequent rainfall stimulate microbial activity. Hence there will be an increase in mineralization or decomposition of organic residues in the soil that releases the CO_2 to the atmosphere (Dotaniya 2013; Dotaniya et al. 2014a, 2015). Increased atmospheric CO_2 increases in plant water-use efficiency (WUE) results in increased biomass production per mm of available water (Kimball 2003). But the decomposition rates are greater than primary production under increased water deficits, so it causes the reduction of biomass accumulation and depletion of soil C and decreases the C:N ratio of soil (Rosenzwig and Hillel 2000; Kirschbaum 2000; Lal 2004b). Drier conditions favor the organic carbon reduction and thus reduction in annual and perennial vegetation. During summer periods increased risk of fires that destroy vegetation cover leads to rapid defoliation and conversion of carbon stores to atmospheric carbon (Hennessy et al. 2005; Meena et al. 2015d, e).

The overall climate change has the negative impact on soil organic carbon and its storage in tropical climate. Globally, the net effect of climate change will be to decrease stocks of organic carbon in soils, thus releasing additional CO_2 into the atmosphere and acting as a positive feedback, further accelerating climate change. In humid tropics and temperate region, climate change-induced increase in rainfall will have a positive influence on C storage in the soil, while some parts of the subtropics where scarce rainfall and higher temperature-induced drought and desertification will reduce the vegetation cover and carbon sequestration potential of the soil (Kushwah et al. 2014). Climate change-mediated soil organic matter

Table 9.3 Pre-industrial and contemporary N depositions (Tg N year^{-1}) onto a different hemisphere

Latitude	Pre-industrial	1990s
Northern Hemisphere	4.03	26.65
Southern Hemisphere	1.87	1.76
Tropical	7.64	15.55

(Source: Adapted from Kundu et al. 2013a, b)

dynamics is a complex process that influenced by many factors and processes will differently interplay on a long-term basis is unpredictable.

In many natural and seminatural ecosystems, the atmospheric N deposition has increased over the years. As compared to estimated inputs of 1–3 kg N ha^{-1} year^{-1} in the early 1900s (Goulding 1998), the atmospheric N deposition rates of 20–60 kg N ha^{-1} year^{-1} in non-forest ecosystems in Western Europe and up to 100 kg N ha^{-1} year^{-1} in forest stands in Europe or the United States have been observed (Bobbink et al. 2002). While there has been an increase in deposition rate across all the biomes at both temperate and tropical latitudes, the increase is the greatest in the Northern Hemisphere temperate ecosystems (Table 9.3). Higher deposition rates are probably driven by biomass burning and soil emissions of NOx and NH_3 as well as lightening production of NOx.

While deposition of N to agricultural lands or croplands could serve as a source of nutrient, it could also adversely affect several processes in the soil. Nitrogen deposition can lead to reduction in biodiversity, soil acidification, and increased nitrous oxide emissions from soils. It also altered the balance of nitrification and mineralization/immobilization and increased NO_3 leaching resulting in contamination of surface and underground water bodies. Increasing temperature in combination with aerobic conditions increases nitrification and increases the mineral nitrogen in the soil. Episodic rainfall in arid and semiarid region increases the risk of nitrate loss due to leaching. In anaerobic conditions, the rate of denitrification will also increase the loss of nitrogen as N_2O or N_2 which reduces the plant-available nitrogen in soil. Depressing effects of temperature on symbiotic nitrogen bacteria also reduce the nitrogen fixation and increase the dependence on artificial nitrogen sources. The models simulate denitrification in soils and indirectly address the impact of climate change by including adjustment functions for substrate availability soil temperature and moisture to calculate actual denitrification rates from the potential rate. The main problems with the models for simulating N regime of agricultural soils are that these do not explicitly include the effect of atmospheric N deposition and changes in ambient CO_2 concentration. So the long-term effect of climate change on nitrogen dynamics is complex and unpredictable.

9.3.4 Soil Enzyme Activity and Microbial Diversity

Factors influencing soil microbial activity exert control over soil enzyme production and nutrient availability (Groenigen et al. 2015). Warming and drought affect the activity of the soil enzymes involved in C and N mineralization in the mid and long term. A recent study conducted in Mediterranean shrublands by Sardans et al. (2008) revealed that warming increased soil urease activity by 10 % during the study period (1999–2005) and increased β-glucosidase activity ~38 %. Soil urease and β-glucosidase activities were positively correlated with soil temperatures in winter and negatively in summer. Drought reduced soil protease activity (9 %) and did not affect β-glucosidase activity. They have observed that warming and drought have changed some soil enzyme activities related to P turnover in some year seasons (Sardans et al. 2008). The effects of warming and drought on soil enzyme activities were due to a direct effect on soil temperature and soil water content (Dotaniya et al. 2014c) and not to changes in soil organic matter quantity and nutritional quality (Sardans et al. 2008; Dotaniya and Kushwah 2013).

Temperature, precipitation, and vegetation changes considerably influence the microbial community and their activity in the soil. Particularly under condition of elevated CO_2 and changed temperature and moisture regimes, soil biodiversity and its function are expected to change. Elevated CO_2 elicited a 47 % average increase in mycorrhizal abundance and that mycorrhizae were stimulated disproportionately more than roots (percentage colonization increased by more than 30 %) (Treseder 2004). Another meta-analysis reported that the mass of ectomycorrhizal (EM) fungi increased by 34 % in CO_2-enriched environments, while that of endomycorrhizal (AM) fungi increased by 21 % (Alberton et al. 2005). Overall, changes in plant growth and rhizodepositions have an impact on soil microbial activity. However, reduced soil moisture creates unfavorable environment for microflora and microfauna as well as macrofuana like earthworms, termites, arthropods, etc.

9.4 Impact of Climate Change on Plant System

Climate change is unavoidable and associated with weather extremes such as high temperature, heat waves, increased frequency of drought, and high intensity of rainfall causing floods. All these factors directly or indirectly affect the plants in various ways. The effect of elevated CO_2 is briefly discussed.

9.4.1 Loss of Crop Yields

Based on many studies covering a wide range of regions and crops, negative impacts of climate change on crop yields have been more common than positive impacts (Ahmed et al. 2015). The smaller number of studies showing positive impacts relates mainly to high-latitude regions, though it is not yet clear whether the balance of impacts has been negative or positive in these regions (Schlenker and Lobell 2010). Climate change has negatively affected wheat and maize yields for many regions and in the global aggregate (Lobell et al. 2013). Many researchers reported detrimental effect of climate change on agricultural crops mainly on cereals in sub-Saharan Africa (Ruane et al. 2013; Waha et al. 2013). Schlenker and Lobell (2010) estimated 15 % decreased mostly in maize and millets in African countries. Effects on rice and soybean yield have been smaller in major production regions and globally, with a median change of zero across all available data, which are fewer for soybean compared to the other crops. Observed impacts relate mainly to the production aspects of food security rather than access or other components of food security. Since AR4, several periods of rapid food and cereal price increases following climate extremes in key producing regions indicate a sensitivity of current markets to climate extremes among other factors.

9.4.2 Physiological Impacts

Increases in atmospheric CO_2 concentration affect photosynthesis, resulting in increases in plant water-use efficiency, enhanced photosynthetic capacity, and increased growth (Steffen and Canadell 2005). Increased CO_2 has been implicated in "vegetation thickening" which affects plant community structure and function (Gifford and Howden 2001). Depending on the environment, there are differential responses to elevated atmospheric CO_2 between major "functional types" of plant, such as C_3 and C_4 plants, or more or less woody species, which has the potential among other things to alter competition between these groups (Jeffrey and Harold 1999). Increased CO_2 can also lead to increased carbon/nitrogen ratios in the leaves of plants or in other aspects of leaf chemistry.

9.4.2.1 CO_2 Fertilization Effect

The atmospheric concentration of CO_2 has two major short-term, but direct effects on plants. It increases photosynthesis in C_3 plants and decreases stomatal conductance and transpiration in C_3 and C_4 plants (Long et al. 2006). The increase in net photosynthetic C accumulation due to increase in atmospheric CO_2 concentration is called "CO_2 fertilization effect." The term literally indicates that with a supply of CO_2, the production increases similar to the production increase observed when nutrients like N and P are added. This effect is pronounced in C_3 plant species. In C_3 plants, the first target of CO_2 is ribulose-1, 5-bisphosphate carboxylase/ oxygenase (RuBisCO). The enzyme is not CO_2 saturated under the current level of atmospheric CO_2 concentration. Under CO_2 enrichment may suppress the oxygen's reactions due to CO_2 competes with O_2 for the same reaction sites on RuBisCO (Long et al. 2006). The estimates from free-air enrichment (FACE) and other studies show 14 % (Lee et al. 2011) to 26–31 % (Nowak et al. 2004; Ainsworth and Rogers 2007) increase in light-saturated photosynthesis. The availability of additional photosynthates enables most plants to grow faster under elevated CO_2, with an average increase in dry matter production by 17 % of the aboveground and more than 30 % of the belowground biomass (de Graaff et al. 2006). The increased growth results in higher yield of crops, with wheat, rice, and soybean accounting for yield increase of 12–14 % in the elevated CO_2 conditions (Ainsworth and Long 2005).

9.4.2.2 Increased Water-Use Efficiency in Plants

Response of C_4 plants, on the contrary, is more as a consequence of increased water-use efficiency (WUE). A reduced stomatal conductance decreases the transpiration loss of water through stomatal opening. Stomatal conductance is reduced by 22 % in field crops and graminaceous species (Ainsworth and Rogers 2007; Lee et al. 2011) and by 33–50 % across a broad range of C_3 and C_4 plant species (Kimball et al. 2002). A possible mechanism may be that the ratio of intercellular and to ambient CO_2 remains approximately constant; hence, under elevated CO_2 conditions, the stomata remains closed as there is no any additional requirement of CO_2 and thus reduces the transpiration loss.

9.4.2.3 Reduction in Leaf N and Protein Concentration

The primary mechanism for reduction of tissue N concentration seems to be from a reduction in RuBisCO content since much of the soluble leaf protein is associated with RuBisCO. The reduction leaf N concentration can be as much as 10–20 % in grasses and 13–31 % in field crops. The reduced leaf N concentration also causes lowering of N in grains by 8–31 % and in protein content in wheat, rice, barley, and potato by 5–14 %. Under elevated CO_2 conditions, leaves tend to accumulate more sugar and starch, resulting in 30–40 % higher nonstructural carbohydrates on the unit leaf area. Overall the leaf C:N ratio increases under elevated CO_2 conditions in both C_3 and C_4 plants. An exception is observed under in case of nodulating legumes where almost no significant effect of elevated CO_2 on C:N ratio is observed. For example, in some comparative studies of legumes with non-legumes, C:N ratio increased to 27–40 % in grain crops, and no changes were observed in legumes. Hence plants demand more N under elevated CO_2 conditions for higher productivity and increasing the protein content (Taub et al. 2008). High-N-fertilized (more than 30 kg N ha^{-1}) treatments show 12 % aboveground and 25 % belowground biomass yield than low-N-fertilized (0–30 kg N ha^{-1}) treatments under elevated CO_2 conditions across different plant species (de Graffe et al. 2006). Another important change in plant tissues is increased in the lignin content. On an average there may be up to 6.5 % increase in lignin content in leaf tissues in elevated CO_2 conditions.

9.5 Soil Organic Carbon Fractions/Pools

Soil organic carbon content has been recognized as one of the important indicators of soil health/soil quality (Jha et al. 2013), because it interacts with other soil components, affecting water retention, aggregate formation, bulk density, pH, buffer capacity, cation exchange properties, mineralization, sorption of pesticides and other agrichemicals, infiltration, aeration, and activity of soil organism (Kundu et al. 2013a). Changes in the levels of organic matter, caused by land use, can be better understood by alterations in the different compartments (Lal 2015b). Soil organic matter can be analyzed on the basis of the different fractions. Soil organic matter in the soil is divided into three pools: living biomass of microorganisms,

Table 9.4 Possible repositories for additional carbon storage in terrestrial ecosystems or their products and approximate residence times for each pool mean residence time are average time spent by a carbon atom in a given reservoir

Repository	Fraction	Examples	Mean residence time
Biomass	Woody	Tree boles	Decades to centuries
	Nonwoody	Crop biomass, tree leaves	Months to years
Soil organic matter	Litter	Surface litter, crop residues	Months to years
	Active	Partially decomposed litter; carbon in macro-aggregates	Years to decades
	Stable	Stabilized by clay, chemically recalcitrant carbon, charcoal carbon	Centuries to millennia
Products	Wood	Structural, furniture	Decades to centuries
	Paper, cloth	Paper products, clothing	Months to decades
	Grains	Food and feed grain	Weeks to years
	Waste	Landfill contents	Months to decades

(Source: IPCC 2000)

fresh and partially decomposed residues, and humus. Generally SOC has been divided into three pools: an active pool consisting of labile C (simple sugars, organic acids, and microbial metabolites) with a mean residential time of days; an intermediate or slow pool consisting of plant residues and physically stabilized carbon; and a resistant carbon pool consisting of lignin and chemically stabilized with a mean resident time of more than 1000 years (Buyanovsky et al. 1994). The possible repositories for additional carbon storage in terrestrial ecosystems and their products with their mean residential time are depicted in Table 9.4.

Researchers across the world were coming up with different concepts and ideas on soil organic carbon pools or fraction by adapting different criterion and norms. Initially, Jenkinson and Rayner (1977) identified five pools in their organic matter cycling model ranging from a decomposable pool, with a radiocarbon age of <1 year, through a biomass pool at 25.9 years to a chemically stabilized pool with a radiocarbon age of 2565 years. Then SOM fractionation has been carried out on the basis of extraction of humic substances (Schnitzer 1982), dissolved organic C (Cook and Allan 1992), particle size (Christensen 1986), natural abundance (Balesdent et al. 1990; Lefroy et al. 1993), microbial biomass C (Sparling 1992), and ease of oxidation of C (Loginow et al. 1987). Based on their degree of oxidation, carbon fractions are grouped into two different pools as labile (oxidizable by $KMnO_4$) and non-labile (not oxidizable by $KMnO_4$) carbon pools (Blair et al. 1995). Recently, Chan et al. (2001) reported that the TOC was apportioned into different pools by modifying Walkley and Black method using 5, 10, and 20 ml of concentrated H_2SO_4 that resulted in three acid-aqueous solution ratios of 0.5:1,

1:1, and 2:1. The amount of C, thus determined, allowed the apportioning of TOC into very labile C (organic C oxidizable by 12.0 *N* H_2SO_4), labile C (the difference in C oxidizable by 18.0 *N* and that by 12.0 *N* H_2SO_4), less labile C (the difference in C oxidizable by 24.0 *N* and that by 18.0 *N* H_2SO_4), and non-labile C (difference in C between TOC and SOC). These concepts are frequently tailored or modified accordingly depending on their necessity and to quantify the changes in soil organic carbon under different land-use management practice systems. Most of the studies use the content of total organic carbon (TOC) to quantify the stocks and changes in organic matter. In many cases, the changes resulting from land use are not duly reflected in TOC values (Roscoe and Buurman 2003), mainly due to the high C concentrations in stable and little variable mineral association (Lal 2006b). However, a minor variation in such big C pool in terrestrial ecosystems may have a significant effect on the C flux (Harrison et al. 1993). Due to environmental complexities, it is difficult to quantify soil C changes (McKenzie et al. 2000; Skjemstad et al. 2000; Clark 2002). However, researchers across the world are trying to estimate the soil carbon dynamics under different land-use and management system.

9.6 Impact of Land-Use and Management Practices on Soil Organic Carbon

The C balance of terrestrial ecosystems can be changed markedly by the impact of human activities, including deforestation, biomass burning, and land-use change (Bhattacharyya et al. 2000). Soils in tropical regions are low in SOC particularly those under the influence of arid, semiarid and subhumid climates (Katyal 2001). Carbon stocks in major soil types and land-use systems in semiarid tropical region of southern India were reported, and the soil organic carbon (SOC) stock was highest in Alfisols (52.84 Mg ha^{-1}) followed by Inceptisols (51.26 Mg ha^{-1}) and Vertisols and associated soils (49.33 Mg ha^{-1}), and in among the different land-use systems, total C stock was highest in forest soils followed by fodder system, paddy, maize, cotton, red gram, intercrop, chili, and permanent fallow and lowest in castor system (Venkanna et al. 2014). Also in different forest system, a significant change in soil organic carbon was observed. For example, soil carbon pool varied in the order from wet temperate forest (165.24 Mg ha^{-1}) > deciduous forest (138.64 Mg ha^{-1}) > tropical thorny forest (135.42 Mg ha^{-1}) > tropical riparian fringing forest (104.94 Mg ha^{-1}). Also tropical thorny and riparian forest had more labile carbon fractions, and wet temperate forest had the greatest quantity of stable, non-labile carbon fractions (Sreekanth et al. 2013). Likewise, it happened in grassland and pasture land-use systems. Maintaining or improving organic C levels in tropical soils is more difficult because of its rapid oxidation under prevailing high temperatures (Lal 1997). The amount and duration of carbon gain within an ecosystem depends on the temporal dynamics of these different pools: transient pools may increase rapidly, but quickly settle down, whereas carbon that is incorporated into more stable pools can produce slow but long-term increases.

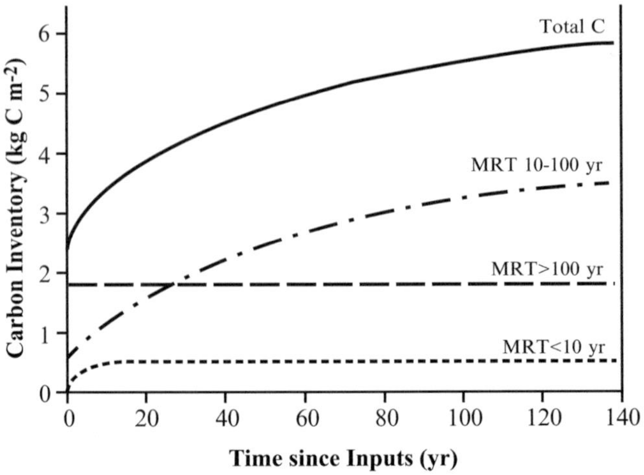

Fig. 9.1 Predicted response of different pools of soil organic matter for an agricultural soil converted to forest in northeastern United States of America (Source: Adopted from Gaudinski et al. 2000)

Consequently, the initial impact of land-use or management change occurs disproportionately in pools with shorter residence times (Huggins et al. 1998), whereas increases in stable soil pools occur slowly over a much longer time period (Fig. 9.1). Early response reflects changes in the relatively small pools with mean residence time (MRT) <10 years (leaf and root residues). Pools with intermediate MRT (10–100 years; including humified organics in litter layers) dominate the overall response because this pool contains most organic matter in this soil. Persistent carbon pools (MRT >100 years) do not change appreciably over a 100-year period. MRT = average time spent by a carbon atom in a given reservoir.

While comparing across the land-use system, grasslands store ~34 % of the global terrestrial stock of carbon, while forests store ~39 % and agroecosystems ~17 % (World Resource Institute 2000). Conversion of grassland or forest land system to agricultural system had a negative impact on soil carbon pools. Therefore, proper management of SOC is important for sustaining soil productivity and protection from land degradation. Among the tillage practices, conservation agricultural practices and long-term recycling of crop residue support the natural systems of storage of more crop residues in soil (Saha and Ghosh 2013). Long-term fertilizer experiments in India also revealed that integrated or balanced nutrient management resulted in building up the organic carbon content of soils (Katkar et al. 2012). Also cereal-based cropping system had low carbon storage when compared to legume-based cropping system. It is obvious from the results that land-use and management systems may influence soil organic carbon storage and dynamics.

9.7 Climate Change Mitigation Strategies and Carbon Sequestration

Global climate change affected the crop growth parameters and various processes in agricultural sector. The agriculture sector contributes ~19 % of the total greenhouse gases in Indian condition. The detrimental reduction in emission of GHGs may be a reduction in the rate of emission from industrial units, burning of fossil fuels, automobiles, and also from various agricultural practices. One of the most important factors for reducing the detrimental effect of global climate change is afforestation, and in other side, we can minimize the effect with the help of various practices in individual or in combination. A variety of options are available for the mitigation of GHG emissions in agriculture sector listed in Table 9.5.

9.7.1 Carbon Capture

Increasing the soil C sequestration (CS) rate by the management practices is very important at grassroots as well as across the globe. Some of the C capturing measures is listed below:

9.7.1.1 Restoration of Degraded Land

Harsh climatic conditions promote the soil degradation. Degradation of soil and vegetation enhanced the greenhouse gas emission into the atmosphere. The change in land-use pattern is the major cause for the vast degradation. In this process, lots of C are lost and converted into CO_2 and accumulated in the atmosphere. As per the National Bureau of Soil Survey and Land Use Planning (NBSS & LUP), Nagpur estimated the 146.82 M ha land as degraded by various means (NBSS and LUP 2005). The indiscriminate use of degraded lands for cultivation, overgrazing, frequent occurrence of drought, overexploitation of groundwater, deforestation,

Table 9.5 Climate change mitigation options in agriculture

Mitigation options in agriculture	
Reduce emissions	Carbon capture
Reduce methane emissions from paddy field	Soil carbon sequestration
Shift to low-carbon and renewable biomass fuels	Soil and water conservation measures
Reduce in energy use in agriculture sector	Afforestation and reforestation
Improved energy management	Conservation agriculture
Reduce crop residue burning	Restoration of degraded lands
Reduce enteric methane formation	
Reduce those subsidies and taxes which enhance C emission	
Drainage of croplands	
Manure management	
Efficient fertilizer N management	

(Source: Adopted from Lenka et al. 2013b)

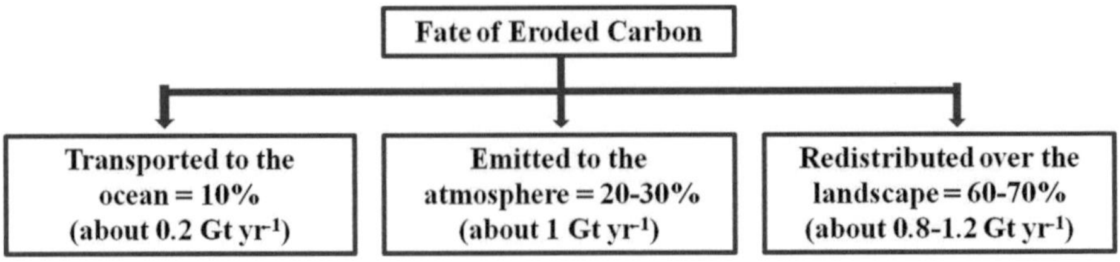

Fig. 9.2 Distribution of C transported by erosion (Source: Adopted from Lenka et al. 2013b)

and soil water erosion are important cause for the problems (Lenka et al. 2013b). Lal (2003a) mentioned that eroded 60–70 % of C gets redistributed over the landscape, and approximately 10 % reached to ocean. Later on, he reported that 20–30 % of C transported by erosion is emitted to the atmosphere, the fate of C transported by erosion mechanism (Fig. 9.2).

9.7.1.2 Afforestation and Reforestation

In the ancient period, the rate of GHGs emission is very low, and the forest cover was higher; this provided a clean environment and healthy soil and crop produce. Speedy growth in industrial sector and increasing population reduced the forest cover and increased the GHGs in the atmosphere. The plants consume CO_2 during the photosynthesis and release the oxygen into atmosphere. Unsustainable exploitation of forest has resulted in vast degradation of forest area which is estimated to ~40 % for the past two decades (FSI 2009). The need of present time is increasing the greenbelt in urban and rural areas and its sustainable management. The sustainable forest management (SFM), having broader goal related to social, economic, and environmental issues, provides a method for balancing the diverse wheel of sustainability regarding the economic and social development in one side and ecological sustainability on the other side. The conservation of existing C sinks through forest conservation and expansion of C sinks through afforestation are important strategies for the Indian conditions. In the year 2007, an action plan was agreed at Bali, Indonesia that is called Reducing Emissions from Avoided Deforestation and Forest Degradation (REDD). This plan has the centerpiece for the mitigation of climate change and integrated sustainable management of forest cover and biodiversity in parallel with improved welfare to rural people mostly in developing countries (IPCC 2000).

9.7.1.3 Soil and Water Conservation Measures

The natural resources are precious, and its proper management provides the sustainability in ecosystem and better quality of life. Use of soil and water conservation mechanisms for reducing the soil and water erosion improves the soil microflora and fauna. Increasing the amount of trees, shrubs, and organic matter input can accumulate more amount of C in the soil. The rainwater harvesting is a common practice in rain-fed areas and the harvested water can be used as a life-saving irrigation during crop production (IPCC 2000, 2007).

Nowadays, many cost-effective bioengineering measures were developed for promoting the cost-effective tools for the soil water conservation. The most common measures are field bunds, contour bounding, trenching, vegetative barriers, agroforestry, water harvesting, and afforestation on undulated/sloppy land. The watershed program is very successful in Indian condition for rational utilization of water and land resources for optimum production without compromising natural resources degradation. It improves the risk-bearing capacity and increases the income of farmers due to social involvement at the grassroots level. The soil and water conservation measures are improving the soil organic matter and reduced the GHGs emissions.

9.7.1.4 Conservation Agriculture

It is an approach to farming that seeks to increase food security and managing agroecosystems for the improved and sustained productivity, long-term environmental and economic sustainability to farming system, and enhancing the resource-based soil and agricultural practices. Principles of conservation agriculture (CA) are not limited to a single set of recommendation and practices for all types of agroecosystems, crops, and soils. As per the CA definitions mentioned, FAO described that the three pillars which are (i) minimum continuous disturbance to soil by using reduced tillage, (ii) permanent soil cover with organic, and (iii) crop diversification may be in sequence or association. In CA, reduced tillage is used to reduce the CO_2 during mechanical operation in agriculture field; burning of organic carbon and release of CO_2 to the atmosphere are well documented by various researchers. The CA practices can contribute to making the agricultural system more resilient to climate change. It has a powerful mechanism to adapt to climate change by increasing resilience to drought and increasing water-use efficiency and thermal stress resistance to crops by changing the hydrothermal regimes of the surface soil. Apart from this it also increased the soil organic matter, reduced the soil erosion, and reduced the soil organic material mineralization rate through better aggregation (Rao and Somasundaram 2013).

9.7.1.5 Soil Carbon Sequestration

Increasing concentration of GHGs enhanced the global temperature in new horizons. To reduce the C concentration in the atmosphere, soil carbon sequestration is an important mechanism for mitigation potential up to 89 % contribution to the technical potential (Smith et al. 2007). It acts as a sink and a key pathway in agriculture through enhancing C capturing from the atmosphere to plant biomass. Sequestering carbon in the soil is a critical element in mitigating climate change. Carbon sequestration is defined as the removal and storage of carbon from the atmosphere in carbon sinks (such as oceans, forests, or soils) through physical or biological processes, such as photosynthesis. The soil carbon pool alone is 3.3 times the size of the atmospheric pool and 4.5 times the size of the terrestrial biotic pool (Lal 2004b). Carbon dioxide cycles globally among the atmosphere, the ocean, and the land biosphere. The atmosphere contains 762 Gt C with nearly all of it as CO_2. The ocean stores about 40,000 Gt, and the quantity of CO_2 exchanged between the ocean and the atmosphere is 90 Gt C year^{-1} (Denman et al. 2007; Lal 2006a). Several strategies have been advocated for stabilizing atmospheric abundance of CO_2. The three main strategies to lower down the CO_2 emissions include reducing the global energy use, developing low- or no-carbon fuel, and sequestering CO_2 through natural and engineering techniques (Kundu et al. 2013a, b). Cropping system can play an important role in mitigating CO_2 emissions through biotic C sequestration in soil and vegetation.

9.7.2 Reduce Emission

In this modification in management techniques, this can reduce the rate of GHGs emission and also enhanced the soil C. Some of the management techniques are mentioned below:

9.7.2.1 Manure/Fertilizer/Nutrient Management

Optimal nutrition and most favorable soil tillage greatly affect water circulation within the plant, which is a highly effective method of combating drought. Under poor nutrient condition, plants have to absorb more amount of water to fulfill nutrient requirement, but in low-moisture conditions, which have a negative effect on plant growth and crop yield (Meena et al. 2014, 2015). Thus, applying the right amount and at the right time can address the problem of environmental stress. Application of silicon as a fertilizer also helps to alleviate negative stress of climate change and its increasing plant resistance to drought stress. One of the alternative methods involves the silicon fertilization for alleviation of negative stress of climate change (Meena et al. 2013a, b). Unfortunately the silica that occurs in soil is in an unavailable polymerized form, and for its absorption by plants it has to be depolymerized and rendered soluble by means of biological or chemical reactions in the soil. Use of proper management techniques for the compost preparation or during storage reduces the production of GHGs from manure (Sejian and Indu 2013).

9.7.2.2 Crop Residue Management Practices

Crop residue management practices are older since the evolution of agriculture crop production to fulfill the human food needs from land. In ancient time peoples were using slash burning or shifting cultivation. In these practices, after taken one or two crops, people were migrated to another place and burn the forest and make the fertile land for cultivation. Such type of activities was dominated in last so many years. The population of global world is increasing with the quantum pace and the land holding decreasing inversely. Nowadays, there are three dominating types of residue management practices: The first practice is crop residue burning, which is mostly used in the highly mechanized belt of agriculture. In this practice farmers are burning the crop residue in situ, to avoid the hindrance during the next crop sowing in the field. During the burning, there is loss of plant nutrient and the environmental pollution increased due to CO_2, CH_4, N_2O, NOx, NMHCs, and SO_2

and particles (Dotaniya et al. 2014b, d). The second practice is crop residue removal, in which crop residue is removed from the field and either used or not in another field. This practice also encourages the loss of plant nutrient from the agricultural field. And the third practice is the crop residue incorporation, in which crop stubbles or residues are incorporated into the field (Dotaniya and Datta 2014). This increased the soil fertility and improved the soil health. The residue is decomposed by soil microbes and converted into ionic form of plant nutrient during plant uptake.

9.7.2.3 Reduce Enteric Methane Fermentation

Methane is a one of the GHGs responsible for a climate change event. Methane and nitrous oxide emissions are projected to further increase 35–60 % by 2030, driven by wide-growing N fertilizer for fulfilling the food requirement of a growing population. Agriculture accounts about 50 % of CH_4 from natural and cultivated wetlands and enteric fermentation. From the rice field, CH_4 emission is emitted and proved by a list of research publication. Apart from this, CH_4 is also emitted from animals during the digestive process by which carbohydrates are broken down by microorganisms. This is mainly happing in four-chambered stomach animals, i.e., cattle, sheep, buffalo, and goat. In general, about 8–12 % of dietary energy is lost during the digestive metabolism in the form of CH_4 in ruminants (Sejian and Indu 2013; Dotaniya 2015). To reduce the CH_4 from ruminant animals, some countries reduced the number of unproductive animals and increased the efficiency of livestock and also animal breeding with low CH_4 emission animals, etc.

9.8 Carbon Sequestration Potential of Global Terrestrial Ecosystem

It is well known that the GHGs especially CO_2 concentration of the global atmosphere have increased during the last few decades, mainly due to energy consumption from fossil fuels. Since the start of the industrial revolution, the atmospheric level CO_2 has increased from 280 ppm to around 385 ppm and continues to rise at approximately 1.8 ppm per year. It is expected that the CO_2 level might reach a concentration of 600–1000 ppm by the end of this century. It appears more likely that greenhouse gases from human activities were the dominant drivers of these global-average temperature changes during the twentieth century. Due to the increase of the CO_2 level, it is expected that the maximum, minimum, and mean global temperatures will also change by 3–4 °C. The IPCC expects a global surface temperature increase, ranging from 1.0 to 3.5 °C from 2100 based on the predictions of the general circulation models (IPCC 2001). Changes in Earth's carbon cycle derived from human activities played an important role in global warming. The interactive effects of global warming and increasing CO_2 levels could especially impact agriculture, affecting both growth and development of crops and ultimately impacting yield and food production. Anthropogenic activities, including combustion of fossil fuels and land-use change, contribute CO_2 emissions to the global C cycle. In 2013, global CO_2 emissions due to fossil fuel use (and cement production) were 36 gigatonnes (Gt CO_2); this is 61 % higher than 1990 (the Kyoto Protocol reference year) and 2.3 % higher than 2012 (CDIAC 2013). Terrestrial sequestration of CO_2 involves the transfer of CO_2 from the atmosphere into soils and vegetation. It includes both the removal of CO_2 from the atmosphere and limiting CO_2 emissions from terrestrial ecosystems into the atmosphere.

Adopting land-use and management practices that enhance CO_2 sequestration (storage) in terrestrial systems can be one important tool for significantly reducing atmospheric concentrations of CO_2. Storage of C in soils and plants has the potential to offset CO_2 emissions to the atmosphere in the coming decades, while new "clean" energy production and CO_2 sequestration technologies are developed and deployed. Because they are economically important, have a rich history of direct research, and can be more easily managed, forests and croplands are best suited for the application of existing and new technology to enhance terrestrial C sequestration in the near term. In addition to proven management approaches, new management and chemical and biological technology have the potential to impact C storage.

The terrestrial ecosystem plays a prominent role in the global C cycle, and its management offers cost-effective ways to enhance C sequestration. Terrestrial sequestration is generally accomplished through the forest and soil conservation practices that enhance the storage of carbon (restoring and establishing new forests, wetlands, and grasslands) or reduce CO_2 emissions (such as reducing agricultural tillage and suppressing wildfires) and also known as biological sequestration. At present, the biosphere constitutes a carbon sink that absorbs about 2.3 gigatonnes of C per year, which represents ~30 % of fossil fuel emissions (IPCC 2000). Carbon sequestration implies transferring atmospheric CO_2 into long-lived pools and storing it securely so it is not immediately reemitted. Soil is the largest reservoir of terrestrial C (about 53 % terrestrial C), and approximately 10 % of the CO_2 in the atmosphere is cycled through the soil every year (Lenka et al. 2013b) Thus, soil C sequestrations mean increasing SOC and SIC stocks through the judicious land-use and recommended management practices (RMPs). The carbon sequestration potential of terrestrial ecosystems depends on the type and condition of the ecosystem, that is, its species composition, structure, and (in the case of forests) age distribution. The ultimate potential for terrestrial C sequestration is not known due to lack of knowledge about the biogeochemical mechanisms (IPCC 1995, 1996) responsible for C fluxes and storage potential at regional and global levels (Metting et al. 2001), which account for two-thirds of global terrestrial organic C stocks.

The primary C sequestration method is rated with high (H), medium (M), and low (L) levels of sustained management intensity required over the long term. Global potential C sequestration (CS) rates were estimated for major ecosystems (Table 9.6) that might be sustained over a long period of up to 50 years (Metting et al. 2001). Enhancing terrestrial C sequestration with proven management practices includes converting marginal land to productive grassland or forest, increasing productivity on crop and forest land with residue management, reduced C loss with modified tillage practices, the efficient use of fertilizer, pesticide, and

Table 9.6 Sustained terrestrial C sequestration potential

Ecosystem	Primary method to increase C sequestration	Potential CS (Gt C/year)
Agricultural lands	Management (H)	0.85–0.90
Biomass croplands	Management (H)	0.5–0.8
Grasslands	Management (H)	0.5
Rangelands	Management (H)	1.2
Forests	Management (H)	1–2
Wetlands	Restoration and maintenance (M)	0.1–0.2
Urban forest lands and grasslands	Creation and maintenance (M)	<0.1
Deserts and degraded lands	Manipulation (H)	0.8–1.3
Sediments and aquatic systems	Protection (L)	0.6–1.5
Tundra and taiga	Protection (L)	0.1–0.3

water, and other technologies. It is difficult to estimate global C sequestration enhancement potential because of inadequate baseline inventories. Enhanced C sequestration in managing lands aims to increase the productivity of crop and forest land. In agriculture, adoption of conservation tillage practices is a viable mechanism that also reduces erosion, increases soil aggregation, and lessens loss of SOM to microbial oxidation (Lal et al. 1998). Degraded lands also represent some potential for C sequestration. Worldwide, there is ~1965×10^6 ha of degraded soils, 4 % from physical degradation, 56 % of water erosion, 28 % from wind erosion, and 12 % from chemical degradation. With proper management this represents a potential to sequester between 0.81 and 1.03 Gt C per year (Lal 2001). Wright et al. (1992) also estimated 14–28 M ha of agricultural cropland is suitable for woody biomass with a sequestration potential between 0.09 and 0.18 Gt C per year.

9.9 Strategies for Terrestrial Carbon Sequestration

Thus, the soil C pool of 2500 Pg is 3.3 times the atmospheric pool and 4.0 times the biotic pool. The SOC pool is at a dynamic equilibrium under a specific land-use and management system. At equilibrium, the C_{input} into a system equals C_{output}. Upon conversion to another land use and management, depletion of SOC pool occurs if $C_{input} < C_{output}$ and sequestration if $C_{input} > C_{output}$ (Eqs. 9.1, 9.2, and 9.3).

Steady state

$$C_{input} = C_{output} \tag{9.1}$$

Depletion

$$C_{\text{input}} < C_{\text{output}} \tag{9.2}$$

Sequestration

$$C_{\text{input}} > C_{\text{output}} \tag{9.3}$$

However, soils of the poorly managed ecosystems have lost 50–75 % of the original SOC pool. The magnitude of SOC depletion is higher in soils prone to erosion and those managed by low-input or extractive farming practices. The loss of SOC pool is also high in soils of coarse texture and those with a high initial pool.

Strategies to compensate the increase in atmospheric CO_2 can be achieved only with improved practices of the C sequestration potential related to land pattern. Modifying the land-use pattern potentially increases C sequestration through increased plant biomass production and/or decreased rates of detritus or soil organic C decomposition (Lal 2003b). Science and technology that might drive enhanced terrestrial C sequestration includes (a) technology for soil, crop, and forest management, (b) exploitation of underutilized land resources and existing biodiversity, (c) plant biotechnology, (d) microbial biotechnology, and (e) innovative chemical technology. In spite of these technologies, there are several opportunities for C sequestration, viz., establishing additional areas of forest for C sequestration, increasing the productivity of existing forest and C sequestration, managing soil systems, agronomic, and desert crop as C sinks, employing the agroforestry as land-use practices to conserve and sequester C, and offsetting fossil fuel combustion with biomass or biomass-derived fuels.

9.10 Challenges and Recommendations

Major challenges and recommendations on mitigation of global climate change effect are listed below:

9.10.1 Major Challenges

- The global climate change is not a regional problem, to remediate it in short periods.
- It requires a lot of fund, which is not equally contributed by developing and developed countries.
- Lots of contradiction in baseline determination and also additionally, which is not formulated due to lack of technical training and data availability, etc.
- The lacking of willingness regarding mitigation of climate change effect across the globe by developed nations and financial limitation in developing nations.

9.10.2 Some of the Recommendations

- The more fund to be invested in the climate change research and its mitigation option for the clear view of mechanisms and measurement of uncertainty.
- The needs of integrating farmers in the C trading process and reduction of emission at block level or district level.
- The joint collaboration of the government sector and nonprofitable institutes for the capacity building program at the grassroots level for monitoring, supervision and enhancement of C sequestration, and reduction in GHG emission with low-cost technologies.
- The developing countries must be aware that they would be required in the future to reduce emissions on their own.
- Promote the afforestation with the help of self-help groups (SHGs) and other agencies.
- Advancement in technological field like clean development mechanism (CDM); supply the appropriate technologies by developed countries to developing countries.

9.11 Conclusions

Increase in the concentration of GHGs in the atmosphere with the pace of industrial development is a major concern particularly to the developing countries. However, fulfilling the need of the growing population in terms of food and nutrition, faster economic development, advancement of science and technology, and welfare of the global community is a prime objective for any nation vis-a-vis sustainable use of resources. But improper governmental policies, mismanagement and selfishness, and greed of human beings all lead to global warming. Therefore, to overcome these issues requires a lot of awareness among the scientific as well as local people, a huge amount of money, the right direction, the teamwork between developed and developing nations (in which the developed nation may provide green technology and monetary help to the developing nations), and also the mass campaign to combat the adverse effect of climate change. In this regard, the government motivates its people for using the green and clean energy and maximum utilization of nonconventional energy (solar and wind); for popularizing these green technologies among the people, it gives subsidy or reduced the cost. Further, the carbon sequestration to counter the increasing CO_2 in the atmosphere through any means, and climate change adaptation and mitigation strategies that are specific to particular agroecological region, should be initiated to overcome the climate change problems.

Acknowledgment The authors are highly thankful to Dr. H. M. Meena, Scientist, Central Arid Zone Research Institute, Jodhpur, India, for the needful help during the writing of the manuscript.

References

Abtew W, Melesse AM (2012) Climate change and evapotranspiration. In: Evaporation and evapotranspiration. Springer, Dordrecht, pp 197–202

Ahmed KF, Wang G, Yu M, Koo J, You L (2015) Potential impact of climate change on cereal crop yield in West Africa. Clim Change 133:321–334

Ainsworth EA, Long SP (2005) What have we learned from 15 years of free-air CO_2 enrichment (FACE)? A meta-analytic review of the responses of photosynthesis, canopy properties and plant production to rising CO_2. New Phytol 165:351–371

Ainsworth EA, Rogers A (2007) The response of photosynthesis and stomatal conductance to rising CO_2: mechanisms and environmental interactions. Plant Cell Environ 30:258–270

Alberton O, Kuyper TW, Gorissen A (2005) Taking mycocentrism seriously: mycorrhizal fungal and plant responses to elevated CO_2. New Phytol 167:859–868

Anderson CR (2013) *The earth surface temperature without greenhouse gases: the shade effect of infra-red active gases*. http://objectivistindividualist.blogspot.in/2013/04/ the-earth-surface-temperature-without.html

Balesdent J, Mariotti A, Boisgontier D (1990) Effect of tillage on soil organic carbon mineralization estimated from ^{13}C abundance in maize fields. J Soil Sci 41:587–596

Benbi DK, Kaur R (2009) Modeling soil processes in relation to climate change. J Indian Soc Soil Sci 57:433–444

Bhattacharyya T, Pal DK, Mandal C, Velayutham M (2000) Organic carbon stock in Indian soils and their geographical distribution. Curr Sci 79:655–660

Blair JG, Lefroy RDB, Leanne L (1995) Soil carbon fractions based on their degree of oxidation, and the development of a carbon management index for agricultural systems. Aus J Agric Res 46:1459–1466

Bobbink R, Ashmore M, Braun S, Fluckiger W, van den Wyngaert IJJ (2002) Empirical nitrogen critical loads for natural and semi-natural ecosystems: 2002 update, Available online http://www.iap.ch

Brinkman R (1982) Clay transformations: aspects of equilibrium and kinetics. In: Bolt GH (ed) Soil chemistry. B. Physicochemical models. Developments in soil science 5B. Elsevier, Amsterdam, pp 433–458

Buyanovsky GA, Aslam M, Wagner GH (1994) Carbon turnover in soil physical fractions. Soil Sci Soc America J 58:1167–1173

CDIAC (2013) *Global carbon project*. doi: 10.3334/CDIAC/GCP-2013-V2.3

Chan KY, Bowman A, Oates A (2001) Oxidizible organic carbon fractions and soil quality changes in an oxic paleustalf under different pasture layers. Soil Sci 166:61–67

Chen R, Senbayram M, Blagodatsky S, Myachina O, Dittert K, Lin X, Blagodatskaya E, Kuzyakov Y (2014) Soil C and N availability determine the priming effect: microbial N mining and stoichiometric decomposition theories. Glob Chang Biol 20:2356–2367

Christensen BT (1986) Straw incorporation and soil organic matter in macro-aggregates and particle size separates. J Soil Sci 37:125–135

Clark DA (2002) Are tropical forests an important carbon sink? Reanalysis of the long-term plot data. Ecol Appl 12:3–7

Cook BD, Allan DL (1992) Dissolved organic carbon in old field soils: compositional changes during the biodegradation of soil organic matter. Soil Biol Biochem 24:595–600

Courty PE, Buee M, Diedhiou AG, Frey-Klett P, Le Tacon F, Rineau F, Turpault MP, Uroz S, Garbaye J (2010) The role of ectomycorrhizal communities in forest ecosystem processes: new perspectives and emerging concepts. Soil Biol Biochem 42:679–698

Das DK, Hati KM (2010) *Impact of climate change and global warming on soil processes and nutrient availability*. Souvenir, 75th annual convention of the Indian society of soil science, pp 39–46 de Graaff MA, van Groenigen KJ, Johan S, Bruce H, van Kessel C (2006) Interactions between plant growth and soil nutrient cycling under elevated CO_2: a meta-analysis. Global Change Biol 12:2077–2091

Denman KL, Brasseur G, Chidthaisong A, Ciais P, Cox PM (2007) Couplings between changes in the climate system and biogeochemisry. In: Climate change: the physical science basis. Contribution of working group I to the fourth assessment report of the intergovernmental panel on climate change. Cambridge University Press, Cambridge, UK, pp 499–587

Dotaniya ML (2013) Impact of various crop residue management practices on nutrient uptake by rice-wheat cropping system. Curr Adv Agri Sci 5(2):269–271

Dotaniya ML (2015) Impact of rising atmospheric CO_2 concentration on plant and soil process. In: Mohanty M, Sinha NK, Hati KM, Chaudhary RS, Patra AK (eds) Crop growth simulation modelling and climate change. Scientific Publisher, New York, pp 69–86. ISBN 978-81-7233-941-8

Dotaniya ML, Datta SC (2014) Impact of bagasse and press mud on availability and fixation capacity of phosphorus in an Inceptisol of north India. Sugar Tech 16(1):109–112. doi:10.1007/s12355-013-0264-3

Dotaniya ML, Kushwah SK (2013) Nutrients uptake ability of various rainy season crops grown in a vertisol of central India. Afr J Agri Res 8(44):5592–5598. doi:10.5897/AJAR2013.7969

Dotaniya ML, Meena VD (2013) Rhizosphere effect on nutrient availability in soil and its uptake by plants –a review. Proc Natl Acad Sci, India Sec B: Biol Sci 85(1):1–12. doi:10.1007/s40011-013-0297-0

Dotaniya ML, Pingoliya KK, Meena HM, Prasad D (2013a) Status and rational use of rock phosphate in agricultural crop production- a review. Agric Sus Devel 1(1):103–108

Dotaniya ML, Prasad D, Meena HM, Jajoria DK, Narolia GP, Pingoliya KK, Meena OP, Kumar K, Meena BP, Ram A, Das H, Chari MS, Pal S (2013b) Influence of phytosiderophore on iron and zinc uptake and rhizospheric microbial activity. Afr J Microbiol Res 7(51):5781–5788. doi:10.5897/AJMR2013.6461

Dotaniya ML, Sharma MM, Kumar K, Singh PP (2013c) Impact of crop residue management on nutrient balance in rice-wheat cropping system in an Aquic hapludoll. J Rural Agric Res 13 (1):122–123

Dotaniya ML, Datta SC, Biswas DR, Meena BP (2013d) Effect of solution phosphorus concentration on the exudation of oxalate ions by wheat (*Triticum aestivum* L.). Proc Natl Acad Sci, India Sec B: Biol Sci 83(3):305–309. doi:10.1007/s40011-012-0153-7

Dotaniya ML, Datta SC, Biswas DR, Meena HM, Kumar K (2014a) Production of oxalic acid as influenced by the application of organic residue and its effect on phosphorus uptake by wheat (*Triticum aestivum* L.) in an Inceptisol of north India. Natl Acad Sci Lett 37(5):401–405. doi:10.1007/s40009-014-0254-3

Dotaniya ML, Datta SC, Biswas DR, Kumar K (2014b) Effect of organic sources on phosphorus fractions and available phosphorus in Typic Haplustept. J Indian Soc Soil Sci 62(1):80–83

Dotaniya ML, Kushwah SK, Rajendiran S, Coumar MV, Kundu S, Rao AS (2014c) Rhizosphere effect of kharif crops on phosphatases and dehydrogenase activities in a Typic Haplustert. Natl Acad Sci Lett 37(2):103–106. doi:10.1007/s40009-013-0205-4

Dotaniya ML, Pingoliya KK, Lata M, Verma R, Regar KL, Deewan P, Dotaniya CK (2014d) Role of phosphorus in chickpea (*Cicer arietinum* L.) production. Afr J Agric Res 9(51):3736–3743. doi:10.5897/AJAR2013.8427

Dotaniya ML, Datta SC, Biswas DR, Meena HM, Rajendiran S, Meena AL (2015) phosphorus dynamics mediated by bagasse, press mud and rice straw in inceptisol of North India. Agrochimica 59(4):358–369

Drigo B, Kowalchuk GA, Knapp BA, Pijl PG, Boschker HTS, van Veen JA (2013) Impacts of 3 years of elevated atmospheric CO_2 on rhizosphere carbon flow and microbial community dynamics. Glob Chang Biol 192:621–636

Emmett BA, Beier C, Estiarte M, Tietema A, Kristensen HL, Williams D, Penuelas J, Schmidt I, Sowerby A (2004) The responses of soil process to climate change: results from manipulation studies of shrub-lands across an environmental gradient. Ecosys 7:625–637

Fiore AM, West JJ, Horowitz LW, Naik V, Schwarzkopf MD (2008) Characterizing the tropospheric ozone response to methane emission controls and the benefits to climate and air quality. J Geophysical Res. doi:10.1029/2007JD009162

FSI (2009) Indian state of forest report 2009, Dehradun FSI, p 172

Gaudinski JB, Trumbore SE, Davidson EA, Zheng S (2000) Soil carbon cycling in a temperate forest: radiocarbon-based estimates of residence times, sequestration rates and partitioning of fluxes. Biogeochemistry 51:33–69

Ghosh BN, Meena VS, Alam NM, Dogra P, Bhattacharyya R, Sharma NK, Mishra PK (2016) Impact of conservation practices on soil aggregation and the carbon management index after seven years of maize–wheat cropping system in the Indian Himalayas. Agr Ecosyst Environ 216:247–257

Ghude SD, Jain SL, Arya BC, Beig G, Ahammed YN, Kumar A et al (2008) Ozone in ambient air at a tropical megacity, Delhi: characteristics, trends and cumulative ozone exposure indices. J Atmos Chem 60:237–252

Gifford RM, Howden M (2001) Vegetation thickening in an ecological perspective: significance to national greenhouse gas inventories. Environ Sci Policy 4:59–72

Gosain AK, Rao S, Basuray D (2006) Climate change impact assessment on hydrology of Indian river basins. Curr Sci 90(3):346–349

Goulding KWT, Bailey NJ, Bradbury NJ, Hargreaves P, Howe M, Murphy DV, Poulton PR, Willison TW (1998) Nitrogen deposition and its contribution to nitrogen cycling and associated soil processes. New Phytol 139:49–58

Groenigen JWV, Huygens D, Boeckx P, Kuyper TW, Lubbers IM, Rütting T, Groffman PM (2015) The soil N cycle: new insights and key challenges. Soil 1:235–256

Grover M, Maheswari M, Desai S, Gopinath KA, Venkateswarlu B (2015) Elevated CO_2: plant associated microorganisms and carbon sequestration. Appl Soil Ecol 95:73–85

Harrison K, Broecker W, Bonani G (1993) A strategy for estimating the impact of CO_2 fertilization on soil carbon storage. Global Biogeochem Cycle 7:69–80

Hennessy K, Lucas C, Nicholls N, Bathols J, Suppiah R, Ricketts J (2005) Climate change impacts on fire-weather in south-east Australia. CSIRO, Melbourne

Huggins DR, Buyanovsky GA, Wagner GH, Brown JR, Darmody RG, Peck TR, Lesoing GW, Vanotti MB, Bundy LG (1998) Soil organic C in the tallgrass prairie-derived region of the corn belt: effects of long-term crop management. Soil Tillage Res 47:219–234

IPCC (1995) Radiative forcing of climate change and an evaluation of the IPCC IS92 emission scenarios. In: Houghton JT (ed) Climate change 1994. Cambridge University Press, Cambridge, p 337

IPCC (1996) Intergovernmental panel on climate change. The science of climate change. In: Houghton JT, Meira Filho LG, Callander BA, Harris N, Kattenberg A, Maskell K (eds) Climate change 1995. Cambridge University Press, Cambridge, p 572

IPCC (2000) In: Robert TW, Ian RN, Bert B, Ravindranath NH, David JV, David JD (eds) Land use, land-use change and forestry. Cambridge University Press, Cambridge, p 375

IPCC (2001) IPCC 3rd assessment report-climate change 2001. Working group II. Impact, adaptation and vulnerability. http://www.ipcc.ch/ipccreports/ index.htm

IPCC (2007) The physical science basis. Summary for policymakers. Intergovernmental panel on climate change. http://www.ipcc.ch/ipccr-wg2.org

Jeffrey SD, Harold AM (1999) Does global change increase the success of biological invaders? Trends Ecol Evol 14(4):135–139

Jenkinson DS, Rayner JH (1977) The turnover of soil organic matter in some of the Rothamsted classical experiments. Soil Sci 123:298–305

Jha P, Lakaria BL, Biswas AK, Saha R, Mohanty SR, Singh M, Rao AS (2013) Stabilization of soil organic matter. In *IISS contribution in frontier areas of soil research*, IISS, Bhopal, pp 83–126

Jones D, Nguyen C, Finlay R (2009) Carbon flow in the rhizosphere: carbon trading at the soil–root interface. Plant Soil 321:5–33

Katkar RN, Kharche VK, Sonune BA, Wanjari RH, Singh M (2012) Long term effect of nutrient management on soil quality and sustainable productivity under sorghum-wheat crop sequence in Vertisol of Akola, Maharashtra. Agropedol 22:103–114

Katyal JC, Rao NH, Reddy MN (2001) Critical aspects of organic matter management in the tropics: the example of India. Nutr Cycl Agroeco 61:77–88

Kimball BA (2003) Response of plants to elevated atmospheric CO_2. Indian J Plant Physio (Special Issue)8:18–24

Kimball BA, Kobayashi K, Bindi M (2002) Responses of agricultural crops to free-air CO_2 enrichment. Adv Agron 77:293–368

Kirschbaum MUF (2000) Will changes in soil organic carbon act as a positive or negative feedback on global warming? Biogeochemistry 48:21–51

Kumar A, Bahadur I, Maurya BR, Raghuwanshi R, Meena VS, Singh DK, Dixit J (2015) Does a plant growth promoting rhizobacteria enhance agricultural sustainability? J Pure Appl Microbiol 9(1):715–724

Kundu S, Dotaniya ML, Lenka S (2013a) Carbon sequestration in Indian agriculture. In: Lenka S, Lenka NK, Kundu S and Rao AS (eds) Climate change and natural resource management. New India Publishing Agency, New Delhi, pp 269–289

Kundu S, Saha JK, Adhikari T, Coumar MV, Biswas AK, Rajendiran S, Hati KM, Bhattacharya T, Rao AS (2013b) Assessment of soil quality and resilience of Vertisols of Madhya Pradesh. In *IISS contribution in frontier areas of soil research*, IISS, Bhopal, pp 83–126

Kushwah SK, Dotaniya ML, Upadhyay AK, Rajendiran S, Coumar MV, Kundu S, Rao AS (2014) Assessing carbon and nitrogen partition in kharif crops for their carbon sequestration potential. Natl Acad Sci Lett 37(3):213–217. doi:10.1007/s40009-014-0230-y

Lal R (1997) Residue management, conservation tillage and soil restoration for mitigating greenhouse effect by CO_2 - enrichment. Soil Till Res 43:81–107

Lal R (2001) Potential of desertification control of sequester carbon and mitigate the greenhouse effect. Clim Change 51:35–72

Lal R (2003a) Global potential of soil C sequestration to mitigate the greenhouse effect. Cri Rev Plant Sci 22:151–184

Lal R (2003b) Offsetting global CO_2 emissions by restoration of degrades soils and intensification of world agriculture and forestry. Land Degrad Dev 14:309–322

Lal R (2004a) Soil carbon sequestration impacts on global climate change and food security. Science 204:1623–1627

Lal R (2004b) Soil carbon sequestration to mitigate climate change. Geoderma 123:1–22

Lal R (2006a) Soil carbon sequestration in Latin America. In: Lal R, Cerri CC, Bernoux M, Etcheves J, Cerri E (eds) Carbon sequestration in soils of Latin America. Food Products Press, New York, pp 49–64

Lal R (2006b) Enhancing crop yields in developing countries through restoration of soil organic carbon pool in agricultural lands. Land Degrad Dev 17:197–209

Lal R (2014) Soil carbon management and climate change. In: Hartemink K, McSweeney K (eds) Soil carbon progress in soil science. Springer International, Cham, pp 339–362

Lal R (2015a) Restoring soil quality to mitigate soil degradation. Sustainability 7:5875–5895

Lal R (2015b) Soil as a sink of atmospheric CO_2 and CH_4. Global soil week 2015, Berlin, 20–23 April 2015

Lal R, Kimble JM, Follett RF, Cole CV (1998) The potential of US cropland to sequester carbon and mitigate the greenhouse effect. Ann Arbor Press, Chelsea

Lee TD, Barrott SH, Reich PB (2011) Photosynthetic responses of 13 grassland species across 11 years of free-air CO_2 enrichment is modest, consistent and independent of N supply. Global Change Biol doi: 10.111/j.1365-2486.2011.02435.x.

Lefroy RDB, Blair GJ, Strong WM (1993) Changes in soil organic matter with cropping as measured by organic carbon fractions and natural isotope abundance. Plant Soil 155 (156):399–402

Lenka S, Lenka NK, Kundu S, Rao AS (2013a) Climate change and natural resource management. New India Publishing Agency, New Delhi

Lenka S, Lenka NK, Chaudhary RS (2013b) Climate change mitigation options in agriculture. In: Lenka S, Lenka NK, Kundu S, Rao AS (eds) Climate change and natural resource management. New India Publishing Agency, New Delhi, pp 195–214

Lobell DB, Hammer GL, McLean G, Messina C, Roberts MJ, Schlenker W (2013) The critical role of extreme heat for maize production in the United States. Nat Climate Chang 3(5):497–501

Loginow W, Wisniewski W, Gonet SS, Ciescinska B (1987) Fractionation of organic carbon based on susceptibility to oxidation. Polish J Soil Sci 20:47–52

Long P, Ainsworth EA, Leakey ADB, Nosberger J, Ort DR (2006) Food for thought: lower than expected crop yield stimulation with rising CO_2 concentrations. Science 312:1918–1921

McKenzie N, Ryan P, Fogarty P, Wood J (2000) Sampling, measurement and analytical protocols for carbon estimation in soil, litter and coarse woody debris, Australian Greenhouse Office, National Carbon Accounting System, Technical report no 14, pp 1–42

Meehl GA, Stocker TF, Collins WD, Friedlingstein P et al (2007) Global climate projections. In: Solomon S, Qin D, Manning M, Chen Z (eds) Climate change 2007: the physical science basis. Contribution of working group I to the fourth assessment report of the intergovernmental panel on climate change. Cambridge University Press, Cambridge, pp 749–784

Meena RS, Yadav RS, Meena VS (2013a) Heat unit efficiency of groundnut varieties in scattered planting with various fertility levels. The Bioscan 8(4):1189–1192

Meena VD, Dotaniya ML, Rajendiran S, Coumar MV, Kundu S, Rao AS (2013b) A case for silicon fertilization to improve crop yields in tropical soils. Proc Natl Acad Sci, India Sec B: Biol Sci 84(3):505–518. doi:10.1007/s40011-013-0270-y

Meena BP, Kumar A, Dotaniya ML, Jat NK, Lal B (2014) *Effect of organic sources of nutrients on tuber bulking rate, grades and specific gravity of potato tubers*. Proc Natl Acad Sci, India Sec B: Biol Sci. doi:10.1007/s40011-014-0398-4

Meena BP, Kumar A, Lal B, Sinha NK, Tiwari PK, Dotaniya ML, Jat NK, Meena VD (2015a) Soil microbial, chemical properties and crop productivity as affected by organic manure application in popcorn (*Zea mays* L. var. everta). Afr J Microbiol Res 9(21):1402–1408

Meena RS, Meena VS, Meena SK, Verma JP (2015b) The needs of healthy soils for a healthy world. J Cleaner Prod 102:560–561

Meena RS, Meena VS, Meena SK, Verma JP (2015c) Towards the plant stress mitigate the agricultural productivity: a book review. J Cleaner Prod 102:552–553

Meena VS, Meena SK, Verma JP, Meena RS, Ghosh BN (2015d) The needs of nutrient use efficiency for sustainable agriculture. J Cleaner Prod 102:562–563

Meena VS, Meena SK, Verma JP, Meena RS, Jat LK (2015e) Current trend to mitigate climate change and sustainable agriculture: a book review. J Cleaner Prod 102:548–549

Meena VS, Verma JP, Meena SK (2015f) Towards the current scenario of nutrient use efficiency in crop species. J Cleaner Prod 102:556–557

Meena RS, Bohra JS, Singh SP, Meena VS, Verma JP, Verma SK, Sihag SK (2016) Towards the prime response of manure to enhance nutrient use efficiency and soil sustainability a current need: a book review. J Cleaner Prod 112(1):1258–1260

Metting FB, Smith JL, Amthor JS, Izaurralde RC (2001) Climatic change. 51:11. Intergovernmental panel on climate change 1995. Impacts, adaptations and mitigation of climate change. Scientific-technical analyses. Cambridge University Press, Cambridge, p 1996

Mohanty M, Sinha NK, Hati KM, Chaudhary RS, Patra AK (2015) Crop growth simulation modelling and climate. Scientific Publisher, Jodhpur

Myhre G, Shindell D, Bréon FM, Collins W, Fuglestvedt J, Huang J, Koch D, Lamarque JF et al (2013) Anthropogenic and natural radiative forcing. In: Stocker TF, Qin D, Plattner GK, Tignor M, Allen SK, Boschung J (eds) Climate change 2013: the physical science basis.

Contribution of working group I to the fifth assessment report of the intergovernmental panel on climate change. Cambridge University Press, Cambridge
NBSS & LUP (2005) Degraded and wastelands of India status and spatial distribution. Indian Council of Agricultural Research, Krishi Anusandhan Bhavan I, p 158
Nowak RS, Ellsworth DS, Smith SD (2004) Tansley review: functional responses of plants to elevated atmospheric CO_2 – do photosynthetic and productivity data from FACE experiments support early predictions? New Phytol 162:253–280
PRB (2015) World population data sheet with a special focus on women's empowerment. Population reference bureau. http://www.prb.org/pdf15/2015-world-population-data-sheet_eng.pdf
Ramaswamy V, Boucher O, Haigh J et al (2001) Radiative forcing of climate change. In: Houghton JT et al (eds) Climate change 2001: the scientific basis. Cambridge Press, Cambridge, pp 349–416
Rao AS, Somasundaram J (2013) Conservation agriculture for carbon sequestration and sustaining soil health, In: Agriculture Year Book 2013, Published by Agriculture Today (The National Agricultural Magazine) pp 184–188
Roscoe R, Buurman P (2003) Tillage effects on soil organic matter dynamics in density fractions of a cerrado Oxisol. Soil Till Res 70:107–119
Rosenzweig C, Hillel D (2000) Soils and global climate change: challenges and opportunities. Soil Sci 165:47–56
Ruane et al (2013) Multi-factor impact analysis of agricultural production in Bangladesh with climate change. Glob Environ Chang 23(1):338–350
Saha R, Ghosh PK (2013) Soil organic carbon stock, moisture availability and crop yield as influenced by residue management and tillage practices in maize-mustard cropping system under hill agro-ecosystem. Nat Acad Sci Lett 36:461–468
Sanyal SK, Majumdar K (2009) Nutrient dynamics in soil. J Indian Soc Soil Sci 57:477–493
Sardans J, Penuelas J, Estiarte M (2008) Changes in soil enzymes related to C and N cycle and in soil C and N content under prolonged warming and drought in a Mediterranean shrubland. Appl Soil Microbio 39:223–235
Schlenker W, Lobell DB (2010) Robust negative impacts of climate change on African agriculture. Environ Res Lett 5:014010
Schnitzer M (1982) Organic matter characterization. In: Page AL, Miller RH, Keeney DR (eds) Methods of soil analysis part 2, chemical and microbiological properties. ASA, SSSA, Madison, pp 581–594
Sejian V, Indu S (2013) Strategies to reduce enteric methane emission in domestic ruminant livestock. In: Lenka S, Lenka NK, Kundu S, Rao AS (eds) Climate change and natural resource management. New India Publishing Agency, New Delhi, pp 235–267
Singh DK, Singh SK, Singh AK, Meena VS (2014) Impact of long term cultivation of lemon grass (Cymbopogon citratus) on post-harvest electro-chemical properties of soil. Annals of Agri Bio Res 19(1):45–48
Skjemstad JO, Spouncer LR, Beech A (2000) Carbon conversion factors for historical soil carbon data. 15, CSIRO Land and Water, Adelaide
Smith P, Martino D, Cai Z, Gwary D, Janzen H, Kumar P, McCarl B, Ogle S, Mara FO, Rice C, Scholes B, Sirotenko O (2007) Agriculture. In: Metz B, Davidson OR, Bosch PR, Dave R, Meyer LA (eds) Climate change 2007: mitigation. Contribution of working group III to the fourth assessment report of the intergovernmental panel on climate change. Cambridge University Press, Cambridge
Sparling GP (1992) Ratio of microbial biomass carbon to soil organic carbon as a sensitive indicator of changes in soil organic matter. Aus J Res 30:195–207
Sreekanth NP, Santhi Prabha V, Babu P, Thomas AP (2013) Soil carbon alteration of selected forest types as an environmental feedback to climate change. Intl J Environ Sci 3:1516–1530
Steffen W, Canadell P (2005) Carbon dioxide fertilization and climate change policy. Australian Greenhouse Office, Department of Environment and Heritage, Canberra, p 33

Taub DR, Wang X (2008) Why are nitrogen concentrations in plant tissues lower under elevated CO_2? A critical examination of the hypotheses. J Integr Plant Biol 50:1365–1374

Treseder KK (2004) A meta-analysis of mycorrhizal responses to nitrogen, phosphorus, and atmospheric CO_2 in field studies. New Phytol 164:347–355

Venkanna K, Mandal UK, Solomon AJ, Sharma KL, Adake RV, Pushpanjali B, Reddy S, Masane RN, Venkatravamma K, Babu BP (2014) Carbon stocks in major soil types and land – use systems in semiarid tropical region of southern India. Curr Sci 106:604–611

Verma JP, Jaiswa DK, Meena VS, Meena RS (2015a) Current need of organic farming for enhancing sustainable agriculture. J Cleaner Prod 102:545–547

Verma JP, Jaiswal DK, Meena VS, Kumar A, Meena RS (2015b) Issues and challenges about sustainable agriculture production for management of natural resources to sustain soil fertility and health. J Cleaner Prod 107:793–794

Waha K, Müller C, Rolinski S (2013) Separate and combined effects of temperature and precipitation change on maize yields in sub-Saharan Africa for mid-to late-21st century. Glob Planet Chang 106:1–12

World Resources Institute (2000) World resources 2000-2001: people and ecosystems: the fraying web of life. World Resources Institute, Canada, p 389

Wright LL, Graham RL, Turhollow AR, English (1992) Opportunities for incresing forest cover. In: Forest and global change by Sampson RN, Hair D (eds) Am Forest Ass 1:123–156

Climate Change Risk Perception, Adaptation and Mitigation Strategy: An Extension Outlook in Mountain Himalaya

10

Anirban Mukherjee, Shantanu Rakshit, Arindam Nag, Mrinmoy Ray, H.L. Kharbikar, Kumari Shubha, Sujit Sarkar, Sudipta Paul, Subhodeep Roy, Aniruddha Maity, Vijay Singh Meena, and R. Roy Burman

Abstract

Climate change is becoming an ever increasing global threat which is difficult to ignore. The major underlying cause is anthropogenic, i.e. excessive use of fossil fuels, destruction of forests for industrialisation and urbanisation with rapid overgrowing population. The danger is such alarming that ecosystem will be irreversibly altered which will lead to suffering of human life by many ways. The overriding appearance of climate change is the increasing average worldwide temperature which is popularly called as global warming, and as a consequence several regions of the Earth are facing visible problems such as melting of glaciers, sea level rising, deviations in precipitation patterns and increase in

---

These views expressed are of the authors' and not necessarily of the organisations where they are attached with.

A. Mukherjee
ICAR-Indian Agricultural Research Institute, New Delhi 110012, India

ICAR-Vivekananda Institute of Hill Agriculture, Almora 263601, Uttarakhand, India

S. Rakshit (✉) • K. Shubha (✉) • S. Sarkar • S. Paul • R.R. Burman
ICAR-Indian Agricultural Research Institute, New Delhi 110012, India
e-mail: rakshitshantanu90@gmail.com; shubha.veg@gmail.com

A. Nag
Bihar Agricultural University, Sabour, Bihar 813210, India

M. Ray
ICAR-Indian Agricultural Statistics Research Institute, New Delhi 110012, India

H.L. Kharbikar • V.S. Meena
ICAR-Vivekananda Institute of Hill Agriculture, Almora 263601, Uttarakhand, India

S. Roy
ICAR-Indian Institute of Vegetable Research, Varanasi, Uttar Pradesh 221305, India

A. Maity
ICAR-Indian Grassland and Fodder Research Institute, Jhansi, Uttar Pradesh 284003, India

J.K. Bisht et al. (eds.), *Conservation Agriculture*,
DOI 10.1007/978-981-10-2558-7_10

plant diseases, and a number of bourgeoning challenges for public health are coming across by many nations. According to the Intergovernmental Panel on Climate Change (IPCC) report, the Indian Himalayan ecosystem (IHE) is one of the extremely vulnerable zones followed by the coastal ecosystem towards the climate change in India, and as per projection the climate change will impart serious environmental, economic and social impacts of the Indian Himalaya agricultural production system. At this juncture, strong adaptation and mitigation strategy is needed for reducing the vulnerability of resource-poor hill farmers and sustainable development of the Himalayan ecosystem. Climate change adaptation involves holistic changes in agricultural and ecological management practices. It comprises a combination of distinct responses, the indigenous knowledge systems, alternative practices and accessible technologies. Adaptation policy should be taking into account the farmers' perspective. In this piece of writing, the focus is to draw an outline of present condition and, furthermore, propose a strategy for effective adaptation and mitigation of climate change suited for Himalayan agricultural system.

Keywords
Climate change • Indian Himalaya • Modelling • Evergreen agriculture • Adaptation and mitigation

10.1 Introduction

The Indian Himalaya is one of the most important 'biological hotspots' (i.e. a natural environment which is rich in biodiversity and containing enormous endangered and endemic species) recognised by conservation international (CI 2011). Rich biodiversity is the prime characteristic feature of this region from biological resources' point of view. Globally various geographical regions as well as domains of knowledge (subject matter) have been explored related to climate change. The Himalayan ecosystem also has been explored with respect to the effect of climate change. But efforts that made at past were much sporadic; dearth of long-term studies, lack of availability of climate data and moreover the authenticity of the data make difficult to comprehend the climate change-related issues in Himalayan agriculture. Various authentic, reliable and peer-reviewed documents were assembled to generate some knowledge base about evergreen agriculture in the Indian Himalayan Region (IHR) vis-a-vis climate change.

The Himalaya are surrounded by the Tibetan Plateau on the north, the Indo-Gangetic Plain (IGP) on the south, the Karakoram and Hindu Kush ranges on the northwest and the Indian states of Assam and Arunachal Pradesh on the east. The hilly and mountain areas are vastly distributed all over the north extending ~2500 km in length and ~250 to 400 km in width. The Himalaya are classified according to the elevation as *Shivalik* (600–1200 m), the Middle Himalaya (~3000 m), the Greater Himalaya (~5200 to 8000 m) and Trans-Himalaya (average altitude ~4500 m expanded over ~60 km).

The Trans-Himalaya is further categorised as:

(a) *Western Himalaya*: Himachal Pradesh and Jammu and Kashmir
(b) *Central Himalaya*: Uttarakhand
(c) *Northeast Himalaya*: all northeastern states along with hilly parts of Assam and West Bengal (parts of Darjeeling district)

The ecology of the Indian Himalayan Region (IHR) varies with soils, climate and altitude. The climate ranges from tropical to highly temperate from the base to the highest elevations, respectively. Rainfall amount increases from west to east along the southern front. It may be due to the diversity of soils, climate and altitude, variability of distinct plant and animal species is seen here. The Himalayan ecosystem has significant importance in terms of species richness, biological diversity and anthropological and sociocultural diversity. The distinctive ecological entity and topographical variability are the elite characteristics of hill and mountain regions in the Himalaya.

The economy of hills is highly dominated by agricultural activities. The overriding physiognomies of hill farming in India are small land holdings, sloppy terrains and rainfed farming. In summer season, crop gets ~75 % amount of the annual rainfall, of which a lot goes to waste through runoff. In spite of this, a variety of cereals, vegetables, flowers, fruits, medicinal and aromatic plants are grown. Rice, wheat, maize, millets, pulses, vegetables and traditional crops like buckwheat (*Fagopyrum esculentum*), black soybean (*Glycine max* L. Merrill), amaranths (*Amaranthus* spp.), etc. are also grown in hills. The major cropping systems in rainfed condition are maize-wheat, rice-wheat, millets-wheat with intercropped pulses and oilseeds, whereas vegetable-based crop sequences along with rice-wheat are dominant under irrigated condition. In the high-hill temperate zone in high-altitude hilly area, only one cropping season is feasible during the summer (April to September).

Hill states are more vulnerable to natural disasters, i.e. landslides, excessive rains, earthquakes, extreme cold and frequent droughts. Climate change is being realised by the hill peoples in the form of untimely flowering in fruit trees, increased incidences of cloud burst and increase in drought-like situations. The major part of agriculture (~90 %) is rainfed which makes the area more vulnerable to climatic variations and changes such as drought and excessive rainfall. Vegetable productivity in the hilly areas of Uttarakhand is declining (Fig. 10.1) because of degradation of resources, light soil with shallow depth, low soil fertility, higher incidences of pest and disease and more over due to climate variation and changes.

The Himalayan mountains are large sources of water for plains (around 3 million hectares or ~ 17 % of the hilly area) due to Himalayan glaciers. The Himalayan glacial snowfields store ~12,000 km^3 of freshwater, and ~15,000 Himalayan glaciers constitutes a distinctive reservoir supply that flows in perennial rivers. Himalayan region has largest snow and ice cover in the world outside the polar regions and is one of the prime mountain systems, referred as ‘third pole’ (Schild 2008) and ‘water tower of Asia’ (Xu et al. 2009).

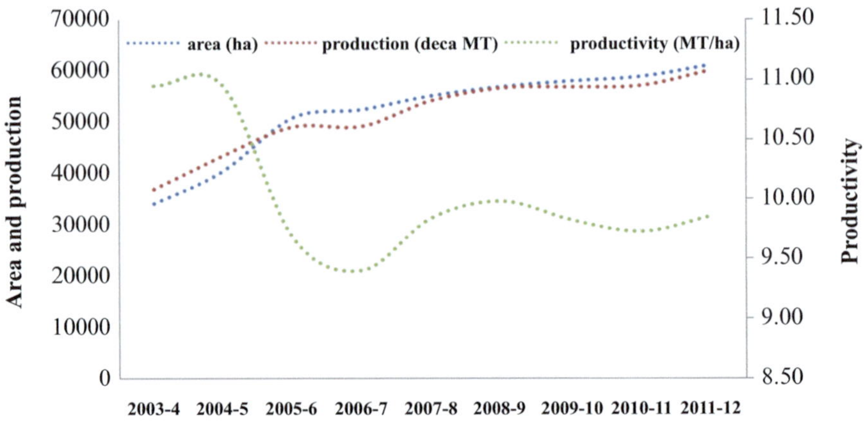

Fig. 10.1 Area production and productivity trends of vegetable crops of the northwest Indian Himalaya

A systematic effort made was made to generalise various projected environmental scenarios of the Himalayan region, in order to depict the climatic picture of the Indian Himalaya. According to Shrestha et al. (2012), the significant changes in temperature, rainfall, and vegetation phenology across have occurred in the Himalaya between 1982 and 2006. During the period of 25 years, an average increase of annual mean temperature and mean annual precipitation is 0.06 °C year^{-1} and 6.52 mm year^{-1}. They have also reported that average start of the growing season and the length of growing season have advanced by 4.7 days or 0.19 days year^{-1} although end of the growing season remained unchanged. The rate of warming in the Himalaya is greater than the global average (Shrestha et al. 2012), confirming about the potential of vulnerability. The Ministry of Environment and Forests, Government of India (2012), projected mean annual temperature will increase from 0.9 ± 0.6 to 2.6 ± 0.7 °C in 2030 and net temperature will increase from 1.7 to 2.2 °C with respect to the 1970s. In this region, 1–4.5 °C increases in minimum temperatures and 0.5–2.5 °C increases in the maximum temperatures are projected. The number of rainy days in the Himalayan region as per projection may increase by 5–10 days on an average in the 2030s, whereas rainfall intensity is likely to increase by 1–2 mmday^{-1} (INCCA 2010).

10.2 Current Status of Climate Change

Agriculture is one of such sectors of the modern economy which is straight way exposed to climate and therefore mostly affected by the climatic variability. The climate-linked instability happened to be one of the prime reasons of vulnerability in rainfed hill region. Long-term changes have been observed in different climatic parameters which lead to question food security issues.

10.2.1 Effect on Temperature

Studies indicate that the Himalayan region appeared to be warmer than the average increase in global temperature (Shrestha et al. 1999; Du et al. 2004; IPCC 2007). Incidences of increase in temperature are more during the winter and autumn rather than the summer, and the increase is greater at higher altitudes (Liu and Chen 2000).Studies have shown the evidences of increase in temperature. In Alaknanda valley of Western Himalaya, the mean annual temperature has raised by 0.15 °C between the years 1960 and 2000 (Kumar et al. 2008); a similar case has been reported in Kashmir valley where the mean temperature has gone up (1.45 °C) over the last 20 years (Sinha 2007). Exponential escalation in greenhouse gases (GHGs) in the atmosphere due to anthropogenic activities has resulted in such incidence (World Climate News 2006). The concentration of CO_2 has reached to 379 ppm in 2005 from 280 ppm after the industrialisation. There is a decreasing trend observed in maximum and minimum temperatures, sunshine hours during June to September (*kharif* season), while increasing trend in maximum temperature was recorded in post-monsoon and winter season, and increasing trend in sunshine hours too was observed during winter (Panday et al. 2014).

Rise in temperature in hill states will favour most of the vegetable and food grain crops in hills. Vegetables help in providing nutritional security and can reduce malnutrition problem prevailing in the hills (Gupta 2003). This temperature rise may bring some negative factors like the temperate fruit crops will be shifted to further higher altitude, change in preference of insect pest scenario, genetic erosion, loss of biodiversity, water scarcity, etc.

10.2.2 Effect on Rainfall and Water Availability

Long-term rainfall and water availability is very important for agriculture in the IHR. Irrigation is limited to ~10 % cultivated area in majority of IHR and confined only in the valleys. In many regions, a major amount of rainfall occurs early. Due to climate change, melting of glaciers begins earlier, and this affects river regimes and water supplies which further affect people's livelihoods. The existence and condition of wetlands in high altitude, reservoirs, rivers, and lakes are also affected (Eriksson et al. 2009).

Due to erratic rainfall, the irrigation systems have been severely affected. In the Kullu valley (Himachal Pradesh), it has been observed that the rainfall has reduced by ~7 cm, and snowfall by ~12 cm, although the average maximum and minimum temperatures have increased by 1 °C and 0.25 °C, respectively, in the last ~110 years (Vishvakarma et al. 2003). However, in Jammu and Kashmir and some other regions of the Indian peninsula, a significant raise in rainfall was noticed (Agarwal 2009).

10.2.3 Effect on Vegetation or Forest Cover

The climate-informatics analysis infers that in the last 10 years, all these climatic zones have shifted upwards in higher altitudes, and the oak and pine forest areas have also decreased. As a result of forest degradation, hydrological hazards (flash floods, high runoff, soil erosion, river-line floods, landslides, etc.) have increased in monsoon time, by which ~22 % of the cultivated area was affected annually during 2005–2010 (Rawat 2012).

Natural forest fire, which is indirectly linked with harsh climate, has also increased nowadays. This has caused insignificant decrease of broadleaves species of trees like *Quercus* spp. (oak), *Rhododendron* (*buransh*), *Myrica esculenta* (kafal), *Alnus* (utis), etc. These tree species are well known for retaining the water and recharging the soil moisture in the area. Hill women are custodians of ITK (indigenous technological knowledge) related to seed conservation, and they preserve a diverse genetic pool of this valuable resource through in situ conservation (Dhakal and Leduc 2010).

10.2.4 Effect on Food Production System

Today, the foremost challenge of the twenty-first century is ensuring food security for an ever increasing population. It is more pertinent to India where the population growth rate is high; a remarkable progress is made in production of food and fibre after 1966 by introducing a green revolution. However, India is still to match the production with a population growth rate to ensure country's food security. The importance of agriculture to food security is not only to provide food and fibre but more importantly to ensure primary sources of livelihood for the majority of the world workforce. Yet livelihood systems that are based on agriculture may face the risk of increased crop failure, frequent incidence of pest and diseases and loss of livestock due to climate change (Howden et al. 2007).

Traditionally, farmers are associated with several farm activities such as crop cultivation, fuel and fodder collection, water for irrigation and day-to-day use with the help of scarce resources available with them such as seeds, manpower, capital, land, water and other agricultural inputs. Such resources are decreasing rapidly and affect the agriculture business that impacted on deterioration of these scarce resources, low productivity, shift of the cropping patterns, shift of farmers towards nonfarm activities and seasonal migration of youth in search of job.

10.3 Vulnerability Assessment

Climate change can affect food production, availability of food grains, energy, fuel, water supply and human health. Some groups of people who live in areas that are more vulnerable to storms, flood, cloud burst, drought, etc. might face greater challenges than others. Agriculture is more vulnerable to climate change.

According to the nature of farming system, farmers' accessibility to the resources and their capacities to cope with the changes, farming community were affected by the climate change (Leduc et al. 2008).

10.3.1 Vulnerability

According to the United Nations IPCC third assessment report, vulnerability is 'the degree to which a system is susceptible to, and unable to cope with, adverse effects of climate change, including climate variability and extremes. The vulnerability is a function of the character, magnitude, and rate of climate change and variation to which a system is exposed, its sensitivity, and its adaptive capacity' (IPCC 2001). The definition quantifies vulnerability as a function of three main components: exposure (E), sensitivity (S) and adaptive capacity (AC) of a particular system (Metzger et al. 2005). The IPCC has figure out the framework for vulnerability assessment (Fig. 10.2). These three components may be combined and measured in many ways with different context-specific indicators (Sehgal et al. 2013; Das et al. 2014).

$$\text{Vulnerability} = f(\text{Exposure, Sensitivity, Adaptive capacity})$$

$$\text{In another way, Vulnerability } = f(\text{Potential Impact} - \text{Adaptive capacity})$$

Where, the exposure is demarcated as 'the nature and degree to which a system is exposed to significant climatic variations'. The sensitivity is clearly defined as 'degree to which a system is affected, either adversely or beneficially, by climate variability or climate change'. While the adaptive capacity is 'the ability of a system to adjust to climate change (including climate variability and extremes) to moderate potential damages, to take advantage of opportunities, or to cope with the consequences'. A system with higher level of adaptive capacity is less vulnerable. Hossain (2001) found that people of the United States are three times more exposed to natural hazards than those of Bangladesh, but they are less affected due to their better coping capacity.

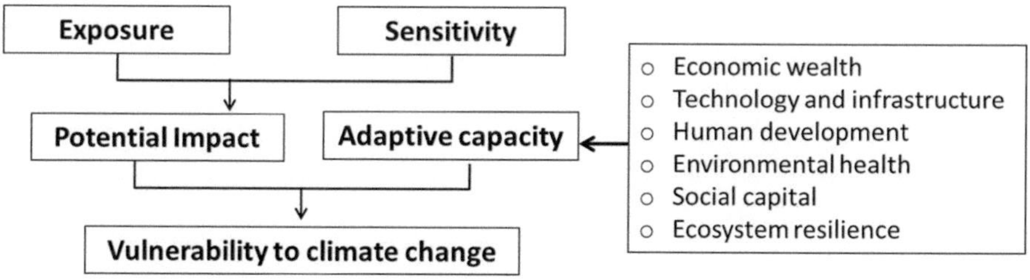

Fig. 10.2 Intergovernmental panel on climate change (IPCC) framework for vulnerability assessment

10.3.2 Components of Vulnerability

Vulnerability assessments provide a guide to design appropriate adaptation and mitigation policies for a specific ecosystem. Assessment can be done at different level, like individual, group, community, region or nation. Social scientists' and climate scientists' view about vulnerability are quite different. In vulnerability assessment social scientists lean towards the set of socio-economic variables that help in determining people's ability to cope with climate variability or climatic alteration (Allen 2003), while climate scientists often emphasise on probability of occurrence and consequences of weather and climate-related aberrations (Nicholls et al. 1999).

10.3.3 Studies on Vulnerability to Climate Change

Scientists of different fields including climate, agricultural sciences, social sciences, geography and environmental sciences have conducted several vulnerability assessment studies in different dimensions like livelihood vulnerability, agricultural vulnerability, social vulnerability and economical vulnerability (Table 10.1).

10.4 Adaptation and Mitigation

The climate change adaptation and mitigation strategies among Himalayan farmers are also reviewed to find some basis for future line of work. Adger et al. (2009) inquired social limits to adaptation and stated that adaptation limits are endogenous to society and hence depend on knowledge, attitudes to risk, ethics and culture of the society in particular. Based on reviewing from various social science perspectives, they have developed four propositions concerning adaptation limits:

(i) The ultimate goals of adaptation reinforced by diverse values
(ii) Uncertainty about future foresight of climate-related risk
(iii) Individual and social factors which limit adaptation actions
(iv) Systematic undervaluation of loss of places and culture disguises as real, experienced but are subjective

Lebel (2013) reviewed 42 studies and addressed three modest questions: (i) How is climate change recognised by people? (ii) What is known about climate change adaptation strategies? (iii) How do people learn about adaptation strategies? Although, traditional coping mechanisms are showing to be gradually insufficient, perfection in adaptation strategies and practices is becoming ever more urgent (Dulal 2014). (a) Efficiency of climate information system, institutions and infrastructure was assessed to explore successful climate change adaptation. Rather than focussing on 'one package fits all' approach, the content and format of climate information should take into consideration the differential information needed and used within vulnerable households and communities. Weak coordination among the government institutions

Table 10.1 Previous studies on vulnerability assessment, conceptual frameworks and indicators employed to assess vulnerability

Study	Concepts	Indicators
Climate change vulnerability	Vulnerability was conceptualised as a function of quality, magnitude and rate of climatic variations to which a system is exposed, its sensitivity and its adaptive capacity	**Exposure**:
Mapping for Southeast Asia (Yusuf and Francisco 2009)	Sensitivity was composed of human and ecological sensitivity	1. Frequency of tropical cyclones
	Adaptive Capacity was considered as a function of socio-economic factors, infrastructure and technology	2. Floods
		3. Droughts
		4. Exposure to landslides
		Sensitivity:
		1. Population density (uses as a proxy for sensitivity)
		2. Percentage of protected areas (as a proxy for ecological sensitivity)
		Adaptive capacity (socio-economic)
		1. Human development index (HDI)
		2. Poverty occurrence
		3. Income disparity
		Adaptive capacity (technology)
		1. Electricity coverage
		2. Extent of irrigation
		Adaptive capacity (infrastructure)
		1. Road density
		2. Communication
Vulnerability of Agriculture to Climate Change: District Level Assessment in the Indo-Gangetic Plains (Sehgal et al. 2013)	Vulnerability (V) of a system is a function of exposure, sensitivity and adaptive capacity	**Exposure**:
	Considered climatic, physical and socio-economic factors together to assess vulnerability	1. Rate of change in maximum temperature (kharif)
		2. Rate of change in maximum temperature (rabi)
		3. Rate of change in minimum temperature (kharif)

(continued)

Table 10.1 (continued)

Study	Concepts	Indicators
		4. Rate of change in minimum temperature (rabi)
		5. Frequency of low rainfall events (kharif)
		6. Severity of low rainfall events (kharif)
		7. Frequency of high rainfall events (kharif)
		8. Severity of high rainfall events (kharif)
		9. Frequency of low rainfall events (rabi)
		10. Severity of low rainfall events (rabi)
		11. Frequency of high rainfall events (rabi)
		12. Severity of high rainfall events (rabi)
		Sensitivity
		1. Net sown area/ geographical area
		2. Productivity of food grains
		3. Organic C content of soil
		4. Available water-holding capacity of soil
		5. Average landholding of farmer
		6. Human population density
		Adaptive capacity
		1. Percentage of irrigated area
		2. Human development index
		3. Cropping intensity
		4. Livestock density
		5. Percentage of villages electrified
		6. Percentage of villages with paved roads
		7. NPK fertiliser consumption

(continued)

Table 10.1 (continued)

Study	Concepts	Indicators
Mapping climate vulnerability and poverty in Africa Thornton et al. (2006)	Vulnerability has been conceptualised as depending on the five livelihood assets: natural, physical, social, human, and financial capital	**Natural capital**
	Country-level data was collected from FAO UNDP (2012), World Bank (2011)	Suitability for crop production
		Soil degradation due to water, wind and anthropogenicerosion sub-basin as internal water resources
		Physical capital
		Accessibility to markets
		Social capital
		Human poverty index
		Governance
		Stunting, poverty
		Human capital
		Infant mortality rate
		% underweight children
		Risk of dengue, malaria
		Public health expenditure
		Prevalence of HIV/AIDS
		Financial capital
		Agricultural GDP
		Global interconnectivity
The vulnerability concept and its application to food security. (Lucas and Hilderink 2004)	Overall vulnerability is calculated by the potential impact (exposure + sensitivity) and the coping capacity	**Exposure**
		Applied on the issue of food and nutritional security, resulting as measures taken for the vulnerability of countries towards food shortages
		Caloric balance index
		Food diversity index
		Sensitivity
		Fraction of agricultural value added
		In total GDP
		GINI coefficient
		GDP per capita (PPP) index
		Water stress index
		Water erosion hazard index
		Desertification risk index

(continued)

Table 10.1 (continued)

Study	Concepts	Indicators
		Land availability index
		Coping capacity
		GINI coefficient
		Literacy rate index
		Infrastructure density index
		Life expectancy index
		GDP per capita (PPP) index
Climate vulnerability index-measure of climate change vulnerability to communities: a case of rural Lower Himalaya, India (Pandey and Jha 2012)	Adaptive capacity (Socio-demographic profile + livelihood strategies + social network)	**Socio-demographic Profile**
	Sensitivity (health + food + water)	1. Family dependency index
	Exposure (Natural disaster + climate variability)	2. House type diversity index
		Family decision index
		Livelihood strategies
		1. % of household with migrated family members
		2. % of household had not changes crop variety
		3. % of household had not introduced new crop
		4. % of household reported change in sowing/planting time
		5. Index of fodder and fuel wood collection
		6. % of household sole dependent on agriculture
		7. Natural resource diversification index
		8. % of household reported depletion in natural resource
		Social network
		1. Percent of household obtained/seek help from social networks
		2. Percent of household provide help to others
		3. Percent of household with money transaction between friends
		4. Profession diversity index

(continued)

Table 10.1 (continued)

Study	Concepts	Indicators
		Health
		1. Percent of household with recent death of infant
		2. Percent of household with recent death
		3. Percent of household reported disease due to climatic factor
		4. Percent of household reported stress due to climatic factor
		5. Percent of household having new disease
		Food
		1. Percent of households dependent on agriculture for food
		2. Percent of households with insufficient food form farm
		3. Percent of households with decreasing food production
		4. Percent of household with reductions in nutrition
		5. Percent of household using pesticides
		6. Percent of household using fertiliser
		7. Percent of household having loss of agricultural land
		Water
		1. Percent of households with problem in availing potable water
		2. Percent of households with problem in availing irrigation water
		3. Percent of households utilising natural source of water
		Natural disaster
		1. Percent of household faced injury or death due to natural disaster

(continued)

Table 10.1 (continued)

Study	Concepts	Indicators
		2. Percent of household losses property due to natural disaster
		3. Percent of household reported high frequency of forest fire
		4. Percent of household reported increase in intensity of forest fire
		Climate variability
		1. Temperature and hot months perception index
		2. Rainfall and rainfall pattern perception index
		3. Hailstorms and cold waves perception index

was found due to contradictory mandates, dysfunctional interagency collaboration and inadequate funding provisions for adaptation. This should change fast or else resilience to climate change will be gloomy. Negi et al. (2013) pointed out protected cultivation using modern poly house technologies. Plug-tray nursery techniques (Shubha and Mukherjee 2015) and plastic mulching are new alternative technologies for vegetable cultivation, in high-altitude region and niche areas of the Himalaya. The adaptation policy should be made in such a way that it is affordable for poor hill farmers (Rawat 2010); policy solely based on only educating and sensitising people shall not be for too long (Vedwan and Rhoades 2001). There are evidences that villagers know well about climate change but could not afford to adapt due to tiny resources and expertise (Rawat 2010). However association of religion with ecosystem management (Joshi 1992) and promotion of adaptation and mitigation strategies could play an important role. Rawat (2013) suggested the following mitigation measures of climate changes among forest farmers:

(a) Social awareness campaigns
(b) Inclusion of subject on climate change and environment in syllabus of school and college students
(c) Concerned effort to save forest by protecting, rejuvenating and planting
(d) Empowering village level institutions for environmental protection
(e) Implementing strict policies for forest protection and management
(f) Rainwater conservation
(g) Encouraging people to reduce, reuse and recycle
(h) Planting plenty of trees

But there is a requirement of convergent measures to deal with the problem of climate change. To deal with this sector, wise potential adaptation options have been compiled and are depicted in following (Table 10.2).

Table 10.2 A sector-wise summary of potential adaptation options for Himalayan region

Sector	Adaptation options
Agriculture	Climate resilient agricultural production systems (short, early-maturing crops, short-duration, resistant and tolerant varieties, appropriate sowing and planting dates, proactive management)
	Diversification of enterprises, improve land cultivation management through adoption of appropriate climate resilient farming systems
	Promote water-use efficiency in agriculture through effective water storage (soil conservation, water harvesting) and optimising water use (drip, micro-jet, sprinkler and root rizospheric irrigation)
	Recycling waste water and solid waste in agriculture
	Adopt agro-ecosystems approach to planning and production methods
	Improve risk management through weather and seasonal forecasting for climatic conditions, crop monitoring and early warning systems and crop insurance
	Adjust cropping sequences and crop rotation to deal with climatic variability and extremes
Water resources	Adopt integrated watershed management and protect ecosystems (restore vegetation cover, prevent and control soil erosion and losses)
	Construct irrigation systems and reservoirs, and adjust the operation of water-supply infrastructure
	Recycling and reusing waste water for different daily activities
	Economic and optimised use of water through rainwater harvesting and contemporary water conservation systems (groundwater recharge, artificial water impoundments) and strengthen the unified management and protection of water resources (efficient water resource systems)
	Flood regulation, protection and mitigation; tapping water sources to increase water-supply capacity
	Monitor water resources to readjust national and sectoral plans (reduce future developments in floodplains); improve preparedness for water-related natural disasters
Human health	Improvise preventative medicinal activities and reinforce disease prevention measures
	Inaugurate epidemic prediction programmes to cope with possible vector-borne diseases, both for humans and domestic animals
	Adopt techno-engineering solutions to prevent vector-borne diseases/ epidemics such as increase infrastructure for waste disposal; improve sanitation facilities; reduce air and water pollution and track water quality, water treatment efficiency, soil quality, etc.
	Strengthen healthcare systems, including surveillance, monitoring and information dissemination
	Improve public education and literacy rates in various communities
Ecosystems and biodiversity	Adopt afforestation and simultaneously prevent deforestation
	Conserve natural habitats in climatic transition zones inhabited by genetic biodiversity with potential for ecosystem restoration and reintroduce endangered species by ensuring local peoples participation
	Initiate awareness programmes and involve school children and local people for the cause

(continued)

Table 10.2 (continued)

Sector	Adaptation options
	Adopt integrated ecosystem planning, strengthen monitoring, and management of vulnerable ecosystems
	Reduce habitat fragmentation and promote development of migration corridors and buffer zones for wild animals
	Enhance forest management policy for the protection of natural forests, prevent desertification processes and mixed-use strategies
	Design a pragmatic programme to develop forestry planting and growth technology for fast growth species to increase share of forest (afforestation and reforestation), to conserve and rehabilitate soil and prevent slope failure and mass wasting, and for forest protection
	Undertake preventive measures for forest hazards
	Preserve gene material in seed banks and explore different ITKs for successful conservation of biodiversity
Natural disasters	Strengthen the early warning system within the national meteorological and hydrological service use of ICT for quick warning delivery
	Undertake full assessment of wildfire-risk zones and increase public awareness
	Initiate different trainings on disasters management for youth

Sources: Documents pertaining to the ICIMOD 2010; Aggarwal 2008, Vision-2050 document ICAR-VPKAS are consulted

10.4.1 Case Study of Adapting Climate Change (Himalayan Perspective)

In Uttarakhand farmers have selected ~40 different crops and ~100 of crop cultivars containing cereals, pseudo-cereals, millets, local pulses and tuber crops (Maikhuri et al. 1997; Agnihotri and Palni 2007). Another example of adopting diversity farming by farmers is mixed cropping of 12 cereals (called *Baranaja*), one of the best example of maintaining agrobiodiversity of that area (Ghosh and Dhyani 2004). These crops has adapted the native environment and developed inherent qualities to resist the environmental risks which has further safeguarded the food and nutritional security of the mountain populace.

10.5 Importance of Indigenous Technological Knowledge (ITK) in Adapting Climate Variability

The most effective means to mitigating climate change should contain the components like sustainability, a vital dimension of climate change adaptation (Cohen et al. 1998; Swart et al. 2003), cost-effectiveness and people's participation. Integrating ITKs into climate change policies may lead to achieve these desired goals (Hunn 1993; Robinson and Herbert 2001). Such strategies comprise the adoption of efficient technologies; environmental resource management practices, viz. soil and water management practices; traditional biodiversity management; changes in cropping pattern; plant protection measures and livestock rearing practices where climate change is visible.

10.5.1 ITKs in Soil and Water Management

In the Indian Himalaya region, one of the valuable and important common natural resources is water. Mountain communities have a traditional basis for effective management of common property resources. In Himachal Pradesh villagers often construct small water storage ponds locally called *chal* to harvest rainwater. They often use the moisture of dew or fog by ploughing the fields in early morning. To collect the spring water, small reservoirs are built at intervals on the high uplands. Ash is used in conserving the rainwater by melting the snow and increase soil porosity. Similarly, mulching is used to conserve soil moisture and maintain the soil temperature. Water drops from earthen pitcher is used as traditional form of modern drip irrigation (Lal and Verma 2006, 2008). In Meghalaya the bamboo-based drip irrigation system is used (Singh 1989). It is widely practised by the farmers in Jaintia Hills of Meghalaya (Jeeva et al. 2006). This irrigation system is mainly used for plantation crops like areca nut (*Areca catechu* L.), black pepper (*Piper nigrum* L.) and betel vines (*Piper betle* L.) (Sharma and Prasad 1994). In Meghalaya soil erosion in agricultural fields is minimised by using bamboo culms, stones and gunny bags filled with soil (Jeeva et al. 2006; Kumar et al. 2005).

Rao et al. (2003) reported different traditional water management structures in the state Uttarakhand of central Himalaya; these are *Gools*, network of channels, used for irrigation; *Gaad*, rainfed small rivers; *Gadhera* rainfed rivulets; *Choyas*, tiny seasonal seepages; *Dhara*, underground water channels that are visible generally on lower or mid hills; and *Naula*, the traditional rocky structures to capture percolated seepage waters and supply adequate water throughout the year. In the western part of Nepal, the size of *Naula* is about $6 \times 4 \times 2\ m^3$ dimension to reserve the spring water and was used for irrigation (Sharma et al. 2009a, b). All these are managed independently by few households and community for agriculture and environment.

10.5.2 ITKs in Traditional Biodiversity Management

The central Himalaya regions of Uttarakhand traditionally use resources from forests and common lands as decided by communities' consensus (Rao et al. 2003). Though timber has potential economic value, timber trade was not practised generally. There are several evidences that villagers of northeastern Himalaya are using plant species (*Milletia pachycarpa*, *Albizzia amara*, *Vitex negundo*, *Xylosma longifolia* Clos., *Cedrela toona* Roxb., *Jatropha curcas* L., etc.) for protection of forest sapling and other crops (Santosh and Chhetry 2012). These villagers know the nature and its wealthy distribution. They share the natural resources together for sustainable development. Villages near the alpine zone are more affluent in summer fodder, medicinal plants and bamboo, whereas those close to foothills are in winter fodder. One village accept other villagers to use their resources not only for achieving social integrity but also for economic gains (Maikhuri et al. 2000).

10.5.3 ITKs in Change in Cropping Patterns

Due to increase in temperature and duration of summer, delayed onset and irregular distribution of southwest monsoon, winter, winter period reduction, reduced and delayed snowfall during winters, unpredictable rainfalls, etc., farmers of the Himalaya have altered the prevailing cropping pattern to suitable one. In the low- and mid-hill regions, basmati rice and sugarcane are replaced by maize and local paddy rice. The apple belt has shifted towards higher altitudes, and former apple growing area are replaced by vegetable crops (Sarkar 2015). Farmers are shifted from traditional native crops (e.g. local rice, wheat and maize) to off-season vegetables, for instance, tomato, cabbage and cauliflower (Rana et al. 2013). In the hills of Nepal, Bhutan, Sikkim and Darjeeling, large cardamom (*Amomum subulatum*)-based agroforestry system have proven to be a sustainable land use. Large cardamom agroforestry system accelerate the nutrient cycling, increases the soil fertility and productivity, reduces soil erosion, conserves biodiversity and restores soil fertility and carbon sequestration (Sharma et al.2009a, b; Mishra and Rai 2013). In the mid- and low-hills of Sikkim, ricebean (*Vigna umbellata*) and urdbean (*Vigna mungo*) are grown in rotation after the maize is harvested (Subba 2009).

10.5.4 ITKs in Plant Protection

A mixture of cow urine, *Vitex negundo* (*Nirgundi*) and *Ferula asafoetida* (*Hing*), is used to control insect pests of wheat crop. Ash, soil mixture and cow urine are used against cabbage aphids. To control insect pests of mustard crops, *Aloe barbadensis* (*Gwarpatha*), *Nicotiana tabacum* (tobacco), *Azadirachta indica* (neem) and *Sapindus trifoliatus* L. (*Aritha*) mixture is used. For management of pod borers in gram crop whey (*lassi*), *Aloe barbadensis* (*Gwarpatha*), *Nicotiana tabacum* (tobacco) are used. The use of *Nirgundi* in paddy is practised to control pest infestation. Spreading of ash in the vegetable field is practised against chewing and sucking type of insects like beetles, leaf-defoliating insects, leaf miners, thrips and aphids. Sharma et al. (2009a, b) reported that in the Himalayan region of Nepal before the rainy season begins, farmers used spores of Yarsagumba on the heads of caterpillars to control underground caterpillars. The pulp of the *Khira* leaf is used to control stem borer (*Chillozonellis*) in rice and wheat. Cactus (locally called as *seewdi* and *khirro*) is used for biofencing.

For protection of seed material, farmers of Uttarakhand commonly used rhizome and leaves of bach (*Acorus calamus* L.), leaves of peach (*Prunus persica* L.), timur (*Zanthoxylum armatum* DC.), neem (*Azadirachta indica* A. Juss), lemon (*Citrus limon* L.), walnut (*Juglans regia* L.), bakayan (*Melia azadirach* L.), turmeric (*Curcuma longa* L.), wooden ash, cow dung ash, cow dung and cow urine mixture, lime powder, mustard oil and red roasted soil (Mehta et al. 2012).

In a study conducted in Bageshwar district of Uttarakhand, Chandola et al. (2011) cited common salt (NaCl) is broadcasted at the rate of 1 kg$Nali^{-1}$ (1 Nali =200 m^2 or 1/50 ha) to control white grub. Pine leaves and leaves of other plants along with ash of cow dung cakes are also used to control white grub in vegetable crop also like chilli. *Chullah* (oven) ash is also used for storing food grains. A mixture of dried leaves of walnut, immature turmeric and mustard oil is properly pasted on the grains of green gram to avoid attack of storage pest (Chandola et al. 2011). Black gram mix with mustard oil is also used to avoid disease and pest attack in stored grains. Some farmers of hills store black gram (*Vigna mungo*) with salt (NaCl) in order to prevent pests of stored grains. Leaves of neem (*Azadirachta indica*), walnut (*Juglans regia* L.) and timur (*Zanthoxylum armatum* DC.tree are used for storing cereals and pulses both grains and seed (Chandola et al. 2011).

10.5.5 ITK in Livestock Disease Control

Devadar tree oil is used for wounds; cow urine and black ash are used to control external parasites; and black cumin powder is used for fever. To control ticks (*Arachnids*) in animals, '*Karoi*' grass (*Striga asiatica*) is rubbed on affected area in the skin. For deworming mustard oil, fermented mixture of neem leaves and buttermilk are used. To get higher milk production, different feed combination is used. '*Pinda*' is prepared by mixing wheat flour in lukewarm water with '*Gur*', butter (extracted from curd), rice, and *Jhingora* (*Echinochola frumentacea*). A combination of '*Bhimal*' leaves, bhatt (black soybean), wheat, *binola* (cotton seeds), *Methi* seeds, *Dhalia* and *Jaggery* is also used to increase the milk production. To recover from mouth ulcers, lemon and salt are used, for mastitis honey and *haldi,* for diarrhoea powder of half ripe *bael* fruit, powder of *Shisam* (*Dalbergia sissoo*) leaves, etc. are used (Subramanyeswari and Chander 2013).

10.6 Statistical Models Predicting Climate Change in the Himalaya

The Indian Himalaya agricultural performance depends on various climatic parameters out of which rainfall, temperature, snowfall and glacier are vital. Hence, forecasting and quantifying the impact of rainfall, temperature, snowfall and glacier through statistical modelling will be serviceable for planning purposes. In this article we have discussed about four models:

(i) Autoregressive Integrated Moving Average model (ARIMA)
(ii) Artificial neural network
(iii) ARIMAX
(iv) ARIMA Intervention

10.6.1 Models for Predicting Temperature and Rainfall

Let

y_t = temperature/ rainfall at time t
y_{t-1} = temperature/ rainfall at time $t-i$, i=1,2,. . .p

10.6.1.1 ARIMA Model Fitting

An ARIMA model is given by $\phi(B)(1-B)^d y_t = \theta(B)\varepsilon_t$ where $\phi(B) = 1 - \phi_1 B - \phi_2 B^2 - \cdots - \phi_p B^p$ (autoregressive parameter)$\theta(B) = 1 - \theta_1 B - \theta_2 B^2 - \cdots - \theta_p B^q$ (moving average parameter)

ε_t = white noise or error term
d = differencing term
B = backshift operator, i.e. $B^a Y_t = Y_{t\text{-}a}$

ARIMA modelling strategy consists of three phases, viz. (i) identification, (ii) estimation and (iii) diagnostic checking. Parameters of this model are experimentally selected at the identification stage, and at the estimation stage the parameters are estimated using iterative least square techniques. The efficacy of the particular model is then tested at the diagnostic checking stage. If the model is found to be insufficient, the three phases are repeated again until satisfactory ARIMA model is generated for the time series under consideration:

(i) *Identification*
Identification of 'd' is imperative to make the nonstationary time series to stationary. A conventional statistical test for the presence of stationarity, commonly known as the test of the unit root hypothesis or augmented Dickey-Fuller (ADF) test can be employed to test the stationarity. Makridakis et al. (1998) discussed a good account on ADF test. In this case generally the null hypothesis is 'the time series is not stationary', and the alternative hypothesis is 'the time series is stationary' (Tables 10.3 and 10.4).
(ii) *Estimation of parameters*
In estimation phase, parameters are estimated for the ARIMA model speculatively selected at the identification stage. Thereafter, estimation

Table 10.3 Primary distinguishing characteristics of theoretical acf's and pacf's for stationary process

Process	acf	pacf
AR	Tails off towards zero (exponential decay or damped sign wave)	Cuts off to zero (after lag p)
MA	Cuts off to zero (after lag q)	Tails off towards zero (exponential decay or damped sign wave)
ARMA	Tails off towards zero	Tails off towards zero

Table 10.4 Summary of stationary conditions for AR coefficients

Model type	Stationary conditions
ARMA(0, q)	Always stationary
AR(1) or ARMA(1, q)	$\lvert\phi_1\rvert < 1$
AR(2) or ARMA (2, q)	$\lvert\phi_2\rvert < 1, \phi_2 + \phi_1 < 1, \phi_2 - \phi_1 < 1$

of parameters for ARIMA model is conventionally done by employing iterative least squares method. Bayesian information criterion (BIC) and Akaike information criterion (AIC) values for ARIMA model are calculated by

$$\mathrm{BIC} = T\log\left(\sigma^2\right) + (p + q + 1)\log T$$

and

$$\mathrm{AIC} = T\log\left(\sigma^2\right) + 2(p + q + 1)$$

where T denotes the number of observations utilised for parameter's estimation and σ^2 represents the mean square error (Box and Tiao 1975; Box et al. 2009).

(iii) *Diagnostic checking*

In diagnostic checking, testing is conducted to check the statistical adequacy of the estimated model, i.e. whether the error terms are 'white noise', i.e. uncorrelated with zero mean and constant variance. Ljung-Box test is employed to the original series or to the residuals after fitting a model. Box et al. (1994) has given a good explanation on Ljung-Box test. The null hypothesis in the series is 'white noise', whereas the alternative hypothesis in one or more autocorrelations up to lag m is not zero. The test statistics is given by

$$Q^* = T(T + 2)\sum_{k=1}^{m} \frac{r_k^2}{T - k}$$

where, T is the observation number deployed for estimation of the model and m is the utmost number of lag. Here statistics Q approximately follows a

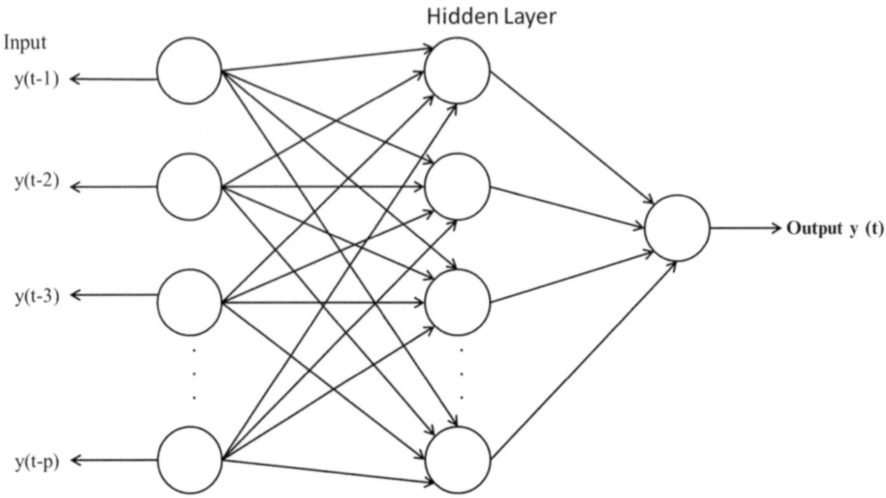

Fig. 10.3 The neural network structure

chi-squared distribution with $(T-k)$ degrees of freedom, where k is the number of parameters estimated in the ARIMA model, and the r_k is the autocorrelation function of residual at lag k. If it is not adequate, we should return back to the identification stage again to speculatively select another model.

10.6.1.2 Artificial Neural Network (ANN) Model

Artificial neural network (ANN) models are well thought out as a class of generalised nonlinear model that are able to capture various nonlinear structures present in the data set. The main advantage of this model is devoid of prior assumption of the data generating process required; instead it largely depends on characteristics of the data popularly known as data-driven approach. Single hidden-layer feedforward network is the most popular for time-series modelling and forecasting (Fig. 10.3). This model is regarded by a network of three layers of simple processing units and thus termed as multilayer ANNs. The first layer is input layer, the middle layer is the hidden layer, and the last layer is output layer.

The relationship between the output (y_t) and the inputs $(y_{t-1}y_{t-2}\cdots y_{t-p})$ can be mathematically represented as follows:

$$y_t = f\left(\sum_{j=0}^{q} \omega_j g\left(\sum_{i=0}^{p} \omega_{ij} y_{t-i}\right)\right)$$

where, $\omega_j (j = 0, 1, 2, \ldots.., q)$ and $\omega_{ij} (i = 0, 1, 2, \ldots.., p, j = 0, 1, 2, \ldots.., q)$ are the model parameters frequently called the connection weights; q is the number of hidden nodes and p is the number of input nodes, and g and f denote the activation function at hidden and output layer, respectively. Activation function defines the relationship between outputs and inputs of a network in terms of degree of the nonlinearity. Most commonly used activation function is logistic function which is frequently used as the hidden-layer transfer function, i.e.

$$g(x) = \frac{1}{1 + e^{-x}}$$

Thus ANN model performs a nonlinear functional mapping between the input and output which is characterised by a network of three layers of simple processing units linked by acyclic links:

$$y_t = f(y_{t-1} + y_{t-2}, \ldots, y_{t-p}, w) + \varepsilon_t$$

where, w is a vector of all parameters and f is a function of network architecture and connection weights. Therefore, the neural network resembles a nonlinear autoregressive model.

The selection of appropriate number of hidden nodes as well as optimum number of lagged observation p for input vector is important in ANN modelling for determination of the autocorrelation structure present in a time series. Though there are no established theories available for the selection of p and q, various training algorithms have been employed for the determination of the optimal values of p and q. The most commonly used training method is the back propagation algorithm. The objective of the training is to minimise the error function that assesses the misfit between the predicted and actual value. The error function which is widely used is mean squared error which can be written as

$$E = \frac{1}{N}\sum_{n=1}^{N}(e_i)^2 = \frac{1}{N}\sum_{n=1}^{N}\left\{y_t - f\left(\sum_{j=0}^{q}\omega_j g\left(\sum_{i=0}^{p}\omega_{ij}y_{t-i}\right)\right)\right\}$$

where N is the total number of error terms. The parameters of the neural network ω_{ij} are changed by an amount of changes in $\Delta\omega_{ij}$ as

$$\Delta\omega_{ij}(t) = -\eta\frac{\partial E}{\partial\omega_{ij}} + \partial\Delta\omega_{ij}(t-1)$$

where, η is the learning rate and $\frac{\partial E}{\partial\omega_{ij}}$ is the partial derivative of the function E with respect to the weight ω_{ij}. δ is the momentum rate.

For sigmoid activation function,

$$\frac{\partial E}{\partial w_{ij}} = -e_i(n) \times \theta'(x) \times y_j(n)$$

where $-e_t(n)$ is the residual at nth iteration

$$\theta'(x) = \frac{\exp(-x)}{(1 + \exp(-x))^2}$$

$y_j(n)$ is the desired output.

10.6.2 Model for Quantifying the Impact of Temperature/Rainfall

10.6.2.1 ARIMAX Model

ARIMAX model is given by $Y_t = \frac{\theta(B)}{\phi(B)}\varepsilon_t + \omega(B)X_t$

where

Y_t = crop yield at time t
X_t = rainfall/temperature at time t
$\omega(B) = \omega_0 + \omega_1 B + \cdots + \omega_s B^s$ (impact parameter)
$\phi(B) = 1 - \phi_1 B - \phi_2 B^2 - \cdots - \phi_p B^p$ (autoregressive parameter)
$\theta(B) = 1 - \theta_1 B - \theta_2 B^2 - \cdots - \theta_p B^q$ (moving average parameter)
ε_t = white noise or error term
B = backshift operator, i.e. $B^a Y_t = Y_{t-a}$

The modelling procedure ARIMAX process is the same as ARIMA model

10.6.3 Model for Quantifying the Impact of Snowfall/Glacier

10.6.3.1 ARIMA Intervention Model

An intervention model is given by $Y_t = \frac{\theta(B)}{\phi(B)}\varepsilon_t - \omega(B)I_t$

Y_t = ARIMA model + [intervention component]*I_t

where

Y_t = crop yield at time t

$$I_t = \begin{cases} 0 \\ 1 \end{cases}$$

1 = if glacier/snowfall occurred
0 = if glacier/snowfall did not occurred
$\omega(B) = \omega_0 + \omega_1 B + \cdots + \omega_s B^s$ (impact parameter)
$\phi(B) = 1 - \phi_1 B - \phi_2 B^2 - \cdots - \phi_p B^p$ (autoregressive parameter)
$\theta(B) = 1 - \theta_1 B - \theta_2 B^2 - \cdots - \theta_p B^q$ (moving average parameter)
ε_t = white noise or error term
B = backshift operator, i.e. $B^a Y_t = Y_{t-a}$

The modelling procedure ARIMA Intervention process is the same as ARIMA model

10.7 Ongoing Projects and Programmes on Climate Risk Mitigation in Himalaya

Agriculture sector plays a key role in socio-economic development and provides employment opportunities to ~51 million people who practise hill agriculture in the IHR. Vulnerability of agricultural sector discourages investments in agriculture sector. The climatic aberrations, consequently decreased productivity of existing crops and the resultant changes in cropping patterns pose major threats to Himalayan agriculture. Therefore agriculture sector involves a lot of risks. Accessibility and efficiency of insurance or agricultural risk management strategies depend on public policies (Walker and Jodha 1986; Raju and Chand 2008a, b). The resource-poor farming communities are highly vulnerable to the risks of climate change due to poor adaptation capacity. Keeping in view of such problems faced by farmers, several programmes has been launched by different organisations for mitigation of climatic risks in agriculture.

10.7.1 National Action Plan on Climate Change

India's first National Action Plan on Climate Change (NAPCC) document was released by the prime minister of India on 30 June 2008 with the objective of demarcation of existing and upcoming policies and programmes for achieving sustainable economic development with a focus on adaptation, mitigation and scientific research (Ghosh 2009). Under this plan, eight national missions were conceptualised:

- The *National Mission for Sustainable Agriculture* for developing of climate resilient agricultural technologies and expanding of weather-based insurance schemes
- The *National Mission for Green India* for afforestation of degraded forestland and the forest area increase
- The *National Solar Mission* for increasing the contribution of solar energy in total energy for power generation and other uses

Likewise several other missions have been planned such as:

- The *National Mission for Enhanced Energy Efficiency*
- The *National Mission on Sustainable Habitat*
- The *National Water Mission*
- The *National Mission for Sustaining Himalayan Ecosystem*
- The *National Mission on Strategic Knowledge for Climate Change*

10.7.2 National Initiative on Climate Resilient Agriculture (NICRA)

The NICRA is a network project of the Indian Council of Agricultural Research (ICAR) launched in the lead institute, Central Research Institute for Dryland Agriculture (CRIDA), on February 2011. The project was implemented at large number of ICAR-Research Institutes, State Agricultural Universities and Farm Science Centres (*Krishi Vigyan Kendras*). The objective of the project is strategic research for improvement of Indian agriculture according to climate change situations, demonstrations of climate resilient technologies and capacity building programmes for scientists, farmers and other change agents in climate resilient agricultural research and its application.

10.7.3 Mechanism of NICRA Project

The development of improved varieties/breeds of crops, livestock and management practises to adapt climate change situations and provide inputs for policymaking in mainstreaming climate resilient agriculture as a part of developmental planning. The project is comprised of four components:

1. 'Strategic research on adaptation and mitigation
2. Technology demonstration on farmers' fields to cope with current climate variability
3. Sponsored and competitive research grants to fill critical research gaps
4. Capacity building of different stakeholders'

ICAR-VPKAS's Farm Science Centre Krishi Vigyan Kendra (KVK) situated at *Chinyalisaur* is implementing National Initiative on Climate Resilient Agriculture (NICRA) project in *Dunda* village in district Uttarkashi of Uttarakhand. The interventions by KVK such as natural resource management, plantation and management of fruit crops, introduction of drought-tolerant variety of pigeon pea, introduction of HYV of oilseeds, pulse, cereals, millets and vegetable crops, etc. are very helpful to the farmers for increasing farm income as well as improvement in their livelihood options. Due to interventions made by KVK, farmers are able to conserve waste water in LDPE tanks and used that water for off-season vegetable cultivation, and net return of Rs 3,500–4,000 per year by each farmer was generated. Farmers produced seventy quintal of vermicompost which had increased soil moisture and fertility of soil (APR 2015).

Since 2011–2012, there are about 17 KVKs working in northeastern states under NICRA project with the aim of sustainable management natural resources in terms of climate risk mitigation on the farmer's fields (Paul et al. 2014). These KVKs have been carrying out demonstrations, capacity building programmes and other activities for sustainable and environment-friendly agricultural technologies adoption by the farmers. Under soil health programmes, various interventions like green manure and mulching for in situ soil moisture conservation, soil reclamation, embankment to check soil erosion, etc. are implemented on farmer's fields, and as a result soil erosion, soil salinity, deterioration of soil health were minimised, and improved soil health as well as maintenance of the soil moisture was reported by Paul et al. (2014). The east Sikkim farmers were inspired for construction of rainwater harvesting tanks, *Jalkund* for conversion of fallow land into cultivated land to cultivate year-round vegetable. Rejuvenation of water bodies resulted a significant impact on crop diversification and increasing cropping intensity. Participatory planning and renovation of drainage channels, recycling of organic wastes in the form of vermicomposting and biogas plants and creation and use of bio-fertilisers can be options for better resource utilisation. These successful efforts by different KVKs managed natural resources in terms of climate risk mitigation on the farmer's fields.

10.7.4 Agricultural Insurance Schemes

Farmers are not able to control the events like natural calamities and weather vagaries. To overcome such problems, there is a necessity of such policies and programmes like insurance schemes for protecting the farmer's production losses (Swain 2014). Insurance can shift the likelihood of risks and enable individuals to be involved in risky activities (Ashan et al. 1982). In India, various programmes and schemes launched time and again for protection and compensation of the farmers. Agricultural insurance schemes played an important role to deal with the risks associated in agriculture sector in effective manner.

10.7.4.1 National Agricultural Insurance Scheme (NAIS)

In India, National Agricultural Insurance Scheme (NAIS) was introduced in the year 1999–2000 (*rabi* season) and implemented by Agricultural Insurance Company of India Ltd. (AIC). This scheme covers all oilseeds, food grains and annual horticultural crops. NAIS is operating both on the basis of 'area approach' and 'individual approach' for extensive calamities and localised calamities, respectively. At present, it is being executed by all the states except four states of northeast, viz., Arunachal Pradesh, Manipur, Mizoram, Nagaland and the state of Punjab. Initially, the payable premium was subsidised at 50 % for small and marginal farmers, now it is 10 %. The subsidised amount equally shared by the central and the concerned state/UTK governments (Economic Survey 2013–2014; NAS 2014).

10.7.4.2 Modified National Agricultural Insurance Scheme (MNAIS)

The MNAIS proposal was made after the study of previous NAIS, and it was implemented in 50 districts on full-fledged basis from Rabi 2013–2014 by approval of the Government of India (MNAIS 2010). The most important enhanced features of MNAIS over NAIS are mentioned below:

- Uniform seasonality norms for both loanee and non-loanee farmers.
- For all farmers, there is higher subsidy in premium ranging up to 75 %.

- Decrease of unit area to all villages in Panchayat for major crops.
- Minimum insurance level was higher (~70 % as compare to 60 % in NAIS).
- An individual assessment of claims in case of specified localised calamity, viz., hailstorm and landslide.
- Addition of insurance amount for sowing risks and for postharvest losses due to cyclones.

10.7.4.3 Weather-Based Crop Insurance Scheme (WBCIS)

A pilot Weather-Based Crop Insurance Scheme (WBCIS) was launched in 2007 in twenty states. The aim of this scheme was to make larger coverage of crop insurance among farmers. The scheme operates on the 'concept of area approach, which implies that a reference area shall be linked to the nearby reference weather station' on the basis of which compensation shall be made. This scheme is existed for all loanee and non-loanee farmers. The WBCIS is envisioned to provide insurance safeguard to the farmers against adverse weather incidences such as excessive and deficit precipitation (mainly rainfall), abnormal temperature, humidity, etc. The scheme has benefit of settlement of claims in a short period (aicofindia.com).

10.7.4.4 Weather Index Insurance

The weather index insurance was introduced in India for farmers in the year 2003 and in 2007; the government of India adopted it as a substitute to crop-yield index insurance. In this scheme by the year 2012, up to ~12 million farmers, growing 40 crops over ~15 million hectares, would be insured against weather-related losses (Report on Evaluation of Pilot Weather Based Crop Insurance Scheme 2011).

10.7.4.5 *Varsha Bima Yojna* (Rainfall Insurance Scheme)

Rainfall insurance scheme (*Varsha Bima*) introduced by Agricultural Insurance Company of India Ltd. (AIC) during the year 2004 (southwest monsoon). The scheme has provided five different options suited for varied requirements of diverse farming community (AIC 2015).

All the interventions made in recent years helped in the minimisation of carbon emissions and climate risk mitigation and benefited to farming community and society at large. Thorough adoption of policies and schemes like National Agricultural Insurance Scheme, Weather-Based Crop Insurance Scheme, *Varsha Bima Yojna* (insurance scheme based on rainfall) and recently in February 2016 Prime Minister Crop Insurance Scheme (*Pradhan Mantri Fasal Bima Yojana*) has been announced keeping the climate change in mind. Through all these programmes, farmers are capable to take risks and able to compensate the losses occurrence in agriculture due to the weather vagaries.

10.8 Strategic Context for Better Climate Resilience in Himalayan Perspective

The resource-poor rural farming communities are the most deprived and highly vulnerable to the risks related to climatic aberrations due to poor adaptation capacity. Moreover, the physiographic and environmental constraints associated with the wide variability in altitude, slope, and aspect limit the adoption of modern agricultural technologies in the IHR. Climatic aberrations has been experienced by the hill farming community through the practise of shift in cropping seasons, untimely flowering in fruit trees, increased incidences of cloud burst and increase in drought-like situations. Most of the farming population in Uttarakhand hills practices rainfed agriculture which is frequently getting affected by localised drought and excessive rainfall. The unique and typical socio-economic characteristics of hill and mountain ecosystem contain unique socio-economic features, ethnicity, climate variability and human activities. Farming community is the ultimate sufferer of climate change. So their part should be addressed through developmental efforts for reducing the effects of climate change.

As the farmers are key stakeholders of agricultural system and critical end users of technology, it is essential to realise their needs, objectives and outlook. Apart from scarcity of water and weather vagaries, today's main issue in agriculture is the migration of youths from agricultural to service sectors. Agriculture no more remains as lucrative as it was before. This issue can be fixed through in-depth quantitative socio-psychological research (Mukherjee and Maity 2015). Presently, young and educated farmers rather than old and traditional are more associating with innovative modern technologies and extension systems delivered by private extension system (Mukherjee et al. 2011). But before that the factors behind association of younger farmers with profitable agriculture through private extension system (Mukherjee et al. 2012a) need to be identified further in hill perspectives. Hill farmers' profit orientation and market orientation (Mukherjee et al. 2012b) are required factors for making agriculture profit oriented along with climate resilient.

There is requirement of superior understanding in the correlation between the Himalayan ecosystem and the climatic factors so that it can provide inputs for sustainable Himalayan development. Currently, there is no solid database available on soil, water, crop genetic diversity and indigenous practices in Himalayan perspective which can frame climate resilient farming models and validate technologies for hill farmers covering aspects related to crops, livestock, agroforestry, farming systems, horticulture and fisheries, soil and water conservation and water efficiency in relation to climate change. Establishment of proper monitoring systems, validation and implementation of generated climate models and capacity building of hill farming community are need of the hour.

10.8.1 Pragmatic Model Addressing the Issues

Rapid environmental degradation as a result of faster rate of deforestation, unobstructed grazing practices destruction of vegetation and unplanned development create threat to hill agriculture. Degradation of aquatic resources, erratic rainfall pattern, and unavailability of inputs and high rate of soil fertility depletion in the absence of recycling of farm residues hinder the sustainability of hill crop production system. Besides these, underdeveloped marketing as well as transport facilities culminates reduction of net return and diminish farmers' interest for establishing high-value commercial crop entrepreneurship. Immediate attention is therefore required to analyse economic impacts of climate change on Himalayan ecosystem, communities and livelihood of hill farmers.

The Himalayan ecosystem is facing problems as consequences of climate change and some other physiographic and socio-economic constraints, which are in culmination affecting hill agriculture and thereto livelihood of hill population. The nature and extent of such adversities are not clearly known. To address the problem, a structure programme on statistical survey on the components of agriculture, ecosystem and livelihood is required (Fig. 10.4). Based on the generated data, simulation and modelling can be done to predict the upcoming future of Himalayan ecology. Thereafter assessing the vulnerability of Himalayan region based on specific lower unit area and ultimately based on the data, we can propose Climate Resilient Farming System (CRFS). Validation is essential before progressing towards mass extension. Continuous monitoring and recording the response at farmers' field are required as the farmers are the end user of the CRFS model. Based on the results, we shall find the empirical solutions in terms of environmental legislation, capacity building of stakeholders and development of strong marketing channels of less energy intensive green inputs, promotion of sustainable energy base creation, along with the cleaner technology on agriculture through a convergence model (Mukherjee et al. 2012c; Mukherjee and Maity 2015). Channelizing all these evergreen agriculture for Himalayan ecosystem is possible.

10.9 Concluding Remarks and Future Perspectives

Anthropogenic activities influence climate change in two means. The first is mitigation, by reducing global emissions of greenhouse, and the second is adaptation, the focus of this paper. By assessing the level and extent of vulnerability, statistical models are very helpful in predicting climate change in Himalaya. There are specific statistical models for predicting temperature and rainfall, quantifying the impact of temperature and rainfall and quantifying the impact of snowfall and glacier, and based on that adaptation, strategy can be made. Side by side vulnerability assessments should be continued to bring out information about the nature and magnitude of probable consequences of climate change and informed decisions about the outline and necessity of adaptation actions and strategies. These assessments help decision-makers to better understand the nature and intricacy of

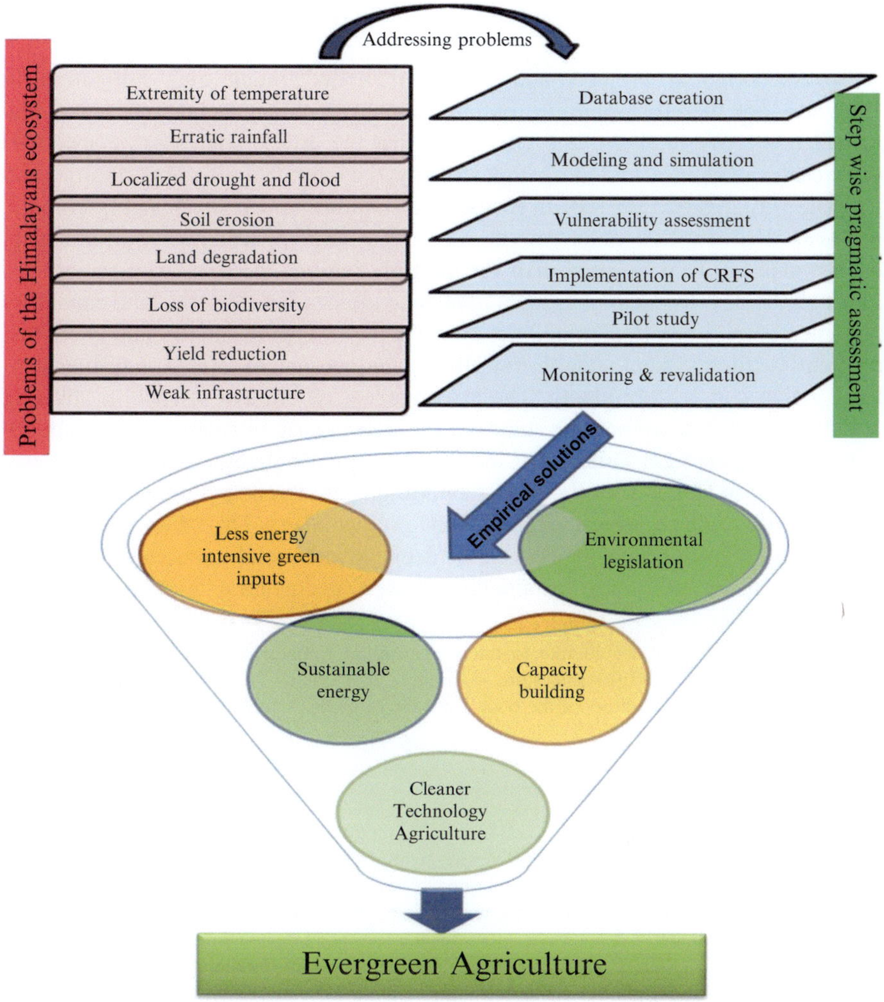

Fig. 10.4 Conceptual model to mitigate the climate change and bring the evergreen agriculture in Indian Himalayans

vulnerability across scales and provide comprehensions that can assist in the development of strategies that could minimise vulnerability and maximise opportunities for adaptation. By providing knowledge about the type of vulnerability of resource users, vulnerability assessments can facilitate policymakers to prioritise their efforts and tailor a choice of adaptation approaches to most effectively accommodate and support the divergent requirements of the different types of resource users. ITKs are the one way to adaptation and mitigation of the climate change. ITKs in soil and water management, traditional biodiversity management, change in cropping patterns, plant protection and livestock disease control are very much crucial for farmers and other stakeholders to survive in the change in time-tested traditional way.

Several ongoing projects and programmes on climate risk mitigation in Himalaya are therefore pointing out the climate resilient technologies and risk management. We analysed and concluded that the article addresses perspectives towards the adaptation and mitigation strategy, the overwhelming opportunities and possible way forward for IHR in focusing agriculture under the changing climatic scenario. We identified major gaps as constraints, and opportunities to fill this gap, but we also put forward road map for the hilly regions in Himalayan ecosystem. Through the development of climate resilience indices for hill soils, land capability classifications, simulation modelling, vulnerability mapping and ultimately climate resilient farming system suitable for Himalayan ecosystem, effective hill agricultural policy can be formulated. The effectiveness of climate science in advising policy formulation can only be achieved through intense inter-institutional collaboration. This paradigm shift in IHR in respect to climate change adaptation requires commitment to stakeholders' capacity building (both inter- and intra-disciplinary) which would bring together the farmers and policymakers. This offers an exclusive opportunity to bridge the gap between the decision and social sciences with the goal of creating more agriculture sustainability under the changing climatic scenarios.

Acknowledgements We would like to thank the Indian Council of Agricultural Research – Vivekananda Institute of Hill Agriculture – Almora, Uttarakhand 263601, India, for providing support and facilities and editors of the book for their valuable comments.

References

Adger WN, Dessai S, Goulden M, Hulme M, Lorenzoni I, Nelson DR, Naess LO, Wolf J, Wreford A (2009) Are there social limits to adaptation to climate change? Clim Chang 93(3–4):335–354

Agarwal PK (2009) Global climate change and Indian agriculture. Indian Council of Agricultural Research, New Delhi

Aggarwal PK (2008) Global climate change and Indian agriculture; impacts adaptation and mitigation. Indian J Agric Sci 78(10):911–919

Agnihotri RK, Palni LMS (2007) On farm conservation of landraces of rice (Oryza sativa L.) through cultivation in the Kumaun region of Indian, Central Himalaya. J Mt Sci 4:354–360

AIC (2015) Varsha Bima. Agriculture Insurance Company of India Limited: available at http://www.aicofindia.com/AICEng/Pages/Default.aspx

Allen K (2003) Vulnerability reduction and the community-based approach. In: Pelling M (ed) Natural disasters and development in a globalizing world. Routledge, London, pp 170–184

APR (2015) Annual progress report of April 2014 to March 2015 of NICRA. ICAR- VPKAS KrishiVigyan Kendra, Chinyalisaur-Uttarkashi (Uttarakhand)

Ashan, Syed M, Ali AAG, Kurian NJ (1982) Towards a theory of agricultural crop insurance. Am J Agric Econ 64(3):520–529

Box GEP, Tiao GC (1975) Intervention analysis with application to economic and environment problems. J Am Stat Assoc 70:70–79

Box GEP, Jenkins GM, Reinsel GC (1994) Time series analysis: forecasting and control. Pearson Education, Delhi

Box GEP, Jenkins GM, Reinsel GC (2009) Time series analysis: forecasting and control, 3rd edn. Holden-Day, San Francisco

Chandola M, Rathore S, Kumar B (2011) Indigenous pest management practices prevalent among hill farmers of Uttarakhand. Indian J Tradit Knowl 10(2):311–315

CI (2011) The biodiversity hotspots. Available at URL Address: http://www.biodiversityhotspots.org/xp/hotspots/himalaya/Pages/default.aspx

Cohen S, Demeritt J, Robinson J, Rothman D (1998) Climate change and sustainable development: towards dialogue. Glob Environ Chang 8(4):341–371

Das MK, Srivastava PK, Rej A, Mandal ML, Sharma AP (2014) A framework for assessing vulnerability of inland fisheries to impacts of climate variability in India. Mitig Adapt Strat Glob Chang 1–18. DOI 10.1007/s11027-014-9599-7

Dhakal TD, Leduc B (2010) Women's role in biodiversity management in Himalayas.Sustainable Mountain Development.International Centre for Integrated Mountain Development (ICIMOD), G.P.O. Box 3226, Khumaltar, Kathmandu, Nepal

Du MY, Kawashima S, Yonemura S, Zhang XZ, Chen SB (2004) Mutual influence between human activities and climate change in the Tibetan plateau during recent years. Glob Planet Chang 41:241–249

Dulal HB (2014) Governing climate change adaptation in the Ganges basin: assessing needs and capacities. Int J Sustain Dev World Ecol 21(1):1–14

Economic Survey (2013–2014) Available at URL address http://www.insightsonindia.com/2014/07/09/download-economic-survey-2013-2014-pdf/

Eriksson M, Jianchu X, Shrestha AB, Vaidya RA, Nepal S, Sandström K (2009) The changing Himalayas – impact of climate change on water resources and livelihoods in the Greater Himalayas. International Centre for Integrated Mountain Development (ICIMOD), G.P.O. Box 3226, Khumaltar, Kathmandu, Nepal

Ghosh P (2009) National Action Plan on climate change. Prime Minister's council on climate change.pp1–31

Ghosh P, Dhyani PP (2004) Baranaaja: the traditional mixed cropping system of Central Himalaya. Outlook Agric 33:261–266

Gupta HS (2003) Agricultural productivity in the hills and mountains of India. In: Pandey RK, Shah D, Kumar A, Sharma R (eds) Sustainable mountain agriculture. Pub. Centre for Development Studies, Uttaranchal Academy of Administration, Nanital, pp 65–80

Hossain SMN (2001) Assessing human vulnerability due to environmental change: concepts and assessment methodologies. Master of Science Degree thesis, Division of land and water resources, Department of Civil and Environmental Engineering, Royal Institute of Technology, KTH, Stockholm, Sweden

Howden SM, Soussana J-F, Tubiello FN, Chhetri N, Dunlop M, Meinke H (2007) Adapting agriculture to climate change. Proc Natl Acad Sci 104:19691–19696

http://documents.worldbank.org/curated/en/2011/11/15643352/weather-index-insurance-agriculture-guidance-development-practitioners

http://mospi.nic.in/Mospi_New/site/inner.aspx?status=3&menu_id=192

http://www.aicofindia.com/AICEng/Generaldocuments/Product Profiles/Varsha rainfall-ENG.pdf

http://www.aicofindia.com/AICEng/GeneralDocuments/Product Profiles/Varsha-rainfall-eng.pdf

http://www.aicofindia.com/AICEng/GeneralDocuments/Product Profiles/WBCIS FAQ.pdf

http://www.aicofindia.com/AICEng/Pages/Default.aspx

http://www.searchschemes.com/rainfall-insurance-varsha-bima) viewed on 20-08-2015

https://en.wikipedia.org/wiki/AgricultureInsurance Company ofIndia

https://www.agriskmanagementforum.org/sites/agriskmanagementforum.org/files/Documents/riskmanagementthroughweatherbasedinsurance.pdf

Hunn E (1993) What is traditional ecological knowledge? In: Williams N, Baines G (eds) Traditional ecological knowledge: wisdom for sustainable development. Centre for Resource and Environmental Studies, ANU, Canberra, pp 3–15

ICIMOD (2010) Climate change vulnerability of mountain ecosystems in the Eastern Himalayas. Climate change impact and vulnerability in the Eastern Himalayas – synthesis report. International Centre for Integrated Mountain Development, Kathmandu, Nepal, June 2010

INCCA (2010) Indian network for climate change assessment. Report 2. Page 164 in Climate Change and India: a 4x4 Assessment – a sectoral and Regional Analysis for 2030s, Ministry of Environment and Forests, Government of India. November 2010

IPCC (2001) Climate change 2001: synthesis report., Intergovernmetal panel on climate change. Available at https://www.ipcc.ch/ipccreports/tar/vol4/english/

IPCC (2007) Climate change 2007: the physical science basis. In: Solomon S, Qin D, Manning M, Chen Z, Marquis M, Averyt KB, Tignor M, Miller HL (eds) Contribution of working group I to the fourth assessment report of the intergovernmental panel on climate change. Cambridge University Press, Cambridge, p 996

Jeeva SRDN, Laloo RC, Mishra BP (2006) Traditional agricultural practices in Meghalaya, North East India. Indian J Tradit Knowl 5(1):7–18

Joshi PC (1992) Afforestation, development and religion: a case from the Himalayas. In: Singh SC (ed) Himalaya: environment economy and people. R. K. Publications, New Delhi, pp 453–465

Kumar R, Singh RD, Sharma KD (2005) Water resources of India. Curr Sci 89(5):811

Kumar K, Joshi S, Joshi V (2008) Climate variability, vulnerability, and coping mechanism in Alaknanda catchment, Central Himalaya, India. AMBIO 37:286–291

Lal C, Verma LR (2006) Use of certain bio-products for insect-pest control. Indian J Tradit Knowl 5(1):79–82

Lal C, Verma LR (2008) Indigenous technological knowledge on soil and water management from Himachal Himalaya. Indian J Tradit Knowl 7(3):485–493

Lebel L (2013) Local knowledge and adaptation to climate change in natural resource-based societies of the Asia-Pacific. Mitig Adapt Strateg Glob Chang 18(7):1057–1076

Leduc B, Shrestha AB, Bhattarai B (2008) Case study: gender and climate change in the Hindu Kush Himalayas of Nepal. WEDO, Kathmandu

Liu X, Chen B (2000) Climate warming in the Tibetan Plateau during recent decades. Int J Climatol 20:1729–1742

Maikhuri RK, Semwal RL, Rao KS, Nautiyal S, Saxena KG (1997) Eroding traditional crop diversity imperils the sustainability of agricultural systems in Central Himalaya. Curr Sci 73:777–782

Maikhuri RK, Semwal RL, Rao KS, Singh K, Saxena KG (2000) Growth and ecological impacts of traditional agroforestry tree species in central Himalaya, India. Agrofor Syst 48:257–272

Makridakis S, Wheelwright SC, Hyndman RJ (1998) Forecasting: methods and applications, 3rd edn. Wiley, New York

Mehta PS, Negi KS, Rathi RS, Ojha SN (2012) Indigenous methods of seed conservation and protection in Uttarakhand Himalaya. Indian J Tradit Knowl 11(2):279–282

Metzger MJ, Leemans R, Schröter D (2005) A multidisciplinary multiscale framework for assessing vulnerability to global change. Int J Appl Earth Obs Geoinf 7:253–267

Mishra PK, Rai SC (2013) Use of indigenous soil and water conservation practices among farmers in Sikkim Himalaya. Indian J Tradit Knowl 12(3):454–464

MNAIS (2010) Modified National Agricultural Insurance Scheme available online at :http://agricoop.nic.in/mnaiso29910.pdf. Viewed on 21-08-2015

Mukherjee A, Maity A (2015) Public–private partnership for convergence of extension services in Indian agriculture. Curr Sci 109(9):1557–1563

Mukherjee A, Bahal R, Burman RR, Dubey SK, Jha GK (2011) Effectiveness of Tata Kisan Sansar in technology advisory and delivery services in Uttar Pradesh. Indian Res J Ext Educ 11 (3):8–13

Mukherjee A, Bahal R, Burman RR, Dubey SK (2012a) Factors contributing farmers' association in Tata Kisan Sansar: a critical analysis. Indian Res J Ext Educ 12(2):81–86

Mukherjee A, Bahal R, Burman RR, Dubey SK, Jha GK (2012b) Constraints in privatized agricultural technology delivery system of Tata Kisan Sansar. J Glob Commun 5(2):155–159

Mukherjee A, Bahal R, Burman RR, Dubey SK (2012c) Conceptual convergence of pluralistic extension at Aligarh District of Uttar Pradesh. J Commun Mob Sustain Dev 7(1&2):85–94

NAS (2014) Ministry of statistics and programme implementation. Available at http://mospi.nic.in/Mospi_New/site/inner.aspx?status=3&menu_id=192

Negi VS, Maikhuri RK, Rawat LS, Parshwan D (2013) Protected cultivation as an option of livelihood in mountain region of central Himalaya, India. Int J Sustain Dev World Ecol 20 (5):416–425

Nicholls RJ, Hoozemans FMJ, Marchand M (1999) Increasing flood risk and wetland losses due to global sea-level rise: regional and global analyses. Glob Environ Chang 9:S69–S87

Panday SC, Mahanta D, Kumar A, Stanely J, Singh S (2014) Climate variation/changes and their impact on crop yields in mid Hills of North- West Himalayas In compendium –cum-Abstracts: International Symposium on "New- Dimensions in Agrometeorology for Sustainable Agriculture (NASA-2014)". 16–18, Oct 2014, G.B.P.U.A. &T, Pantnagar, India pp73

Paul S, Jat PC, Singha AK, Bhalerao AK, Bordoloi R, Pasweth A, Gogoi AK (2014) Sustainable management of natural resources for climate resilient agriculture in North Eastern Region: successful cases of KVK Interventions. Zonal Project Directorate, Zone-III, Umiam- 793103, Meghalaya

Raju SS, Chand R (2008a) Agricultural insurance in India problems and prospects. NCAP Working Paper No. 8. pp 1–78

Raju SS, Chand R (2008b) Study on performance of National agricultural insurance scheme and suggestions to make it more effective. Agric Econ Res Rev 21:11–19

Rana RS, Bhagat RM, Kalia V, Lal H, Sen V (2013) Indigenous perceptions of climate change vis-a-vis mountain agricultural activities in Himachal Pradesh, India. Indian J Tradit Knowl 12 (4):596–604

Rao KS, Semwal RL, Maikhuri RK, Nautiyal S, Sen KK, Singh K, Saxena KG (2003) Indigenous ecological knowledge, biodiversity and sustainable development in the central Himalayas. Trop Ecol 44(1):93–111

Rawat M (2010) People's perception on effect of climate change in Garhwal Himalayas region – a case study from Kunjapuri Hills. Reflections of climate change leaders of Himalaya, Leadership for Environment and Development (LEAD) Report, India, New Delhi

Rawat PK (2012) Impacts of climate change and hydrological hazards on Monsoon crop patterns in the Lesser Himalaya: a watershed based study. Int J Disaster Risk Sci 3(2):98–112

Rawat VS (2013) People's perception on climate change and their influence on various aspects around Tones Valley of Garhwal Himalaya. Environ Ecol Res 1(3):150–154

Report on impact evaluation of pilot weather based crop insurance study (WBCIS) (2011) Submitted by Agricultural Finance Corporation Ltd., Mumbai to Department of Agriculture and Cooperation Ministry of Agriculture Government of India on January 2011. 5–237

Robinson J, Herbert D (2001) Integrating climate change and sustainable development. Int J Glob Environ Issues 1(2):130–148

Santosh T, Chhetry GKN (2012) Agro-biodiversity management related ITKs in North-Eastern India. J Biol Agric Healthcare 2(6):83–92

Sarkar S (2015) Assessment of climate change led vulnerability and simulating the adaptive behaviour of farmers in the Himalayan and Arid Ecosystem. Ph. D. thesis. Indian Agricultural Research Institute, New Delhi

Schild A (2008) The case of the Hindu Kush-Himalayas: ICIMOD's position on climate change and mountain systems. Mt Res Dev 28(3/4):328–331

Sehgal VK, Singh MR, Chaudahry A, Jain N, Pathak H (2013) Vulnerability of agriculture to climate change: district level assessment in the Indo-Gangetic plains. Available at http://www.nicra-icar.in/nicrarevised/images/Books/Vulnerability%20of%20agriculture%20to%20climate%20change.pdf

Sharma UC, Prasad RN (1994) Potential indigenous farming systems of Northeastern Hill region, (ICAR Research Complex for NEH Region, Shillong) 48

Sharma G, Sharma R, Sharma E (2009a) Traditional knowledge systems in large cardamom farming: biophysical and management diversity in Indian mountainous regions. Indian J Tradit Knowl 8(1):17–22

Sharma S, Bajracharya R, Sitaula B (2009b) Indigenous technology knowledge in Nepal–a review. Indian J Tradit Knowl 8(4):569–576

Shrestha AB, Wake CP, Mayewski PA, Dibb JE (1999) Maximum temperature trends in the Himalayas and its vicinity: an analysis based on temperature records from Nepal for the period 1971–94'. J Clim 12:2775–2787

Shrestha UB, Gautam S, Bawa KS (2012) Widespread climate change in the Himalayas and associated changes in local ecosystems. PLoS One 7(5):e36741. doi:10.1371/journal.pone.0036741

Shubha K, Mukherjee A (2015) Plug tray nursery: an innovative technique for off-season vegetable cultivation. Agrobios News Lett 13(9):79–80

Singh A (1989) Bamboo drip irrigation system. ICAR Research Complex for NEH Region, Barapani

Sinha S (2007) Impact of climate change in the highland agroecological region of India, Sahara Time Magazine, (http://www.saharatime.com/Newsdetail.aspx?newsid=2659)

Subba JR (2009) Indigenous knowledge on bio-resources management for livelihood of the people of Sikkim. Indian J Tradit Knowl 8(1):56–64

Subrahmanyeswari B, Chander M (2013) Integrating indigenous knowledge of farmers for sustainable organic farming; an assessment of Uttarakhand State of India. Indian J Tradit Knowl 12(2):259–264

Swain M (2014) Crop insurance for adaptation to climate change in India. Asia research centre working paper 61:1–40

Swart R, Robinson J, Cohen S (2003) Climate change and sustainable development: expanding the options. Climate Policy. 19–40

UNDP (2012) Managing climate risks in the Himalayas: a community centric-approach. Regional climate risk reduction project. Available at URL: http://www.undp.org/content/undp/en/home/librarypage/crisis-prevention-andrecovery/managing-climate-risks-in-the-himalays.html

Vedwan N, Rhoades RE (2001) Climate change in the western Himalayas of India: a study of local perception and response. Clim Res 19:109–117

Vishvakarma SCR, Kuniyal JC, Rao KS (2003) Climate change and its impact on apple cropping in Kullu Valley, North-West Himalaya, India. In: 7th international symposium on temperate zone fruits in the tropics and subtropics, 14–18 October, Nauni-Solan (H.P.)

Walker TS, Jodha NS (1986) How small farm households adapt to risk. In: Hazell P, Pomareda C, Valdes A (eds) Crop insurance for agricultural development. The Johns Hopkins University Press, Baltimore

World Bank (2011) Weather index insurance for agriculture: guidance for development practitioners. Agriculture and rural development discussion paper; no. 50. World Bank, Washington, DC, pp 1–116

World Climate News (2006) Homing in on rising sea-levels. WMO, Geneva, No. 29, June 29: 1–12

Xu J, Grumbine R, Shrestha A, Eriksson M, Yang X, Wang Y et al (2009) The melting Himalayas: Cascading effects of climate change on water resources, biodiversity, and human livelihoods in the Greater Himalayas. Conserv Biol 23:520–530

11 Toward the C Sequestration Potential of Agroforestry Practices to Combat Climate Change in Kumaon Himalaya, India

R.P. Yadav, Jaideep Kumar Bisht, B. Gupta, and T. Mondal

Abstract

The Himalayan region has a long tradition of tree-based smallholder agroforestry practices, which are rich in the different species of trees. Several indigenous agroforestry systems based on people's needs and site-specific characteristics have been developed over the years by the smallholders for various uses. These agroforestry practices have attractive, wide, and promising potential to store carbon (C) and remove atmospheric carbon dioxide (CO_2) through enhanced growth and development of trees. From the recent study results reported as the maximum biomass (1,170 Mg ha^{-1}), C stock (526.5 Mg ha^{-1}), and biomass C equivalent CO_2 (1,932.2 Mg ha^{-1}) was found in high density of oak plantation, followed by pecan nut-based agrihorticulture system biomass (48.7 Mg ha^{-1}), C stock (21.9 Mg ha^{-1}), and biomass C equivalent CO_2 (80.0 Mg ha^{-1}). However, minimum biomass (8.6–28.4 Mg ha^{-1}), C stock (3.8–12.7 Mg ha^{-1}), and biomass C equivalent CO_2 (13.9–46.6 Mg ha^{-1}) reported in fruit tree-based agrihorticulture system. The C storage in plant biomass is only feasible in abovementioned type of perennial agroforestry systems, and smallholder farmers are the real practitioner of these agroforestry systems. Thus, agroforestry systems are not only remunerative to the farmers from livelihood point of view but also contributing toward trapping of atmospheric CO_2 vis-à-vis mitigation of climate change.

R.P. Yadav (✉)
ICAR-Vivekananda Parvatiya Krishi Anusandhan Sansthan, Almora 263601, India
e-mail: rams34052@gmail.com

J.K. Bisht • T. Mondal
ICAR-Vivekananda Institute of Hill Agriculture, Almora 263601, Uttarakhand, India

B. Gupta
Dr. Y S Parmar University of Horticulture and Forestry, Nauni, Solan 173230, India

J.K. Bisht et al. (eds.), *Conservation Agriculture*,
DOI 10.1007/978-981-10-2558-7_11

Keywords
Agrihorticulture systems • Carbon stock • High-density plantation • Himalaya • Pecan nut • Oak

11.1 Introduction

The C sequestration potential of agroforestry systems has attracted attention from developing countries in recent years following the recognition of agroforestry as a greenhouse gas (GHG) mitigation strategy under the Kyoto Protocol (Albrecht and Kandji 2003). The sale of C sequestered through agroforestry could be an attractive economic opportunity for subsistence farmers in developing countries, who are the major practitioners of agroforestry, and it will be an environmental benefit to the global community at large as well. The Himalayan region has a long tradition of agroforestry; several indigenous agroforestry systems based on people's needs and site-specific characteristics have been developed over the years (Yadav et al. 2016a). In addition to already existing indigenous agroforestry systems, improved practices and technologies are now being promoted in the region for perceived benefits such as wasteland management, reducing erosion hazards, improving biodiversity and food security, and a variety of other benefits (Chavan et al. 2015; Verma et al. 2015b; Meena et al. 2016; Ghosh et al. 2016).

Agroforestry practices play a vital role in removal of atmospheric CO_2 through enhanced growth and development of trees and thus have wide and promising potential to store C in its biomass. Tree components in agroforestry systems can be significant sink of atmospheric CO_2 due to their fast growth and sustainable productivity. By including trees in agricultural production systems, agroforestry can, arguably, increase the amount of C stored in lands devoted to agriculture, while still allowing for the growing of food crops (Kursten 2000). The average potential of agroforestry has been estimated to be 25 t C/ha over 96 Mha (Sathaye and Ravindranath 1998; Basu 2014). Sequestration of C in biomass is presently considered as the most promising approach to mitigate GHG effect (Yadav et al. 2016a). However, little has been reported regarding C sequestration potential of agroforestry systems in Indian Himalaya. In this scenario, it is imperative that the C sequestration potential for agroforestry practices in the region is also investigated. In this article, we have reviewed biomass production, C stock, and CO_2 mitigation potential of agroforestry systems, viz., pecan nut in agrihorticulture system, fruit trees in agrihorticulture system, and high-density plantation of oak in Kumaon Himalaya, India.

11.1.1 Area and Climate

The Indian Himalaya cover an area of 53.7 Mha, which is ~17 % of total geographical area of the country. Out of 21 agroecological regions of the country, four regions are covered exclusively and one partly in the hill and mountain agroecosystem. These five agroecological zones have a wide variation in their climate, i.e., from cold arid to warm per humid. The average annual rainfall in the region varies from 150 to 4,000 mm. The mean annual temperature varies from 8 to 22 °C. The growing period of different crop ranges from 90 to ~270 days in a year. Skeletal and calcareous to brown forest podzolic are the major soil groups of the region (Tulachan 2001); these are alkaline to acidic in nature. The natural vegetation in the Indian Himalayan region is tropical deciduous, alpine, temperate, and wet evergreen forest.

11.1.2 Population and Land Use Pattern

The population of the Himalayan region is ~39 million, which is ~4 % of the total population of India; the primary source of income here is agriculture, which contributes ~45 % of the total regional income. This region is thinly populated (population density 627/1,000 ha). However, the actual pressure on agricultural land in the hills is much higher since the net cultivated area is only ~12 % of the reporting area. The per capita availability of cultivated land in hill region is just ~0.17 ha. Forest is the major land use in the Himalaya, and it covers ~59 % of total reporting area of hills, the permanent pastures and grazing land region having ~12 % area.

The Himalayan ecosystem has relative advantages for horticulture because of its specific environmental conditions and several micro-situations. It has a subtropical to temperate climate, and a wide range of fruits (citrus, banana, mango, apple, walnut, plum, peach, and cherry), vegetables (potato, pea, cabbage, cauliflower, tomato, etc.), spices (ginger, turmeric, chilies, etc.), and flowers (orchids, gladiolus, marigold, chrysanthemum, etc.) are being here. The total area under fruits and vegetable is ~13–15 % of the gross cropped area (Tulachan 2001).

11.1.3 The Forests

The nonarable land constitutes about ~84 %, and the forest occupies ~60 % of the total geographical area (Alam 2016). The agrarian economy today depends on forest for energy supply, fodder, non-timber product, and livestock rearing. These all forests are under rapid pace of degradation. In hill region, farming has been intricately integrated with forestry. However, to increase the production of fuel wood, fodder, and other minor products and environmental protection, greater thrust is needed on agroforestry (Yadav and Bisht 2013).

11.2 Agroforestry Ecosystem Services and Climate Change

Agroforestry landscapes provide a variety of ecosystem services which can either be enhanced or protected by management (Nair et al. 2009). Ecosystems can store C in trees and other biomass as well as in the soil. In climate change terms, this is called mitigation because better management results in lower emissions or in increased removal of C from the atmosphere, thus lowering the CO_2 in the atmosphere (Yadav et al. 2016a, b). At the same time, these same landscapes can provide benefits that increase resiliency to climate change. Such adaptation benefits include improved storage and release of water, maintained local and regional rainfall patterns, and diversified economic opportunities through agricultural products, timber, non-timber products, and tourism (Rahman et al. 2015). These same landscapes can also provide services such as biodiversity maintenance and places where local people uphold cultural or religious values. Some of the important uses of agroforestry ecosystems are given below:

11.2.1 Biomass Production

Agroforestry produces more biomass per unit area, which is the main objective of any landscape, and for this purpose, we can use farm bunds, canal and road sides, farm ponds, lakes, water-logged area, and ravines in an economic way by adopting agroforestry.

11.2.2 Soil and Water Conservation

It can help in maintenance and conservation of soil, water, and biodiversity. This loss may be in terms of soil loss through sedimentation and water loss. This way, a huge amount of valuable top fertile soil is being lost. It has been estimated that on an averages soil is displaced at the rate of ~16 t ha^{-1} $year^{-1}$ and washed in to the sea, which is much higher than the permissible limit of 4.5 t ha^{-1} $year^{-1}$. This loss can be reduced by using agroforestry for the land cover in barren lands of a watershed (Meena et al. 2013, 2015a; Singh et al. 2014a; Kumar et al. 2015).

11.2.3 Improvement of Microclimate

Agroforestry will help in the moderation of microclimate around the trees; for this purpose, we can use boundary plantation of trees on farm bunds.

11.2.4 Agro-based Cottage Industry

With the help of agroforestry in watershed, we can promote the agro-based cottage industry. This will include herbal drugs, paper pulp, fiber, poultry, piggery, aquaculture, dairy, beekeeping, sericulture, and mushroom production. This will help in the income generation of the marginal farmers.

11.2.5 Ecological Balance

Pressure on forest is increasing due to population explosion and increased number of industries day by day and because of this increasing pressure encroachment on forest area and deforestation is taking place to meet food requirement of the people. Through agroforestry, we can grow trees on marginal lands with crops and grasses; it will be helpful in restoring the ecological balances.

11.2.6 Watershed Management

With the adoption of agroforestry, we can utilize the entire land of a watershed in a proper way. As per need of the watershed, we have to decide how to manage the land, i.e., cropping area, pasture developments, forest plantation, etc.

11.3 Classification of Agroforestry Systems

Agroforestry systems prevalent in the study area were classified on the basis of structure (nature and arrangement) and function (role of output) of components (Nair 1987). Stratified classification of agroforestry practices as given by Zou and Sanford (1990) was used to indicate the system types and their system units. A system type was named by considering the major components, whereas system unit was termed on the basis of major functional unit in combination with specific tree species or other related components. Hence, functional units like food grains, vegetable and pulses in agriculture, specific fruit trees in horticulture, grasses in grasslands, and trees/grass species in silvopasture were considered to identify systems and system units.

11.3.1 Components

The different components of agroforestry systems were identified as (a) primary components, the components occupying the larger area of the total unit area and serving the major function, i.e., production of primary output needed by the farmers, and (b) secondary components, the component occupying relatively lesser

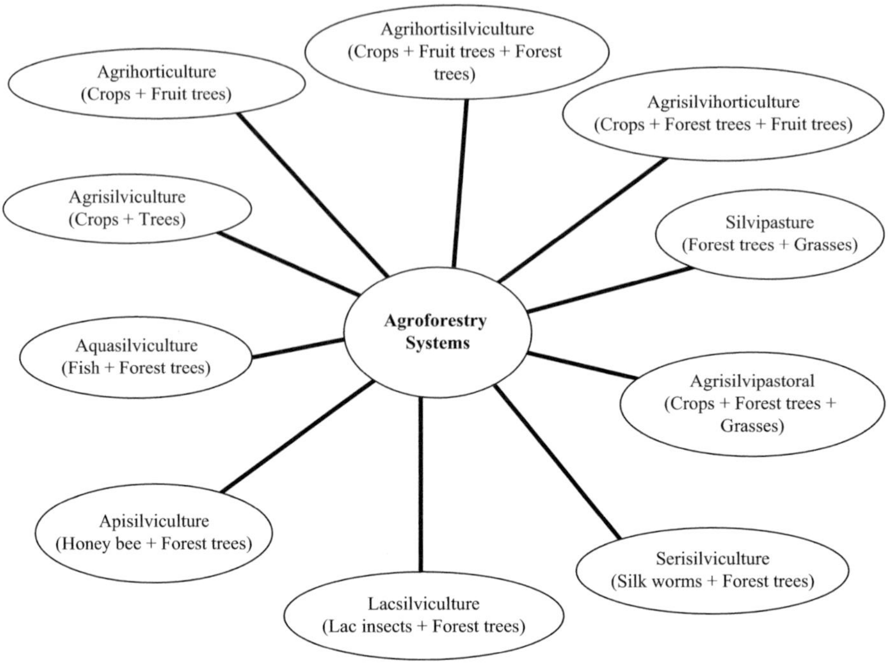

Fig. 11.1 Classification of agroforestry system based on structure and function

area of the total unit area compared to area under primary component and yielding secondary needs of the farmers(Fig. 11.1). Using the abovementioned framework for identifying agroforestry systems, it can be classified as mentioned in Fig. 11.1.

11.4 Agroforestry and Mitigation

The mitigation measures address the causes of the problem increasing GHG concentrations. For example, mitigation could be measures to reduce energy consumption and promote clean technologies. Mitigation is a global issue, when a mitigation agroforestry project reduces emissions of GHGs, it benefits the whole world. This is why mitigation has been addressed extensively at the international level, and financial mechanisms have been created (IPCC 2006). Today, the major sources of emissions from deforestation are located in developing countries, especially in South Asia and tropical America. The trend is now to increase tree cover, which contributes to the absorption of carbon. At the agroforestry scale, two concepts of stock and flux are very important. An agroforestry or any ecosystem is a set of C stocks and C is everywhere, from the leaves to the soil. A good way to visualize a stock of C is to think of the biomass stored in the ecosystem. Biomass is the mass of living biological organisms in a given area or ecosystem at a given time (Meena et al. 2015b, e; Verma et al. 2015a).

Almost 50 % of the dry biomass is C, if the dry biomass of a tree is 2 tonnes, then it contains ~1 tonne of carbon. Agroforestry or any ecosystem is a set of C fluxes. Using the daylight as a source of energy, the leaves absorb CO_2 from the atmosphere and transform it through the process of photosynthesis. The products of this process are distributed to the plant and move to the litter and soil when branches or leaves fall and decompose. Other fluxes are emitting CO_2 back in to the atmosphere through respiration and soil mineralization. Products exported from the ecosystem, such as wood, are also responsible for C fluxes. The important fluxes are those between the atmosphere and the biosphere.

The two concepts of stock and flux are directly linked. If the C stock increases, it means that the ecosystem absorbs more C and it is due to the law of conservation of mass. For example, if your bank account is growing, it means there is more money entering than going out.

In a growing ecosystem, the net balance of flux is an inbound flux; it means that CO_2 is removed from the atmosphere, the atmospheric concentration of GHGs is decreased, and climate change is reduced. This process is called C fixation, absorption, or removal and the ecosystem is called a C sink. If the C stock decreases, for instance, in a removal of trees from the agroforestry system, an outbound flux will increase the atmospheric concentration of GHGs and increase climate change. The process is called C emission and the ecosystem is called a C source. If we measure a C stock of 20 tonnes of carbon at a given time and a stock of 120 tonnes of C after 10 years, it means that, on average, the ecosystem has removed or absorbed 10 tonnes of C per hectare per year from the atmosphere.

Many agroforestry activities contribute to climate change mitigation (Fig. 11.2). C stocks can be increased by creating agroforestry through plantations. The benefit of creating agroforestry plantations is the difference between the present growing stock and the baseline, as show on the graph. Existing growing stocks can be conserved through reducing felling or no felling at all of trees from agroforestry farmlands. In this situation, the benefit of conserving is estimated with reference to the degradation or tree felling scenario. Emissions caused by agroforestry activities can be reduced, for example, by using less energy or fertilizers in various agroforestry operations. Biomaterials and bioenergy can be produced from agroforestry

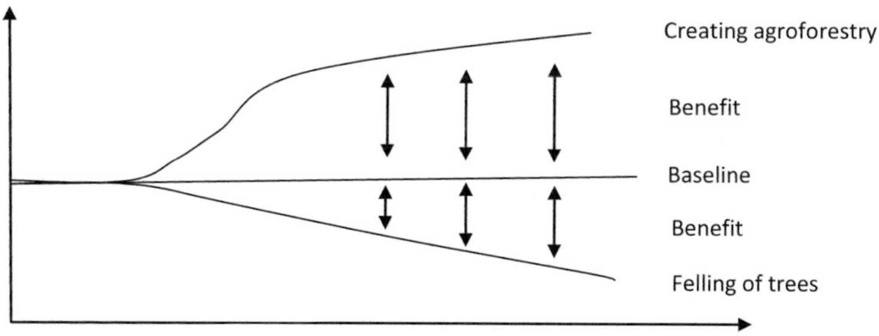

Fig. 11.2 C sequestration through creating agroforestry plantation and reduced felling of trees from agroforestry farmlands

products to substitute materials or energy that generates GHGs. The first two activities refer to C sequestration in the agroforestry or any other ecosystem, while the last two refer to energy-related emissions (Meena et al. 2015c, d).

11.4.1 Fruit Tree-Based Land Use Systems

Agrihorticulture, i.e., fruit tree-based land use systems are a unique and very common practice in the Himalayan region of India (Yadav et al. 2015). Trees are deliberately incorporated in the land use systems by the farmers to fulfill their varied need, namely, fodder, fuel, fiber, fruits, small timber, and agricultural implements along with the agricultural produce (Yadav and Bisht 2013). In Indian Himalaya ~5, 36,800 ha area under fruit tree plantation and considerable area of these fruit tree plants is being used for growing of agricultural crops in the form of agroforestry. These fruit tree-based land use systems which include the cultivation of agricultural crops in association of horticultural trees on the same piece of land provide the stable and better output to the farmers (Yadav and Bisht 2014a). These fruit tree-based land use systems are the backbone for the existence of the hill populations which are the major source of livelihood besides conservation of soil and water and help in protecting agricultural crops from climatic abnormality; improvement in the soil quality takes place due to favorable effects of the trees in addition to C sequestration.

11.4.2 Methodology

The study was conducted during 2011–2012 at experimental farm Hawalbagh (29°36′ N and 79°40′ E, 1,250 m amsl altitude) of ICAR-Vivekananda Institute of Hill Agriculture, Almora, Kumaon Himalaya. A long-term (11 years) experiment on the fruit trees, i.e., *Citrus lemon* (hill lemon), *Prunus persica* (plum), *Pyrus communis* (pear), and *Prunus armeniaca* (apricot), was used for study, which were planted at a spacing of 6.0 × 6.0 m (277 trees ha^{-1}). The different fruit tree-based land use systems include hill lemon + wheat, plum + wheat, pear + wheat, apricot + wheat, and wheat (*Triticum aestivum* L.) monocropping.

All the fruit trees within the land use systems was measured for diameter at breast height and for understory biomass of wheat crop clipped near the ground and converted on hectare basis. Plants of fruit trees were assumed to contain ~45 % C of their biomass (Schroth et al. 2002) and C stock estimated using equation CS = 0.4545 × B (Magnussen and Reed 2004), where CS is the C stock and B is the biomass. The fruit tree biomass was calculated with an equation $\left(\text{Biomass} = \pi D^2 hs/4\right)$ given by Hairiah et al. (2001), where biomass is expressed in kg, D = tree diameter (cm), h = height (cm), and s = density (g cm^{-3}). The data on wood density was extracted from a wood density database created by ICRAF. To get an estimate of equivalent CO_2 assimilation, C stock was multiplied by a factor of 3.67 (44/12), and this factor was used for agroforestry tree species by Chauhan et al. (2009). The C stock of each plot was divided by the number of years since establishment of the plot, to get aboveground biomass C accumulation rate for each land use system. The data on biomass, C stock, and C accumulation rate were analyzed after one-way analysis of variance (ANOVA) using SAS 9.3 statistical software. Significant differences were tested at $P \leq 0.05$ using Tukey's least significant difference test.

11.4.3 Biomass, Biomass C, and C Stock Equivalent CO_2

The aboveground biomass, biomass C, and C stock equivalent CO_2 of the fruit tree-based five land use systems including monocropping of wheat crop varied significantly (10.8–37.8 Mg ha^{-1}, 4.8–17.0 Mg C ha^{-1}, and 17.6–62.3 Mg CO_2 ha^{-1}) as given in Table 11.1. The highest biomass, biomass C, and C stock equivalent CO_2 were observed in the pear + wheat (37.8 Mg ha^{-1}, 17.0 Mg C ha^{-1}, 62.3 Mg CO_2 ha^{-1}) followed by apricot + wheat (26.5 Mg ha^{-1}, 11.9 Mg C ha^{-1}, 43.6 Mg CO_2 ha^{-1}) > plum + wheat (22.4 Mg ha^{-1}, 10.0 Mg C ha^{-1}, 36.5 Mg CO_2 ha^{-1}) > hill lemon + wheat (18.8 Mg ha^{-1}, 8.4 Mg C ha^{-1}, 30.9 Mg CO_2 ha^{-1}), whereas the lowest was in the wheat monocropping (10.8 Mg ha^{-1}, 4.8 Mg ha^{-1}, 17.6 Mg CO_2 ha^{-1}).

Aboveground biomass, biomass C, and C stock equivalent CO_2 accumulation rates (Table 11.1) also varied among the fruit tree-based land use systems, but it was not significant. The highest rate of biomass accumulation was found in the pear + wheat (12.0 Mg ha^{-1} year^{-1}) followed by apricot + wheat (11.5 Mg ha^{-1} year^{-1}) > hill lemon + wheat (11.0 Mg ha^{-1} year^{-1}) > plum + wheat (10.8 Mg ha^{-1} year^{-1}), and the lowest (10.7 Mg ha^{-1} year^{-1}) rate of biomass accumulation was noted for the wheat monocropping. The highest rate of C and CO_2 accumulation was found in the pear + wheat (5.34 Mg C ha^{-1} year^{-1}, 19.6 Mg ha^{-1} year^{-1}) followed by apricot + wheat (5.16 Mg C ha^{-1} year^{-1}, 18.9 Mg ha^{-1} year^{-1}) > hill lemon + wheat (4.90 Mg C ha^{-1} year^{-1}, 18.0 Mg ha^{-1} year^{-1}) > wheat (4.79 Mg C ha^{-1} year^{-1}, 17.6 Mg ha^{-1} year^{-1}), and the lowest (4.72 Mg C ha^{-1} year^{-1}, 17.3 Mg ha^{-1} year^{-1}) rate of C accumulation was noted for the plum + wheat. The contributions of tree and crop in aboveground biomass of different land use systems were calculated, and the tree species in each plot of the sampled land use system was plotted against C stock. In general, aboveground carbon stocks were varied in different land use systems with different tree species (Fig. 11.3). From the cropping system, pear + wheat, apricot + wheat, plum + wheat, and hill lemon + wheat were found to contain a fairly high C stocks due to tree species. Conversely, the land use systems without tree species had low C stocks, i.e., wheat monocropping.

Fruit tree-based land use systems such as pear + wheat, apricot + wheat, plum + wheat, and hill lemon + wheat showed more biomass, biomass C, and C stock equivalent CO_2 accumulation rates in comparison to wheat monoculture. This study has shown that the presence of diverse fruit tree species in land use systems has high C and C stock equivalent CO_2 accumulation rates in aboveground biomass in Indian Himalayan fruit tree-based land use systems. In comparison to only annual crops,

Table 11.1 Aboveground biomass, C stocks, C stock equivalent CO_2, and their accumulation rate in fruit tree-based land use systems in Kumaon Himalaya (Yadav et al. 2015)

Treatment	Aboveground biomass		Aboveground C stock		Aboveground C stock equivalent CO_2	
	Biomass (Mg ha^{-1})	Accumulation rate (Mg $ha^{-1}year^{-1}$)	C Stock (Mg ha^{-1})	Accumulation rate (Mg $ha^{-1}year^{-1}$)	C stock equivalent CO_2 (Mg ha^{-1})	Accumulation rate (Mg $ha^{-1}year^{-1}$)
Hill lemon + wheat	18.8 ± 1.09	11.0 ± 0.30	8.4 ± 0.45	4.9 ± 0.06	30.9 ± 1.65	18.0 ± 0.21
Plum + wheat	22.5 ± 0.67	10.8 ± 0.25	10.0 ± 0.33	4.7 ± 0.01	36.5 ± 1.21	17.3 ± 0.05
Pear + wheat	37.8 ± 1.81	12.0 ± 0.21	17.0 ± 0.92	5.3 ± 0.20	62.3 ± 3.36	19.6 ± 0.72
Apricot + wheat	26.5 ± 0.99	11.5 ± 0.23	11.9 ± 0.65	5.1 ± 0.24	43.6 ± 2.38	18.9 ± 0.87
Wheat	10.8 ± 0.30	10.7 ± 0.30	4.8 ± 0.16	4.8 ± 0.16	17.6 ± 0.60	17.6 ± 0.60
C.D.	3.7	NS	1.9	NS	6.9	NS

C.D critical difference, *NS* nonsignificant (0.05 %), and values after ± sign is standard error

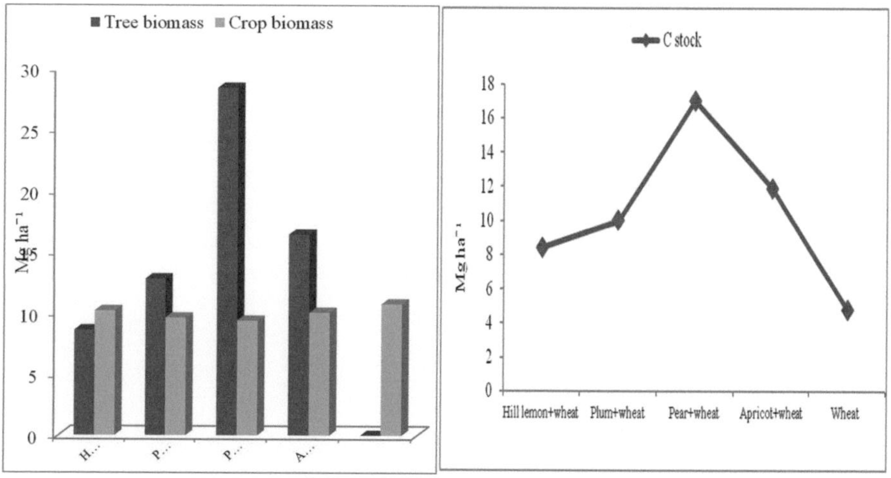

Fig. 11.3 Aboveground biomass (Mg ha^{-1}) of trees and crop and C stocks in relation to different fruit tree-based land use systems in Kumaon Himalaya (Yadav et al. 2015)

which accumulated limited C, tree-crop systems sequestered C at a higher rate. Therefore, by moving away from only annual crops to tree-based systems like fruit tree-based land use systems in the form of agroforestry and fodder tree plantations, significant quantities of C can be sequestered (Yadav and Bisht 2014c; Nair et al. 2009; Yadav et al. 2016a, b). Annual crops will only accumulate C through the roots and retention of crop residues, whereas the tree crops will accumulate C through roots, litter, and aboveground biomass. Geographical regions, plant species, spacing, and age of plants are the major determinants of biomass, C stock, and C accumulation rate in vegetation, and therefore, cultivating agricultural crops with fruit tree plantation, therefore, may be an attractive option for storage of atmospheric CO_2 in the Kumaon Himalaya (Yadav et al. 2015).

In addition to the accumulation of high average C stock, agroforestry systems have several advantages over monocultures including crops for household consumption. Agroforestry also may provide a viable combination of C storage with minimal negative effects on food production (Pandey 2002; Yadav and Bisht 2014b). High and long-term biomass accumulation with early generation of income from annual and semi-perennial intercrops was characteristic feature of agroforestry systems. This paper shows that fruit tree-based land use systems in Kumaon Himalaya have higher potential for C stock and C accumulation in aboveground biomass. For instance, tree-crop systems pear + wheat, apricot + wheat, plum + wheat, and hill lemon + wheat sequestered higher C than those containing monocropping of wheat, which has limited accumulation of C. In general, C accumulation rates varied with diverse tree species in land use systems with high C accumulation. Cultivating agricultural crops with fruit tree plantation, therefore, may be an attractive option for storage of atmospheric CO_2 in the Kumaon Himalaya.

11.5 Pecan Nut-Based Agrihorticulture System

In the Kumaon Himalayan mid-hill situation, pecan nut [*Carya illinoinensis* (Wangenh.) K. Koch] trees were planted at a spacing of 6.0 × 7.0 m with a density of 238 trees ha^{-1}. Intercropping patterns of pecan nut and crop were designed at the time of planting, and the orientation of the plots was managed in the East–West. The plots were managed in the rainfed system according to farmers' normal practice; two crops wheat (*Triticum aestivum*) and lentil (*Lens esculentum*) were grown under the pecan nut tree as intercropping which is the intensive activity in an agroecosystem.

11.5.1 Methodology

The study was conducted during 2012–2013 at experimental farm Hawalbagh (29° 36′ N and 79° 40′ E, 1,250 m amsl) of ICAR-Vivekananda Institute of Hill Agriculture, Almora, Kumaon Himalaya, India. The pecan nut plantation was well-managed tree-crop intercropping as an agrihorticulture system at the time study and was 11-year-old. Diameter at breast height and height of the trees within the plot were measured and recorded. The regression equations given by Ares et al. (2006) for pecan nut in silvopastures have been used to assess the tree biomass. The equation to estimate wood volume of pecan nut tree as a function of DBH (m) and height (m) was $V = 0.6134\ \mathrm{DBH}^{1.7775}\ \mathrm{H}$, where V = wood volume (m^3), DBH = mean diameter at breast height (1.37 m) above the ground level (m), and H = height (m). Aboveground biomass of crops was measured by cutting, drying, and weighing all the plant material within the 1 m square from replicated plots at full maturity. Fresh weights of aboveground components were determined in the field, and subsamples for each component were collected for moisture and C analysis. From the biomass sampling measurement, crop biomass was expanded to an area basis in the field as a whole.

In the present study, we followed the equation $C = 0.45 \times B$, (Magnussen and Reed 2004), where C is the C content and B is biomass to assess the carbon content in different components of pecan nut plantation. All plant materials were dried at 70 ± 1 °C in oven and ground and stored in airtight containers for C concentration analysis of plant samples, and the mass of C stored in crop compartments was estimated by multiplying their measured mass by the C concentrations and was expanded to an area basis. The values of pecan nut trees and crops were added to get total C stock in the each intercropping configuration. The values of C stock were multiplied with a factor of 3.66 to obtain C stock equivalent CO_2 mitigation. Long-lived C storage, heat from biomass combustion, and C storage from coal substitution were estimated by the formula (Wang and Feng 1995). The total amount of C sequestration in woody component was estimated by adding long-lived C storage in wood products and the C storage due to substitution biomass for coal and expressed in Mg ha^{-1}.

11.5.2 C Stock and CO_2 Mitigation Through Biomass

The data in Fig. 11.4 shows the variation in biomass, biomass C, and biomass C equivalent CO_2 level for lentil + pecan nut, wheat + pecan nut in pecan nut-based agrihorticulture systems, and lentil and wheat as a pure crop management system. The aboveground biomass ranged between 56.3 and 2.75 Mg ha^{-1} in which the maximum biomass was recorded 56.5 Mg ha^{-1} in wheat + pecan nut system followed by 53.2 Mg ha^{-1} in lentil + pecan nut system, while the minimum value of biomass was recorded 2.75 Mg ha^{-1}in lentil as a pure lentil production system. Aboveground biomass C stocks ranged between 25.30 and 1.17 Mg ha^{-1}, in which the maximum C stock was observed in pecan nut + wheat system (25.30 Mg ha^{-1}), followed by pecan nut + lentil system (23.92 Mg ha^{-1}), and the minimum was recorded 1.17 Mg ha^{-1} in lentil system as a pure crop. The biomass C equivalent CO_2 mitigation potential ranged between 92.5 and 4.29 Mg ha^{-1}, in which the maximum was recorded in pecan nut + wheat system (92.5 Mg ha^{-1}) followed by pecan nut + lentil system (87.78 Mg ha^{-1}). The minimum mitigation of CO_2 (4.29 Mg ha^{-1}) was recorded in lentil system as a pure crop. Biomass C stocks and biomass C equivalent CO_2 through biomass are dependent on the tree species, density, age, climatic region, and C concentration in different components of the system.

The aboveground biomass obtained in this study is in the range of those reported by Pandey et al. (1987), Lodhiyal et al. (1995), Yadava (2010), and Kanime et al. (2013) for *Populus deltoides* and *Eucalyptus* species, respectively. The results of the biomass C stocks are comparable with the findings of Albrecht and Kandji (2003), Pandey et al. (2010), Wani et al. (2010), Rizvi et al. (2011), Yadav and Bisht (2014a, c), Rajput et al. (2015), and Yadav et al. (2016a) which reported that tree-based land use systems can store C in the range of 12–228 Mg ha^{-1}. Gera et al. (2011) attributed that the variations in the C sequestration potential relate to the mean annual increment, which varied with site, age, density, and plantation, as well as the quality of planting stock. Biomass C equivalent CO_2 mitigation by

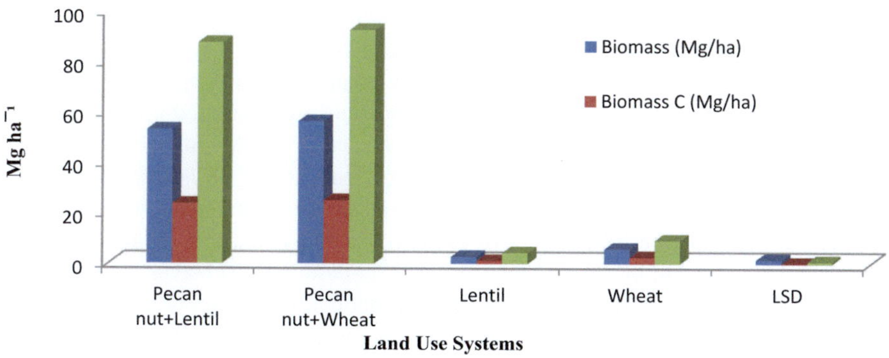

Fig. 11.4 Aboveground biomass, biomass C, and biomass C equivalent CO_2 of pecan nut-based agrihorticulture systems in Kumaon Himalaya (Yadav et al. 2016c)

Table 11.2 Biomass, C storage, and C sequestration in pecan nut tree in agrihorticulture system in Kumaon Himalaya

Biomass (Mg ha^{-1})	48.75
C storage (Mg ha^{-1})	21.93
Long-lived C storage (Mg ha^{-1})	9.21
Heat from biomass combustion ($\times 10^{10}$)	54.42
C storage from coal substitution (Mg C ha^{-1})	9.14
Total C sequestration (Mg C ha^{-1})	18.35
C sequestration (Mg ha^{-1} $year^{-1}$)	1.67

Yadav et al. 2016c

aboveground parts varied from 4.07 Mg ha^{-1} in lentil with pecan nut to 83.71 Mg ha^{-1} in pecan nut biomass (Fig. 11.3). Due to the presence of trees, higher CO_2 mitigation value of pecan nut + wheat system and pecan nut + lentil system can be attributed to more biomass and more C stock in tree-based agrihorticulture system as compared to sole agriculture system.

C stock refers to the absolute quantity of C held at the time of inventory, whereas C sequestration refers to the process of removing C from the atmosphere and depositing it in a reservoir (Takimoto et al. 2008); large C stock does not necessarily mean a large C sequestration potential. The long-lived C storage (9.21 Mg C ha^{-1}), heat from biomass combustion (54.4×10^{10}), and C storage from coal substitute through branches and twigs/leaves (9.14 Mg C ha^{-1}) were observed in pecan nut (Table 11.2). The C sequestration was recorded 18.53 Mg C ha^{-1} and 1.67 Mg C ha^{-1} $year^{-1}$ by pecan nut tree. Tree stem sequesters the C for longer time after felling as compared to the C stored in leaves and branch biomass (Wang and Feng 1995; Kaul et al. 2010; Gera et al. 2011; Yadav and Bisht 2014a).

11.6 High-Density Plantation of Oak

Oaks (*Quercus* spp.) are the dominant, climax tree species of the moist temperate forests of the Indian Himalayan region (Troup 1921), where ~35 species of *Quercus* are extensively distributed between 1,000 and 3,500 m elevations. *Q. leucotrichophora* is an evergreen tree, locally known as “banj oak,” and it is a valuable keystone species with great societal relevance (Ramakrishnan 2001). Banj is among the main forest-forming species in the densely populated mid-altitudinal zones of the Central Himalaya and provides a variety of ecosystem services (Zobel and Singh 1997; Sharma et al. 2009). The oaks, particularly *Q. leucotrichophora*, are intricately associated not only with agroecosystems but also with the life support systems of the inhabitants of the hills in the Himalaya (Makino 2011). High calorific value of banj wood makes it an excellent fuel; its leaves are palatable which makes them a major source of cattle fodder in the winter and dry seasons (Negi et al. 1997; Yadav and Bisht 2013) and can be correlated with natural springs and wildlife (Singh 1981).

11.6.1 Methodology

Q. leucotrichophora in the year of 1983 was planted at the study site at a spacing of 1 × 1 m, and this high-density plantation was managed properly for many years. For fodder removal, four cutting methods, i.e., coppicing at 1 m height, pollarding at 2 m height, local practice, and keeping above one-third part undisturbed practices, are introduced in the plantation in the year of 1997. Fodder was removed from trees during winter season which is the scarcity period for fodder due to drying of grasses and rainfed farming, and there is no source of fodder than trees in mid-hills of Himalaya. Terms used are "lesser disturbed tree" which means the minimum biomass was removed from the trees such as local practice and keeping above one-third part undisturbed of the trees and "increased disturbance" which means the maximum biomass of the trees was removed such as coppicing at 1 m height and pollarding at 2 m height of the trees.

In plots of different cutting methods, trees were measured for the girth at breast height except for basal area in 1 m height cutting methods plots and height with the help of measuring tape and Ravi multimeter, respectively, a total of 240 trees, comprising 60 trees from each plot of cutting methods. The diameter at breast height (dbh) and basal diameter were obtained from gbh and basal area by using the equation $G = \pi D$, where G is the girth and D is the diameter. The volume was obtained by using the equation $\text{Volume}/d^2 = 5009470 + 0.00563/d^2$ (FSI 2009), and the volume was multiplied with tree density to obtain the biomass. Belowground biomass of trees was calculated by Cairns et al. (1997) and as per the IPCC (2006). By considering C content 0.45 (Magnussen and Reed 2004), C stock was estimated, and this C stock was multiplied with a factor 3.67 (Chauhan et al. 2009) to get biomass equivalent CO_2. The mean tree values were multiplied with the corresponding density of trees to obtain standing stock of biomass and C on area basis (per ha).

11.6.2 Biomass, Biomass C, and Biomass C Equivalent CO_2

Among growth parameters, the mean height, diameter, and volume per tree are significantly highest (10.14 m, 12.21 cm and 0.0833 m^3) for the trees kept one-third top portion undisturbed followed by other practices and lowest (2.02 m, 5.35 cm and 0.0202 m^3) for trees which were coppiced at 1 m height which was on par with the pollarding at 2 m height among different cutting methods (Fig. 11.4).

With increasing disturbance to trees in the form of cutting methods, the mean height, diameter, and volume which decreased the mean growth were less in more disturbed trees by cutting methods, i.e., cutting at 1 m and 2 m height as compared to the trees which were managed with less disturbance in the form of cutting, i.e., one-third top portion undisturbed and local practice. In general, during cutting

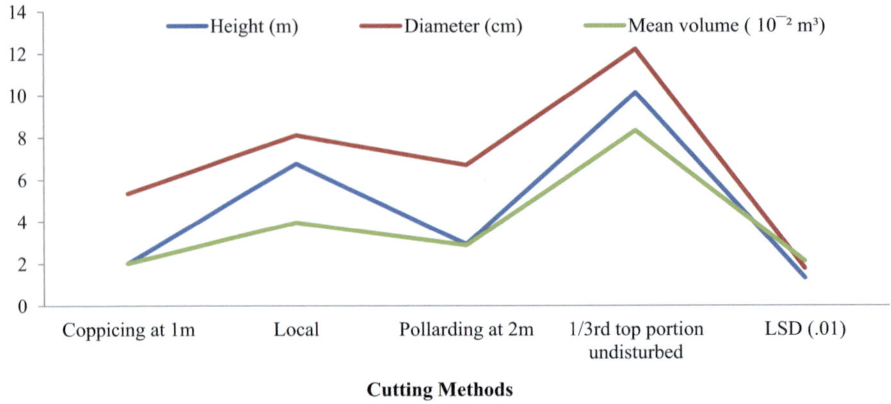

Fig. 11.5 Cutting method influence on biomass, biomass C, and biomass C equivalent CO_2 of Oak in Kumaon Himalaya (Yadav et al. 2016a)

methods, removal of higher proportion of leaves may adversely affect photosynthetic rates and would depress tree growth including decrease in annual diameter growth (Gyenge et al. 2010).

According to Fig. 11.5 in total biomass, biomass C, and biomass C equivalent CO_2, a highly significant ($p < 0.01$) variation was observed for different cutting methods. Total (aboveground and belowground) tree biomass, biomass C, and biomass C equivalent CO_2 ranged between 192.6–793.2, 86.7–356.9, and 317.2–1,306.5 Mg ha^{-1}. The least disturbed trees in the form of cutting methods, i.e., one-third top portion undisturbed practice, recorded the significantly highest (793.2, 356.9, and 1,306.5 Mg ha^{-1}) total tree biomass, biomass C, and biomass C equivalent CO_2, and declined with increasing tree disturbance, i.e., coppicing at 1 m, the significantly lowest (192.6, 86.7, and 317.2 Mg ha^{-1}) total tree biomass, biomass C, and biomass C equivalent CO_2(Singh et al. 2014b; Yadav and Bisht 2014d, 2016a). Roots may restrain resource acquisition potential of trees leading to reductions in their biomass production, and it is a well-known fact that limitations imposed by cutting methods of trees on the lateral expansion affects resource acquisition from soil and environment.

In the pollarding at 2 m (123.4 Mg ha^{-1}) compared to that of the cutting at 1 m (86.7 Mg ha^{-1}), almost one and a half-fold increase in C stocks has been observed as well as in the one-third top portion undisturbed (356.9 Mg ha^{-1}) as compared to local practice (169.6 Mg ha^{-1}) twofold. Twofold and fourfold C stock was depressed due to cutting at 1 m height and one and a half-fold, and almost threefold C stock was reduced pollarding at 2 m height of the tree in comparison to local and one-third top portion undisturbed practices, respectively. C stocks being a function of the overall tree growth (Fig. 11.6) are not surprising. Majid and Paudyal (1992) noted reductions in tree growth when crown length removal from below exceeded ~40 % in the *A. mangium* plantation of Peninsular Malaysia. According to Singh

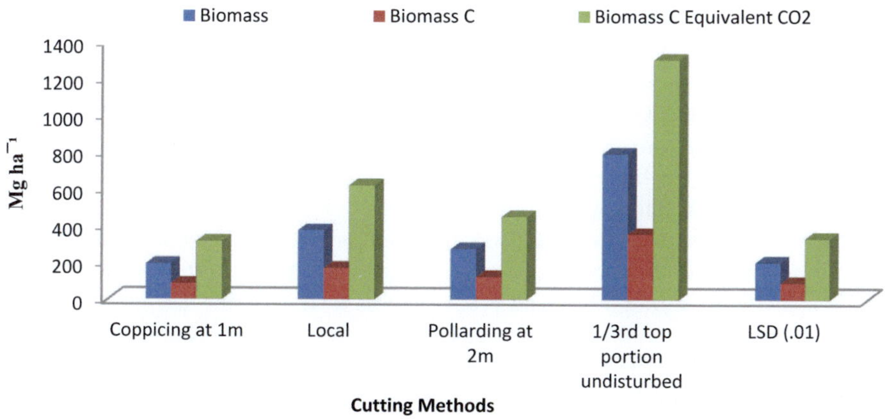

Fig. 11.6 Effects of cutting methods on the height, diameter, and volume of oak in Kumaon Himalaya (Yadav et al. 2016a)

Table 11.3 C sequestration of oak as influenced by cutting methods in Kumaon Himalaya

Treatment	C sequestration rate (Mg ha⁻1 year⁻1)
Coppicing at 1 m	2.9
Local	5.7
Pollarding at 2 m	4.1
One-third top portion undisturbed	11.9
LSD ($p = 0.01$)	3.4

Yadav et al. 2016a

et al. (2014a, b), the impact of human-induced small-scale disturbances (lopping of branches and leaf removal) adversely affects the functioning of banj oak (*Quercus leucotrichophora* A. Camus) forests of Central Himalaya. The biomass C CO_2 mitigation by the Oak high-density plantation followed the trend of biomass and C stock and varied between 317 and 1,306 Mg ha^{-1} for different cutting methods (Fig. 11.6).

The least /minimum disturbance in the form of cutting methods, i.e., in one-third top portion undisturbed (11.9 Mg ha^{-1} year^{-1}) obtained higher rate of total C sequestration followed by other methods and in the highest disturbed trees in the form of cutting methods obtained least in cutting at 1 m height (2.9 Mg ha^{-1} year^{-1}) (Table 11.3). Results of the C sequestration rates were almost within range to those obtained in earlier studies (KTGAL 2004; Yadav and Bisht 2014c, d; Yadav et al. 2016a). The varying rate of C sequestration of trees can be attributed to their differential cutting methods.

11.7 Conclusion and Future Prospective

The agroforestry practices are mainly practiced by the smallholder farmers and are rich in various species of trees. Plantation of multipurpose tree species (MPTs) on farmlands serves a variety of ecosystem services. These agroforestry landscapes can also increase resiliency to climate change, through such ecosystem services include soil and water conservation and quality improvement, improvement in microclimate, increased economic opportunities for livelihood security, and biodiversity maintenance besides the C sequestration and affect rainfall patterns. This landscape store C through biomass production as well as in the soil, and it comes under mitigation to combat climate change due to lower emissions or increased removal of CO_2 from the atmosphere and trapping it either in the biomass or in the soil. The farmland productivity increases due to integration of trees, which interacts positively with other farm components, i.e., crops, vegetables, grasses, and animals. Due to these positive interactions, the biomass production increases per unit area and uses soil water and nutrients efficiently in such agroforestry practices. Here in this article we have discussed three land use systems, i.e., fruit tree-based land use systems, pecan nut-based agrihorticulture system, and oak high-density plantation. All these systems store a good amount of C ranging from 3.8 to 526.5 Mg ha^{-1} through biomass production. On the conversion of this C stock in the equivalent CO_2, it comes to be in the range of 13.9–1,932.2 Mg ha^{-1}. This variation in the C stock and equivalent CO_2 amount is due to well-known fact that it depends on the tree species, spacing or density, age, climatic region, etc. In the hills of the Kumaon Himalaya, these agroforestry landscapes exist everywhere on the farmlands, which are rich in the species of trees. And these agroforestry practices play a major role in the livelihood security of local inhabitant besides climate change mitigation and adaptation through various ecosystem services. There is an urgent need for the research on the influence of climate change on such agroforestry practices and on the smallholders and their livelihood security. These people not only help to combat climate change but also conserve biodiversity so they should get benefits of C trading in economic terms. The smallholders should be motivated and rewarded for their contribution in reducing climate change vulnerability.

The review of past studies showed high potential for C sequestration in the agroforestry practices in Kumaon Himalaya. It is suggested that the influence of factors affecting C sequestration dynamics of agroforestry such as biological, physical, social, and economical needs to be analyzed. Agriculturally marginal lands and range lands can be diverted to agroforestry land through careful consideration. Therefore, scientific management needs to be implemented to protect and improve C pools as well as C sequestration rates of agroforestry practices against threats. The large-scale studies are required not only for understanding the C storage and C sequestration potential of agroforestry systems but also for scientific management and effective implementation of agroforestry systems at various levels for the development of Himalayan region. Hence, these agroforestry systems have potential for significant C accumulation which needs to be protected to maintain and improve C sequestration for mitigation of climate change.

Acknowledgments The authors are indebted to Director, Indian Council of Agricultural Research- Vivekananda Institute of Hill Agricultural Research, Almora for funding the research work in the Kumaon Himalaya.

References

Alam A (2016) Long-term strategies and programmes for mechanization of agriculture in agro climatic zone–I: Western Himalayan region, pp 78–86 http://farmech.gov.in/06035-04-ACZ1-15052006.pdf

Albrecht A, Kandji ST (2003) C sequestration in tropical agroforestry system. Agric Ecosyst Environ 99:15–27

Ares A, Reid W, Brauer D (2006) Production and economics of native pecan silvopastures in central United States. Agrofor Syst 66:205–215

Basu JP (2014) Agroforestry, climate change mitigation and livelihood security in India. N Z J For Sci 44 (Suppl 1), doi: 10.1186/1179-5395-44-S1-S11

Cairns MA, Brown S, Helmer EH, Baumgardner GA (1997) Root biomass allocation in the world's upland forests. Oecologia 111:1–11

Chauhan SK, Gupta N, Ritu, Yadav S, Chauhan R (2009) Biomass and C allocation in different parts of agroforestry tree species. Indian Forester 135:981–993

Chavan SB, Keerthika A, Dhyani SK, Handa AK, Newaj R, Rajarajan K (2015) National agroforestry policy in India: a low hanging fruit. Curr Sci 108:1826–1834

FSI, [Forest Survey of India] (2009) Forest survey of India-annual report. Ministry of Environment and Forest, Dehradun

Gera M, Mohan G, Bisht NS, Gera N (2011) C sequestration potential of agroforestry under CDM in Punjab state of India. Indian J For 34:1–10

Ghosh BN, Meena VS, Alam NM, Dogra P, Bhattacharyya R, Sharma NK, Mishra PK (2016) Impact of conservation practices on soil aggregation and the carbon management index after seven years of maize–wheat cropping system in the Indian Himalayas. Agric Ecosyst Environ 216:247–257

Gyenge JE, Fernańdez ME, Schlichter TM (2010) Effect of stand density and pruning on growth of ponderosa pines in NW Patagonia, Argentina. Agroforest Syst 78:233–241

Hairiah K, Sitompul SM, VanNoordwijk M, Palm C (2001) Methods for sampling C stocks above and below ground, ASB lecture note 4B. International Center for Research in Agroforestry (ICRAF), Bogor, Available at http://www.worldagroforestrycentre.org/Sea/Products/Training/Materials/lecture%20notes/ASB-LecNotes/ASBLecNote%204B.pdf. Accessed in Aug 2013

IPCC, [Intergovernmental Panel on Climate Change] (2006) Guidelines for national greenhouse gas inventories. http://www.ipcc-nggip.iges.or.jp/public/2006gl/index.html. Accessed on 26 Feb 2014

Kanime N, Kaushal R, Tewari SK, Raverkar KP, Chaturvedi S, Chaturvedi OP (2013) Biomass production and C sequestration in different tree-based systems of Central Himalayan Tarai region. Forests Trees Livelihoods 22:38–50

Kaul M, Mohren GMJ, Dadhwal VK (2010) C storage and sequestration potential of selected tree species in India. Mitig Adapt Strat Glob Clim Chang 15:489–510

KTGAL, [Kyoto Think Global Act Local] (2004) Project report: 2003–2004 [submitted to International Centre for Integrated Mountain Development, Kathmandu, Nepal]. Central Himalayan Environment Association, Nainital

Kumar A, Bahadur I, Maurya BR, Raghuwanshi R, Meena VS, Singh DK, Dixit J (2015) Does a plant growth promoting rhizobacteria enhance agricultural sustainability? J Pure Appl Microbiol 9(1):715–724

Kursten E (2000) Fuel wood production in agroforestry systems for sustainable land use and CO_2 mitigation. Ecol Eng 16:69–72

Lodhiyal LS, Singh RP, Singh SP (1995) Structure and function of an age series of poplar plantation in Central Himalaya-I, dry matter dynamics. Ann Bot 76:191–199
Magnussen S, Reed D (2004) Modelling for estimation and monitoring (FAO-IUFRO, 2004)
Majid NM, Paudyal BK (1992) Pruning trial for *Acacia mangium* wild. Plantation in Peninsular Malaysia. For Ecol Manag 47:285–293
Makino Y (2011) Lopping of oaks in Central Himalaya, India. Mt Res Dev 31:35–44
Meena RS, Yadav RS, Meena VS (2013) Heat unit efficiency of groundnut varieties in scattered planting with various fertility levels. Bioscan 8(4):1189–1192
Meena RS, Meena VS, Meena SK, Verma JP (2015a) The needs of healthy soils for a healthy world. J Clean Prod 102:560–561
Meena RS, Meena VS, Meena SK, Verma JP (2015b) Towards the plant stress mitigate the agricultural productivity: a book review. J Clean Prod 102:552–553
Meena VS, Meena SK, Verma JP, Meena RS, Ghosh BN (2015c) The needs of nutrient use efficiency for sustainable agriculture. J Clean Prod 102:562–563
Meena VS, Meena SK, Verma JP, Meena RS, Jat LK (2015d) Current trend to mitigate climate change and sustainable agriculture: a book review. J Clean Prod 102:548–549
Meena VS, Verma JP, Meena SK (2015e) Towards the current scenario of nutrient use efficiency in crop species. J Clean Prod 102:556–557
Meena RS, Bohra JS, Singh SP, Meena VS, Verma JP, Verma SK, Sihag SK (2016) Towards the prime response of manure to enhance nutrient use efficiency and soil sustainability a current need: a book review. J Clean Prod 112(1):1258–1260
Nair PKR (1987) Agroforestry systems inventory. Agrofor Syst 5:301–317
Nair PKR, Kumar BM, Nair VD (2009) Agroforestry as a strategy for C sequestration. J Plant Nutr Soil Sci 172:10–23
Negi AK, Bhatt BP, Todaria NP, Saklani A (1997) The effects of colonialism on forests and the local people in the Garhwal Himalaya, India. Mt Res Dev 17:159–168
Pandey DN (2002) C sequestration in agroforestry systems. Clim Pol 2:367–377
Pandey MC, Tandon VN, Rawat HS (1987) Organic matter production and distribution of nutrient in eucalyptus hybrid plantation ecosystem in Karnataka. Indian Forester 114:713–724
Pandey U, Kushwaha SPS, Kachhwaha TS, Kunwar P, Dadhwal VK (2010) Potential of Envisat ASAR data for woody biomass assessment. Trop Ecol 51:117–124
Rahman SA, Foli S, Pavel MA, Mamun MA, Sunderland TCH (2015) agroforestry landscape management, ecosystem services livelihoods. Int J Dev Sustain 4:479–491
Rajput BS, Bhardwaj DR, Pala NA (2015) CO_2 mitigation potential and C density of different land use systems along an altitudinal gradient in North-Western Himalayas. Agroforest Syst 89:525–536
Ramakrishnan PS (2001) Ecology and sustainable development. National Book Trust of India, New Delhi
Rizvi RH, Dhyani SK, Yadav RS, Singh R (2011) Biomass production and C stock of popular agroforestry systems in Yamunanagar and Saharanpur districts of northwestern India. Curr Sci 100:736–742
Sathaye JA, Ravindranath NH (1998) Climate change mitigation in the energy and forestry sectors of developing countries. Ann Rev Energy Environ 23:387–437
Schroth G, D'Angelo SA, Teixeira WG, Haag D, Lieberei R (2002) Conversion of agroforestry and monoculture plantations in Amazonia: consequences for biomass, litter and soil C stocks after 7 years. For Ecol Manag 163:131–150
Sharma CM, Gairola S, Ghildiyal SK, Suyal S (2009) Forest resource use patterns in relation to socioeconomic status. Mt Res Dev 29:308–319
Singh JS (1981) Rural ecosystems and development in the Himalaya. In: Singh JS, Singh SP, Shastri C (eds) Science and rural development in mountains. Gyanodaya Prakashan, Nainital, pp 74–87

Singh DK, Singh SK, Singh AK, Meena VS (2014a) Impact of long term cultivation of lemon grass (Cymbopogon citratus) on post-harvest electro-chemical properties of soil. Ann Agri Bio Res 19(1):45–48

Singh V, Thadani R, Tewari A, Ram J (2014b) Human influence on Banj oak (*Quercus leucotrichophora* A. Camus) forests of central Himalaya. J Sustain For 33:373–386

Takimoto A, Nair PKR, Nair VD (2008) C stock and sequestration potential of traditional and improved agroforestry systems in the West African Sahel. Agric Ecosyst Environ 125:159–166

Troup RS (1921) Silviculture of indian trees. Vol. I–III. Clarendon Press, Oxford

Tulachan PM (2001) The state of mountain agriculture in India. In: Rana GM, Shrestha B, Joshi SM (eds) State of mountain agriculture in the Hindu Kush-Himalayas: a regional comparative analysis. ICIMOD, Kathmandu, pp 25–31

Verma JP, Jaiswal DK, Meena VS, Meena RS (2015a) Current need of organic farming for enhancing sustainable agriculture. J Clean Prod 102:545–547

Verma JP, Jaiswal DK, Meena VS, Kumar A, Meena RS (2015b) Issues and challenges about sustainable agriculture production for management of natural resources to sustain soil fertility and health. J Clean Prod 107:793–794

Wang X, Feng Z (1995) Atmospheric C sequestration through agroforestry in China. Energy 20:117–121

Wani N, Velmurugan A, Dadhwal VK (2010) Assessment of agricultural crop and soil carbon pools in Madhya Pradesh, India. Trop Ecol 51:11–19

Yadav RP, Bisht JK (2013) Agroforestry: a way to conserve MPTs in North Western Himalaya. Res J Agric For Sci 1:8–13

Yadav RP, Bisht JK (2014a) Agroforestry to combat climate change. Kheti 66:11–14 (In Hindi)

Yadav RP, Bisht JK (2014b) Litter fall and potential nutrient returns from pecan nut (*Carya illinoinensis*) in agroforestry system in Indian Himalaya. Int J Herb Med 2:51–52

Yadav RP, Bisht JK (2014c) C sequestration potential of agroforestry practices in the mid hills of central Himalaya. In: Book of Abstracts, 3rd World Congress on Agroforestry, Trees for life-Accelerating the impact of agroforestry. Nairobi: World Agroforestry Centre, pp. 149

Yadav RP, Bisht JK (2014d) Effect of cutting management on the biomass and C stock of oak under high density plantation in Indian Himalaya. In: Verma KS, Pawar P, Kaushal R, Chauhan S, Chander J, Chandel RS (eds) Souvenir cum abstracts of ISTS-IUFRO Conference on "Sustainable Resource Management for Climate Change Mitigation and Social Security". Solan: Indian Society of Tree Scientist, p 53

Yadav RP, Bisht JK, Pandey BM (2015) Aboveground biomass and C stock of fruit tree based land use systems in Indian Himalaya. Ecoscan 9:779–783

Yadav RP, Bisht JK, Pandey BM, Kumar A, Pattanayak A (2016a) Cutting management versus biomass and C stock of oak under high density plantation in central Himalaya, India. Appl Ecol Environ Res 14:207–214

Yadav RP, Bisht JK, Mondal T, Mishra PK (2016b) Agroforestry practices for C sequestration in the mid hills of central Himalaya. In: Peshin R, Dhawan AK, Bano F, Resam KS (eds) Natural resource management: ecological perspectives, proceeding of the Indian Ecological Society: international conference-2016, SKUAST-Jammu, 18–20 Feb 2016 1:447

Yadav RP, Bisht JK, Bhatt JC (2016c) Biomass and C allocation in different production system in the mid hills of central Himalaya. Trop Ecol (in press)

Yadava AK (2010) C sequestration: underexploited environmental benefits of Tarai agroforestry systems. Report Opin 2:35–41

Zobel BD, Singh SP (1997) Himalayan forests and ecological generalizations. Bio Sci 47:735–745

Zou X, Sanford RLJ (1990) Agroforestry system in China: survey and classification. Agrofor Syst 11:85–94

12 Soil Conservation of Northwestern Himalayas (NWH): Their Constraints and Potentials for Sustainable Hill Agriculture

Gurjant S. Sidhu

Abstract

The Northwestern tract of Himalayan system was located between 28° 43′ N to 37° 05′ N latitude and 72° 02′ E to 81° 02′ E longitude and covers an area of ~33.12 M ha. The climate ranges from cold arid zone to temperate, subtropical temperate transitional, and low-altitude subtropical zone. The temperature regime ranges from cryic to hyperthermic and moisture regime from aridic cold in Greater Himalayas to humid in Lesser Himalayas and subhumid in Siwaliks. This tract is broadly divided in to three distinct landform units, i.e., Greater Himalayas, Lesser Himalayas, and Siwaliks. These units are subdivided into nine subunits. Soils of Greater Himalayas mostly pertain to Leh and Ladakh, Lahaul and Spiti, and high reaches of Uttarakhand. The moisture regime is cryic and temperature regime is aridic. These soils are coarse textured, nearly neutral to mildly alkaline, deep, low in organic matter, and base saturated. Flood plain of Indus river system is mostly calcareous, neutral to slightly alkaline. Some of high-altitude soils are classified as cryic Eutrochrepts or cryollic Eutrochrepts and typic Cryorthents and cryic Udifluvents. Soils under Lesser Himalayas cover the valley of Kashmir and Pir Panjal ranges in J&K comprising parts of the district of Punch, Rajouri, Udhampur, Doda, and Kishtwar of Jammu region and all the districts of Kashmir region. The soils of Lesser Himalayas are highly variable depending on the vegetation and geology of the area and physiography. They are mostly shallow to medium deep and deep, well to excessively drained, sandy, loamy-skeletal, coarse-loamy, and fine-loamy calcareous as well as noncalcareous soils. They are slightly acidic to neutral on high reaches and neutral to slightly alkaline on lower hills. The organic status of the soils is high to very high. They have been classified as lithic/typic Udorthents, typic/dystric

G.S. Sidhu (✉)
ICAR-National Bureau of Soil Survey and Land Use Planning, New Delhi 110012, India

Flat # 82, Evergreen Apts., Plot # 9, Sector # 7, Dwarka, New Delhi 110075, India
e-mail: gssidhu_ps@yahoo.com

J.K. Bisht et al. (eds.), *Conservation Agriculture*,
DOI 10.1007/978-981-10-2558-7_12

Eutrochrepts, typic Hapludolls, and typic Udifluvents. Typic Hapludalfs are also reported in Lesser Himalayas of Kashmir region. Soils of Siwalik regions cover areas of southwestern part of Himalayas comprising Jammu region of J&K, Himachal Pradesh, Uttarakhand, Haryana, Uttar Pradesh and Punjab states. The soil-scape represents moderately to gently sloping situated in lower elements of topography comprising Hapludalfs, Hapludolls, and Eutrochrepts, whereas moderately to steeply sloping pediments at lower elevations are represented by Udorthents, Haplumbrepts, and Argudolls. The nearly level to level flood plain of Indus river system is mostly calcareous and neutral to slightly alkaline and represented by Udifluvents and Haplaquepts. The dominant soil great groups covering the area are Ustochrept, Udorthents, Ustochrept, Udorthents, Haplustalfs, Hapludalfs, Entrochrept, and Hapludolls. The other great groups found in pockets are Ustochrepts, Ustorthents, and Hapludalfs. The major soil management options for improving soil health in northwestern Himalayas have been discussed in the paper.

Keywords
Soil classification • Landform units • Soil management

12.1 Introduction

Soil is the most precious natural resource. Maintaining soils in a state of high productivity on sustainable basis is important for meeting the basic needs of the people. Therefore, knowledge of soils with respect to their extent, distribution, characteristics, and potential use is important for optimizing land use. It is still more important for hilly areas like Himalayas, which is highly prone to tremendous loss of soil and plant nutrients due to steeply sloping landscape and exploitation of natural resources like soils and forest. Himalayas endowed with varied physiography offer niches for a variety of cereal, horticultural plantation, and aromatic and medicinal crops. The Himalayan climate varies according to the elevation. The productivity of the majority of crops in the region is low due to a number of production constraints. The impaired soil health influenced by increased soil erosion, deficiencies and toxicities of nutrients, frequent moisture stress and low soil biological activity, etc. is one of the major factors for low productivity. The future desired gains in productivity and agricultural transformation in the region cannot be realized on a deteriorating natural resource base. It is natural that there is a dire need to identify the factors involved and take suitable remedial measures (Meena et al. 2013, 2016; Singh et al. 2014; Kumar et al. 2015; Verma et al. 2015b; Ghosh et al. 2016). Thus, the present paper focused on the kinds and distribution of soils indifferent physiographic units with reference to agroecological regions in one of the most important and fragile northwestern regions of the Himalayas, India.

12.2 Soil Resource Database of NW Himalayas

12.2.1 Historical Background

The earliest investigation on soils in India by Voelcker dates back to 1893 and by Leather to 1897. They differentiated the soils into four major soil groups, namely, the Indo-Gangetic alluvium, the black cotton or regur soils, the red soils, and the laterite soils. The first soil map of India was published in 1932 by Schokalskaya based on the Russian concept, in a total of 16 soil groups based on climate, vegetation, soil forming material, salinity, alkalinity, swamps, and peats. Based on geological information, Wadia compiled the soil map of India in 1936, which was further improved in 1945 by him. The soils were classified as red soils, black soils (regurs), laterite and lateritic soils of peninsular India, delta soils, desert soils, bhabar and Tarai soils, and alkali soils of the Indo-Gangetic plains. Vishwanath and Ukil in 1944 published a soil map by placing the soils of India into different climatic zones on the basis of N.S. quotients. These studies were based on sporadic information compiled to a soil map of the country. With the advent of soil taxonomy of USDA and its wide applicability throughout the world, the soil mapping and classification in India took new turn in the early 1980s. Prior to it, systematic studies on soil based on genetic factors had been instituted by Raychaudhuri during the 1950s. Raychaudhuri et al. (1963) divided India into 27 major soil groups. Govindarajan and Dutta Biswas in 1968 have grouped the soils of India into 23 major groups and accordingly prepared the soil map, under the scheme of FAO/UNESCO World Soil Map project, which was later on revised by Govindarajan in 1971 and published on 1:7 million scales with 25 broad soil classes. USDA seventh approximation soil classification system was adopted for the first time by him for classifying the soils at great group level. Murthy and Pandey (NBSS & LUP 1985) prepared soil map of India showing associations of suborders as mapping units on 1:6.3 million scales in 1985. The soil associations were assigned equivalent names under FAO/UNESCO system. In all abovementioned studies, extensive extrapolation was utilized and no uniform pattern was adopted.

Soil resource mapping of country on 1:250,000 scales was in initiated by J.L. Sehgal, the then director of NBSS & LUP in 1986. The methodology was finalized after number of workshop meetings of working groups held from time to time. The soil resource information as association of soil families was generated at state level on 1:2, 50,000 scales, along with the executive summary reports. The soil resource data of states compiled to bring out soil map of India on 1:1 million scales as associations of subgroups. A three-tier approach (using remote sensing and aerial photo interpretation techniques, ground truth observation and soil analysis, cartography and printing) was considered appropriate for achieving the task of mapping than the conventional methods (Sarma et al. 1987; Sehgal et al. 1987). The LANDSAT MSS data on bands 2 and 4 and false color composite (FCC) TM/RS on 1:2, 50,000 scales were used.

The soil maps of all the states that were published on 1:500,000 scales were printed. The soil data at 1: 500,000 scales were reduced to 1:1 million scales through cartographic and categorical generalization to prepare the soil map of India as association of subgroups (Staff NBSS & LUP 2002). In total 1649 soil mapping units have been identified. Besides, the soil survey at district, village, watershed, ORP villages, etc. was taken up as per the demands of stakeholders, SAUs, ICAR, and other organizations from time to time. The status of different types of soil surveys completed in different states of the country has been presented in the form of various state soil survey reports of Sidhu et al. (1997), Sidhu et al. (2007a, b) for HP and Punjab states; Rana et al. (2000) for J&K state and Ram et al. (2013) for Uttarakhand state. But the regional wise data is lacking. Some attempts have been initiated by Sidhu and Surya (2014),to compile soils of northern region of India; Sidhu (2012) to compile soils of Indo-Gangetic plains (IGP); Suri et al. (2013), to characterize soils of western Himalayas (Sidhu et al. 2000); Sidhu and Sharma (2008) and Rao et al. (1997), to characterize soils of Siwaliks; and Mahapatra et al. (2000a, b) to characterize and classify soils of J&K state (Sidhu et al. 1998) of Jammu region and to characterize and classify soils of cold desert of the country (Walia et al. 1999; Sehgal et al. 1993). In addition to above studies, the soil surveys of some districts (NBSS & LUP (1992) for Anantnag and Pulwama districts of J&K, NBSS & LUP (1998), soils of Bilaspur district HP and NBSS LUP (1993), soils of ORP villages of Siwaliks, Punjab) have been prepared to characterize and classify soils for perspective land use planning.

12.2.2 Soil Resources

The Indian Himalayan Region (IHR) is spreading on ten states (administrative regions), namely, Jammu and Kashmir, Himachal Pradesh, Uttaranchal, Sikkim, Arunachal Pradesh, Meghalaya, Nagaland, Manipur, Mizoram, and Tripura and hill regions on two states, viz., Assam and West Bengal of Indian Republic (Fig. 12.1). Starting from foot hills in the south (Siwaliks), the region extends to Tibetan plateau in the north (trans-Himalayas) comprising about 95 districts of the country. The region occupies the strategic position of entire northern boundary (northwest to northeast) of the nation and touches almost all the international borders (seven countries) with India. It contributes ~16 % of India's total geographical area, and most of the area is covered by snow-clad peaks, glaciers of Greater Himalayas, and dense forest cover of mid-Himalayas. The IHR shows a thin and dispersed human population as compared to the national figures due to its physiographic condition and poor infrastructure development, but the growth rate is much higher than the national average. In the present paper, soil resources of three northern states representing northwestern Himalayan region, i.e., Jammu and Kashmir, Himachal Pradesh, and Uttarakhand, have been described.

The northwestern tract of Himalayan system was located between 28° 43′ N to 37° 05′ N latitude and 72° 02′ E to 81° 02′ E longitude and covers an area of approximate 33.12 Mha. The climate ranges from cold arid zone to temperate,

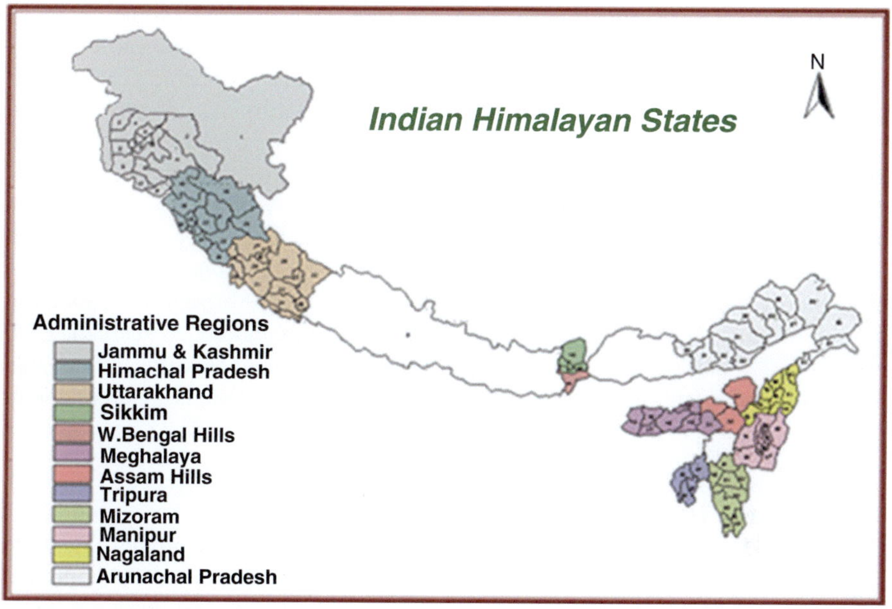

Fig. 12.1 Himalayan system of the country (Maps of India 2015 online)

subtropical temperate transitional, and low-altitude subtropical zone. The temperature regime ranges from cryic to hyperthermic and moisture regime from aridic cold in Greater Himalayas to humid in Lesser Himalayas and subhumid in Siwaliks. This tract is broadly divided in to three distinct landform units, i.e., Greater Himalayas, Lesser Himalayas, and Siwaliks. These units are subdivided into nine subunits.

12.3 The Brief Description of Soils of Himalayan System in Northern Region

12.3.1 Soils of Western Himalayas

The soils of this region comprise of mostly Himalayan ranges in states of Himachal Pradesh (HP), Jammu and Kashmir (J&K), and Uttarakhand (Sidhu et al. 1995; Rana et al. 2000; Singh et al. 2004). These soils occupy an area of 25.9 Mha. The dominant soils belong to typic Cryopsamments, typic Cryofluvents, typic Udifluvents, lithic/typic Cryorthents, lithic/typic Udorthents, typic Endoaquepts, typic Eutrocryepts, dystric/typic Eutrudepts, typic Dystrudepts, typic Hapludolls, aquic/typic Hapludalfs, glaciers, and rocky lands (NBSS & LUP 2002). The soils on Greater Himalayas are mostly skeletal and calcareous in nature. These soils are less weathered and subject to only physical weathering and therefore are loosely fragmented. These soils are subject to severe soil erosion due to landslides and movement of glaciers and lead to form moraines and screes. However, soils of Lesser Himalayas are relatively more weathered and high in organic carbon.

The soils of western Himalayas mostly confined to agroecological regions (AERs) 1 and 14. AER-1 region is characterized my warm and dry summers and cold continental type winters. The mean annual temperature (MAT) is 11 °C, which ranges from 9 to 14°C; the mean annual precipitations range from 100 to 115 mm. The length of growth period (LGP) is <90 days. While AER-14 region represents warm subhumid to cool humid (with inclusion of per humid) ecosystem, the region is characterized by mild summers and cold winters. The mean annual temperature in general varies from 12 to 20 °C. The soil moisture control section (SMCS) in most of the areas belongs to humid or per humid moisture regime. As such, the LGP is 180 to +210 days (Sehgal et al. 1992). The distribution and classification of soils for Jammu and Kashmir, Himachal Pradesh, and Uttarakhand states are presented in Tables 12.1, 12.2, and 12.3, respectively.

12.3.1.1 Land Use

On Greater Himalayas, the soils are mostly under thin forests and alpines. They are sporadically cultivated to vegetables, wheat, barley, high-value crops like hops, and some fruit plants such as apple, walnut, apricot, etc. Among the livestock, mule dominates the cattle population followed by sheep, goat, and yak. The Lesser Himalayan region is rich in moist temperature, subtropical pines, and subalpine forests. The higher reaches are occupied with alpine which are one of the best grasslands during summer months. The agriculture confined to less than 20 % areas (Sidhu et al. 1995; Singh et al. 2004) mostly on the terrace lands and valley areas. The major crops grown are wheat, barley, maize, rice, and minor millets like *madua*, *lahi*, *thingora*, *sewa*, etc. The region is famous for apple, plums, pear, peaches, etc.

12.3.1.2 Constraints

- Shallow, skeletal, and coarse-textured soils with low water-holding capacity and low nutrient retentive capacity
- Severe erosion on steeply to very steeply sloping lands
- Extreme cold climate which hampers plant growth
- Moderately to highly calcareous soils in Lahaul and Spiti areas which cause nutrient imbalances
- Soil acidity in small pockets in Lesser Himalayas which causes nutrient imbalance
- Lack of soil and water conservation measures in Siwaliks and doughtiness during post-rainy season

12.3.1.3 Potentials

- The area has high potential for recreational purposes, i.e., tracking and winter sports, etc; as catchment for rivers and rivulets; and space for glaciers etc.

Table 12.1 Area distribution of different orders, suborders, great groups, and families (particle-size class) in Himachal Pradesh

Suborder	Great group	Subgroup	Sr. No.	Soil family (particle-size class)	Area
Orthents (2854.3, 51.3)	Cryorthents (830.8, 14.9)	Lithic Cryorthents (165.5, 2.9)	1	Sandy-skeletal	(113.1, 2.0)
			2	Loamy-skeletal	(11.0, 0.2)
			3	Loamy	(41.4, 0.7)
		Typic Cryorthents (665.3, 12.0)	4	Sandy-skeletal	(176.3, 3.2)
			5	Loamy-skeletal	(460.8, 8.3)
			6	Coarse-loamy	(28.2, 0.5)
	Udorthents (1862.4,33.5)	Lithic Udorthents (1679.9, 3.1)	7	Loamy	
			8	Loamy-skeletal	(80.8, 1.5)
		Typic Udorthents (1694.5, 30.4)	9	Sandy-skeletal	(166.6, 3.0)
			10	Sandy-skeletal over fragmental	(44.9, 0.8)
			11	Fragmental	(15.7, 0.3)
			12	Loamy-skeletal	(742.3, 13.3)
			13	Sandy	(12.9, 0.2)
			14	Coarse-loamy	(530.7, 9.5)
			15	Fine-loamy	(181.4, 3.3)
	Ustorthents (27.8, 0.5)	Typic Ustorthents (27.8,0.5)	16	Sandy-skeletal	(1.2, 0.02)
			17	Loamy-skeletal	(6.4, 0.1)
			18	Coarse-loamy	(18.5, 0.3)
			19	Fine-loamy	(1.7, 0.03)
Fluvents (126.0, 3.2)	Udifluvents (69.4, 1.2)	Typic Udifluvents (69.4, 1.2)	20	Sandy	(13.0, 0.2)
			21	Coarse-loamy	(56.4, 1.0)
	Ustifluvents (56.6, 2.0)	Typic Ustifluvents (56.6, 2.0)	22	Sandy-skeletal	(26.1, 0.5)
			23	Coarse-loamy	(25.6, 0.5)

(continued)

Table 12.1 (continued)

Suborder	Great group	Subgroup	Sr. No.	Soil family (particle-size class)	Area
			24	Fine-loamy over sandy	(4.9, 0.09)
Psamments (7.3, 3.2)	Ustipsamments (7.3, 0.1)	Typic Ustipsamments (7.3, 0.1)	25	Sandy	(7.3, 0.1)
Cryepts (17.4, 0.3)	Eutrocryepts17.4, 0.3)	Typic Eutrocryepts (17.4, 0.3)	26	Loamy-skeletal	(0.6, 0.01)
			27	Fine-loamy	(16.8, 0.3)
Udepts (959.4, 17.2)	Eutrudepts (959.4, 17.2)	Fluventic Eutrudepts (15.3, 0.2)	28	Loamy-over-fragmental	(1.9, 0.03)
			29	Coarse-loamy	(13.4, 0.2)
		Dystric Eutrudepts (713.1, 12.8)	30	Loamy-skeletal	(8.1, 0.1)
			31	Coarse-loamy	(94.5, 1.7)
			32	Fine-loamy	(610.5, 11.0)
		Typic Eutrudepts (231.0,4.1)	33	Loamy-skeletal	(8.0, 0.1)
			34	Coarse-loamy	(1.1, 0.02)
			35	Fine-loamy	(221.9, 4.0)
Ustepts (123.1, 2.23)	Dystrustepts (42.4, 0.8)	Typic Dystrustepts (42.4, 0.8)	36	Coarse-loamy	(8.6, 0.2)
			37	Fine-loamy	(33.8, 0.6)
	Haplustepts (80.7, 1.5)	Udic Haplustepts (80.7, 1.5)	38	Fine-loamy	(53.1, 1.0)
			39	Coarse-loamy	(27.6, 0.5)
Udolls (40.0, 0.8)	Hapludolls (40.0, 0.8)	Lithic Hapludolls (6.0, 0.1)	40	Loamy	(6.0, 01)
		Typic Hapludolls (34.0, 0.7)	41	Loamy-skeletal	(4.6, 0.08)
			42	Coarse-loamy	(8.5, 0.2)
			43	Fine-loamy	(20.9, 0.4)

(continued)

Table 12.1 (continued)

Suborder	Great group	Subgroup	Sr. No.	Soil family (particle-size class)	Area
Udalfs (23.8, 0.4)	Hapludalfs (23.8, 0.4)	Typic Hapludalfs (23.8, 0.4)	44	Fine-loamy	(23.8, 0.4)
Rocky outcrops (1538.2, 27.7)					

*Area of unit in 000 ha and **% of area for figures in ( )

Table 12.2 Area distribution of different orders, suborders, great groups, and families (particle-size class) in Jammu and Kashmir

Suborder	Great group	Subgroup	Sr. No.	Soil family (particle-size class)	Area
Entisols (7572.6, 34.1)					
Orthents (6846.2, 30.8)	Cryorthents (4842.0, 21.8)	Lithic Cryorthents (3521.7, 15.9)	1	Fragmental	666.3, 3.0
			2	Sandy-skeletal	747.8, 3.4
			3	Sandy-skeletal, calcareous	1159.4, 5.2
			4	Loamy	199.4, 0.9
			5	Loamy, calcareous	748.9, 3.4
		Typic Cryorthents (1320.2, 5.9)	6	Sandy-skeletal	599.2, 2.7
			7	Sandy-skeletal, calcareous	92.0, 0.4
			8	Loamy-skeletal	359.2, 1.6
			9	Loamy-skeletal, calcareous	31.2, 0.1
			10	Coarse-loamy, calcareous	52.0, 0.2
			11	Fine-loamy	186.6, 0.8
	Udorthents (1910.0, 8.6)	Lithic Udorthents (596.6, 2.7)	12	Loamy -skeletal	236.3, 1.1
			13	Loamy	360.3, 1.6
		Typic Udorthents (1293.6, 5.8)	14	Sandy-skeletal	16.0, 0.1
			15	Loamy-skeletal	272.9, 1.2

(continued)

Table 12.2 (continued)

Suborder	Great group	Subgroup	Sr. No.	Soil family (particle-size class)	Area
			16	Clayey-skeletal	10.1, 0.1
			17	Coarse-loamy	669.9, 3.0
			18	Fine-loamy	292.8, 0.9
			19	Fine-loamy, calcareous	31.8, 0.1
		Mollic Udorthents (19.7, 0.1)	20	Fine-loamy	1.9.7, 0.1
	Ustorthents (94.3, 0.4)	Typic Ustorthents (94.3, 0.4)	21	Loamy-skeletal	63.7, 0.3
			22	Coarse-loamy	29.7, 0.1
			23	Fine-loamy	0.9, neg
Fluvents (269.5, 1.2)	Cryofluvents (243.0, 1.1)	Typic Cryofluvents (214.4, 1.0)	24	Coarse-loamy	120.6, 0.5
			25	Coarse-loamy, calcareous	93.7, 0.4
		Mollic Cryofluvents (28.6; 0.1)	26	Coarse-loamy, calcareous	28.6, 0.1
	Udifluvents (11.4, 0.1)	Typic Udifluvents (11.4, 0.1)	27	Fine-loamy	11.4, 0.1
	Ustifluvents (15.0, 0.1)	Typic Ustifluvents (15.0, 0.1)	28	Coarse-loamy	7.0, neg
			29	Fine-silty	8.0, neg
Aquents (15.5, 0.1)	Fluvaquents (15.5, 0.1)	Typic Fluvaquents (15.5, 0.1)	30	Coarse-loamy, calcareous	15.5, 0.1
Psamments (441.4, 2.0)	Cryopsamments (441.4, (2.0)	Typic Cryopsamments (441.4, 0.2)	31	Sandy	144.6, 0.6
			32	Sandy, calcareous	296.7, 1.3
Inceptisols (1427.2, 6.4)					
Ochrepts (1360.1, 6.1)	Cryochrepts (145.7, 0.7)	Typic Cryochrepts (91.3, 0.4)	33	Coarse-loamy	34.3, 0.2
			34	Coarse-loamy, calcareous	7.8, neg
			35	Fine-loamy	3.2, neg
			36	Fine-loamy, calcareous	45.9, 0.2
		Fluventic Cryochrepts (54.4, 0.3)	37	Coarse-loamy	32.8, 0.2
			38	Coarse-loamy, calcareous	21.6, 0.1

(continued)

Table 12.2 (continued)

Suborder	Great group	Subgroup	Sr. No.	Soil family (particle-size class)	Area
	Dystrochrepts (134.2, 0.6)	Typic Dystrochrepts (134.2, 0.6)	39	Fine-loamy	116.9, 0.5
			40	Fine	17.3, 0.1
	Eutrochrepts (922.3, 4.1)	Typic Eutrochrepts (865.4, 3.0)	41	Fine-loamy	587.2, 2.6
			42	Fine-silty	45.3, 0.2
			43	Fine	31.2, 0.1
			44	Coarse-loamy	201.7, 0.9
		Aquic dystric Eutrochrepts (2.8, neg)	45	Fine-loamy	2.8, neg
		Fluventic Eutrochrepts (21.0, 0.1)	46	Coarse-loamy, calcareous	18.3, 0.1
			47	Coarse-loamy	2.5, neg
		Typic Eutrochrepts (29.8; 0.1)	48	Coarse-loamy	13.3, 0.1
			49	Fine-silty, calcareous	6.3, neg
			50	Fine, calcareous	10.2, 0.1
		Vertic Eutrochrepts (2.9, neg)	51	Fine	2.9, neg
	Ustochrepts (157.8, 0.7)	Udic Ustochrepts (94.4, 0.4)	52	Coarse-loamy	9.5, neg
			53	Fine-loamy	84.9, 0.4
		Fluventic Ustochrepts (63.4; 0.3)	54	Coarse-loamy	42.5, 0.2
			55	Coarse-loamy, calcareous	19.3, 0.1
			56	Fine-loamy, calcareous	1.6, neg
Aquepts (67.2, 0.3)	Haplaquepts (67.2, 0.3)	Typic Haplaquepts (52.5, 0.2)	57	Fine-loamy	10.3, neg
			58	Fine-loamy, calcareous	22.2, 0.1
			59	Fine-silty, calcareous	20.1, 0.1
		Aeric Haplaquepts (14.6, 0.2)	60	Fine-loamy, calcareous	12.4, 0.1
			61	Fine, calcareous	2.2, neg

(continued)

Table 12.2 (continued)

Suborder	Great group	Subgroup	Sr. No.	Soil family (particle-size class)	Area
Mollisols (47.7, 0.2)					
Udolls (47.7, 0.2)	Hapludolls (47.7, 0.2)	Typic Hapludolls (47.7, 0.2)	62	Fine-loamy	47.7, 0.2
Alfisols (100.0, 0.5)					
Udalfs (100.0, 0.5)	Hapludalfs (100.0, 0.5)	Typic Hapludalfs (80.8, 0.4)	63	Fine-loamy	36.4, 0.2
			64	Fine-silty	8.0, neg
		Aquic Hapludalfs (19.2, 0.1)	65	Fine	19.2, 0.1

*****neg* negligible

*Area of unit in 000 ha and **% of area for figures in ( )

- The deep soils in valleys have high potential for wheat and fruit crops like apple, apricot, etc. These areas can also be exploited for seed multiplication of crops and off-season high-value crops like vegetables, hops, etc.

12.3.2 Soils of Western Siwaliks

These soils occur in the foot hills of Himalayas, i.e., Siwaliks, mostly in the southern parts of the states of J&K, HP, and Uttarakhand and northern parts of Punjab and Haryana (Rana et al. 2000; Sidhu et al. 1995, 1997; Sachdev et al. 1995; Singh et al. 2004). These soils cover an area of 2.5 Mha. The dominant soils are typic Udorthents, followed by typic Ustorthents, typic Haplustepts, dystric/typic/fluventic Dystrudepts, and typic Dystrudepts (NBSS & LUP 2002). These soils are mostly light textured having low to medium organic carbon (OC) contents. These soils pertain to AER-14 (warm subhumid to humid with inclusions of per humid). The region represents warm subhumid to cool humid (with inclusion of per humid) ecosystem. Due to high variability in relief, the mean annual temperature ranges from 12 to 20 °C. The mean annual rainfall ranges from 1200 to 2000 mm (Sehgal et al. 1992).

12.3.2.1 Land Use

The natural vegetation consists of subtropical pine forests. Some species of Khair, ber, and wild pomegranate are also found in these areas. These soils mostly fall under rainfed agriculture. The main crops grown are wheat, mustard, sunflower, pulses, and rice in some low-lying areas having assured irrigation facilities. These soils also support some plantation crops like ber, kinnow, mango, etc. Some trees of medicinal importance and some horticultural crops like, pine apple, citrus, peach, and pear are also grown in some areas.

Table 12.3 Soils on different physiographic units and their characteristics and taxonomic classification of Uttarakhand state

Physiographic unit	No. of profile observations	Soil characteristics	Soil taxonomy	Area	
				(000 ha)	[a](%)
I Soils of Greater Himalayas:					
(i) Summits and ridge tops	10	Shallow to medium deep, excessively drained, sandy-skeletal to loamy-skeletal, highly calcareous with low O.C. and AWC, slightly alkaline, and severely eroded	Lithic Cryorthents and typic Cryorthents	259	4.6
(ii) Mountain and valley glaciers	10	Similar in (i) above except that they are covered with snow for longer period	Lithic Cryorthents and typic Cryorthents	199	3.6
(iii) Side/reposed slopes	77	Shallow to medium deep, excessively drained, loamy-skeletal and coarse-loamy, both calcareous and noncalcareous, slightly acidic to slightly alkaline, severely eroded soils with low O.C. and AWC, and moderate to strong stoniness	Lithic/typic Udorthents, typic Eutrocryepts, dystric Eutrocryepts	2008	36.1
(iv) Glaciofluvial valleys	38	Medium deep to deep, well to excessively drained, loamy-skeletal to coarse-loamy and fine-loamy soils having low AWC. Dominantly slightly acidic to neutral with medium to high contents of O.C. Some soils are highly calcareous and slightly to moderately alkaline	Lithic/Typic Cryorthents, Typic Eutrocryepts and Dystric Eutrudepts	124	2.3
(v) Fluvial valleys	13	Medium deep to deep, well to somewhat excessively drained, loamy-skeletal, slightly acidic to neutral soils with low AWC	Typic Udorthents/Udifluvents and dystric/typic Eutrudepts	29	0.5

(continued)

Table 12.3 (continued)

Physiographic unit	No. of profile observations	Soil characteristics	Soil taxonomy	Area	
				(000 ha)	[a](%)
II Soils of Lesser Himalayas:					
(vi) Summits/ridge tops	7	Shallow to medium deep, well to excessively drained, coarse-loamy and fine-loamy, neutral to slightly alkaline soils with low to medium AWC and severe erosion	Typic Udorthents and dystric Eutrudepts	3	0.7
(vii) Side/reposed slopes	75	Shallow to medium deep and deep, well to excessively drained, sandy, loamy-skeletal, coarse-loamy and fine-loamy calcareous as well as noncalcareous soils. They are slightly acidic to neutral on high reaches and neutral to slightly alkaline on lower hills. The O.C. status of these soils is high	Lithic/typic Udorthents, typic/dystric Eutrudepts and typic Hapludolls	1605	28.9
(viii) Fluvial valleys	26	Medium deep to deep, well to excessively drained, stratified coarse-loamy and fine-loamy, mostly noncalcareous and neutral to slightly alkaline soils with moderate to high O.C. status and medium AWC	Typic Udifluvents/Udorthents and dystric Eutrudepts	108	1.9
III Soils of Siwaliks:					
(ix) Summits and ridges	5	Medium deep to deep, well to excessively drained, loamy-skeletal soils with severe to very severe erosion. They are slightly acidic to slightly alkaline with moderate to high O.C. status and low AWC	Lithic/typic Udorthents	25	0.4

(x) Side/reposed slopes	94	Medium deep to deep, well to excessively drained, loamy-skeletal, coarse-loamy, and fine-loamy soils with moderate to severe erosion. They are slightly acidic to slightly alkaline with low to moderate O.C. and AWC status.	Lithic/typic Udorthents, typic/dystric Eutrudepts and typic Hapludolls	972	17.6
(xi) Fluvial valleys	21	Medium deep to deep, well-drained, sandy, sandy-skeletal, coarse-loamy, and fine-loamy soils with moderate erosion. They are neutral to slightly and moderately alkaline with low to medium O.C. status and AWC.	Typic Ustorthents, typic Ustifluvents, typic Haplustepts, and typic Hapludalfs	81	1.5
(xii) Piedmont plains	22	Medium deep to deep, well drained, coarse-loamy soils with severe erosion. They are calcareous as well as noncalcareous and slightly to moderately alkaline with low to medium O.C status and medium AWC	Typic Ustorthents and udic Haplustepts	67	1.2
IV Soils of alluvial plains:					
(xiii) Alluvial piedmont plains	7	Medium deep to deep, well drained, coarse- and fine-loamy soils with slight to moderate erosion. They are calcareous as well as noncalcareous, neutral to slightly and moderately alkaline with medium O.C. status and medium to high AWC	Typic Eutrudepts, Dystric Eutrudepts and Typic Udifluvents	14	0.3
(xiv) Active flood plains	5	Deep, well to excessively drained, mostly stratified coarse-loamy and fine-loamy soils with slight erosion and stoniness, prone to slight and moderate flooding	Typic Ustifluvents and typic Ustipsamments	15	0.3

[a]% TGA of the state

12.3.2.2 Constraints

- Draftiness is experienced due to excessive runoff and coarser texture of soils.
- Shallow soil depth on steeply sloping hills.
- Very severe soil erosion during rainy season.
- Floods due to seasonal streams (Choes) during rainy season.

This zone (western Himalayan region) encompasses three hilly states, viz., Himachal Pradesh, Jammu and Kashmir, and Uttarakhand. The region has much diversity. The valleys receive good rainfall and have rich soils, though the temperate zones of Ladakh face rather harsh climatic barriers against agricultural growth. This zone in particular is highly prone to enormous loss of soil and plant nutrients due to its sloppy landscape and also unplanned and indiscriminate use of land and water resources. It is thus imperative to study the soils in this area for properly classifying them according to their characteristics and climate and landform conditions for rationalizing land use.

12.4 Land Capability Classification of Western Himalayas

Data land capability classification (LCC) in HP (Sidhu et al. 2007a, b) revealed that ~78 % area belongs to land capability classes VI, VII, and VIII which are nonarable lands and not suited for agriculture but suitable for forest, orchards, pasture, and recreational purposes. Classes III and IV lands covering an area of 21.4 % are moderately to marginally suitable for growing climatically adopted crops. Class II lands occupy negligible area, and there being no class I lands. Soils, especially in mid- and high hills, are not being used as per their capability and suffer from severe soil degradation. Therefore, appropriate land use planning is an absolute must to check further land degradation. This requires detailed soil survey of each micro-watershed, land capability classification, and preparation of soil/land capability maps of the area. Owing to undulating topography in the hilly region, suitable soil and water conservation measures are urgently warranted to combat the problems of soil erosion and land degradation. On cultivated lands, bench terracing, trenching, contour farming, bunding, strip cropping, and mixed cropping strategies need to be adopted extensively. Wastelands, including cultural wastes, grazing lands, degraded agricultural lands, fallows, and the areas unsuitable for cultivation are highly erodible and therefore need special management practices for soil and water conservation and permanent vegetative cover to check further degradation. Current pastoral or agricultural land use needs to be replaced by more effective soil and water resource conservation system like forestry, agroforestry, silvopasture, controlled and rotational grazing, and enclosure of severely degraded sites for natural regeneration (Singh and Prakash 2000).

12.5 Soil Resource Problems of Western Himalayas

Soil resource assessment for developmental planning requires easily understandable information by the concerned user agencies. This information will be useful for diagnosis of problems and potentials of soils. The deep and shallow soils occupy about 18 % of the area. Medium deep soils occur on 35 % of the TGA. The rest of the area (30 %) is occupied by rock outcrops and glaciers. Loamy-skeletal soils occupy the maximum area (23.9 %), followed by fine-loamy (20.4 %), coarse-loamy (17.6 %), sandy-skeletal (9.3 %), and fragmental (1.1 %) particle-size classes. Majority of the soils (91.2 %) belong to somewhat excessively drained (49.5 %) and well-drained (41.7 %) classes, whereas moderately well to excessively drained soils occupy only 5.4 and 3.3 % area, respectively. Variation in soil depth, texture, and clay mineralogy gives considerable range in available water-holding capacity. It was found that the major area (67.7 %) of the state was very low to low ($<$100 mm m^{-1}) in available water-holding capacity and the remaining 32.3 % area belongs to medium available water-holding capacity (100–150 mm m^{-1}). It is further observed that 44.9 % area is occupied by slightly alkaline soils followed by slightly acidic (27.4 %), neutral (21.5 %), and moderately acidic (5.7 %) soils. The slightly alkaline soils are mostly confined to Lahaul and Spiti, Kinnaur, and Siwaliks (Una, Hamirpur, Bilaspur, and southern parts of Chamba, Solan, Mandi, and Sirmaur districts). The pH seldom exceeds 8.0 in these areas. Therefore, these soils may be treated as good as neutral soils for the purpose of crop management. The pH of up to 8.0 may be ascribed to development of soils on uplifted marine deposits rich in alkaline materials in Lahaul and Spiti region, and the presence of alkaline ranges in Siwaliks and Lesser Himalayas. Shallow and light-textured soils occurring on steep to very steep slopes on different landforms are problematic soils for growing the climatically adapted crops. These soils in Greater Himalayas support xerophytic plants and minor millets. In Lesser Himalayas, these soils support thick forest vegetation and grassy meadows. Soils on side/reposed slopes in Lesser Himalayas are highly suitable for growing the climatically adapted crops/fruit crops.

These soils have sufficient moisture throughout the year to sustain plant growth; their potentiality can be further enhanced by adopting suitable soil and water conservation measures and introducing improved package of practices. Soils of fluvial valleys, having relatively warmer temperatures than the higher reaches, can be brought under vegetables, fruit crops (like plums, pears, etc.), and cereal and cash crops. Soils of Siwaliks are mostly denuded and require effective soil and water conservation measures to protect them from erosion losses (Meena et al. 2015a, b, c, d, e; Verma et al. 2015a). These soils are suitable for plantation of tropical dry deciduous and thorny forests. The brief abstraction of land use options in different physiographic unit is given in Table 12.4.

Table 12.4 Main physiographic regions: geology, climate, and present land use

Physiographic region	Elevation (m)	Geology	Bio-climate type	Length of growing period (LGP)	Present land use	Area	
						(000 ha)	[a](%)
Greater Himalayas	5000–6500	Sedimentary rocks, granite, granite-gneiss	Arid to semiarid	<60–180 days	Barren or rock outcrops covered with permanent glaciers and alpines. Cultivated for barley, potato, wild gram, wheat, coarse millets, and vegetables in small areas	2620	47.1
Lesser Himalayas	1500–5000	Granite-gneiss, phyllites, slates, shale, sandstone, etc.	Temperate and humid to per humid	180 to >300 days	Highly suitable for growing wide range of fruit and cereal crops, viz., apple, peas, plum, grapes, wheat, maize, barley, rice, etc. The area is also supported with deciduous and coniferous thick forests	1751	31.5
Siwaliks	600–1500	Sedimentary rocks of tertiary formation (sand stone, shales, clays, and conglomerates)	Semiarid, subhumid to humid	270–300 days	Denuded hills with thin forests pertaining to tropical dry deciduous and thorny trees and plants. The main crops grown are wheat, maize, rice, oil seeds, etc.	1146	20.0
Alluvial plains	300–600	Alluvium	Semiarid, subhumid to humid	270–300 days	Most of the areas are prone to floods due to overflow of seasonal streams during monsoon. The major crops grown are wheat, paddy, maize, vegetables, etc.	29	0.6

[a]Percentage of the total geographical area (TGA) of the state

12.6 Soil Degradation in Western Himalayas and Its Management

12.6.1 Greater and Lesser Himalayas

Since sloppy landscape is the main cause of land degradation, proper engineering measures as well as soil and water conservation practices should be adopted to arrest further degradation of soils. Land ~33 % slopes may be cultivated with agricultural crops. Sloping lands can be benched as terraces and may be improved by providing shoulder bunds, leveling, and inter-terrace drains. Agricultural crops should be raised using good agronomic practices like contour farming, mulching, intercropping with legumes, high-yielding improved seeds, fertilizers, manures, pesticides, etc. and also on watershed management basis for harnessing the water resources. Landforms with slopes from 33 % to 50 % may be brought under suitable horticultural plantation and orchards after constructing proper engineering structures and land shaping, for example, orchard terraces may be restricted to less slopes i.e. $<$ 15 % slopes. Land with slopes $>$50 % should be brought under permanent vegetation of suitable fuel, fodder, tree species, and grasslands with soil conservation measures like contour trenching, staggered trenching, and contour furrows (Singh et al. 1990). The area of shallow soil which is not suitable for tree growth may be treated with engineering measures like gully plugs, check dams, spurs, etc. and rehabilitated by proper vegetative measures. Partial development with improved seeds may be taken up in areas where no agricultural practices are possible. Sidhu et al. (2009, 2010) and Yadav and Sidhu (2010) comprehensively presented soil erosion status of Himachal Pradesh. Extent, types, and severity of soil degradation in the state have been deduced explicitly.

12.6.2 Siwaliks

As far as the northwestern Siwaliks are concerned, about 36 % area is affected by severe soil erosion, 49 % by moderate, and 13 % by slight erosion. Stoniness covering 25 % area is also a major problem in severely eroded areas. These areas should be treated by adopting suitable soil conservation measures to minimize soil erosion losses. In brief, it can be concluded that shallow soil depth, severe to very severe erosion hazards, stoniness, and rapid runoff leading to low water storage, low available water-holding capacity, and low nutrient status are the main constraints. These soils need intensive soil conservation measures along with introduction of erosion-resistant varieties of grasses like *bhabbar grass* which can also be utilized for paper manufacturing. Some of these soils are well suited for raising horticultural crops, protected pastures, forestry, and recreational activities. Soil degradation status of Himachal Pradesh and J&K generated through soil and climatic data by different studies is presented in Tables 12.5 and 12.6 (Sidhu et al. 2009, 2010; Yadav and Sidhu 2010; Rana et al. 2000).

Table 12.5 Soil degradation (Mha) status of Himachal Pradesh

Nature of soil degradation	Degree of degradation (%)	Total area				
	Slight	Moderate	Strong	Extreme	(M ha)	(%)
Water erosion	0.11 (2.0)	0.88 (15.9)	1.87 (33.5)	0.13 (2.4)	2.99	53.8
Stoniness	0.16 (2.9)	0.91 (16.3)	0.22 (3.9)	–	1.29	23.1
Flooding	–	0.02 (0.3)	–	–	0.02	0.3
Land not fit for agriculture (rocky outcrops, glaciers)	–	–	–	–	1.27	22.8
Total	0.27 (4.9)	1.81 (32.5)	2.09 (37.4)	0.13 (2.4)	5.57	100

Figures within parentheses show the percentage of the total geographical area Source: Sidhu et al. (2007a, b)

Table 12.6 Soil degradation (Mha) status of Jammu and Kashmir

Nature of soil degradation	Degree of degradation (%)				Total area	
	Slight	Moderate	Strong	Extreme	(Mha)	(%)
Water erosion	0.43 (1.9)	1.30 (5.8)	3.21 (14.5)	0.52 (2.4)	5.46	2.46
Wind erosion	0.03 (0.1)	0.78 (3.6)	0.55 (2.5)	–	1.36	17.2
Flooding waterlogging	0.09 (0.4)	0.10 (0.5)	0.01 (1.5)	–	0.20	0.9
Land not fit for agriculture (rocky outcrops, glaciers)					12.59	56.7
Total					9.6	53.3

Figures within parentheses show the percentage of the total geographical area Source: Rana et al. (2000)

12.7 Soil Health Under Changing Climate

The potential impacts on soil health resulting from climate change would be chiefly in the organic matter supply, temperature regimes, hydrology, and salinity.

The following are the major likely consequences of global climate change on soil health:

- Soil carbon levels are expected to decrease due to decreased net primary production.
- Any gains by increased plant water-use efficiency, due to elevated CO_2, are likely to be outweighed by increased carbon mineralization after episodic rainfall and reduced annual and growing season rainfall.
- The quality of soil organic matter may also shift where the more inert components of the carbon pool prevail. The residues of crops under elevated

CO_2 concentration will have higher C:N ratio, and this may reduce their rate of decomposition and nutrient supply.

- Increase of soil temperature will increase N mineralization, but its availability may decrease due to increased gaseous losses through processes such as volatilization and denitrification.
- With increased temperature and higher rate of evapotranspiration, soil may be drier. This may also result in lowering of groundwater table at some places. The melting of glaciers in the Himalayas will increase water availability in the Ganges, Brahmaputra, and their tributaries in the short run, but in the long run, the availability of water will decrease considerably.
- The soil-water balance in different parts of India may be disturbed, and the quality of groundwater along the coastal tract may be more affected due to intrusion of seawater.
- Change in rainfall volume and frequency and increasing events of intense wind may alter the severity, frequency, and extent of soil erosion. Therefore, an increased risk of soil erosion and nutrient loss due to reduced vegetation cover in combination with torrential rainfall and greater wind intensities is expected.
- Soil biodiversity and its functions are expected to change under conditions of elevated CO_2 and changed moisture and temperature regimes.

As soil biodiversity regulates nutrient dynamics and many disease risks, nutrient availability to crops and pastures could change as per the exposure to soilborne diseases.

12.8 Future Prospects

The present information may be helpful in identifying the location specific constraints and their mitigation by adopting suitable corrective measures through adaptation of soil conservation practices. The demarcation of areas under different land capability potentials will help to identify the priority areas for development of horticulture and other high-value crops.

12.9 Conclusions

The present study on soil resources of NW Himalayas focused on status of soil resources in three hilly states, viz., Himachal Pradesh, J&K, and Uttarakhand. The soil resources and land use greatly were influenced by the landform and climates. Soils on Greater Himalayas are immature and loose and subjected to soil erosion due to landslides and glacier movements. The area is denuded having scanty vegetation due to less rainfall and extreme cold. The soils developed on Lesser Himalayas are productive, and OC content is high and best suited for plantation of fruit crops. The soils in the Siwaliks region again are subjected to erosion and need proper and adequate soil conservation measures on watershed basis to achieve sustainable crop yields. Climate change may affect land use and ecology of Himalayan system resulting in shifting of cultivation of fruit crops toward higher reaches.

Acknowledgments The author is highly thankful to the staff of National Bureau of Soil Survey and Land Use Planning with whose untiring efforts the data on soil resources is generated especially in high altitudinal and difficult areas. The valuable comments of reviewers and editors to improve the quality of manuscript are also highly appreciable.

References

Ghosh BN, Meena VS, Alam NM, Dogra P, Bhattacharyya R, Sharma NK, Mishra PK (2016) Impact of conservation practices on soil aggregation and the carbon management index after seven years of maize–wheat cropping system in the Indian Himalayas. Agric Ecosyst Environ 216:247–257

Govindarajan SV (1971) Soil map of India 1:7 million. In: Kanwar JS, Raychaudhary SP (eds) Review of soil research in India

Govindarajan SV, Dutta-Biswas NR (1968) Classification of Indian soils and their mapping. Int Geogr Congr India 1:374–376

Kumar A, Bahadur I, Maurya BR, Raghuwanshi R, Meena VS, Singh DK, Dixit J (2015) Does a plant growth promoting rhizobacteria enhance agricultural sustainability? J Pure Appl Microbiol 9(1):715–724

Leather JW (1897) Reclamation of reh or usar lands. Agric Ledger Nos. 7 and 13

Mahapatra SK, Rana KPC, Sidhu GS, Walia CS, Tarsem Lal (2000) Characterization and classification of the soils of different physiographic units in the sub-humid eco-system of Kashmir region. J Indian Soc Soil Sci 48(3):572–576

Mahapatra SK, Rana KPC, Sidhu GS, Walia CS (2000) Assessment of soil degradation status of Jammu and Kashmir State for their amelioration. J Indian Soc Soil Sci 48(3):577–580

Meena RS, Yadav RS, Meena VS (2013) Heat unit efficiency of groundnut varieties in scattered planting with various fertility levels. Bioscan 8(4):1189–1192

Meena RS, Meena VS, Meena SK, Verma JP (2015a) The needs of healthy soils for a healthy world. J Clean Prod 102:560–561

Meena RS, Meena VS, Meena SK, Verma JP (2015b) Towards the plant stress mitigate the agricultural productivity: a book review. J Clean Prod 102:552–553

Meena VS, Meena SK, Verma JP, Meena RS, Ghosh BN (2015c) The needs of nutrient use efficiency for sustainable agriculture. J Clean Prod 102:562–563

Meena VS, Meena SK, Verma JP, Meena RS, Jat LK (2015d) Current trend to mitigate climate change and sustainable agriculture: a book review. J Clean Prod 102:548–549

Meena VS, Verma JP, Meena SK (2015e) Towards the current scenario of nutrient use efficiency in crop species. J Clean Prod 102:556–557

Meena RS, Bohra JS, Singh SP, Meena VS, Verma JP, Verma SK, Sihag SK (2016) Towards the prime response of manure to enhance nutrient use efficiency and soil sustainability a current need: a book review. J Clean Prod 112(1):1258–1260

NBSS & LUP (1985) Soil map of India 1:7 million scale. NBSS & LUP, Nagpur

NBSS & LUP (1992) The Soils of Anantnag and part Pulwama district (Jammu & Kashmir) for Land Use Planning. Report No. 26, NBSS & LUP Publ. 53 p

NBSS & LUP (1993) Soil survey & evaluation of ORP villages Boothgarh, Jhandupur & Majhot, district Hoshiarpur, Punjab. Report No. RCD 91, NBSS & LUP Publ.30+i-xix

NBSS & LUP (1998) Soils of Bilaspur district (Himachal Pradesh) for Land Use Planning. Report No.536 (ICAR) NBSS & LUP Publ. 119

Ram J, Singh SP, Yadav RC, Mahapatra SK, Sidhu GS, Sarkar D, Sharda VN (2013) Soil erosion in Uttrakhand. Technical bulletin no. 156, NBSS&LUP, Nagpur. (ISBN 978-81-89043-32-2)

Rana KPC, Walia CS, Sidhu GS, Singh SP, Velayutham M, Sehgal J (2000) Soils of Jammu and Kashmir for Optimising Land Use. NBSS Publ. 62b (Soils of India Series). National Bureau of Soil Survey and Land Use Planning Nagpur, India. 71+ XVI pp. + 4 sheets soil map (1:500 000 scale)

Rao RVS, Mahapatra SK, Verma TP, Sidhu GS, Rana KPC (1997) Characterization and classification of some soils of Siwalik Hills of H.P. Agropedology 7(1):14–21

Raychaudhuri SP, Agarwal RP, Dutta Biswas NR, Gupta SP, Thomas PK (1963) Soils of India. Indian Council of Agricultural Research, New Delhi

Sachdev CB, Lal Tarsem, Rana KPC, Sehgal J (1995) Soils of Haryana for optimising land use. NBSS Publ. 44b (Soils of India Series). National Bureau of Soil Survey and Land Use Planning Nagpur, India. 59 pp. + 2 sheets soil map (1:500 000 scale)

Sarma VAK, Krishnan P, Budhihal SL (1987) Laboratory manual. Tech Bull 14, NBSS & LUP. Nagpur, India

Sehgal J, Saxena RK, Vadivelu S (1987) Field manual, bulletin no. 13, NBSS & LUP, Nagpur

Sehgal J, Mandal DK, Mandal C, Vadivelu S (1992) Agro-ecological regions of India. 2nd edn. Tech Bull, NBSS & LUP, Publ.24, 130 p, NBSS &LUP, Nagpur -440010, India

Sehgal J, Singh K, Ram J, Sidhu GS, Gajbhiye (1993) Cold desert soils of India – their taxonomic problems. Agropedology 3:13–28

Sidhu GS, Sharma.JP (2008) Soil resource data for perspective land use planning in Shiwaliks of Himachal Pradesh. In "58(2): 212–220 Dr YSP UH & F, Nauni (Solan), NRM Div. of Min. of Agric. and H. P. State Department of Forests. (ISBN 978-81-211-0676-4), pp. 79–100

Sidhu GS (2012) Chapter 5: Natural soil resources of North Western part of the Indo-Gangetic Plains and their management for rainfed crops. In: Bhattacharyya T, Pal DK, Sarkar D, Wani SP (eds) Impact of climate change in soils and rainfed agriculture of tropical ecosystem. Published by Studium Press LLC, pp 91–100. ISBN 978-93-80012-40-7

Sidhu GS, Surya JN (2014) Soils of North-Western Himalayan eco-system and their land use, constraints, productivity potentials and future strategies. Agropedology 24(01):1–19

Sidhu GS, Walia CS, Lal Tarsem, Rana KPC, Sehgal J (1995) Soils of Punjab for optimising land use. NBSS Publ. 45b (Soils of India Series). National Bureau of Soil Survey and Land Use Planning Nagpur, India. 75 p. + 2 sheets soil map (1:500 000 scale)

Sidhu GS, Rana KPC, Sehgal J, Velayutham M (1997) Soils of Himachal Pradesh for optimising land use. NBSS Publ. 57b (Soils of India Series). National Bureau of Soil Survey and Land Use Planning Nagpur, India. 44pp. + 2 sheets soil map (1:500000 scale)

Sidhu GS, Rana KPC, Walia CS, Mahapatra SK, Tarsem L (1998) Characteristics and classification of dominant soils of Jammu region for land use planning. Agropedology 9(1):22–29

Sidhu GS, Walia CS, Sachdev CB, Rana KPC, Dhankar RP, Singh SP, Velayutham M (2000) Soil resource of N-W Shiwaliks for perspective land use planning. In: Mittal SP, Aggarwal RK, Samra JS (eds) Fifty years of research on sustainable resource management in Shivaliks. CSWCRTI, Research Centre Chandigarh, Chandigarh, pp 23–34

Sidhu GS, Rana KPC, Tarsem Lal, Mahapatra SK, Verma TP, Rao RVS, Ram J, Singh SK, Singh SP (2007a) Soils of Himachal Pradesh: inventory for land use planning. J Indian Soc Soil Sci 55(2):203–208

Sidhu GS, Rana KPC, Tarsem Lal, Mahapatra SK, Verma TP, Rao RVS (2007b) Soils of Himachal Pradesh–II: land capability classification and assessment of soil degradation status for suggested land use. J Indian Soc Soil Sci 55:335–339

Sidhu GS, Yadav RP, Sharma JP (2009) Soil erosion status in Himachal Pradesh for land use planning. In: Bhan Suraj, Karale RL, Singh Shamsher, Bharti VK, Bali JS (eds) Conservation farming, post publication of 2nd International Conference. Soil Conservation Society of India, New Delhi, pp 259–268

Sidhu GS, Yadav RP, Singh SP, Sharma JP, Aggarwal RK, Tiwari AK, Gajbhiye KS, Sarkar D, Sharda VN (2010) Soil erosion status of Himachal Pradesh. NBSS & LUP Publ. Bulletin No. 132 (ISBN -81-89043-11-0), 50 p

Singh HN, Prakash O (2000) Characteristics and erodibility of some degraded soils of hill region of Uttar Pradesh. Agropedology 10:101–107

Singh G, Venkataramanan C, Sastry G, Joshi BP (1990) Manual of soil and water conservation practices, Oxford. IBH Publishing Company, Pvt. Ltd, New Delhi

Singh SP, Ram J, Walia CS, Sachdev CB, Dhankar RP, Rana KPC, Sehgal J, Velayutham M, Gajbhiye KS (2004) Soils of Uttar Pradesh for optimising land use. NBSS Publ. 68b (Soils of India Series). National Bureau of Soil Survey and Land Use Planning Nagpur, India. 91 + XIII p. + 6 sheets soil map (1:500 000 scale)

Singh DK, Singh SK, Singh AK, Meena VS (2014) Impact of long term cultivation of lemon grass (Cymbopogon citratus) on post-harvest electro-chemical properties of soil. Ann Agri Bio Res 19(1):45–48

Soil Survey Staff NBSS & LUP (2002) Soils of India. NBSS Publ. 94, National Bureau of Soil Survey and Land Use Planning, Nagpur, India, pp.130 + 11 sheets soil maps

Suri VK, Sidhu GS, Kumar A (2013) Physical attributes: soil and landscape characteristics of western Himalayan region of India. In: Chopra VL (ed) Climate change and its ecological implications for the Western Himalaya. Scientific Publishers, Jodhpur, pp 1–48. ISBN 9788172338091

Verma JP, Jaiswa DK, Meena VS, Meena RS (2015a) Current need of organic farming for enhancing sustainable agriculture. J Clean Prod 102:545–547

Verma JP, Jaiswal DK, Meena VS, Kumar A, Meena RS (2015b) Issues and challenges about sustainable agriculture production for management of natural resources to sustain soil fertility and health. J Clean Prod 107:793–794

Walia CS, Rana KPC, Sidhu GS, Mahapatra SK, Tarsem L (1999) Characterization and classification of some soils of Ladakh Region for land use. Agropedology 9(1):16–21

Yadav RP, Sidhu GS (2010) Assessment of soil erosion in Himachal Pradesh. J Indian Soc Soil Sci 58(2):212–220

Towards the Sustainable Management of Problematic Soils in Northeast India

13

Rakesh Kumar and Vijay Singh Meena

Abstract

The northeast region is unique in its diversity of climate, crops and people. Jhum cultivation is prevalent in the region for food production and sustaining livelihood. One major constraint in sustainable crop production is problem soils, those soils which have serious physical and chemical limitations to crop cultivation. The North Eastern Hill (NEH) region is having monocropping with low levels of production in agriculture for sustaining the livelihood of poor tribal farmers. Reports revealed that 76.6 t/ha/year of soil is lost with practising of traditional jhum farming. Soil erosion with jhum cultivation having slope (60–70 %) in the first, second and third year as an abandoned was recorded 147, 170 and 30 t/ha/year, respectively. The region is having mostly problems of land degradation, acidity and severe erosion because of existing jhum farming by tribal farmers. The severe soil acidity problems (pH <5.5) are prevalent in the region, which accounts ~54 % of acid soils of India. Soil acidity is having a prominent role in determining the nutrient availability to crops and in many instances causes mineral stress problems also. In jhum cropping, problems of soil erosion are highly erratic from year to year depending on rainfall attributes. This region accounts ~37.1 % of total geographical area in severe threat of land degradation. The problem of waterlogging especially in tea plantation areas has also been alarmingly increasing and turned out to be an acute problem in the plains of Northeast (NE) India. Presently ~2.0 lakh ha of plantations is highly affected with waterlogging, and ~50,000 ha area is actually now available there for its cultivation. The proper management and conservation measures with respect to problematic soils are needed for the sustainable development of the region.

R. Kumar (✉)
ICAR-Research Complex for Eastern Region, Patna 797 106, Bihar, India
e-mail: rakeshbhu08@gmail.com

V.S. Meena
ICAR-Vivekanand Institute of Hill Agriculture, Almora 263 601, Uttarakhand, India

J.K. Bisht et al. (eds.), *Conservation Agriculture*,
DOI 10.1007/978-981-10-2558-7_13

Keywords
Jhum farming • Problem soils • Land degradation • Soil acidity

13.1 Introduction

The North Eastern Hill Region refers to the eastern parts of India and is known as the seven sister states, i.e. Arunachal Pradesh, Assam, Manipur, Nagaland, Mizoram, Meghalaya, Tripura and Sikkim. This region is having a population of ~39 million and covering a geographical area of ~26 m ha, which is having ~4 % and 8 % in terms of population density and area coverage of the country, respectively. Out of the total geographical area, ~28 % has an elevation >1200 m, ~18 % between 600 and 1200 m and ~11 % between 300 and 600 m amsl. The two-third of area (~72 %) of this region lies in hilly terrains. The plains area of the region mainly falls in Brahmaputra and Barak valleys, which are subjected to extreme inundation. Manipur has an area of 1853 km^2 and Tripura plains with 3500 m^2, and both are facing the problems of erosiveness. Annual rainfall of ~2450 mm is received in the region, which is mainly contributed by the southwest monsoon. The farmers of this region are mostly practising jhum and some extent of settled cultivation, which is permanent in nature. Permanent or settled cultivation is mostly adopted in plains, valleys, foothills and terrace but jhumming is prevalent in the hilly terrain of the north-eastern states except Sikkim. Tribal farmers of the region are practising crop production with an altitude of ~3600 m and having slopes of ~60 %. Most of the agricultural area is used for food grain, and rice contributes ~90 % of the total food grain acreage.

The agriculture ecosystem in the region is mainly at subsistence level, and the region as a whole faces a deficit of 1.6 MT of food grains. This region is characterised by heavy soil erosion, loss of soil fertility and deforestation causing acute environmental degradation and severe ecological imbalance. Over 600 MT of soil accompanied by 1.5 MT of nutrients (NPK) gets eroded every year due to steep slopes in the region and negligence of rainwater. A greater frequency of high-intensity rainfall can further increase these losses. Soil acidity is also one of the biggest impediments to crop productivity in the region that are prevalent. Thus, there is an urgent need to prioritise development and implementation of various ways to manage the soil, water conservation and best management practices such as rainwater harvesting, improvement of irrigation potential, soil erosion control, balanced or adequate fertilisation, soil organic matter improvement, acidity amelioration and conservation agriculture to improve climate resilience of north-eastern agriculture. The physical limitations can be managed by irrigation, drainage, mulching, manuring, tillage and soil conservation measures, viz. terracing, contouring and cover crops, whichever is appropriate. Use and disturbance of soils on steep slopes and in wetlands should be avoided because of the risk of these areas to degradation.

13.2 An Overview of the Soils of Northeast India

The northeast region is becoming more and more palpable to climate vulnerability. The maximum and mean temperature from 1901 to 2003 has amplified by 1.02 °C and 0.60 °C, respectively (Deka 2009), but at the same time temperatures are anticipated to rise by further 3–5 °C in the second third of this century (Cline 2007). The most prominent confirmation of changes in precipitation comes from considerably sinking the amount of rainfall in Cherrapunji, one of the wettest places in the world. The annual precipitation there has now fallen rudely to less than one-third of the amount in the 1860s. The recurrent deficit in rains and periodic drought in the region added validation to the climate change-induced variation in rain prototype. Predominantly notable is the planned decline (~10 %) in winter rainfall and increase (~20 %) in the rest of the year in the second half of this century, signifying risk of rising measures of high amount of rainfall in prospect, which is directly linked to the agricultural production system in the region. In spite of sharing ~8 % of geographical area and ~13 % of rains, the region contributes merely 1.5 % to the total food grain basket of the country, leading to a continuous food deficit in the region. This is predominantly owing to soil productiveness- and water accessibility-associated problems. As per latest information by the Ministry of Environment and Forests, government of India, yield of paddy, wheat and maize in the region possibly will turn down up to 35, 20 and 40 %, respectively, in the climatic change scenario of 2030. In view of the existing increase rate in atmospheric warming, coupled with climatic changes and their consequence on agriculture interrelated to water accessibility and soil richness, definite impacts on food production might be additionally stern than expected. The elevated evapotranspiration victims in atmosphere warming will exacerbate more water scarcity in the region. With river feeding by Himalayan glaciers diminishing, water strain will amplify in the Brahmaputra basin, where large residents depend on agriculture and their allied activities. The usual augmentation in water requisite for agriculture (20 km^3 in 2001 to 28 km^3 in 2025) together with topsoil acidity, a main barrier to crop production in the region, possibly will exaggerate more the rising of atmospheric CO_2 concentrations.

This prospect takes from recurrent trial and study of improved CO_2 production in soil due to better soil respiration in elevated CO_2 conditions (Gill et al. 2002). It leads to additional carbonic acid (H_2CO_3) production in soil water, which upon leaching, removes base cations from soil, thereby causing soil acidification. The doubling of distinctive CO_2 might increase acid input from H_2CO_3 leaching ~50 %. Actually, it is understood that increases in atmospheric CO_2, since the interglacial epoch, have slowly acidified soil all throughout the world (Oh and Richter 2004), and will carry on further, so a high-CO_2 atmosphere is in the prospect. The soil organic matter (SOM), a key determinant of soil health, possibly will endure quantitative and qualitative turndown in the global warming era (Kumar 2010). At elevated temperatures, there is better breakdown of SOM, mostly labile fractions, which supply nutrients with rising proportion of rather inert carbon fraction. Such reduction in soil organic matter can trim down water and nutrient holding capacities, decline buffering capacity, degrade soil aggregate firmness rendering it more liable to erosion, decrease biological activity and cause many other related changes that lead to progressive decline in soil health and their productivity. More than 600 MT of soil accompanied by 1.5 MT NPK get eroded per annum owing to vertical slopes in the region and rains negligence. A larger incidence of high amount of rains can augment more of these losses.

The risk of high temperature-induced temporary salinity, denitrification, nutrient losses owing to volatilisation and also loss of nutrients through enhanced leaching in high amounts of precipitation rainfall cannot be ignored (Kumar 2010). The yield profit of elevated CO_2 is realised only when there is no inadequacy of water and nutrient in the soil (Kumar 2011a). Nevertheless, a continuous liquidation of soil owing to serious nutrient removal (258,000, 73,000 and 179,000 t of N, P_2O_5 and K_2O), not enough fertilisation and deteriorating water accessibility strikes out every risk of elevated CO_2-induced yield remuneration agriculture in the region.

Obviously, climate change-induced aggravations in water shortage and marked decline in soil fertility are liable to be the main factors for aggressive prospects of food production in the region. Therefore, there is an urgent need to prioritise expansion and execution of different soil and water management practices, viz. rainwater harvesting, enhancement of irrigation potential, soil erosion control, balanced fertilisation, SOM improvement, soil acidity amelioration and conservation agriculture, to recover the climate resilience of the region.

National Bureau of Soil Survey and Land Use Planning, Regional Centre, Jorhat and Calcutta, surveyed the soil of mountainous states of these regions, based on a multi-tiered approach, and a soil resource map has been prepared for optimising land use and forms an important base for delineated agroclimatic zones at state levels for comparable transfer of improved technology. The soils of these states are categorised into five orders, 22 great groups and 45 subgroups through allocation of soil orders in different states (Table 13.1). Soil depth at different places varies much because of differences in physiographic position and slopes. The geological erosion normally exceeds soil formation except in forest cover. Mass movements of soil in the form of slips, glides and mud flows are common due to high amount of rains in the region.

Soil acidity affects ~50 % of the world's total potential cultivable land especially in humid regions. In India, one-third of area of total arable land is highly affected by soil acidity, and most of these soils are dominant in Northeast India. An estimate reveals that about 65 % of total area falls in extreme form of soil acidity in the region (Sharma and Singh 2002). In acid soil, potential productivity of crop is mainly affected by Al and Fe toxicity, P deficiency, low base saturation, impaired biological activity, acidity-induced fertility and other nutritional problems (Kumar et al. 2012). The intensity of acidity and its associated impact on the fertility status of soil and agricultural productivity intensified in context of climate change (Kumar 2011b). Hence, reclamation of soil acidity through eluviations and toxicity of Al and Mn and improving productivity of crop on such soil is an important aspect for enhancing food security. Northeast India is a potentially agriculturally based economic state and is having acute problems of soil acidity coupled with high

Table 13.1 Area and population of the northeast region

States	Area (km^2)	Population	Density (pop./km^2)	Forest cover (km^2) in hills
Arunachal Pradesh	87,743	109,117	13	68,045 (81.3 %)
Assam	78,438	26,638,407	340	27,714 (35.3 %)
Manipur	22,327	2,388,634	107	12,230 (69.9 %)
Meghalaya	22,429	2,306,069	103	16,926 (75.8 %)
Mizoram	22,081	891,058	40	15,584 (69.5 %)
Nagaland	16,579	1,988,636	120	17,494 (79.2 %)
Sikkim	7096	540,493	76	13,345 (80.5 %)
Tripura	10,486	3,191,168	304	3193 (45 %)
Total NE India	263,179	38,053,582	145	7065 (67.4 %)
India	3,287,263	1,027,015,247	312	–
% of India*	8.00	3.71	–	–

*Indicates % geographical area
2001 census and forest cover of hills Patiram and Ramesh (2008)

rainfall. Acidity-induced soil fertility problems coupled with no use of inorganic fertilisers are responsible for low crop productivity in this region.

The soils are formed in situ except at foothills and near rivers and stream, which are colluvial and alluvial in origin. Soils on steep slope in the upper part of hills are varying from shallow to very shallow. In the hilly terrain, soils are light in colour, highly leached and poor in bases and have low cation exchange capacity (CEC) with low-activity clays. Soils are quite rich in organic carbon with decreasing trend as depth of soil increases. Soils in valleys and thick forests are fairly rich in organic matter and are well drained. Most of acid soils possess a high amount of exchangeable Al^{3+} throughout soil column with an almost increasing trend, indicating that they still have some preserved weatherable minerals, which liberate Al^{+3} ions from edges of clay mineral through weathering processes. Jhum cultivation has resulted in heavy loss of organic matter from soil that suffers from erosion hazard. The amount of clay and soil pH tend to decrease as elevation increases with an increase in sand, SOM and CEC of soil. Steep slopes accelerate the removal of soil separates and exchangeable cations through various agencies, viz. high intensity of rainfall, hill agriculture and movement of human being and animals, etc. The clay content is invariably low in the narrow valleys, because runoff sediments do not get the chance for settling and are ultimately diverted to the main drainage (Table 13.2). However, in wide valleys, poorly permeable or poorly drained soils are heavy in texture and of colluvial or alluvial origin and formed from the adjoining hills' eroded material as noticed in Manipur valley with shale and sedimentary parent rock. The adjoining hills are acidic in reaction, while valley soils are acidic to neutral in reaction due to deposition of bases brought by surface runoff water and eroded materials.

Table 13.2 Major soil types of the NEH region

	State (% of geographical area)						
Soil order	Meghalaya	Manipur	Sikkim	Tripura	Nagaland	Mizoram	A. P.
Alfisols	3.6	0.2	–	5.0	4.8	2.6	0.3
Entisols	10.7	23.1	43.0	8.0	4.0	21.5	35.6
Inceptisols	45.7	38.4	33.4	80.0	76.0	37.3	37.3
Mollisols	–	–	23.6	–	–	38.6	–
Ultisols	40.0	36.4	–	7.0	17.2	–	14.2

Adapted from Velayutham and Bhattacharya (2000)

13.3 Effect of Shifting Cultivation

13.3.1 Change in Forest

Forest cover in the region is 141,652 km^2, which is ~54 % of geographic area besides the national average of 19.4 % (Anonymous 2001). The hilly states, i.e. Manipur and Meghalaya, have thick forest cover of 25.5 % and 25.3 %, respectively. Similarly Nagaland, Sikkim, Tripura and Mizoram have opaque forest cover of 32.5 %, 33.7 %, 33.0 % and 42.4 %, respectively. Among the seven sister states of the North Eastern Hill region, Arunachal Pradesh is having opaque forest cover of 64 %. While jhum cultivation is still prevalently practised, every year, the dense forest is rehabilitated into jhum fields.

13.3.2 Effect of Burning on Soil Fertility

Burning process linked to jhum practices has a marvellous result on the soil ecosystem of the region. The impact of fire is intense; its penalties are dependent on the amount and incidence of fire, quantity of biomass burned, onset of monsoon setting and total rains. The degree to which organic matter is changed into ash depends on a number of factors, viz. intensity or duration of fire, fuel load, moisture content in fuel, weather and topography. The burning of overground vegetation resulted to increase in pH and cations, diminishing carbon and nitrogen contents in surface soil (Ram and Ramkrishanan 1988). The rapid release of nutrients especially cations after burning in jhum farming has been reported by Kellman et al. (1985). Organic carbon of soil declined drastically after burning of crop residues which resulted to soil degradation (Meena et al. 2013, 2015a; Singh et al. 2014; Kumar et al. 2015; Ghosh et al. 2016).

13.3.3 Soil Erosion and Nutrient Loss

Soil erosion under shifting cultivation is highly erratic from year to year depending on rainfall characteristics. Studies on steep slopes (44–53 %) have indicated soil loss to the tune of 40.9 t/ha, and corresponding nutrient losses per ha are 702.9 kg of organic carbon, 63.5 kg of P and 5.9 kg of K (Ram and Singh 1993). The soil loss from hill slopes (60–79 %) under first year, second year and abandoned jhum was estimated to be 147, 170 and 30 t/ha/year, respectively. In general, tolerable soil loss value is 11.2 Mg/ha/year (5.0 t/ac/year), while it is between 5 and 12.5 Mg/ha/year (2.2 and 5.6 t/ac/year) in north-west Himalayas (Mandal et al. 2006). During the first few years of clearing of jhum, carbon and nitrogen levels decrease rapidly. According to one estimate, annual loss of topsoil, N, P and K due to shifting cultivation is 88,346, 10,669, 0.372, and 6051 thousand tonnes in the region (Sharma 1998). Singh et al. (1996) also reported nutrient loss to the tune of 6.0 MT of organic carbon, ~10 tonnes of available P and 5690 tonnes of K from the NEH region. Nutrient losses from jhum field through runoff and percolation are rather heavy during the cropping.

13.4 Problem in Soils of the NEH Region

Problem soils are those soils that have serious physical and chemical limitations to the crop cultivation, and these soils need special management for satisfactory agricultural production. Major soil struggles of the region are land degradation, acidity and erosion owing to jhumming. These problems of land degradation are rigorously exaggerated in states like Mizoram (89 %), Nagaland (60 %) and Meghalaya (54 %) mostly with water erosion and acidity, and 60 % soils of Tripura face degraded erosion, waterlogging or flooding along with soil acidity (Table 13.3).

Table 13.3 Statewise soil degradation status in north-eastern states

States	Water erosion	Waterlogging/ flooding	Soil acidity	Complex problems	Total deg. area	TGA of state	Deg. area (%)
Assam	688	37	612	876	2213	7814	28.2
A.P.	2372	176	1955	–	4503	8374	53.8
Manipur	133	111	481	227	952	2233	42.6
Meghalaya	137	07	1030	34	1208	2243	53.9
Mizoram	137	–	1050	694	1881	2108	89.2
Nagaland	390	–	127	478	995	1658	60.0
Tripura	121	191	203	113	628	1049	59.9

Gajbhiye (2006)

13.5 Soil Acidity Problem and Its Management Measures

Soil acidity is one of the most important factors affecting the agricultural production system of the region. Soil acidity dwells in ~54 % of total country soil in the region (Table 13.3). In such soil, acidity has a key role in decisive nutrient accessibility to crop plants and in many instances specific mineral stress.

13.5.1 Management Measures of Problematic Soils

13.5.1.1 Liming

Although growing of acid-tolerant plant species and varieties is suggested for acid soils, liming is recognised as a standard management practice for increasing the productivity of such soils. Liming means adding to soil any calcium-and magnesium-containing compound that is able of dipping soil acidity towards the neutrality of soils for better crop production. Lime is added to acid soils primarily for three purposes:

- To supply calcium and magnesium as a plant nutrient
- To reduce the toxicity of aluminium, manganese and iron
- To increase the pH of acid soils thereby making other plant nutrients more available

Northeast India has vast reserves of limestone totalling about 6686 MT (Table 13.4). The limestone found in the region is mostly organic and originated from nummilite shells in those areas, which was in intermediated sea condition during the Mesozoic period. These sources can be utilised for the amendment of acid soils.

Table 13.4 Extent of acid soil in the northeast region (m ha)

States	pH <5.5	pH 5.5–6.5	Total acid soil	Geo. area	Geog. area acid soil (%)
Arunachal Pradesh	6.52	0.27	6.79	7.786	81.08
Assam	2.33	2.33	4.66	7.844	59.41
Manipur	1.87	0.32	2.19	2.233	98.07
Meghalaya	1.19	1.05	2.24	2.243	99.87
Mizoram	1.27	0.78	2.05	2.208	97.20
Nagaland	1.60	0.05	1.64	1.658	99.50
Sikkim	0.60	–	0.60	0.710	84.51
Tripura	0.81	0.24	1.05	1.049	100.00
Total NE India	16.19	5.04	21.23	26.219	80.97
India	30.00	58.94	89.95	328.726	27.36

Sharma et al. (2006)

Table 13.5 Limestone reserves in Northeast

States	Reserves (MT)
Arunachal Pradesh	1503
Assam	135
Nagaland	450
Meghalaya	4665
Manipur	4.6
Sikkim	2.1
Total reserves of the region	6686

India Patiram and Ramesh (2008)

Barrenness in acid soil is a syndrome problem, which affects plant growth due to low pH. Soil acidity is the result of low availability of P to crop plants and more P fixation or less solubility of Fe and Al phosphates. These soils are often considered by low CEC, intermediate texture ranging from sandy loam to loam and low organic matter content. These soils contain relatively higher amounts of Fe and Al ions in soil solution and exchangeable H^+ and Al^{3+} ions, and a high capacity of solid phase to adsorb anions, especially P ions, and very low activities of organic biological soil fraction are found (Table 13.5).

It has been further found that in high rainfall condition of this region, optimum rates of lime applied once are sufficient for 2 years only because precipitated exchangeable Al started to reappear as cropping number increased, and after 2 years a half rate of limestone is needed for sustained productivity. In this hilly terrain, inputs are carried to distant agricultural field by head-load; therefore, furrow application at lower rates was tried, which suits their requirement and can be carried to their distant field. It has been found that applying only 250–500 kg limestone/ha in furrow per year as compared to high rates of broadcasted limestone can optimise the crop yields (Patiram 1994, 1996). The highly acidic soil (pH <5.5) would be limed to lift soil pH to 5.5 to eliminate toxicity of Al^{+3} to provide favourable soil conditions for plant root development and nutrient uptake. The ready reckoner of lime dose to get the soil pH to 5.5 for acid soil has been prepared (Table 13.6) taking into account pH, organic matter and clay content, because exchangeable Al^{+3} did not improve the certainty. This reckoner can be utilised by state soil testing laboratories to advice farmers how much lime would be needed to raise soil pH to 5.5 for getting the optimum production. There is an alternative to liming of acid soils, and that is to grow those plants which grow well in acid soils without liming such as rice, buckwheat, potato, sweet potato, ginger, turmeric, etc.

13.5.1.2 Crop Response to Liming

The direct and indirect effects of liming generally lead to increase the crop production, the magnitude of which varies widely under different soil conditions and cropping pattern. The beneficial effects of lime may be attributed to amelioration of toxic effect of exchangeable Al, Fe and Mn, enhanced availability of plant nutrients and increased efficiency of applied fertilisers. On the basis of degree of response to lime, crops have been grouped into the following three classes:

Table 13.6 Lime requirement of soil to achieve pH 5.5 as affected by pH, organic matter and clay content of soils (t/ha)

	Organic matter, 2 %				Organic matter, 4 %				Organic matter, 6 %				Organic matter, 8 %			
	Clay (%)				Clay (%)				Clay (%)				Clay (%)			
Soil pH	10	15	20	25	10	15	20	25	10	15	20	25	10	15	20	25
4.5	3.56	3.83	4.11	4.39	3.72	4.00	4.28	4.56	3.89	4.17	4.44	4.72	4.05	4.33	4.61	4.89
4.6	3.15	3.42	3.70	3.98	3.32	3.59	3.87	4.14	3.49	3.77	4.04	4.32	3.66	3.93	4.21	4.49
4.7	2.74	3.01	3.29	3.60	2.91	3.18	3.34	3.62	3.08	3.36	3.63	3.91	3.24	3.52	3.80	4.08
4.8	2.25	2.53	2.81	3.09	2.42	2.70	2.98	3.26	2.59	2.87	3.14	3.42	2.76	3.03	3.31	3.59
4.9	1.93	2.22	2.49	2.75	2.10	2.38	2.66	2.93	2.27	2.54	2.82	3.10	2.43	2.71	3.00	3.27
5.0	1.52	1.80	2.08	2.35	1.69	1.97	2.24	2.52	1.86	2.13	2.41	2.69	2.02	2.30	2.58	2.86
5.1	1.11	1.40	1.68	1.94	1.29	1.56	1.84	2.12	1.46	1.73	2.01	3.29	1.62	1.90	2.18	2.46
5.2	0.71	1.00	1.27	1.56	0.90	1.16	1.43	1.71	0.06	1.33	1.61	1.90	1.22	1.50	1.78	2.06
5.3	0.31	0.59	0.86	1.13	0.48	0.76	1.03	1.31	0.64	0.92	1.20	1.48	0.81	1.09	1.37	1.64
5.4	–	0.18	0.44	0.72	0.07	0.34	0.62	0.90	0.23	0.52	0.82	1.10	0.40	0.68	0.96	1.23

Patiram (1991)

- High-response class: pigeon pea, soybean and cotton
- Medium-response class: chickpea, lentil, pea, groundnut, maize and sorghum
- Low-response class: barley, finger millet, little millet, rice and mustard

Although rice is highly tolerant to soil acidity but responds to liming in strong acid soils, it is known that the effect of lime applied in full dose lasts for 4–5 years, which decline gradually in the last 2 years. Hence, in rainfed conditions, crops like groundnut, soybean, cotton and pigeon pea are highly responsive to liming (pH 6.5–7.0) and as such may be grown in the first 2 years of liming followed by medium-response (pH 6.0–6.4) crops such as maize, sorghum, black gram and green gram in the fifth and sixth years; when the effect of lime is greatly reduced (pH <6.0), low-responsive crops like rice, small millet and potato may be grown.

13.5.1.3 Lime Requirements

Soil testing for pH is essential for finding out approximately the correct dose of liming material. Table 13.6 provides pH values of soil and recommended doses of pure calcium carbonate ($CaCO_3$) to be used in an acre or ha of land. Some of the methods used for determining lime requirements of a soil are as follows:

13.5.1.4 Indirect Methods Based on pH and Texture

The lime requirement increases with increase in clay content. Soils with less than average amount of organic matter will require less lime, and soils with more than average amount of organic matter will require more lime than the amount indicated in Table 13.7.

- By actual and direct titration and formation of titration curves
- By using modified pH meters (also called lime meters)
- Rapid lime requirement methods involving equilibrium extraction of soil with strongly buffered system followed by determination of the exchange acidity in the extract by direct titration or by measuring the change in pH of the extract

Soils of this region are rich in sesquioxides especially as iron fixates ~80 % of added water-soluble phosphate like SSP or TSP, and thus efficiency of phosphate fertilisers is very much reduced in such soils. The use of water-insoluble phosphate like rock phosphate has been recommended in these soils for increasing efficiency of phosphatic fertilisers. The rock phosphate can be applied in three ways:

- Double dose of P_2O_5 as rock phosphate may be applied in rows.
- A mixture of rock phosphate and SSP in ratio of two-third to one-third of phosphate is recommended in these soils in rows at the time of sowing of crops.
- Rock phosphate can be applied 20–25 days before sowing, and in such case phosphate dose is the same as SSP.

Table 13.7 Lime requirement to bring soil to indicated pH

pH of soil buffer suspension (field soil sample)	Lime required to bring pH down to indicated level ($CaCO_3$) in t/acre		
	pH 6.0	pH 6.4	pH 6.8
6.5	1.8	2.2	2.5
6.4	2.3	2.7	3.1
6.3	2.7	3.2	3.7
6.2	3.1	3.7	4.2
6.1	3.5	4.2	4.8
6.0	3.9	4.7	5.4
5.9	4.4	5.2	6.0
5.8	4.8	5.7	6.5
5.7	5.2	6.2	7.1
5.6	5.6	6.7	7.7
5.5	6.0	7.2	8.3
5.4	6.5	7.7	8.9
5.3	6.9	8.2	9.4
5.2	7.4	8.4	10.0
5.1	7.8	9.1	10.6
5.0	8.2	9.6	11.2

Methods of Lime Application

The following general procedure should be observed in applying lime:

- Lime should be applied before ploughing or applied on ploughed land and then disced or mixed into the soil.
- In case of large quantity of lime, it is advisable to apply small quantities two to three times rather than giving one heavy application.
- On strong acid soils, when large doses of lime are required, half the quantity should be applied before ploughing and remaining half doses applied and disced in after ploughing.
- When only one application of lime is to be given, it should never be given before ploughing with a furrow-turning plough.

13.5.1.5 Alternative to Liming

An acceptable alternative to liming are:

- To grow acid-tolerant crops and cultivars laying emphasis on diversification.
- Agroforestry for resource management utilising lowest one-third of hill slopes put to the agriculture crops, middle one-third of hill slopes to hortipasture and top one-third to agroforestry.
- The importance of an integrated nutrient management system (INMS) applied to acid soils, such as organic manure, crop residues like rice straw, rhizobia, blue green algae and azotobacter, cannot be underestimated.

- Organic matter complexes aluminium and manganese and thereby decreases their amount in soil solution. Where available, a large amount of organic matter, e.g. easily decomposable plant residues, could be incorporated into acid soils.
- Humified material should decrease the concentration of aluminium and manganese in soil solution because of formation of organo-complexes.

Application of lime with farmyard manure is suggested to augment phyto-accessibility of essential nutrient and improve acidity-induced fertility problems in acid soil (Kumar et al. 2012; Meena et al. 2015b, c, 2016; Verma et al. 2015b). Hence, a field trial was conducted to study the effect of liming on production and economics of maize by Kumar et al. (2012) on sandy loam soil at Medziphema during two consecutive *rabi* seasons of 2010–2012. The results revealed that application of 0.6 t lime ha^{-1} in furrow recorded significantly higher grain yield (3.91 t/ha) and stover yield (4.25 t/ha). This might be due to the effect of liming in lowering of exchangeable Al^{3+} and H^{+} and to an increase in Ca, Mg, CEC and pH. Chatterjee et al. (2005) also reported that incorporation of lime at 10, 50 and 100 % lime requirement of soil in furrows application increased dry pod and haulm yield in groundnut over treatment receiving recommended dose of NPK (25:50:50 kg/ha) alone, thereby signifying response to liming. It may conclude that application of lime in furrow (at 0.6 t/ha) proved the best management practice to achieve the sustainable production of maize under the foothill condition of Nagaland.

Lime application and integrated nutrient management is often recommended to increase the crop productivity on acid soils. To ascertain the individual and synergistic effects of lime, NPK and FYM application on maize, a field experiment was undertaken on an acid Alfisol (pH 4.6) of Meghalaya by Kumar et al. (2012), and results indicated that application of recommended dose of NPK (80, 60 and 40 kg/ha of N, P_2O_5 and K_2O) resulted in 53 % yield improvement, while liming at 300 kg/ha (furrow) caused 32 % yield increase over control. Combined application of NPK + lime resulted in 147 % yield increase, while application of FYM at 5 t/ha along with NPK + lime further boosted the yield improvement up to 291 % over control. Results of this study suggest that liming along with INM practices, if adopted properly, can lead to more than three fold increase in maize productivity on acid soils of Meghalaya and other north-eastern states of India with similar soils.

13.6 Soil Erosion and Land Degradation and Its Management Measures

Nearly 37 % of total geographical area, which is almost double than national average of 20.17 % (Anonymous 2000) in Northeast India, is under the threat of land degradation, where erosion is a major land degradation process. In NEH region, there is high percentage of area under wasteland due to jhum farming, which is caused by degraded forest. On such lands, soil erosion through running water is the main agent of land degradation. The existing community or private land system has been excessively exploited for survival and realisation of short-term

Table 13.8 Effect of liming on yield and economics of *rabi* maize

Treatment	Grain yield (t/ha)	Stover yield (t/ha)	Gross return ($\times 10^{3\prime}$/ha)	Net return ($\times 10^{3\prime}$/ha)	Benefit: cost ratio
Liming in furrow application					
Control	2.61	3.41	28.95	18.53	1.35
0.2	2.99	3.74	33.03	21.61	1.44
0.4	3.44	4.00	37.70	25.28	1.52
0.6	3.91	4.25	42.56	29.14	1.59
SEm±	0.04	0.05	0.36	0.34	0.02
CD ($P = 0.05$)	0.11	0.14	1.10	1.06	0.06

Kumar et al. (2014)

objective without taking care of soil health (Table 13.8). The major crop land areas of hill agriculture are eroding faster than natural processes and have been significantly degraded (Meena et al. 2015d, e; Verma et al. 2015a).

Jhum cultivation is the most customary and main land use system in the northeast region. This kind of cultivation is confined to the village border and done repeatedly after 2–3 years, then the cultivated area is left vacant, and a new site is selected to be cultivated again. But this farming became indefensible now a days mainly owing to the increase in population pressure that led to amplify food demand. The practices of jhum cycle on the same site, which were extensive to 20–30 years in the past, have now been condensed to 3–6 years (Borthakur 1992). The land degradation problems are much severe in Manipur, Nagaland and Sikkim, where >50 % of total geographical area is distinct as wastelands. Of various degradation types, water erosion, reduced infiltration, acidification, nutrient leaching, burning of vegetation and decline in vegetative cover and biodiversity are important in context to the region. Soil erosion in jhum cultivation is extremely unpredictable from year to year and it depends on rainfall type.

The further studies on steep slopes (44–53 %) have noticed that soil loss is to the tune of 40.9 t/ha and nutrient losses are 702.9 kg of organic carbon, 63.5 kg of P and 5.9 kg of K, respectively, on per ha basis (Ram and Singh 1993). The loss of soil from hill slopes (60–79 %) in the first year, second year and abandoned jhum was computed to 147, 170, and 30 t/ha/year, respectively. Generally, average soil loss value is 11.2 Mg/ha/year (5.0 t/ac/year), while it is between 5 and 12.5 Mg/ha/year (2.2 and 5.6 t/ac/year) in NW Himalayas (Mandal et al. 2006). It is to be reported that annual loss of topsoil, N, P and K due to jhum cultivation is 88,346, 10,669, 0.372 and 6051 thousand tonnes in the northeast region (Sharma 1998). Singh et al. (1996) also revealed the loss of nutrients to the tune of 6.0 MT of organic carbon, 9.7 t of available P and 5690 t of K in the region.

13.7 Long-Term Strategies for Step-Up of Problematic Soil in the NEH Region

Main arable area of total geographical area in the region is at risk of land degradation, where erosion is a key land derivative practice. With huge anxiety of poor soil health and stern land degradation, there is a need of a practicable alternative for eco-renovation and upholding of soil resources, which could maintain long-term soil productivity in the region. The following strategies, suitable for altered land condition, elevation and topography existing in the region, are discussed herewith.

13.7.1 Multipurpose Trees (MPTs)

This trees form an integral part of various agroforestry interventions in an agricultural production system with special reference to the northeast region. These components also furnish several outputs, viz. fuel, fodder, timber and miscellaneous products, which help in soil health improvement. The farmers in the region put together different tree species in various land uses, though the main concern of suitable species varies from place to place within a state based on racial diversity and food habits of tribal communities. There are ~40 capable species that are cultivated in the tropical and subtropical regions and 30 in the temperate zone, which fit well in different farming systems.

In addition, 28 bamboo species and 2 genera of cane also were found prominently in various agroforestry programmes. However, tree thickness is also a decisive factor on sloppy land. Usually, optimal tree compactness in case of agri-horticulture system is 400 trees/ha, while in agri-silviculture, it is 200 plants/ha, so as to minimise the effect of shade and biochemical interactions on growth and production of agricultural production system (Bhatt 2003; Umashankar 2005). The long-term consequence of different multipurpose tree species on soil physical performance has been studied by Saha et al. (2007). The MPTs with higher surface cover, steady leaf litter fall and wide spread root system augmented soil organic carbon (SOC) by 96.2 %, porosity by 10.9 %, aggregate stability by 24 %, soil moisture by 33.2 %, and altogether abridged bulk density and erosion ratio by 15.9 % and 39.5 %, respectively. Among MPTs experienced, *P. kesiya*, *M. oblonga* and *Alnus nepalensis* were established appropriately as bio-ameliorant in hilly topography with regard to organic matter rise through occurrence of leaf litter, better soil aggregation, transmissivity and infiltrability through an extensive root system and improved soil conservation with constant surface cover with leaf biomass. Such upgrading in soil hydro-physical properties in a tree-based farming system has a direct bearing on long-standing sustainability in agricultural production system in hills.

13.7.2 Agroforestry Interventions in Degraded Land

The northeast region is having an extremely high rate of land degradation. It is to be reported that 7.85 mha of areas are degraded, which necessitate remedy with establishing different agroforestry models (Anonymous 2000). These systems today become recognised ways of integrated land management not only for renewable resource production but also for ecological concern. It represents incorporation of agriculture and forestry to boost the production of a farming system.

13.7.2.1 Soil Fertility Build-Up

Study reveals that organic carbon, available P and exchangeable cations content in surface soil ranged in between 2.0 % and 2.5 %, 10.4–13.2 ppm and 5.9–8.4 cmol (p +)/kg, respectively, in jackfruit-based AFS, while 1.5–1.8 % organic carbon, 3.8–6.7 ppm available P and 3.9–5.9 cmol (p+)/kg total cations were noted in areca nut/khasi mandarin-based farming system approach (Singh and Dhyani 1995). In a study of the Farming System Research Project carried out at Barapani, effect of different AFS like silvipasture, silvi-horticulture and agri-horti-silvipasture was evaluated for long-term studies with respect to soil fertility indices by Majumdar et al. (2002). The natural fallow and neglected jhum land at Umiam were taken for comparison. The soil organic carbon content amplified in all the agroforestry systems including natural fallows. The highest build-up of these cations was noted with agri-horti-silvipasture, silvi-horticulture and subsequently natural fallow and silvi-pastoral agroforestry. However, increase of exchangeable K was higher in silvi-horticulture followed by agri-horti-silvipasture. Available N, P and S were recorded maximum in agri-horti-silvipastoral and silvi-horticulture as compared to natural fallow and silvipastoral systems.

13.7.2.2 Soil Physical Health

The effect of different land use systems on soil physical properties showed that the highest decline in bulk density over jhum was observed in forest (17.6 %) followed by agri-horti-silvipastoral (14.3 %), livestock-based (13.4 %), natural fallow and agriculture systems (12.6 %). A higher percentage of macroaggregates (54.5 %), soil organic C content (2.95 %) and biotic movement were noted in the forest ecosystem. The soil biota influences on its properties all the way through formation of stable aggregates, development of organo-mineral complexes by recuperating macroporosity and continuity of pores from surface to subsoil, which finally augment water transmission and reduce runoff.

The maximum transmission and storage pore volume together with lower value of residual pores coupled with modified land use systems as compared to jhum cultivation were therefore a sign of maintaining pore geometry of soil in these systems. The improved soil aggregation in natural forest, multi-storeyed and silvi-hortipastoral systems keeping rigorous vegetative cover round the year could be attributed to the response of higher proportion of organic matter, clay content and high amount of Al and Fe oxides in soil.

13.7.2.3 Soil and Water Conservation

Several latent farming systems, viz. crop production on bench terraces, horticulture and agri-horti-silvipasture, have been evaluated by Satapathy (2005) at Barapani for long-standing runoff, soil and nutrient losses, production behaviour and biotic and abiotic changes. Results revealed that diverse land use systems with suitable soil management measures, i.e. bench terraces and contour trenches, were the most efficient in retaining 90–100 % rains and simulated special effects of natural forest. The watersheds have continuous rivulet flow, which generated a base flow level of 70–90 % of total water productivity. Watershed treated with jhum farming yielded maximum peak runoff, while one left intact with natural vegetation gave lowest peak runoff, and hence results showed that agroforestry and mixed land use systems were found to be most efficient in conserved soil moisture and significantly abridged peak runoff. The low erosion percentage values in silvi-horti-pastoral and multi-storeyed AFS (3.07 and 3.06, respectively) indicated that these systems were most appropriate for soil and water conservation in the region (Saha et al. 2005). This could be attributed to consequence of profound litter fall, which may have augmented cohesiveness in soil after decomposition and bind soil strongly in the lower horizon by their deep roots.

13.7.2.4 Soil Carbon Sequestration Potential

The consideration of soil feature is a valuable tool in decisive sustainability and environmental impact of agricultural production system. Soil excellence in different agroecosystems using soil organic carbon (SOC) and soil microbial C (SMBC) as soil features suggest jhum-cultured areas had the lowest SMBC value of 192 mg/kg, while soil in *Michelia oblonga* plantation had the significantly highest value of 478 mg/kg. The amount of SMBC to total soil organic carbon was in the range of 0.76–1.96 % across all systems. MPTs, viz. *P. kesiya*, *A. nepalensis*, *P. roxburghii*, *M. oblonga* and *G. arborea*, with greater surface cover, regular leaf litter fall and widespread root systems increased SOC by 96.2 %, which helped improve the aggregate stability by 24 %, enhance the available soil moisture by 33.2 % and in turn abridge the soil erosion by 39.5 % (Saha et al. 2007, 2010).

13.7.3 Resource Conservation Techniques (RCTs)

13.7.3.1 Conservation Tillage

It is a system of management in which crop residue is left on the soil surface with minimum disturbance. Stubble mulch or reduced tillage, no tillage and direct drill are the mechanisms of conservation tillage. The main objectives are:

i. To leave adequate crop residues on soil surface for managing the water and wind erosion
ii. To protect the soil and water
iii. To reduce the energy use in agricultural production system (Ghosh et al. 2010)

Some of the conservation tillage practices followed in hill ecosystems are discussed here.

13.7.3.2 In Situ Residue Management

Little native soil N and P together with the lack of concern of farmers to the use of inorganic fertiliser are key constraints to restrictive crop production in the region. Production and nutrient recycling prospective in rice-vegetables cropping system with low-input in situ residue management was evaluated at Umiam. After harvesting of paddy, five vegetable crops, viz. tomato, potato, frenchbean, cabbage and carrot, was grown without using any chemical inputs. Only the economic parts of crop were harvested, and left-out portions including weed residues are chopped and integrated into the soil. Significant amounts of nutrients were used during in situ weed biomass inclusion.

13.7.3.3 Incorporation of Local Grass

In long-standing effects of various locally accessible grasses and weeds on soil hydro-physical properties, rice yield during 5-year field trial at Barapani, Meghalaya, resulted to integration of jungle grass (*Ambrosia spp.*) in puddle rice improved soil organic carbon (SOC) by 21.1 % and stability of microaggregates, moisture retention capacity and infiltration rate by 82.5 %, 10 % and 31.3 %, respectively, and soil bulk density diminished by 12.6 % (Saha and Mishra 2009). However, locally accessible jungle grasses are similarly good as an organic amendment, which would ease the difficulty of removal of these grasses in peak monsoon. Hence, these organic sources might provide as an option to FYM and have a remarkable outcome on continuing agricultural production system in the region.

13.7.3.4 Zero Tillage (ZT)

This practice in rice-based cropping system improves physical properties of soil, viz. soil structure, increases relative proportion of bio-channels and macrospores and decreases vulnerability to crusting. It has been noticed that bulk density of soil decreased by ~25 %; total porosity and soil aggregates enlarged by 29 % and 32 %, respectively, over usual tillage. It also augments SOC by 12.5 %, available P by 14.3 % and K by 29.4 % over conventional tillage. ZT practice saved ~20 energy and fertiliser requirements as compared to usual tillage devoid of jeopardising output (Mishra et al. 2005).

13.7.3.5 Integrated Plant Nutrient Supply (IPNS)

Integrated use of a balanced fertiliser blended with lime and organic manure improved soil health for sustainable crop production in the region. The report revealed that adding of NPK fertilisers together with organic manure, lime and bio-fertilisers has improved SOC content, aggregate stability, moisture retention capacity and infiltration rate of soil although tumbling bulk density (Saha et al. 2010).

13.7.3.6 Pastoral Development

The resource management and ecological friendly production strategies are attractive for agrarian economies. The grass cover is a key factor in recuperating soil physico-chemical health by assuring constant addition of organic matter, therefore tumbling surface runoff and soil erosion. Some capable perennial grasses, viz. *Setaria*, Congo signal, Guinea, Napier and broom grass, are available for maintaining grass cover and managing soil health properly. Ghosh et al. (2010) reported that long-term (15-year) grass covers appreciably augmented SOC, with *Setaria* having the highest SOC content (2.24 %). Likewise, soil microbial biomass carbon, soil aggregation and infiltration rate in a mixture of grass covers were also noted to be high as compared to a plot without grass covers.

13.7.3.7 Hedgerow Intercropping

Growing of a variety of hedge-row species even in a jhum field could be a viable alternative for them as this species has a short growth period. Hedgerows alone abridged soil loss by 94 % and runoff by 78 %. As soon as twigs and tender stems of hedge plants are used for mulch, it conserved ~83 % of oil and 42 % of rains. In an experimental trial conducted at Changki, Nagaland, soil loss was abridged by 22 % with integration of hedgerow species in jhum plots as compared to traditional jhum site (38.1 t/ha/year). Therefore, contour hedgerow technology provides an alternative farming on hill slopes on a sustainable basis. The raising of nitrogen-fixing hedgerow species on field bunds helps in build-up of atmospheric N and decreases leaching losses. Their dynamic root system mobilises P, K and other micronutrients. ICAR Research Complex for NEH Region, Barapani, screened a variety of hedgerow species for plantation, and *Cajanus cajan*, *Crotalaria tetragona*, *Desmodium rensonii*, *Flemingia macrophylla*, etc. have been found the most appropriate for growing in the region. Nutrient content of N, P and K in the litters of hedgerow species ranged from 3.23 to 3.86 %, 0.32–0.81 % and 1.26–1.67 %, respectively. The results also reported that total leaf biomass creation on dry weight basis after 1 year was found maximum in *C. tetragona* (22.98 q/ha) followed by *G. maculata* (20.75 q/ha), *I. tinctoria* (16.99 q/ha), and *T. candida* (15.3 q/ha). Among the hedgerow species, *C. tetragona* enriched soil fertility more resourcefully as it accumulated high amounts of N, P and K (79.7, 11.1 and 37.4 kg/ha) during its leaf incorporation. However, recycling of bases in debris of hedgerow could be viable to neutralise the acidification (Laxminarayana et al. 2005).

13.7.4 Organic Farming

Organic farming is primarily in operation in areas under shifting cultivation and traditional land use systems in Northeast India. Nearly 57.1 % of total geographical area (TGA) in India is under threat of land degradation mainly by water erosion. On an average, 37.1 % TGA in Northeast India is under the degraded state.

13.7.5 Management Measures

Soil conservation measures are the best tools for managing erosion and land degradation. The practical methods of soil and water conservation fall into two important classes, biological and mechanical. The basis of all conservation measures should be patterned as closely as possible after nature's own way of conservation. The more emphasis should, therefore, be laid on biological method of erosion control, which is discussed below:

13.7.5.1 Contour Farming

A contour line connects all points of the same level and it runs across the slope of land. Contour farming consists in doing all the tillage operation including sowing of seeds and planting in rows along contour lines instead of straight lines or in rows running up and down the slope. All terrace lands should be formed on contour. For slopes up to 1–2 %, contour farming alone is effective. On steeper slopes and gullied lands, it ought to be supported by terracing, strip cropping and other measures to be effective.

13.7.5.2 Contour Tillage

It is an essential gradient of conservation farming. It helps in creating obstructions to flow of water at every furrow and results in a more uniformed distribution of water, more infiltration of water and less runoff and erosion. The conservation tillage practices such as no tillage, reduced tillage, mulch tillage, conventional tillage and traditional tillage can also be practised where ~30 % of soil is covered by residues after planting to reduce water erosion or where wind erosion is a primary concern.

13.7.5.3 Mulching

Crop residue mulch is one of the most effective measures against soil erosion. Mulch prevents direct impact of raindrops on soil aggregates, maintains pore space continuity and high infiltration rate and enhances crop growth to provide an early ground cover through improvement in soil temperature and moisture regimes and other physico-chemical properties. Tillage that leads to surface soil clod and mulched with crop residue (stubble) is an effective measure to minimise the soil erosion and land degradation.

13.7.5.4 Water Harvesting and Runoff Recycling

The portion of rainfall, which moves down to a stream, channel, river or ocean as surface or sub-surface flow is termed 'runoff'. The runoff can be harvested from the field, stored in a pond or tank and recycled for life-saving or supplemental irrigation. The technology of water harvesting consists of collecting and storing water from land treated to increase runoff. Runoff from catchment can be increased by building contour ditches to collect hillside runoff, by clearing vegetation and smoothening land surface, by treating soil surface to reduce infiltration rate and by covering soil surface with impermeable sheeting. Rainfall harvesting consists of in situ and ex situ measures, whereas in in situ measures, infiltration rate is augmented by deep ploughing, profile modification and vertical and mulching. In case of ex situ water harvesting technology, it includes rooftop water collection, dugout ponds or storage tanks, gully control structures or check dams, percolating tanks or ponds and sub-surface dams or barriers.

13.7.5.5 Choice of Crops and Cropping Sequence

Raising of close growing crops like grasses and good canopy-producing crops like legumes controls soil erosion and improves soil structure due to good canopy, higher root mass and root secretions helping in binding soil particles. Cropping sequences that provide early and continuous ground cover permit less erosion. A suitable cover crop such as pigeon pea (*Cajanus cajan*), stylo (*Stylosanthes guianensis*) and velvet beans (*Psophocarpus palustris*) used in crop rotation helps to improve soil properties.

13.7.5.6 Strip Cropping

It can also be practised wherein crops can be grown in relatively narrow strips across a slope of land in such a way that rows of erosion-permitting crops are put alternatively with rows of erosion-resisting crops. This practice reduces the velocity of runoff water, and silt gets deposited within erosion-resisting crops. The dense foliage of these strips prevents detachment of soil by splash caused by raindrops.

13.7.5.7 Nutrient Management

Organic manures and green manures not only supply plant food like fertilisers but help in improving the soil's physical condition. Residues of crops spread over in fields after harvesting have a remarkable impact on soil and water management. The maintenance of residues is a very useful tool for the decline of soil erosion.

13.7.5.8 Organic Sources

Organic sources of nutrients, if pooled collectively, can provide 13.1 kg N/ha, 7.18 kg P/ha and 7.34 kg K/ha in the region. Micro-nutrients made available from organic sources may be in adequate amounts. A considerable quantity of potash can be obtained from crop residues if it was managed properly and added in the soils. Similarly if the crop seed is treated with bio-fertilisers efficiently, it can enhance crop yield of 5–30 %. Another approach is the use of vermicompost, which is made up of rural wastes, and this holds a vast promise in extenuating nutrient starvation of soils. Soil amelioration with the use of limestone deposit accessible in the region can be brought in use. Ultimately watershed-based technology with proper soil and water conservation measures can be useful avenue to rear the soil health for sustainable food production (Saha et al. 2012).

13.7.5.9 Growing Grasses

Grasses are considered to be one of the best tools in soil conservation. It acts in same way as legumes in controlling soil and water loss. Strips of grasses laid in between growing field crops are highly beneficial in conserving the soil and soil moisture and improving the structure and productive capacity of soil. Some of the grasses that can be grown are tropical grasses, such as *Pennisetum* grasses, several species of vetiver, lemon grass, citronella, Napier grasses, Rhodes grass and tropical panic grass, and temperate grasses, such as switchgrass, wheat grass, pampas grass, bamboos, giant reed and ribbon grass.

13.7.5.10 Agroforestry

Afforestation makes barren lands into fertile land by minimising erosion and restoration of nutrient account. In the initial stage, harshly eroded land needed full forest cover and protection from free grazing systems. The local perennial tall tufted grass species Amliso (*Thysanolaena agrostis*) may be capable to recoup degraded land, terrace risers, waterways and land between trees and vulnerable points and to provide fodder to animals in winter and spikes for brooms for livelihood improvement of tribal community of the region. The region has an extremely high rate of land degradation, which needs to be cured through agroforestry models (Anonymous 2000). Agroforestry has been a long-standing custom in the region, where cereals, rhizomes, pineapple, coffee, tea, spices and vegetable crops are grown along with fruits and other trees such as pine, pear, plum, areca nut, mandarin, guava, coconut, jackfruit, banana and large cardamom with alder trees.

13.7.6 Waterlogging and Its Management Measures

Another major problem that results in crop loss in the hills of Northeast India is soil waterlogging. The reasons of waterlogging are:

i. Human interference: The growing human population, which affects the drainage system.
ii. Riverine flood waterlogging: In the rainy season, the flood comes to the nearby land from the river having excess floodwater. Due to rising water level in nearby river Brahmaputra and its tributary and subtributary, waterlogging occurs.
iii. Excessive rainfall: Seasonal waterlogging due to runoff water accumulating in the lowlands and depression in the rainy season.
iv. Perennial waterlogging: Deep water swamp accumulates rainwater; runoff water or seepage water of canal causes perennial waterlogging.

Due to various factors, the problem of waterlogging especially in tea plantation areas has been alarmingly increasing and has turned out to be an acute problem in the plains of Northeast India. At present, ~2 lakh ha of plantations endured with waterlogging and ~50,000 ha of lands are existing for cultivation (Bordoloi 1993). Excessive rainfall can cause waterlogging of soil, and under this condition soil become hypoxic or anoxic. Either localised accumulation of water or high water table to root zone (~90 cm depth) for longer period of time causes stunted growth, defoliation or death of plants. In Brahmaputra and Barak valleys in Northeast India, major tea-growing area has been suffering from a rising situation of waterlogging. The best management measure for waterlogging is to provide proper drainage. It provides desirable environment in crop zone by removing excess salts and water.

The drainage system starts at field level and ends at outfall. Though field drainage system is supposed to be constructed and maintained by farmers, disposal system and outlets or outfalls are constructed and maintained by the government. The accomplishment of surface drainage of agricultural lands includes land farming, land smoothening, land grading or levelling, bedding system, open ditches and levels. Sub-surface drainage, i.e. bio-drainage including tree species of *Eucalyptus*, poplar and *Acacia*, can be helpful, which involves growing certain plants that habitually draw their main water supply directly from groundwater or capillary fringe just below it.

13.8 Future Prospects

The northeast region includes eight states covering ~8 % of total geographical area and 3.8 % population density of the country. Two-third of area of this region lies in hills of varied elevation from subtropical to alpine zones. The productivity of agricultural production system particularly in rainfed ecosystem of the region is regulated with population growth in spite of rich natural resources. Most of the soils of this region are highly acidic, which also are suffering from Al and Fe toxicities and deficiencies of phosphorus accompanied by low use of fertilisers. These can be addressed through proper soil fertility management and field trials on farmers field through frontline demonstration of diverse extension agencies involved for increasing the agricultural productivity. The balanced nutrition of crops, which depends upon soil productivity maintenance either through inorganic fertilisers or organic sources of nutrients taking into environmental degradation and use of soil amendments, can increase the productivity through farmer's participatory approach with the use of fertilisers. However, jhum cultivation is measured as a foremost basis of rural economy in the region and will stay as an imperative one as it is associated with socio-economic and cultural systems of the poor tribal community. Due to this practice, land degradation will persist in years to come and may reach out of control, if proper attention is not taken immediately. Therefore, to reduce the degradation level, a complete forest policy is required as a continuing approach in the region for sustainability and increase of food, fuel, fodder and timber. In this trend, agroforestry together with viable resource conservation approaches needs to be strengthened for long-standing sustainable agricultural production and environmental conservations in delicate ecosystem, which will contribute to improve food security and income generation for resource-poor tribal farmers. To achieve these desired goals, a convergent approach is needed (Mukherjee et al. 2012; Mukherjee and Maity 2015).

Integrated farming system has emerged as a growingly accepted, single-window and sound approach for congruent concurrently joint management of land, water, vegetation, livestock and human resources. Actually this strategy was developed for hill areas, which reduced risk of land degradation, turned soils into prolific potential and trimmed down the risk of environmental degradation. Moreover, these interventions have a tree crop with a good quality of leaf litter and root-binding ability to diminish erodibility from rainfall or runoff and improve the physico-chemical condition. The effort should also be prepared to manage soil health through addition of organic materials. In addition to these, other ways like agroforestry, growing acid-tolerant crops and INM system are an alternative to liming. Soil conservation measures are found to be the best tools for managing erosion and degradation of land, and biological measures such as contour farming, contour and conservation tillage, agroforestry, watershed management, etc. are found to be of greater significance. Also for waterlogging, which is prevalent in the plains of NEH region, introduction of proper drainage and development of resistant crop can help in mediating the effects of waterlogging.

In addition to agricultural crop production, the varied physiography and elevation offer the scope of horticultural crops according to the suitability of soils:

- Higher altitudinal zones of Greater Himalayas (Sikkim and Arunachal Pradesh) are usually covered with alpine, sub-alpine and temperate (coniferous) forests. Soils of side slopes of hills, which are acidic and rich in organic matter together with high rainfall and cold climate, favour plantation of temperate fruits like apple, pear, plum and walnut. Some areas have good potential for spice cultivation.
- Soils in the lower Himalayas, Siwalik hill and Purvanchal hills have the potential for sub-tropical fruits like orange, lemon and jackfruit for social forestry. The foothill soils are suitable for pineapple, banana, coffee and tea. In Tripura, soils on moderately sloping to steep slopes with land capability classes VIe and IIIe are under different plantation crops like coffee and rubber. In Meghalaya, main horticultural crops are citrus fruits and khasi mandarin. Pineapple and banana are successfully grown in the lower slopes of hills.
- Foothills and undulating upland of Assam are highly suitable for tea plantation. Besides, various tropical fruits, i.e. orange, guava, litchi, jackfruit and lemon, are also grown in these soils.
- Alluvial plains of Assam mainly central and lower Brahmaputra valley and narrow valleys in states are suitable for a wide range of agriculture. If these valleys can be made productive at par with the northern plains of India, this can become not only self-sufficient in food production but also export to other regions.

13.9 Conclusions

The problem of soil acidity affects ~50 % of the world's potentially cultivable land, mostly in humid tropic regions; in India, one-third of cultivated land is exaggerated with soil acidity, and mostly these soils are concerted in the northeast region of India. The crop productivity on such soils is inhibited by Al and Fe toxicity, P deficiency, low base saturation, impaired biological activity, related acidity-induced soil fertility and nutritional problems. Problems of soil acidity management and improving crop productivity on such soils are therefore, imperative for enhancing food safety worldwide. Acidity-induced soil fertility exertion together with usually less application of inorganic nutrients is often accountable for poor agricultural production system in the region. However, liming along with INM is repeatedly suggested to augment phyto-accessibility of essential nutrients and improve acidity-induced fertility restriction. It is therefore, very important to make assured yield remuneration of being as well as collective appliance of lime, inorganic nutrient and organic manure in a meticulous edapho-climatic condition of the northeast region of India.

Acknowledgements The author would like to thank the Indian Council Agricultural Research, Research Complex for North Eastern Hill Region, Jharnapani, Medziphema, for providing a platform to bring out this book chapter.

References

Anonymous (2000) *Wastelands Atlas of India*. Ministry of Rural Development, Government of India and National Remote Sensing Agency, Hyderabad, India

Anonymous (2001) *State of forest report*. Forest survey of India. Ministry of Environment and Forests, Government of India

Bhatt BP (2003) Agroforestry for sustainable mountain development in NEH region. In: Rawat MSS (ed) Central Himalaya environment & development: potentials, actions and challenges. Transmedia Publisher, Uttaranchal, pp 206–223

Bordoloi PK (1993) Drainage in tea of north east India: A Review. Proc. 58th Ann. Res. & Dev, Central Board of Irrigation and Power, Apr 19–22, Bangalore, pp 143–148

Borthakur DN (1992) Agriculture of the North Eastern region with special reference to hill agriculture. Beecee Prakashan, Guwahati

Chatterjee A, Doshni AAK, Talashikar SC, Mehta VB (2005) Effect of lime on yield, quality and nutrient uptake by six groundnut varieties and properties of an Alfisol. J Indian Soc Soil Sci 53 (3):128–132

Cline WR (2007) Global warming and agriculture: impact estimates by country. Peterson Institute for International Economics, Washington, DC

Deka RL (2009) ISPRS archives XXXVIII-8/W3. Workshop proceedings: impact of climate change on agriculture

Gajbhiye KS (2006) Land utilization. In: Hand book of agriculture, 5th edn (revised and expanded). Indian Council of Agricultural Research, New Delhi

Ghosh PK, Das A, Saha R (2010) Conservation agriculture towards achieving food security in North East India. Curr Sci 99(7):915–921

Ghosh BN, Meena VS, Alam NM, Dogra P, Bhattacharyya R, Sharma NK, Mishra PK (2016) Impact of conservation practices on soil aggregation and carbon management index after seven years of maize–wheat cropping system in Indian Himalayas. Agric Ecosyst Environ 216:247–257

Gill RA, Polley HW, Johnson HB, Anderson LJ, Maherali H, Jackson RB (2002) Non-linear grassland response to past and future atmospheric CO_2. Nature 417:279–282

Kellman M, Miyanishi K, Hiebert P (1985) Nutrient retention by savanna ecosystems II. Retention after fire. J Ecol 73(3):953–962

Kumar M (2010) Soil responds to climate change: is soil science in India responding to? Curr Sci 99:892–894

Kumar M (2011a) North East India: soil and water management imperatives for food security in a changing climate. Curr Sci 101:1119

Kumar M (2011b) Evidences, projections and potential impacts of climate change on food production in northeast India. Indian J Hill Farming 24(2):1–10

Kumar M, Khan MH, Singh P, Ngachan SV, Rajkhowa DJ, Kumar A, Devi MH (2012) Variable lime requirement based on differences in organic matter content of iso-acidic soils. Indian J Hill Farming 25(1):26–30

Kumar R, Deka BC, Baishya LK, Pandey A, Ngachan SV (2012) Effect of liming, mulching and INM on production potential of *rabi* maize under rainfed condition of Nagaland. Extended summaries Vol. 2: 3rd International Agronomy Congress, Nov 26–30, New Delhi, India

Kumar A, Bahadur I, Maurya BR, Raghuwanshi R, Meena VS, Singh DK, Dixit J (2015) Does a plant growth promoting rhizobacteria enhance agricultural sustainability? J Pure Appl Microbiol 9(1):715–724

Laxminarayana K, Bhatt BP, Tulsi R (2005) Soil fertility build-up through hedgerow intercropping in integrated farming system: a case study. In: Bhatt BP, Bujarbaruah KM (eds) Agroforestry in North East India: opportunities and challenges. ICAR RC for NEH Region, Umiam, pp 495–506

Majumdar B, Venkatesh MS, Satapathy KK, Kumar K, Patiram (2002) Effect of alternative farming systems to shifting cultivation on soil fertility. Indian J Agric Sci 72(2):122–124

Mandal D, Dadhwal KS, Khola OPS, Dhyani BL (2006) Adjusted T values for conservation planning in Northwest Himalayas of India. J Soil Water Conserv 61(6):391–397

Meena RS, Yadav RS, Meena VS (2013) Heat unit efficiency of groundnut varieties in scattered planting with various fertility levels. The Bioscan 8(4):1189–1192

Meena RS, Meena VS, Meena SK, Verma JP (2015a) The needs of healthy soils for a healthy world. J Clean Prod 102:560–561

Meena RS, Meena VS, Meena SK, Verma JP (2015b) Towards the plant stress mitigate the agricultural productivity: a book review. J Clean Prod 102:552–553

Meena VS, Meena SK, Verma JP, Meena RS, Ghosh BN (2015c) The needs of nutrient use efficiency for sustainable agriculture. J Clean Prod 102:562–563

Meena VS, Meena SK, Verma JP, Meena RS, Jat LK (2015d) Current trend to mitigate climate change and sustainable agriculture: a book review. J Clean Prod 102:548–549

Meena VS, Verma JP, Meena SK (2015e) Towards the current scenario of nutrient use efficiency in crop species. J Clean Prod 102:556–557

Meena RS, Bohra JS, Singh SP, Meena VS, Verma JP, Verma SK, Sihag SK (2016) Towards prime response of manure to enhance nutrient use efficiency and soil sustainability a current need: a book review. J Clean Prod 112(1):1258–1260

Mishra VK, Saha R, Bujarbaruah KM (2005) Zero tillage technique for transplanted rice in high rainfall eco-system, *Scientific leaflet*, ICAR RC for NEH Region, Umiam

Mukherjee A, Maity A (2015) Public–private partnership for convergence of extension services in Indian agriculture. Curr Sci 109(9):1557–1563

Mukherjee A, Bahal R, Roy Burman R, Dubey SK (2012) Conceptual convergence of pluralistic extension at Aligarh district of Uttar Pradesh. J Community Mobil Sustain Dev 7(1&2):85–94

Oh NH, Richter DD (2004) Soil acidification induced by elevated atmospheric CO_2. Glob Chang Biol 10:1936–1946

Patiram (1991) Liming of acid soils and crop production in Sikkim. J Hill Res 4:6–12

Patiram (1994) Efficacy of furrow-applied limestone on maize production on an acid inceptisol of Sikkim. J Indian Soc Soil Sci 42:309–313

Patiram (1996) Efficacy of furrow application of limestone on soybean and wheat grown in sequence on an acid soil. Indian J Agric Sci 66(2):81–85

Patiram, Ramesh T (2008) Technologies for soil and soil fertility management of North Eastern Region. ICAR RC for NEH Region, Umiam

Ram SC, Ramakrishnan PS (1988) Hydrology and soil fertility of degraded grasslands at Cherrapunji in North Eastern India. Environ Conserv 15:29–35

Ram M, Singh BP (1993) Soil fertility management in farming systems, Lectures notes, off-campus training on farming system, Aizawl, Mizoram, India

Saha R, Mishra VK (2009) Effect of organic residue management on soil hydro-physical characteristics and rice yield in eastern Himalayan region, India. J Sustain Agric 33 (2):161–176

Saha R, Mishra VK, Tomar JMS (2005) Effect of agroforestry systems on erodibility and hydraulic properties of Alfisols in eastern Himalayan region. Indian J Soil Conserv 33:251–253

Saha R, Tomar JMS, Ghosh PK (2007) Evaluation and selection of multipurpose tree for improving soil hydro-physical behaviour under hilly ecosystem of north east India. Agrofor Syst 69(3):239–247

Saha R, Mishra VK, Majumdar B, Laxminarayana K, Ghosh PK (2010) Effect of integrated nutrient management on soil physical properties and crop productivity under a Maize (*Zea mays*)-mustard (*Brassica campestris*) cropping sequence in acidic soils of Northeast India. Commun Soil Sci Plant Anal 41(18):2187–2200

Saha R, Chaudhary RS, Somasundaram J (2012) Soil health management under hill agroecosystem of North East India. Appl Env Soil Sci

Satapathy KK (2005) Runoff production on hill slopes under different land use systems. In: Bhatt BP, Bujarbaruah KM (eds) Agroforestry in North East India: opportunities and challenges. ICAR RC for NEH Region, Umiam, pp 451–459

Sharma UC (1998) Methods of selecting suitable land use system with reference to shifting cultivation in NEH region. Indian J Soil Conserv 26(3):234–238

Sharma UC, Singh RP (2002) Acid soils of India: their distribution, management and future strategies for higher productivity. Fertilizer News 47:45–52

Sharma PD, Baruah TC, Maji AK, Patiram (2006) Management of acid soils of NEH region. ICAR – Krishi Anunsandhan Bhawan, Pusa Campus , New Delhi

Singh BP, Dhyani SK (1995) Significance of jackfruit in restoration of soil fertility. Annual report, ICAR Research Complex, Umiam, Meghalaya

Singh NP, Singh OP, Jamir NS (1996) Sustainable agriculture development strategy for NEH Region of India. Shukla SP and Sharma N (eds) Mittal, New Delhi

Singh DK, Singh SK, Singh AK, Meena VS (2014) Impact of long term cultivation of lemon grass (*Cymbopogon citratus*) on post-harvest electro-chemical properties of soil. Ann Agric Biol Res 19(1):45–48

Umashankar (2005) Indigenous agro-forestry tree species for conservation and rural livelihood. In: Bhatt BP, Bujarbaruah KM (eds) Agroforestry in North East India: opportunities and challenges. ICAR RC for NEH Region, Umiam, pp 149–174

Velayutham M, Bhattacharya T (2000) Soil resource management. In: Yadava JSP, Singh GB (eds) Natural resource management for agriculture production in India. Alfa Printers, New Delhi, pp 1–135

Verma JP, Jaiswa DK, Meena VS, Meena RS (2015a) Current need of organic farming for enhancing sustainable agriculture. J Clean Prod 102:545–547

Verma JP, Jaiswal DK, Meena VS, Kumar A, Meena RS (2015b) Issues and challenges about sustainable agriculture production for management of natural resources to sustain soil fertility and health. J Clean Prod 107:793–794

The Indian Himalayan Ecosystem as Source for Survival

14

Indu Bhushan, Ashok Kumar, Jai Singh Patel, R.P. Yadav, Sher Singh, Ramawatar Meena, Sunita Kumari Meena, and Vijay Singh Meena

Abstract

The Indian Himalayan Region (IHR) covers ~95 districts of the Indian union, which starts from the foothills in the south (Siwalik); the region extends to the Tibetan Plateau in the north (trans-Himalaya). The IHR occupies the strategic position of the entire northern boundary (northwest to northeast) of the country and touches almost all the international borders of seven countries with India. The contribution of India is ~16 % of total geographical area, out of which ~17 % area is under permanent snow cover and ~35 % is under seasonal snow cover. The IHR is responsible for providing water to a large part of the Indian subcontinent and contains varied flora and fauna; it was estimated that ~40 million of the population reside in this region. The Indian Himalayan rivers

I. Bhushan
Department of Biotechnology, Shri Mata Vaishno Devi University, Katra 182320, Jammu & Kashmir, India

A. Kumar
Department of Botany, MMV, Banaras Hindu University, Varanasi 221005, India

J.S. Patel
Department of Botany, Banaras Hindu University, Varanasi 221005, India

R.P. Yadav
ICAR-Vivekananda Parvatiya Krishi Anusandhan Sansthan, Almora 263601, India

S. Singh • V.S. Meena (✉)
ICAR-Vivekananda Institute of Hill Agriculture, Almora 263601, Uttarakhand, India
e-mail: sharmasmvdu92@gmail.com; vijay.meena@icar.gov.in

R. Meena
Department of Soil Science and Agricultural Chemistry, Institute of Agricultural Sciences, Banaras Hindu University, Varanasi 221005, India

S.K. Meena
Division of Soil Science and Agricultural Chemistry, ICAR-Indian Agriculture Research Institute, New Delhi 110012, India

J.K. Bisht et al. (eds.), *Conservation Agriculture*,
DOI 10.1007/978-981-10-2558-7_14

run off ~1,600,000 million m^3 of water annually for drinking, irrigation, hydropower, etc. The IHR has been a potential source of important medicinal herbs and shrubs. This region is extremely rich in plant life and abounds in genetic diversity of all types of fauna and flora. The medicinal virtues of the northwest (NW) Himalayan plants are well known from the early times of the great epics of Ramayana and Mahabharata and are mentioned in the oldest Hindu scriptures, viz., Rigveda, which is said to be the source of the Ayurvedic medicine system. These high hills are the storehouse of numerous herbs and shrubs, which are exploited not only for the pharmaceutical industries worldwide. In fact, a large percentage of crude drugs in the Indian market come from this Himalayan region. Besides this, the Himalayan regions remain as a source of many cereal crops, pulses, vegetables, fruits, and animal husbandry. The climate change impact is at a global level, and this Himalayan region is no exception. Due to the climatic changes, a lot of disturbances happening like flooding, drought, wildfire, and other global changes derive from pollutions and overexploitation of resources. These changes drastically degrade our natural resources, and nowadays it challenges a need to adopt a comprehensive master plan for conservation of these resources for the survival in the future.

Keywords
Indian Himalayan Region • Flora • Fauna • Ecosystem and biodiversity

14.1 Introduction

14.1.1 Himalaya

The Indian Himalayan Region (IHR) occupies a unique place in the mountainous ecosystems of the globe. This region is a paramount standpoint of climate and, as a life provider, supplies water to huge areas of the Indian subcontinent, and it also nourishes a great biodiversity (Kala 2013). Among the worldwide mountain ecosystem, the Himalayan ecosystem is the most diversified and multifarious, and it divides South Asia from the northern portion of Asia. This IHR is a separate ecological and geographical entity and records significantly in key biophysical establishments of the world. The IHR ranges ~2500 km in length, from 80 to 300 km wide, shaped a characteristic climate of its own, and influences the climatic conditions similarly as Asia (Meena et al. 2015c, d, 2016; Verma et al. 2015b). The enormous distinction in topographical appearance causes huge biological diversity in climatic and local conditions within IHR. The variations with respect to time and space caused by variety in orogeny of geology have resulted in a noticeable distinction in environment and physiography and as a result in the circulation pattern of biotic elements. This unique place and heterogeneous diffusion of biological diversity basics have led to complications in biogeographical patterns of the IHR. The eastern part of Himalaya (EH) and the western Himalaya (WH) support ~8000 and ~5000 species of flowering plants, respectively (Rao et al. 1994). This IHR all together supports ~50 % of the total flowering plants in

India, of which ~30 % flora is endemic to the IHR. A total of ~816 tree species, ~675 edibles, and nearly ~1743 species of medicinal value is found in the IHR.

This IHR is the most susceptible ecological system to climate change, and the communities mostly dependent on livestock, farming, and fruit products, which include vegetables, are influenced (Dash and Hunt 2007). The IHR currently is facing the problems shaped due to rising aridity, warmer cold season, inconsistency in rainfall, and unanticipated storms and frosts (Renton 2009), which mostly influence the whole range of biological diversity (Renton 2009; Meena et al. 2015e; Verma et al. 2015a). Though the IHR is loaded with biological diversity and has the most susceptible hill ecosystems to climate change, there is a scarcity of organized scrutiny of climate change and its effects on the IHR (Xu et al. 2009; Bawa et al. 2010; Shrestha et al. 2012).

14.1.2 Biodiversity of Medicinal Plants of the Indian Himalayan Region

The natural world has been a resource of medicinal healing for centuries, and the plant-originated system still plays a vital role in the primary well-being of ~80 % of the global population (Gupta 2001). The whole region is gifted with a broad range of physical geography, type of weather, soil, and biodiversity. The IHR is one of the wealthiest reservoirs of biodiversity in the planet and is known as a storeroom of priceless medicinal crop species (Samant et al. 2007). The IHR supports ~18,440 species of plants, of which ~25 % are endemic (Singh and Hajra 1996), ~1748 medicinal plants (Samant et al. 1998), and ~675 wild edibles (Samant and Dhar 1997; Samant et al. 1998). This biological diversity is being used by the residents of the IHR in a variety of forms, i.e., medicine, feed, fuelwood, forage, construction timber, making farming tools, filament, spirituality, and a variety of other purposes (Samant and Dhar 1997). The plants with medicinal characteristics enjoyed the maximum status in the native systems of medicine globally and still comprise one of the most important sources of drugs in current as well as conventional systems of medicine in spite of remarkable progress in the area of man-made antibiotics and drugs (Aswal 2001). In the inaccessible areas of IHR, where primary well-being centers are unavailable, inhabitants generally depend on plants for well-being and treatment and prevention of different ailments/diseases, such as tumor, kidney problems, asthma, skin disorders, piles, etc. (Samant et al. 1998).

14.1.3 Impact of Climate Change on the Properties of Medicinal and Aromatic Plants

Climate change and global warming are the warning calls and threats today worldwide, and almost all the species of world biodiversity are affected by the same (Badola 2010). Climate change refers to variation in the earth's climate over time. It was estimated that there will be a further increase in temperature of

1.4–5.8 °C by the twenty-first century. It will have an effect on the biological diversity, forcing one to adapt in the changing habitation and change in life phase and develop a new combating approach for climate change (Anon 2008). In the recent years, the market for substitute medicine and herbal output has also been increasing exponentially, and a lot of the Himalayan medicinal and aromatic plants (MAPs) are extremely valued as inputs for these products (Rodgers and Panwar 1988; Samant et al. 2011). The significance of the IHR MAPs in markets as well as local lives comprises a prospect as well as a menace that requires a tactical approach and supervision (Anon 2007, 2008). The majority of the plant outputs in use is extracted from the uncultivated, and the damage of their habitations due to growth pressures along with the harmful effects of climate change has also contributed to their dwindling numbers. The numerous IHR species of MAPs have suffered reduction rates of ~80 % in the previous decade. Apart from biological diversity and environmental impacts, the reducing plant wealth in the uncultivated state also has unfavorable impacts on the IHR poor who rely on them for their well-being and food supplements. Simultaneously, this blowing-up demand for MAPs substance in the global markets is a chance that should be apprehended to improve the financial status of IHR farmers. Through the cultivation of high-value case crops, they could confirm to be economically sustainable. However, unavailability of farming packages, marketing troubles and quality guarantee issues are a few of the bottlenecks (Singh and Rawat 2011; Tiwari et al. 2011; Ved and Goraya 2015).

The IHR has lost ~70 % of its unique habitation and is one of the 34 biological diversity hotspots of the globe. The gradually growing human populace in the region has led to extensive alteration of the forests, wetlands, and grasslands for farming. The road building has contributed to the disintegration of habitation and also facilitates the extend of invasive species, diseases, and injurious insects in the IHR. Without planning, inadequately managed tourism along with global instability has led to enlarged habitat demolition. The residual wild areas are a basis of minor forest products for the countryside people, and extraction is growing with the rising population, contributing to the degradation of IHR sustainability. The environmental change as a consequence of worldwide warming was anticipated, which is particularly disastrous for IHR MAPs because of their habitation specificity and thin range of distribution (Anon 2008). The investigation of tree ring samples of *Taxus baccata*, *Abies pindrow*, and *Abies spectabilis* from a variety of forest stands has provided precious information toward the plant expansion and climatic association. Thus, a larger understanding and susceptibility evaluation of a variety of species and niche are essential, with a spotlight on their adaptability range (Anon 2007, 2008). The plants of alpine climate may also be predominantly impacted by climate change; advancing tree lines and extinctions of montane plant populations have become gradually more apparent and recognized by researchers globally (Salick et al. 2009). The researchers from IHR have found that some cold-adapted plant species in alpine environments have begun to slowly advance up mountain summits, an occurrence associated with warm temperatures (Yoon 1994; Zobel and Singh 1997; Dash and Hunt 2007). In a few cases, these plants travel upward until there are no elevated areas to occupy; at that point they may face extinction

(Salick et al. 2009; Yoon 1994). As per the projection of the Intergovernmental Panel on Climate Change (IPCC), the IHR is likely to experience some of the most extreme climate changes in the globe outside of the polar regions, with the temperature increases of 5–6 °C and precipitation increases of 20–30 % (Salick et al. 2009).

14.1.4 Unsustainable Extraction of Medicinal Plants

The global market for MAPs material is of the order of over US$ 60 billion and rising rapidly; closer homes, numerous herbal industries, pharmacies, and TM practitioners use MAPs for helping the healthcare requirements of a large population base in India, while a lot of countryside households are supported by incomes from the collection of MAPs for export and family use. The higher economic use significance of MAPs leads to overextraction and resultant endangerment of the species. The IHR MAPs are much in demand due in part to their rarity and small populations; ~18 % of the MAPs material are traded in India (Valdiya 1980).

The value trends of most of the IHR species of MAPs traded in the market, such as *Picrorhiza*, *Aconitum*, and *Jatamansi*, have been on a repeated upward trend, pinpointing to the surge in demand and value for these species in markets globally. To meet the high demand of raw materials in MAPs trade, large-scale unlawful harvesting from the uncultivated is resorted in order to meet the demand-supply gap. Close to ~90 % of the plant material used locally and in the herbal industry is collected from the wild, and as much as ~70 % of it is destructively harvested (India's Forests 2007). The open access to medicinal plants combined with the small rates paid to commercial collectors leads to the removal of the high-value species from their usual habitations in the IHR. The society norms that had conventionally ensured the sustainable utilization have been replaced by forest acts of the nation, whose execution and enforcement, especially in the inaccessible landscape of the IHR, is tremendously weak. The outcome of the study conducted on IHR MAPs indicated that ~41 % of the initial-level traders source their material by collection and ~45 % follow a diverse sourcing. A huge bulk of the plants in the trade is utilized for parts that involve destructive harvesting; ~63 % of the MAPs material in trade include roots and ~5 % include whole MAPs. The only twenty out of ~800 species that are currently used in industry for large-scale production of herbal products are under commercial cultivation (Banerji and Basu 2011).

According to the Kuniyal et al. (2015) report, the growing domestic and worldwide demand of MAPs products has put the local herbal resources under noteworthy stress. The uncultivated populations of numerous MAPs species, in the forests as well as outside the forests, have been exhausted to such an degree that their very endurance has become a cause of apprehension. Numerous species are facing the danger of extinction on account of their thin distribution or endemicity and degradation of their specific habitat. It is a disappointment that the majority of these species have never been the focal point of any devoted management. This lack of management practices focus has been a cause, as well as a consequence, of the inadequacy of data related to the biodiversity of IHR. The Foundation for

Revitalization of Local Health Traditions (FRLHT), Bangalore, a center of excellence of medicinal plants and traditional knowledge of ministry of environment, forests and climate change of GOI, has undertaken two kinds of efforts to identify and prioritize wild medicinal plant species needing urgent management. The Conservation Assessment and Management Plan (CAMP) process has been utilized by FRLHT, over the past ~13 years, to undertake rapid assessment of the threat status of wild medicinal plant species of conservation concern in different states of India. During this period, FRLHT has facilitated 14 such CAMP workshops covering 17 states of India. The second effort relates to the NMPB-sponsored nationwide study to assess the demand and supply of MAPs in the country undertaken by FRLHT during 2006–2007. A total of ~960 medicinal plant species, identified as sources of ~1289 botanical raw drugs, has been recorded in trade in this study. Further scrutiny and analysis of this list has resulted in short-listing of 178 species in high trade.

14.2 Interdependence of Ecological and Socioeconomic Activities

The problems in the Himalayan region are complex, having intricate linkages between social, economic, and ecological concerns. The solutions, therefore, cannot be addressed in isolation. To cite an example, the agro- and forest ecosystems are so intricately interrelated and interdependent that it is futile to talk of forest management in isolation without considering the cropland. In the Central Himalayan region, it is estimated that the cost of subsistence agriculture on the forest ecosystem is high. For example, for each unit of energy obtained in agronomic production, seven units of energy are expended from the forest through the use of firewood, fodder, and vegetal manure (Pandey and Singh 1994). Similarly, pasture development or revegetation of wasteland without solving the problems of animal husbandry, fodder, and fuel is not possible. The traditional agri-silvi-pastoral mode of subsistence living of the inhabitants of the region is not anymore sustainable, both ecologically and economically (Singh et al. 1992). It is apparent that sectoral practices of management (or development) will not work, and therefore, the only approach which will work is a holistic one consistent with ecological and social principles.

This approach also implies that the hill and adjoining plains must be taken as the macro-planning unit, with smaller structurally and functionally definable units for micro-level planning. The various ecosystems should be categorized into protective, productive, and waste-dissipative systems and should be managed according to their roles (Singh and Singh 1991). Therefore, the basis of any planning for sustainable development in mountain areas has to be centered on man's relationship with nature. The relationship is desired to be governed by a sense of justice and equity. Each culture is the result of the people trying to survive within their environment and indeed of an attempt to optimize the use of its resources (Agrawal 1992). Lifestyle and production systems develop steadily by experimentation and observations over centuries, till they become so culturally incorporated that they are like genetic knowledge. This has been inherent in many tribal societies, but in the modern acquisitive society, "economy" gets priority over "ecology."

There is need to evolve a new paradigm to restore balance between economic interests and ecological imperatives. Although the ecological and economic systems have a myriad of interconnections, the most simple and most obvious is this: ecological system provides raw materials to the economic system and absorbs the waste generated by the economic system. Therefore, the system will be constrained by the production and waste-absorption capacities of the ecological system. When one or both these capacities are exceeded, ecological backlashes are bound to occur. Once the waste-disruptive capacity of the Ganga was exceeded, a severe pollution problem emerged, which is now costing the government a huge sum of money, with still a doubtful level of the final outcome. Similar is the case for water bodies in the hills. When timber extraction or biomass extraction exceeded the limit of harvestable productivity of the forest, the latter began to diminish (Singh 1998).

The ecological and economic considerations are therefore to be combined to attain ecologically sustainable development. Both ecological and economic values can be served individually in a variety of ways, but combining ecological and economic considerations adds geometrically to the complexity of development programs (Caldwell 1984). When sociocultural systems are added onto the ecological-economic relationships, the situation becomes further complicated. However, development driven solely by economic considerations has changed the aspirations, value systems, and management priorities. Demographic and legal factors further complicate the application of ecological considerations in the development of goals and processes.

14.3 Factors Involved in the Development of Indian Himalayan

14.3.1 Crop Production

Agricultural crop production is one of the major occupations of the farmers, which is mainly dependent upon the animal organic manure. In the Himalayan region, the land use reveals various crops sown under the 3.7 M ha area (Table 14.1). The average is 1.4 T/ha grain production of sown areas, and the total of 5.8 MT of grains is available for the human being consumption. The crop production is the leading source of residues availability, which is the basic need for the animals feed. In general, 1 ha of agricultural crop area produces almost 3.55 T of crop residues, and the total grown area provides up to 14.5 MT of crop residues. This one of the major contributions of crop production that cannot be achieved without the crucial inputs such as irrigation provided through ecological factors and farmyard manure (Misri 2002; Meena et al. 2013, 2015a, b; Singh et al. 2014; Kumar et al. 2015; Ghosh et al. 2016).

Table 14.1 The major land use pattern in the Himalayas (thousand ha)

States	Total geographical area	Reporting area	Forest	Permanent pastures	Net area sown	Others
Arunachal Pradesh	8374	5495	5154	–	185	156
Manipur	2233	2212	602	–	140	1469
Meghalaya	2243	2241	937	–	206	1098
Mizoram	2108	2088	1598	–	109	381
Nagaland	1658	1546	863	–	211	472
Sikkim	710	710	257	69	95	289
Tripura	1049	1049	606	–	277	166
Assam	1532	1521	548	21	350	602
West Bengal	315	242	117	–	99	26
Total	20,322	17,103	10,682	90	1672	4659
Western Himalaya						
Himachal Pradesh	5567	3404	1049	1194	568	593
Jammu and Kashmir	22,224	4505	2747	216	734	898
U.P. Hills	5113	5060	3430	254	660	716
Punjab	383	383	38	11	116	218
Haryana	213	213	123	–	68	22
Total	33,500	13,565	7387	1585	2146	2447
Total Himalaya	53,822	30,668	18,069	1675	3818	7106

14.3.2 Animal Husbandry

Animal production is the second most important component of the system. Misri (2002) reported that 21.4 M is the total livestock population in the Indian Himalayan Region. Animals are reared as one of the most important supplements of the family income and also to sustain the crop production. There are 2.16 MT annual milk production in these area and almost 5.90 million kg of total wool production per annum. The animal rearing is mainly dependent upon grazing lands and other pastures. Grazing in grazing lands, forests, and other pastures contribute up to 61 % of fodder, while crop residues and tree leaf fodder compensate the rest. The animals excreta are the major contribution for recycled and agricultural crop production. Ravindranath and Hall (1995) reported that each bovine provides up to 4.5 kg dung/day, while 0.4 kg dung/day is provided by every goat or sheep. However at this rate, up to 9.5 M goat and sheep and 11.8 M bovines produce 56.9 M kg of organic matter per day for agricultural crop production, while about 25–30 % of this dung has been used as fuel.

14.3.3 Grasslands, Forests, and Ecology

All of these grasslands, forests, and ecology systems are interdependent to each other to form a sequential chain that provides their crucial input to sustain the system. There are different ecological factors that play a very crucial role in the establishment of growth and survival of the forests and grasslands and in turn provide tree fodder and herbage to the animal production. Both forests and grassland systems have been used to be the most common property resources (CPRs) in the Himalayan region, while nowadays most of these are under the government body control, and access to the farmers for these resources is declining. However, still their contribution is the substantial system. Thus overall the important inputs of the system coexist in harmony and balance, which leads to the sustainability of the agropastoralism in the Himalayan region. The major outputs of the system are sustainability, food security, and employment, which are the most basic and essential for any systems.

14.4 Outputs of the System

There are different employments generated through various agricultural activities including animal husbandry, which is enormous. Out of the total workers in the Himalayan region, 59.17 % is engaged in different agricultural enterprises, which supplemented the various other avenues such as seasonal migration to the plains area, private enterprises, and recruitment in the government offices. In this region, the major crop rotations are rice-wheat, rice-maize, and rice-barley. About 1.4 t/ha average grain production has been recorded in these systems and 5.8 MT/annum total grain production. The availability of enough food grains for landless, small, and marginal farmers may still be a problem, and extra cash is generated through other components of the system such as animal production CPRs which compensate via this and exist through the large food security by the agropastoralists.

The sustainability is directly proportional to the assured and continuous contributions through internal resources of a system. In case of agropastoralism, all the internal crucial resources are assured and continuously available that might lead to the sustainability of this system. These resources are available in abundance, but at the later stage, imbalance has started to creep into the system that could affect the sustainability of the system. The following needs are required in order to remove this imbalance:

- Develop integrated watershed for the management of the natural resources
- Replenishment of vegetation covers, which minimize the water and soil loss
- Mixed population of trees require plantation in the wastelands region
- Requirement of water conservation resources.
- Provision of health cover for livestock

Table 14.2 Source of household energy in the Indian Himalayan Region (G-SHE 2009)

Source of cooking energy	Total IHR household (%)
Firewood	64.50
LPG	23.80
Kerosene	4.60
Cow dung cakes	3.60
Crop residues	2.10

14.5 Energy Security

- In the Himalayan region, biomass is one of the major sources of energy at the household level and majority of which is provided through fuelwood sourced from forests (Table 14.2).
- The shortage of fuelwood and the increase of the price of imported conventional fuels as a result of high energy vulnerability.
- Requirement of ecosystem-based approach for energy security.
- Improved cookstoves for decreasing the BC and other non-CO_2 gas emission.

14.6 The Urgent Need to Protect the Himalayan Region

The Indian Himalayan region and its people must require economic growth, while the planners and policy makers in New Delhi and also the Himalayan state capitals should take necessary action to care for their development and economic growth, which takes into account the fragility and ecology of the region. The current state of indiscriminate construction of dams and roads and also rapid urbanization have led to short-term gains, which may cause irreparable damage in the future.

Toward the Possibilities of the Solutions

- We develop to relook our current development of the model that will be very effective to the survival of the Himalayas and, as a result, our future generations.
- To look for an alternate source of energy such as solar and micro-hydro power (Sulgaon model).
- A scientific study should be required in the major key areas of the Himalayan region that should be declared as eco-sensitive zones where no tunnels, infrastructure, unscientific ways of building roads, power projects, and different hotels have been constructed for tourism.
- To encourage the indigenous knowledge and different methods of animal husbandry and organic agriculture and promote and support the ecotourism and also local village industries.
- Ownership of forests and products of forest to be given to the village communities level for management, safeguard, and also conservation of forest ecosystems.

- Afforestation as per the unique local fruit tree plantation and flora that will provide and support an alternative source of livelihood to the people.
- The create policies that will be helpful in sustaining urbanization and decreasing its impact like better urban planning process, garbage and sewage disposal system, strict implementation of construction rules, etc.

14.7 Why Is This Important?

- Heavy deforestation due to the construction of roads, dams, etc. has started to affect the rich biodiversity of this region, leading to a decrease of the unique flora and fauna. Blasting, digging, excavation of the mountainside, and dumping of debris can cause the ecological devastation in the region and drying up of natural springs.
- Many dams on rivers impact aquatic life and decrease water flow, which will impact the lives and livelihood of people and animals downstream not only of our country but also our neighboring country of Bangladesh as well. According to some scientists, a reduction of even up to 10 % can dry up more swathes of farmland in Bangladesh.
- The ecological disturbance in this region can aggravate microclimatic changes thereby again enhancing its impact on the lives and livelihood of people. The effect of climate change has already started in the form of untimely hail, warmer climates, erratic rainfall, less snowfall, retreating glaciers, soil erosion, movement of plant species upward, etc.

14.8 Concluding Remark and Future Prospective

Himalayas is a very sensitive zone for natural disasters. Constructions that do not caution for the geology of this region will cause huge damage and destruction to lives, property, and livelihood. A scientific study should be conducted, and the necessary areas across the Himalayas should be declared as eco-sensitive zones where no infrastructure, tunnels, power projects, unscientific ways of hotels, and building roads should be allowed to be constructed. The states can be incentivized or compensated for ecosystem services provided through the standing forests. Afforestation, as per the unique fruit tree plantation and local flora, will support an alternative source of livelihood to the people. The unscientific exploitation of natural resources is leading to enhancing the environmental degradation and aggravating the natural hazards. There is a necessity to evolve a new paradigm to restore the balance between ecological imperatives and economic interest due to the sociocultural principles.

Acknowledgment I am grateful to Mr. Arvind Kumar Yadav and Ms. RashmiSaraswat for critically reviewing this work and providing their valuable suggestions and comments, which were of great help in improving this chapter.

References

Agrawal A (1992) Community participation in restoration of environment. In: Singh JS (ed) In restoration of degraded land: concepts and strategies. Rastogi Publication, Meerut, pp 291–310

Anon (2007) Inventorying and threat assessment of medicinal and aromatic plants of the high altitude Himalayas; presentation at 'Management of the Herbal Wealth of the High Altitude Himalayas': a MultiStakeholder consultative workshop hosted by NMPB and Pragya; NRM Team, Pragya

Anon (2008) Management of the herbal wealth of India – report on the proceedings of a MultiStakeholder Consultative Workshop hosted by NMPB and Pragya; NRM Team, Pragya

Aswal BS (2001) Biologically active medicinal plants of the Himalaya. In: Samant SS, Dhar U, Palni LMS (eds) Himalayan medicinal plants: potential and prospects. GyanodayaPrakashan, Nainital, pp 377–380

Badola SK (2010) Sikkim Himalayan biodiversity under changing climate emerging issues and possible solutions, abstract book, International workshop mountain biodiversity and impact of climates change with special reference to Himalayan biodiversity, Hot spot G. B. Pant Institute of Himalayan and Development

Banerji G, Basu S (2011) Sustainable management of the herbal wealth of the Himalayas: prioritising biodiversity for conservation and development. Pre-Congress Workshop of 1st Indian Forest Congress

Bawa KS, Koh LP, Lee TM, Liu J, Ramakrishnan P (2010) China, India and the environment. Science 327:1457–1459

Caldwell LK (1984) Political aspects of ecologically sustainable development. Environ Conserv 11:299–308

Dash SK, Hunt JCR (2007) Variability of climate change in India. Curr Sci 93:782–788

G-SHE (2009) Governance for sustaining Himalayan ecosystem (guidelines and best practices). MoEF/GoI/GBPIHED, New Delhi/Kosi-Katarmal, Almora

Ghosh BN, Meena VS, Alam NM, Dogra P, Bhattacharyya R, Sharma NK, Mishra PK (2016) Impact of conservation practices on soil aggregation and the carbon management index after seven years of maize–wheat cropping system in the Indian Himalayas. Agric Ecosyst Environ 216:247–257

Gupta SK (2001) Pharmacology and therapeutics in the new millennium. Narosa Publishing House, New Delhi

India's Forests (2007) Ministry of Environment and Forests, New Delhi

Kala CP (2013) Climate change and challenges of biodiversity conservation. In: Kala CP, Silori CS (eds) Biodiversity, communities and climate change. The Energy and Resources Institute, New Delhi, pp 259–269

Kumar A, Bahadur I, Maurya BR, Raghuwanshi R, Meena VS, Singh DK, Dixit J (2015) Does a plant growth promoting rhizobacteria enhance agricultural sustainability? J Pure Appl Microbiol 9(1):715–724

Kuniyal PC, Bisht VK, Negi JS, Bhatt VP, Bisht DS, Butola JS, Sundriyal RC, Singh SK (2015) Progress and prospect in the integrated development of medicinal and aromatic plants (MAPs) sector in Uttarakhand, Western Himalaya. Environ Dev Sustain 17(5):1141–1162

Meena RS, Yadav RS, Meena VS (2013) Heat unit efficiency of groundnut varieties in scattered planting with various fertility levels. The Bioscan 8(4):1189–1192

Meena RS, Meena VS, Meena SK, Verma JP (2015a) The needs of healthy soils for a healthy world. J Clean Prod 102:560–561

Meena RS, Meena VS, Meena SK, Verma JP (2015b) Towards the plant stress mitigate the agricultural productivity: a book review. J Clean Prod 102:552–553

Meena VS, Meena SK, Verma JP, Meena RS, Ghosh BN (2015c) The needs of nutrient use efficiency for sustainable agriculture. J Clean Prod 102:562–563

Meena VS, Meena SK, Verma JP, Meena RS, Jat LK (2015d) Current trend to mitigate climate change and sustainable agriculture: a book review. J Clean Prod 102:548–549

Meena VS, Verma JP, Meena SK (2015e) Towards the current scenario of nutrient use efficiency in crop species. J Clean Prod 102:556–557

Meena RS, Bohra JS, Singh SP, Meena VS, Verma JP, Verma SK, Sihag SK (2016) Towards the prime response of manure to enhance nutrient use efficiency and soil sustainability a current need: a book review. J Clean Prod 112(1):1258–1260

Misri B (2002) Impact of agro-pastoralism on development of the Indian Himalayan

Pandey U, Singh JS (1994) Energy-flow relationships between agro- and forest ecosystem in Central Himalaya. Environ Conserv 11:45–53

Rao R, Singh B, Singh MP (1994) Biodiversity in India: floristic aspect. Bishen Singh Mahendra Pal Singh, Dehradun

Ravindranath NH, Hall DO (1995) Biomass, energy and environment. A developing country perspective from India. Oxford University Press, Oxford

Renton A (2009) Suffering the science: climate change, people and poverty. Oxfam briefing paper number 130. Oxfam International, Oxford

Rodgers WA, Panwar HS (1988) Planning a wildlife protected area network in India. Wildlife Institute of India, Dehradun

Salick J, Zhendong F, Byg A (2009) Eastern Himalayan alpine plant ecology, Tibetan ethnobotany, and climate change. Glob Environ Change 16:488–495

Samant SS, Dhar U (1997) Diversity, endemism and economic potential of wild edible plants of Indian Himalaya. Int J Sustain Dev World Ecol 4:179–191

Samant SS, Dhar U, Palni LMS (1998) Medicinal plants of Indian Himalaya: diversity, distribution, potential values. GyanodayaPrakashan, Nainital

Samant SS, Pant S, Singh M, Lal M, Singh A, Sharma A, Bhandari S (2007) Medicinal plants in Himachal Pradesh, north western Himalaya, India. Int J Biodivers Sci Ecosyst Serv Manag 3:234–251

Samant SS, Vidyarthi S, Pant S, Sharma P, Marpa S, Sharma P (2011) Diversity, distribution, indigenous uses and conservation of the medicinal plants of Indian Himalayan region used in cancer. J Biodivers 2:117–125

Shrestha UB, Gautam S, Bawa KS (2012) Widespread climate change in the Himalayas and associated changes in local ecosystems. PLoS One 7:36–41

Singh JS (1998) Sustainable development: an ecological viewpoint. In: Sundriyal RC, Uma S, Upreti T (eds) Perspectives for planning and development in Northeast India, vol 11. Himavikas Occasional Publication, Institute of Himalayan Environment and Development, Kosi-Katarmal, Almora, pp 5–16

Singh DK, Hajra PK (1996) Floristic diversity. In: Gujral GS, Sharma V (eds) Changing perspectives of biodiversity status in the Himalaya. British Council, New Delhi, pp 23–38

Singh G, Rawat GS (2011) Ethnomedicinal survey of kedarnath wildlife sanctuary in Western Himalaya, India. Indian J Fundam Appl Life Sci 1:35–46

Singh SP, Singh JS (1991) Analytical conceptual plan to reforest Central Himalaya for sustainable development. Environ Manag 15:369–379

Singh SP, Negi GCS, Pant MC, Singh JS (1992) Economic considerations in the Central Himalayan agro-ecosystems. In: Agrawal A (ed) The price of forest. Centre for Science and Environment, New Delhi, pp 291–296

Singh DK, Singh SK, Singh AK, Meena VS (2014) Impact of long term cultivation of lemon grass (Cymbopogon citratus) on post-harvest electro-chemical properties of soil. Ann Agri Biol Res 19(1):45–48

Tiwari N, Purohit M, Nautiyal AR (2011) Cultivation: a promising method for conserving *Picrorhiza kurroa*. Int J Med Aromat Plants 1:14–17

Valdiya KS (1980) Geology of kumaun lesser Himalaya. Wadia Institute of Himalayan Geology, Dehradun, Uttarakhand, India

Ved DK, Goraya GS (2015) Prioritization of wild medicinal plant species for different states for guiding conservation action at the state level

Verma JP, Jaiswa DK, Meena VS, Meena RS (2015a) Current need of organic farming for enhancing sustainable agriculture. J Clean Prod 102:545–547
Verma JP, Jaiswal DK, Meena VS, Kumar A, Meena RS (2015b) Issues and challenges about sustainable agriculture production for management of natural resources to sustain soil fertility and health. J Clean Prod 107:793–794
Xu J, Grumbine R, Shrestha A, Eriksson M, Yang X (2009) The melting Himalayas: cascading effects of climate change on water, biodiversity, and livelihoods. Conserv Biol 23:520–530
Yoon CK (1994) Warming moves plants up peaks, threatening extinction. New York Times. June 21, C4
Zobel DB, Singh SP (1997) Himalayan forests and ecological generalizations. Bioscience 11:735–745

Soil Degradation in North-West Himalayas (NWH): A Case Study of Himachal Pradesh

15

Gurjant S. Sidhu and R.P. Yadav

Abstract

Soil erosion is the main soil degradation type; therefore, the extent of water-induced soil erosion in the state has been computed on the basis of soil resource map of the state. The data indicated that about 22 % of TGA of the state has annual soil loss less than 5 t ha^{-1}, and this can be termed as very well within the tolerance limit. This area does not require any specific soil conservation measures. Improved land and crop production technologies need to be adopted for improving the productivity on sustainable basis. These areas are distributed in small patches in the entire state. About 7 and 5 % areas are experiencing annual soil loss in the range of 5.0–10.0 and 10.0–15.0 t ha^{-1}, respectively. While slight erosion (5–10 t ha^{-1}) class has been termed as within safe limit, yet this along with moderate erosion (10.0–15.0 t ha^{-1}) class needs appropriate conservation measures to sustain the agricultural production from these areas of the state. These areas can be rehabilitated by adoption of low-input soil conservation measures, viz. land shaping, contour farming, field bunding, conservation tillage and introduction of erosion-resistant crops or cropping sequences. About 27 % area of the state has soil loss >15 t ha^{-1}, and it includes moderately severe (15–20 t ha^{-1}), severe (20–40 t ha^{-1}), very severe (40–80 t ha^{-1}) and extremely severe class (>80 t ha^{-1}) having 3.75 %, 7.40 %, 5.74 % and 10.08 % area, respectively. These areas are concentrated in patches in the entire state, but no area of extremely severe erosion class is present in Hamirpur and Bilaspur

G.S. Sidhu (✉)
ICAR-National Bureau of Soil Survey and Land Use Planning, New Delhi 110012, India

Flat # 82, Evergreen Apts., Plot #9, Sector 7, Dwarka, New Delhi 110075, India
e-mail: gssidhu_ps@yahoo.com

R.P. Yadav
ICAR-National Bureau of Soil Survey and Land Use Planning, New Delhi 110012, India

J.K. Bisht et al. (eds.), *Conservation Agriculture*,
DOI 10.1007/978-981-10-2558-7_15

districts. It has been deduced that the state experiences a total soil displacement to the tune of 258 M tonnes, out of which 83 % is contributed by 10 % area subjected to extremely severe erosion class. This area has been delineated in the map, and effective conservation measures here can not only cause marked reduction in soil loss but can pave way for regeneration of areas under other erosion classes.

Keywords
Soil degradation • Soil erosion • North-west Himalayas

15.1 Introduction

Management of natural resources – soil, water, vegetation, biodiversity and weather – is vital to sustain life support system on the earth. Among the natural resources, soil is a finite resource, which is declining both in quantity and quality as far as its availability for biomass production is concerned. While decrease in land quantity is due to the shift in use to more remunerative and essential sectors like housing, transport and industrial sector, land-quality decline is owing to the various processes of land degradation. Among them water-induced erosion is the single most destructive process. Although soil erosion is the physical process of soil detachment, transportation and deposition, it causes deterioration of physical, chemical and biological properties of soils.

Human population of our country has already crossed one billion mark, and by 2015 AD, it already crossed 1.25 million human and 600 million livestock, requiring 275 MT of food grains, 1083 MT of green fodder and 235 mm^3 of firewood as against our current supplying capacity of 200 MT of food grains, 513 MT of green fodder and 40 mm^3 of firewood. Such a level of production can be attained only if processes of land degradation are checked and processes of restoration of soil productivity are initiated through appropriate soil and water conservation measures and suitable land uses. The first step towards this goal requires quantitative database on the intensity/severity of soil erosion and its spatial distribution in all the states of our country.

The soil resources of the world are finite and non-renewable. They are fragile too. Its misuse and mismanagement especially in harsh environment and ecologically sensitive ecoregions lead to scarcity of good-quality soils. Besides, an accelerated soil erosion leads to changes in soil quality especially that of surface horizon (Lal 1998). It has been a problem of numerous bygone civilisations (McCracken 1987). Siltation of irrigation canal and fields was a major problem in the ancient Mesopotamian agriculture (Jacobson and Adams 1958). Silt deposition has raised the level of fields by almost 1 m over a 500-year period. This rise in the land surface has reduced the groundwater level. Further, siltation of the canal bed sharply reduced water flow and reduces its supply to branches of the main canal. The cost of desilting was enormous and its consequences disastrous. Olson (1981,

1985) also postulated that soil erosion was the cause of downfall of several cultures and civilisations. In the Middle East, deforestation of cedar forests from the countries surrounding the Mediterranean caused severe erosion, toppled the Phoenicians and destroyed granaries of the Roman Empire. Agriculture that was thriving about 10,000 BC in Mesopotamia, present-day Iraq, was destroyed by accelerated erosion (Lowdermilk 1953). It was also the cause of extinction of many ancient cultures in south Asia (Harappan-Kalibangan civilisation in western India), Negev in the Middle East and the Maghreb region of north-west Africa (e.g. Tunisia, Algeria and the Mayan civilisation in Central America) (Olson 1985).

The present global civilisation is also plagued by accelerated soil erosion. Some soils and regions are more vulnerable than others, and the degradative effects are specific to soils and ecoregions depending on the management and environmental factors (Wolman 1985). Slow rate of growth in the world's food output in the 1980s and 1990s has been attributed partly to global soil degradation (Brown 1991a. b; Kazgan 1991) and especially to accelerated soil erosion (Pimentel et al. 1993, 1995). The per capita cropland is decreasing day by day due to population pressure. It is apparent, however, that climatic, topographic and hydrologic constraints effecting water-induced degradation are less manageable than soil-related constraints (El-Swaify 1991).

In India, out of the total geographical area of 328.73 M ha and 147 M ha suffers from degradation of one or other forms (NBSS & LUP 2004–2005) which are reconciled recently to 120.4 M ha (ICAR-NAAS 2010). Among various forms of degradation, water-induced erosion contributes ~93 M ha area. Among the various regions of the world, the Himalayan-Tibetan ecosystem is the global hot spot of the erosion-induced soil degradation (Gurung 1989). Himachal Pradesh is situated in the greater, middle and lower Himalayas and Shivaliks, thus among the most severely erosion-affected states of India. Many dams and reservoirs have been constructed across the rivers flowing through the state, and their life span is dependent on the sediment load of the water channels. Water erosion, though a physical process of land degradation, causes physical, chemical and biological deterioration of soil properties. The resultant effect is the nonreversible loss in soil productivity. The loss in productivity depends upon the severity of erosion, soil type and crop grown. Velayutham and Bhattacharya (2000) reported a loss in productivity in the range of 5–50 % due to moderate erosion (10–20 t ha^{-1} $year^{-1}$). The removal of 2.5, 5.0, 7.5, 10.0, 15.0, 22.5 and 30.0 cm topsoil resulted in grain yield reduction of maize by 11 %, 17 %, 33 %, 39 %, 45 %, 56 % and 65 %, respectively, in alluvial soil of Doon valley (Khybri et al. 1980). Besides loss in productivity, its off-site effects may result in siltation of multipurpose dams, flood and sediment deposition in fluvial plains below and damage to the infrastructures (Meena et al. 2013; Singh et al. 2014; Kumar et al. 2015; Ghosh et al. 2016).

Thus, conservation of land resources needs to be taken up on priority to support and sustain the livelihood on account of its food requirement. Biotechnological, genetic engineering and soil and water management technologies offer potential to improve agricultural productivity. However, the benefits of these technologies are not realised to the desired extent. Improved agricultural technology cannot always and effectively launched on severely degraded soils (Wallace 1994). The proper assessment of the problem is the first requisite for control of soil degradation. The early estimate by Dhruv Narayan and Ram Babu (1983) revealed that about 5334 MT of soils are lost annually through water erosion. This works out to be 16.35 t ha^{-1} $year^{-1}$, which is higher than the permissible limit. NBSS & LUP, Nagpur, during soil resource mapping of various states, has delineated various states into four categories of erosion, viz. slight, moderate, severe and extreme on the basis of field observations. However, these estimates are qualitative in nature. Various workers (Rao et al. 1997; Mahapatra et al. 2000; HPKVV 1991; Anonymous 2008; Yadav and Sidhu 2010; Suri et al. 2013; Sidhu and Jaya 2014) have attempted to characterise the soils and demarcate the soil degradation and soil erosion status at different scales in some areas of the state.

Recently, efforts are made to quantify the average annual soil loss from Himachal Pradesh using universal soil loss equation (USLE) (Sidhu et al. 2010; Meena et al. 2015a, b, c, 2016; Verma et al. 2015b). It is based on comprehensive data about landscape, vegetation, soil, crop cover management and soil conservation practices adopted. In spite of the inherent limitations of USLE and the scale and the peculiarity of the state, the scientific database on soil loss in Himachal Pradesh will be very useful to planners, administrators and scientists (Sidhu et al. 2000, 2007a).

15.2 Geographical Setting of Himachal Pradesh

Himachal Pradesh is surrounded by Jammu and Kashmir in north, Punjab and Haryana in south, Uttar Pradesh in south-east and Tibet in the east and lies between 31°22′ 40″ and 33°12′ 20″ N latitudes and 75° 45′ 55″ and 79°04′ 20″ E longitudes. It has an area of 55,673 km^2 (5,567,300 ha) and a population of 6.08 million, distributed in 16,807 inhabited villages and 57 towns. The state is divided into 12 districts, 52 subdivisions and 75 tehsils (Fig. 15.1). The state capital is at Shimla. The state has been divided into five major physiographic regions, namely, (1) greater Himalayas, (2) lesser Himalayas, (3) outer Himalayas or Shivaliks, (4) piedmont plains and (5) flood plains. The Himachal region is veiled from the Punjab plains by the Shivalik hills. It is a hilly and mountainous tract with altitude ranging from 350 to 6975 m above mean sea level. The region presents an intricate mosaic of mountain ranges, hills and valleys. The white snow-clad peaks are the most prominent landmarks.

The Dhauladhar range looks in supreme majesty over the Kangra Valley, while the Pir Panjal and the Zanskar ranges stand guard over Chamba, Lahaul-Spiti, Kullu and Kinnaur. There is general increase in elevation from west to east and south to north (Fig. 15.1). The state has unique distinction of providing water to both the Indus and Ganga basins. The major river systems of the region are the Chandrabhaga or the Chenab, the Ravi, the Beas, the Satluj and the Yamuna. The catchment of these rivers is fed by snow and rainfall in the state (Sidhu et al. 2009, 2010).

Fig. 15.1 Administrative units of Himachal Pradesh (Source maps of India)

Geologically, the area is more complex and complicated. The region falls into four broad stratigraphical zones, viz. outer Himalayas/Shivaliks, lower Himalayas, higher/greater Himalayas and Tibetan/Tethys Himalayas. The outer Himalayas have great thickness of detrital rocks, sandstones, clays and conglomerates, while lower Himalayan zone is composed of granites and other crystalline rocks. The granitic rocks and granite gneisses are well outcropped intermittently within the metamorphic in the greater Himalayan zone. The Tibetan or Tethys Himalayas have great variation in the rocks.

Climatically, the state is much diversified due to variation in elevation (350–7000 m) and aspect. The climatic conditions vary from hot and subhumid tropical in the southern low tracts to temperate, cold alpine and glacial in the northern and eastern high mountains. Temperature varies from (–) 25 °C in January to 42 °C in June. The relative humidity is generally higher in Himachal region than adjoining plains during pre-monsoon and monsoon period. The mean annual rainfall ranges from 350 to 3800 mm. About 70 % of the annual rainfall is received during July to September (rainy season), ~20 % from October to February (winter season) and ~10 % from March to June (summer season).

15.2.1 Soils

Soils of H. P. are formed from a large variety of igneous (granite, granite gneisses), sedimentary and metamorphic rocks, the details of important soil families occupying more than 1 % of geographical area of the state (Table 15.1). The soils belong to 4 orders, 6 suborders, 17 subgroups and 44 families. Fourteen families occupy more than 1.0 % of geographical area; these families occupy an area of 3.6 M ha, which is 64.4 % of TGA and 89.2 % of non-rocky area of the state (Sidhu et al. 1997, 2007a, b). The Entisols are dominant soils and cover about 51 % of TGA, followed by Inceptisols, Mollisols and Alfisols, which cover about 20.0 %, 0.8 % and 0.4 % of TGA, respectively.

15.2.2 Land Use

The Himachal Pradesh is a mountainous state having 19.9 % of TGA under forests, 21.4 % under pastures and 9.7 % as net sown area. The remaining 50 % area is lying as waste or fallow land or as nonagricultural area (Table 15.2). Gross-irrigated area is only 3.3 % of TGA of the state.

Table 15.1 Important soil families occupying more than 1 % of geographical area of Himachal Pradesh

	Area	
Soil family	'000' ha	(%)
Sandy-skeletal, lithic Cryorthents	113.1	2.0
Sandy-skeletal, typic Cryorthents	176.3	3.2
Loamy-skeletal, typic Cryorthents	460.8	8.3
Loamy, lithic Udorthents	87.1	1.6
Loamy-skeletal, lithic Udorthents	80.8	1.5
Sandy-skeletal, typic Udorthents	166.6	3.0
Loamy-skeletal, typic Udorthents	742.3	13.3
Coarse-loamy, typic Udorthents	530.7	9.5
Fine-loamy, typic Udorthents	181.4	3.3
Coarse-loamy, typic Udifluvents	56.4	1.0
Coarse-loamy, dystric Eutrudepts	94.5	1.7
Fine-loamy, dystric Eutrudepts	610.5	11.0
Fine-loamy, typic Eutrodepts	221.9	4.0
Fine-loamy, udic Haplustepts	53.1	1.0
Total	3575.5	64.4

Source: Sidhu et al. (1997)

Table 15.2 Land use in Himachal Pradesh

Land use	Details	Area	
		'000' ha	(%)
Forests (all types)		1106.1	19.9
Land not available for agriculture		1132.1	20.3
Other uncultivated lands excluding current fallows		1700.4	30.5
Present fallow land		82.1	1.5
Cultivated area	Area sown more than once	402.2	7.2
	Net sown area	538.4	9.7
	Gross sown area	940.6	16.9
Irrigated area	Net-irrigated area	105.6	1.9
	Gross-irrigated area	185.9	3.3

Source: Anonymous (2014)

The main crops grown in the state are wheat (6.7 % area), barley (0.3 % area), gram (negligible area) during rabi and maize (5.2 % area), rice (1.4 % area), ragi (negligible area) and millets (0.08 % area) during kharif season and pulses (1.16 % area) in both kharif and rabi seasons. The yield of crops in the state is comparatively very less as compared to other states of the country, for example, 1.73, 2.3, 1.45, 1.29 and 1.16 t ha^{-1} for rice, maize, wheat, barley and pulses, respectively. Considerable area (218,303 ha) is under fruit crops and the total fruit production is 866,340 tonnes. Another important crop is potato, the production of which is about 68,800 MT, mostly cultivated as seed production.

15.2.3 Soil Degradation

The main soil degradation problem is due to soil erosion. Total erosion in the form of sheet, rill and gully and specialised form of erosion, viz. landslides, ravines, stream banks, etc., are termed as gross erosion. Erosion from sheet and rill from a unit area of field at a specific slope is termed as soil loss. Several models are available to estimate soil loss from a field. Models vary from simple empirical to complex process based on the large number of input variables. Universal soil loss equation (USLE) developed by Wischmeier and Smith (1978) is a simple empirical model that estimates the average soil loss from a field. The USLE has been used extensively in various parts of the world. The USLE was used for estimation of soil loss of Himachal Pradesh.

15.2.4 Computation of Factors

The six factors of universal soil loss equations were calculated by standard methods/equations developed at international and national levels. These were calculated as defined below (Table 15.3):

Table 15.3 Factors responsible for different classes of soil erosion in Himachal Pradesh

Causative factors for varying rates of soil erosion	Soil erosion class				
	Very slight	Slight	Moderate and moderately severe	Severe and very severe	Extremely severe
Flat lands with varying crops, rainfall erosivity and soil erodibility	√	–	–	–	–
Cultivation in valleys of medium-to-high erosivity and erodibility	√	–	–	–	–
Good forest cover on moderately sloping lands	√	–	–	–	–
Erosion-promoting crops on soils having high erodibility	–	√	–	–	–
Practice of fallowing during kharif/monsoon season	–	√	√	–	–
Soils with low erodibility under dense forest in high-rainfall areas	–	√	√	–	–
Cultivation in areas not suitable for crops without conservation measures	–	–	√	–	–
Erosion -promoting crops on gentle slopes (3–10 %) in high-rainfall areas	–	–	√	–	–
Degraded forest/pasture in medium-to-high-rainfall areas	–	–	–	√	–
Denuded hills with no vegetation in low-rainfall areas	–	–	√	√	√
Cultivated fallow on moderate slopes	–	–	√	√	√
Crop cultivation on moderate slopes without conservation measures	–	–	–	√	√
Poorly managed forest cover in high-rainfall areas	–	–	–	√	√
Cultivation of paddy and potato in terraced fields on steep slopes without risers	–	–	–	–	√

R factor: The method of Wischmeier and Smith (1958) is used for estimating the erosion index of each storm greater than 12.5 mm. In India, rainfall data from 45 stations having self-recording rain gauges were collected (RamBabu et al. 1978), and R-factor was calculated.

K factor: The soil-erodibility factor 'K' is the soil loss from a unit plot per erosive index (EI) unit. It should be measured in the field but can also be calculated from soil and profile characteristics. For each soil family, it was calculated by the regression equation (Wischmeier and Smith 1978).

L and S factors: LS is the expected ratio of soil loss per unit area from a field slope to that from a 22.13 m length of uniform 9 % slope under otherwise identical conditions. In the modified universal soil loss equation, the LS factor has been combined as single topographic factor. The single value of LS factor has been compiled (Renard et al. 1994) for different slope gradients, and an LS value was obtained from these tables relating to the given slope length and gradient.

C factor: 'C' is the expected ratio of soil loss from land cropped under specified conditions to soil loss from clean-tilled fallow on identical soil and slope and under the same rainfall. It reflects the combined effect of cover, crop sequence, productivity level, length of growing season, tillage practices, residue management and the expected time distribution of erosive rainstorm with respect to seeding and harvesting date in the locality. To obtain the value of C, the ratio of soil loss in each crop stage period is combined; these values for different crops have been calculated by various workers as compiled (Singh et al. 1981; Wischmeier and Smith 1978).

P factor: The supporting conservation practice factor 'P' is the ratio of soil loss with a specified supporting practice to the corresponding loss with up-and-down cultivation. The specified practices include contour cultivation, contour bunding, strip cropping, terracing and values of these factors (Singh et al. 1981; Wischmeier and Smith 1978). Yadav and Sidhu (2010) summarised the factors responsible for different classes of soil erosion in Himachal Pradesh (Table 15.4) and basic framework of measures to conserve soil under different erosion classes in Himachal Pradesh (Table 15.5).

15.2.5 Soil Erosion Status

Water-induced soil erosion refers to detachment, transportation and deposition of soil particles through erosive power of raindrops and surface flow of runoff. Erosive power of water detaches the different textural fractions preferentially and also carries organic matter and nutrients with them, forming terrains, thus causing physical and chemical degradation. However, deposition in lower elevations may cause improvement in soil productivity. Erosion is primarily responsible for variations of soil characteristics in a topographic sequence as it erodes elevated topo-sequences and simultaneously deposits in depressions. Ecological imbalance caused by human interventions has aggravated the problem of soil erosion due to deforestation, mining, cultivation of marginal lands, overgrazing in pastures, construction activities, etc.

The extent of soil erosion in Himachal Pradesh was qualitatively assessed by NBSS & LUP, Nagpur, during soil resource mapping. Accordingly, four erosion classes, viz. slight, moderate, severe and very severe, were established. The area under rock outcrops has not been included under any erosion class as no soil is present, and it has been shown as separate entity. Only about 2.0 % of TGA of the

Table 15.4 Basic framework of measures to conserve soil under different erosion classes in Himachal Pradesh

Soil conservation and land management techniques	Soil erosion class					
	Slight	Moderate	Moderately severe	Severe	Very severe	Extremely severe
Use of organics to improve soil organic carbon and lower soil 'K'	√	√	–	–	–	–
Cultivation of deep-rooted and erosion-resistant crops	√	√	–	–	–	–
Incorporation of soil-binding/nitrogen-fixing legumes in rotation	√	√	–	–	–	–
Agronomic measures like intercropping, strip cropping and contour farming	–	√	√	√	–	–
Tree-based perennial vegetation in common and panchayat wastelands	–	√	√	√	–	–
Proper utilisation of current fallow, wasteland and old terraces	–	√	√	√	–	–
Land levelling with bunding, contour bunding and contour ditching to reduce slope length in moderately sloping lands	–	–	√	√	√	–
Safe disposal mechanism for removal of excess run-off	–	–	√	√	√	–
Silvi-pastoral, horti-pastoral or forest cover improvement techniques	–	–	√	√	√	–
Bench terracing for field crops and contour trenching for other land uses	–	–	–	√	√	√

(continued)

Table 15.4 (continued)

Soil conservation and land management techniques	Soil erosion class					
	Slight	Moderate	Moderately severe	Severe	Very severe	Extremely severe
Rainwater harvesting, run-off diversion and gully bed and slope stabilisation measures	–	–	–	√	√	√
Integrated participatory watershed management programmes for environmental rehabilitation and sustaining productivity	–	–	–	√	√	√

Table 15.5 Total soil loss in Himachal Pradesh and geographical distribution under different erosion classes

S. no.	Range of soil loss (t ha^{-1} $year^{-1}$)	Erosion class	Area '000' ha	Area (%)	Av. soil loss (Mt $year^{-1}$)
1.	<5	Very slight	1202.09	21.59	3.00
2.	5–10	Slight	379.84	6.82	2.85
3.	10–15	Moderate	302.41	5.43	3.78
4.	15–20	Moderately severe	208.62	3.75	3.65
5.	20–40	Severe	412.20	7.40	12.37
6.	40–80	Very severe	319.64	5.74	19.18
7.	>80	Extremely severe	561.14	10.08	213.23
8.	Rock outcrops		2181.36	39.19	–
Total			5567.30	100.0	258.06

state is under slight erosion class. Moderate, severe and very severe form of erosion occurred in 0.88, 1.87 and 0.13 M ha, respectively, constituting 16 %, 34 % and 2 % of the state area.

Quantification of soil loss in the bulletin has been done with the computation of various USLE factors. The average annual soil loss through water erosion has been calculated from these data in tonnes per hectare per year. The values of annual soil loss in the state have been categorised into seven different erosion classes. The rocky outcrops and snow-bound area have been shown as separate entity. Spatial distribution of soil loss in the state into different erosion classes is generated using GIS techniques in the form of the soil erosion map (Fig. 15.2). The area, calculated from the map, under different erosion classes is given in Table 15.3. In the entire

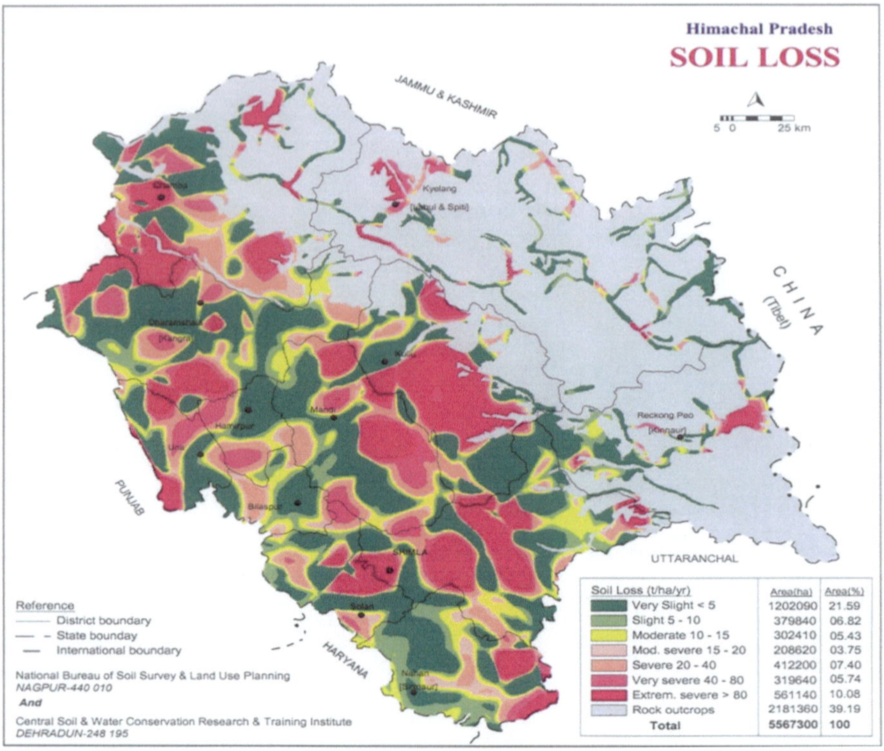

Fig. 15.2 Soil loss map of Himachal Pradesh

state, annual soil loss ranged from as low as 0.08 t ha^{-1}year^{-1} to as high as 683.10 t ha^{-1} year^{-1}. It is the cumulative effect of wide variations in R, K, LS, C and P values in the state.

About 22 % (1.2 M ha) of TGA of the state produces annual soil loss <5 t ha^{-1} year^{-1}. Annual soil loss up to 5 t ha^{-1} year^{-1} can be termed well within safe limit and is, thus, designated as very slight. Slight (5–10 t ha^{-1} year^{-1}) erosion occurs in 0.4 M ha constituting 7 % area of the state. Annual soil loss less than 10 t ha^{-1} year^{-1} has been included within threshold limit for alluvial soils. Thus, 34 % area of Himachal Pradesh has soil loss within the safe limit. The quantity of soil loss in this area can be reduced further by scientific agronomic and tillage practices. Moderate erosion (10–15 t ha^{-1} year^{-1}) is experienced in about 5 % area, and this range can also be termed as sustainable for very deep soils and steep landscape of the Shivaliks (Fig. 15.2).

About 27 % of TGA of the state is suffering a severe form of erosion (>15 t ha^{-1} year^{-1}) and requires bioengineering and other soil and water conservation measures for its management. The annual soil losses in the range of 15–20, 20–40, 40–80 and >80 t ha^{-1} year^{-1} occur in 4 %, 7 %, 6 % and 10 % of TGA of the state, respectively. These areas are scattered in patches in almost all the districts of the state. Kullu, Lahaul and Spiti and Kinnaur districts have small areas under moderately severe and severe erosion classes. Extremely severe erosion is also experienced in the entire state except Hamirpur and Bilaspur districts. The annual soil losses up to 683 t ha^{-1} have been estimated from several grids, and all these have been grouped under extremely severe erosion class.

The total annual soil loss in Himachal Pradesh is about 258 M tonnes, out of which 213 M tonnes, more than 80 % of the total, comes from 10 % area experiencing extremely severe erosion. On the other hand, 28 % area experiencing very slight and slight erosion (<10 t ha^{-1} $year^{-1}$) contributes only 2 % (6 MT $year^{-1}$) towards the total soil displacement in the state. The soil loss besides causing land degradation on the site causes problem of terrain formation; landslides; landslips; clogging of drains; siltation of ponds, lakes and reservoirs; and overloading of rivers and channels. The average annual soil loss per hectare for Himachal Pradesh works out to 46.35 tonnes, which is beyond the safe limit and requires implementation of integrated watershed development programmes in priority areas, bioengineering measures to control torrential form of erosion as well as sound crop and land management practices in low-priority areas. Shivalik region, a large part of which lies in H. P., is the most fragile ecosystem. Adoption of soil and water conservation measures in the delineated regions will not only help in concentrating the conservation measures in the problem area but will also bring better returns per unit investment. The entire state thus needs to be regenerated starting from the most severely affected areas for sustainable biomass production from this Himalayan state.

15.2.6 A Site-Specific Example

To determine the average soil loss in t ha^{-1} $year^{-1}$ for a particular grid point, values of six factors of USLE are first determined as discussed here for the site at 32° 25′ N latitude and 76° 00′ E longitude near Dalhousie in Chamba district of Himachal Pradesh:

1. *Rainfall factor* (*R*): *600.0 MJ-mm/ha-hr-yr*
   The seasonal R value for the monsoon months was taken for this location from the iso-erodent map of Himachal Pradesh prepared by calculating R values of 18 locations in Himachal Pradesh.
2. *Soil factor* (*K*): *0.31 t-ha-hr/ha-MJ-mm*
   The soil at the site is coarse-loamy, dystric Eutrudepts having 39.5 % silt, 16.3 % very fine sand and 9.0 % clay. It has 1.33 % organic matter and very fine subangular blocky structure. The soil profile is excessively drained and has rapid permeability rate. From these properties, K value is determined as per procedure described under 3.2.2.
3. *Topography factor* (*LS*): *8.00*
   The location of the grid point has gradient (slope) in the range of 30–40 %, and the length of slope is in the range of 300–600 m. From these factors, LS value of 8.00 is calculated as per procedure described under 3.2.3.

4. *Crop factor* (*C*): *0.50*
   The land, where the grid is located, is degraded culturable waste having natural grasses and bushes. As the site is a culturable wasteland, C value of 0.50 is taken from Table 15.4.
5. *Conservation factor* (*P*): *0.50*
   The region around the grid point has the abandoned terraces. These terraces are not properly maintained at present; otherwise, a value of 0.35 would have been given as per Table 15.3. The abandoned terraces with degraded forests are still effective up to some extent in reducing erosion. A value of 0.50 has been given for (P) conservation factor.

The annual soil loss 'A' is estimated from the above-mentioned values using USLE equation:

$$A = \mathrm{RKLSCP}.$$

Putting these values, we get

$$A = 600 \times 0.31 \times 8.00 \times 0.50 \times 0.50 = 372\ \mathrm{t\ ha^{-1} year^{-1}}.$$

So, the average annual soil loss from this specific site will be 372 t ha^{-1} $year^{-1}$, and it will be placed under extremely severe erosion class in the map. Similarly, A for all grids was determined and placed in the map.

15.2.7 Soil Loss Tolerance

Soil tolerance is the limit that denotes the maximum level of soil erosion that will permit soil productivity to be sustained economically and indefinitely. Based on the work carried out, tolerance limits have been found ranging from 4.5 to 11.2 t ha^{-1} $year^{-1}$ (Mannering 1981). The magnitude of soil erosion that can be tolerated without affecting agricultural production is based on the soil depth prior to erosion and other factors affecting soil erodibility.

Soil loss in excess of 11.2 t ha^{-1} $year^{-1}$ affects the effectiveness of soil conservation structures; at this stage, the gully formation starts and obstructs the cultural operations (Singh et al. 1981). Yield reduction in crops depends upon soil type and crop grown (Velayutham and Bhattacharya 2000) and is generally low in alluvial soils and high in well-developed soils. Slight erosion in the range of 5–10 t ha^{-1} $year^{-1}$ caused loss in soil productivity to the tune of less than 5 %, 5–10 % and 10–25 % in alluvial, black and red soils, respectively. Considering these data, soil loss tolerance has been fixed to 10 t ha^{-1} $year^{-1}$ for soils of Himachal Pradesh, majority of which are alluvial in nature.

As per the erosion data of the state, 1.8 M ha of area constituting 32 % of TGA of the state has an annual soil loss higher than 10 t ha^{-1} $year^{-1}$. These areas have been delineated in the map and need proper soil conservation measures to manage our land resources for sustainable production.

Cost estimation of land degradation is a difficult task as it includes both on-site and off-site cost. Further, it has local and global impacts. The local costs include decline in soil fertility and agricultural productivity, increased floods, siltation of reservoirs, dust pollution, etc., while global costs are due to loss of biodiversity and carbon sequestration capacity of the land. In Nepal, available information on crop yield in the mountains and hills over a 25-year period show continuously declining trends of yields of major crops such as paddy, maize and millet with loss in productivity estimated at 17.2 %, 17.0 % and 9.5 % in the mountains and 12.7 %, 21.7 % and 8.8 % in the hills, respectively, over the period. The total estimation using the productivity loss approach alone ranged from 395 million in 1984–1985 to 285 million Nepalese rupees. The loss ranged from 0.89 % of GDP in 1984–1985 to 0.37 % in 1995–1996 with an average loss of 0.55 % $year^{-1}$ (Tiwari 1998). The unit cost of carbon abatement is now considered between Rs 480 and 5760 per tonne of carbon. Looking into the severity of erosion in the state and monetary losses associated with it, there is a need to adopt appropriate soil and water conservation measures in severely eroded areas of Himachal Pradesh.

15.2.8 Soil and Water Conservation Planning

The severity of soil erosion determines the soil and water conservation measures to be adopted. Accordingly, the state has been divided into seven erosion classes, viz. very slight, slight, moderate, moderately severe, severe, very severe and extremely severe having average annual soil loss of <5, 5–10, 10–15, 15–20, 20–40 and 40–80 and >80 t ha^{-1} $year^{-1}$, respectively. Details of area under each class and major factors contributing to erosion are discussed class-wise. Although, the type of soil conservation measures to be adopted will depend upon the socio-economic condition, level of knowledge and awareness and attitude of farmers and contribution of the state departments (Meena et al. 2015d, e; Verma et al. 2015a). The basic frame-ups of technologies to be adopted under each class are discussed as under.

15.2.8.1 Very Slight Erosion Class

The land experiencing annual soil loss less than 5 t ha^{-1} falls under very slight erosion class. It covers an area of 1.20 M ha constituting 22 % of TGA of the state. Soil loss of 3.00 M tonnes occurs from this class. This class contributes only ~1 % of the total soil loss in the state. All the 12 districts of the state fall under this class of soil erosion having soil losses less than 5 t ha^{-1} $year^{-1}$. Majority of the area in Kangra, Mandi, Solan and Una districts is under this class. Significant parts of Pangi, Salooni, Chauraha, Dalhousie and Bhatog tehsils of Chamba; Ghumarwin and Jhandutta tehsils of Bilaspur; Rajgarh, Pachhad, Nahan, and Paonta Sahib tehsils of Sirmaur; Tira Sujanpur and Hamirpur tehsils of Hamirpur; Rampur, Seoni, Theog, Shimla, Jubbal and Kumharsain tehsils of Shimla; Kullu, Anni and

Nirmand tehsils of Kullu; Lahaul tehsil of Lahaul; and Spiti and Kinnaur, Kalpa and Sangla tehsils of Kinnaur districts also experience soil loss of less than 5 t ha^{-1} year^{-1}.

The land allowing less than 5 t ha^{-1} year^{-1} soil losses is due to combination of two to three favourable USLE factors in different sets. Various combinations found in Himachal Pradesh responsible for lower soil loss are:

1. Cultivation of paddy ($C = 0.2$) on nearly levelled lands in the valley (LS = 0.10) with medium to higher rainfall erosivity and soil erodibility
2. Cultivation of wide-spaced crops like maize, sorghum, etc. on nearly levelled land under low-rainfall condition (R <250 MJ-mm/ha-hr-yr) with medium to higher soil erodibility
3. Cultivation of wide-spaced crops like maize, sorghum, etc. on very gentle sloping land with medium soil erodibility and medium to high-rainfall erosivity
4. Cultivation of crops like sugarcane, potato, vegetables or orchards on very gentle sloping land having medium to high soil erodibility and rainfall erosivity
5. Good forest cover on moderately sloping lands in high-rainfall areas

The soil loss less than 5 t ha^{-1}year^{-1} is well within the safe limits. However, adopting simple agronomic practices and improving soil organic matter and structure can further curtail it. Intercropping of legumes, integrated nutrient management, organic farming in agricultural fields and good vegetation cover management in forestlands and pastures are suggested for further reducing the erosion rates. These practices will both decrease the soil erodibility (K) or crop factor (C) and, thus, reduce soil losses.

15.2.8.2 Slight Erosion Class

Land producing annual soil loss in the range of 5–10 t ha^{-1}year^{-1} is classified under slight erosion class. It covers an area of 0.38 M ha constituting 7 % of the state area. The area produces 2.85 M tonnes of soil loss, which is 1 % of the total soil loss observed in the state. These areas are well scattered in small zones around slight erosion regions. Majority of the area experiencing slight erosion is in the districts of Chamba, Kangra, Una, Bilaspur, Solan, Sirmaur, Hamirpur, Mandi, Shimla, Kullu, Lahaul and Spiti and Kinnaur districts.

Small areas under Pangi, Chaurah, Salooni, Dalhousie, and Bhatog tehsils of Chamba; Dehra and Jaisinghpur tehsils of Kangra; Una, and Amb tehsils of Una; Ghumarwin and Jhandutta tehsils of Bilaspur; Nalagarh, Arki and Kasauli tehsils of Solan; Rajgarh, Nahan, Ponta Sahib and Pachhad tehsils of Sirmaur; Hamirpur and Barsar tehsils of Hamirpur; Jogindernagar, Sarkaghat, Mandi, Sundernagar and Karsog tehsils of Mandi; Rampur, Seoni, Theog, Shimla, Jubbal and Kumharsain tehsils of Shimla; Kullu, Ani and Nirmand tehsils of Kullu; Lahaul tehsil of Lahaul and Spiti; and Kinnaur, Kalpa and Sangla tehsils of Kinnaur districts fall under this class of soil erosion experiencing annual soil loss in the range of 5–10 t ha^{-1}.

Although soil loss $<10\,\text{t ha}^{-1}\text{year}^{-1}$ is below the tolerance limit, adoption of soil conservation measures will further reduce the soil erosion losses and will enhance the soil productivity. The soil losses in this range are due to combination of erosion assisting and resisting factors of the USLE. These combinations in case of Himachal Pradesh are:

- Cultivation of wide-spaced erosion-promoting crops like maize, sorghum, etc. on nearly levelled land having highly erodible soils ($K > 0.40$)
- Current fallow on nearly levelled lands
- Cultivation of wide-spaced erosion-promoting crops like maize, sorghum, etc. on very gentle sloping land with moderately erodible soils
- Cultivation of crops like maize, legumes, etc. on terraces constructed across medium hills in high-rainfall areas
- Soils with low erodibility under dense forests on steep slopes in high-rainfall areas

Once the reasons are found, it is easier to evolve conservation strategy. The conservation planning in these areas should revolve around:

- Introducing close-spacing erosion-resistant crops, intercropping or strip cropping to reduce C factor
- Adoption of soil management practices to improve the organic matter and soil structure, thus reducing K factor
- Good pasture/forest development on community and wasteland
- Land levelling and bunding in gentle sloping fields
- Bringing current fallows under pasture, forest or horticultural crops
- Proper management of terraces with live hedges on the outer ridge and grass waterways

15.2.8.3 Moderate Erosion Class

Land experiencing annual soil loss in the range of 10–15 $\text{t ha}^{-1}\text{year}^{-1}$ is categorised as moderate erosion class. It covers an area of 0.30 M ha constituting about 5 % of the state area. Annually, 3.78 M tonnes of soils are lost from this area, which is about 1.5 % of the total soil loss in the state.

The area experiencing moderate soil losses is very small but distributed over the entire state of Himachal Pradesh. It is mainly concentrated in Bhatog and Brahmaur tehsils of Chamba; Dehra and Jaisinghpur tehsils of Kangra; Amb tehsil of Una; Ghumarwin and Bilaspur tehsils of Bilaspur; Nalagarh, Solan and Kasauli tehsils of Solan; Rajgarh, Nahan and Ponta Sahib tehsils of Sirmaur; Hamirpur and Barsar tehsils of Hamirpur; Jogindernagar, Sarkaghat, Mandi, Sundernagar and Karsog tehsils of Mandi; Rampur, Seoni, Theog, Shimla, Jubbal and Kumharsain tehsils of Shimla; Kullu, Ani and Nirmand tehsils of Kullu; Lahaul tehsil of Lahaul and Spiti; and Kinnaur, Kalpa and Sangla tehsils of Kinnaur districts.

The soil losses produced from these areas are higher than the tolerance limits and, thus, require corrective measures to stop degradation of the land. The major causes of water-induced erosion in this range are due to:

- Moderate erosion due to slope in the range of 3–10 % in high-rainfall areas.
- Poor fertility status of soils resulting in initial poor crop stand during high erosive rainfall storms.
- Crop cultivation on area suitable for perennial plantations without soil conservation measures on moderate slope.
- Soils with low K value under dense forest on fragile hills in high-rainfall regions.
- Cultivation without appropriate conservation measures to check erosion and surface cover to check soil fertility depletion on medium sloping landscape.
- Practice of old traditional cropping system like keeping fallow during kharif for moisture conservation.
- Poor rainfall with low erosivity in the range of 100–250 MJ-mm/ha-hr-yr in steep hills.
- Specific soil and water conservation practices need to be adopted to check water-induced soil erosion. Following practices could prove effective in reducing soil erosion.
- Crop diversification with emphasis on erosion-resistant crops and cropping patterns, viz. intercropping, strip cropping, etc.
- Change of cropping system from old traditional to new scientific cropping sequences.
- Safe disposal of excess run-off to water-harvesting bodies, viz. ponds, dams, etc.
- Introduction of silvi-pastoral or horti-pastoral systems on community and degraded lands.
- Appropriate engineering measures like land levelling, terracing in agricultural fields and contour trenching in other land uses.
- Contour farming to check soil erosion and conserve soil moisture in soil profile in low-rainfall regions.

Moderate erosion around 10–15 t $ha^{-1}year^{-1}$ seldom occurs in isolation; rather, it is a transition phase from low to severe erosion in a landscape or watershed. The conservation measures can neither be applied in isolation, but these measures will form part of integrated approach of watershed treatment.

15.2.8.4 Moderately Severe Erosion Class

About 4 % of the state area (0.21 M ha) lies under this erosion class. The land under this class experiences 15–20 t ha^{-1} soil losses annually. The area under this erosion class produces ~4 % of the total soil loss. The areas experiencing 15–20 t $ha^{-1}year^{-1}$ of soil loss are distributed in almost the whole state in small patches. Moderate-to-very severe classes generally occur in continuum. The excess run-off from these areas may cause gully erosion and silt deposition in downslope fields (Fig. 15.3).

The area experiencing moderately severe soil erosion is spread over in Bhatog and Brahmaur tehsils of Chamba; Dehra and Khundian tehsils of Kangra;

Fig. 15.3 Exposure of tree roots as a result of soil erosion

Ghumarwin and Bilaspur tehsils of Bilaspur; Nalagarh, Solan and Kasauli tehsils of Solan; Renuka and Ponta Sahib tehsils of Sirmaur; Hamirpur and Barsar tehsils of Hamirpur; Sundernagar and Karsog tehsils of Mandi; Rampur and Chaupal tehsils of Shimla; Manali, Kullu, Ani and Banjar tehsils of Kullu; and Kalpa, Puh and Sangla tehsils of Kinnaur districts.

The major factors responsible for causing water-induced erosion of this magnitude are:

1. Moderate erosion due to slope in the range of 3–10 %in high-rainfall areas.
2. Poor fertility status of soils resulting in initial poor crop stand during high erosive rainfall storms in June to July months.
3. Crop cultivation on area suitable for perennial plantations without soil conservation measures on moderate slope.
4. Soils with low K value under dense forest on fragile hills in high-rainfall regions.
5. Adoption of inappropriate conservation practices like contour farming and degraded pastures with live hedges on medium sloping landscape instead of terraced contour farming.
6. Practice of old traditional cropping system like keeping fallow during kharif for moisture conservation.
7. Poor rainfall with low erosivity in the range of 100–250 MJ-mm/ha-hr-yr in medium to high hills.

Specific soil and water conservation practices need to be adopted to check water-induced soil erosion. Following practices could prove effective measures in reducing soil erosion.

1. Conservation farming with emphasis on erosion-resistant crops and cropping patterns, viz. intercropping, strip cropping, etc.

2. Shifting from erosion-susceptible old traditional cropping system to erosion-resistant and more remunerative land uses like horticultural and forestry-based cropping systems.
3. Contour farming to check soil erosion and conserve moisture in soil profile in gentle sloping lands.
4. Adoption of integrated watershed management approach in the vast areas which include land experiencing varying levels of erosion.
5. Appropriate engineering measures like land levelling, terracing in agricultural fields and contour trenching in other land uses.

15.2.8.5 Severe Erosion Class

Land experiencing annual average soil loss in the range of 20–40 t ha^{-1} is classified under severe erosion class. It covers an area of 0.41 M ha, which constitutes >7 % of the state area. Annually, 12.37 M tonnes of soils are eroded from this area, which is about 5 % of the total soil loss in the state (Fig. 15.4).

The area experiencing severe erosion is spread over the entire state although the area is small but covers all the districts of the state. Severe erosion class includes tehsils of Dalhousie, Bhatog and Bharmour of Chamba; Indora, Jawali, Fatehpur, Dehra and Jaisinghpur of Kangra; Una and Bangana of Una; Bilaspur and Ghumarwin of Bilaspur; Nalagarh, Solan and Kasauli of Solan; Renuka, Shillai and Ponta Sahib of Sirmaur; Nadaun, Hamirpur and Barsar of Hamirpur; Jogindernagar, Mandi, Sunder Nagar, Karsog and Paddar of Mandi; Seoni, Shimla, Jubbal, Theog, Chaupal and Chirgaon of Shimla; Manali, Kullu, Banjar and Anni of Kullu; Spiti of Lahaul and Spiti; and Sangla, Morang and Puh of Kinnaur districts.

The major reasons behind such high rate of soil erosion in these areas are combinations of more than two unfavourable factors of universal soil loss equation and these combinations are as under (Fig. 15.5):

1. Degraded forest or wastelands on moderate to moderately steep slopes in high-rainfall regions.

Fig. 15.4 Man-made soil erosion (**a**, **b**) due to road cut in Shivalik. (**c**) Severely eroded Shivalik hills on Bilaspur–Shimla road

Fig. 15.5 The control of soil erosion through vegetation in Shivaliks. (**a**) Soil binder (bhabbar grass) in Shivaliks and (**b**) soil binder (cactus-prickly pair)

2. Medium-managed forests or pastures on highly degradable hills in medium-to-high-rainfall areas.
3. Non-utilisation of current fallow for biomass production in high-rainfall areas.
4. Cultivation of annual crops in high-rainfall regions on soil of medium erodibility and moderate slopes up to 8 %.
5. Indiscriminate damage to forest trees either through overgrazing or illicit cutting.
6. Conservation measures in these regions should include combination of biotic and abiotic measures coupled with suitable policy decisions.

1. Integrated watershed management approach with emphasis of catchment's treatment to reduce sediment flow.
2. In situ soil moisture conservation for establishment and growth of forest ecosystem.
3. Diversion of excess run-off in monsoon season to water-harvesting structures for supplemental irrigations to plants during the first 2–3 years of establishment.
4. Establishment of permanent cover on the current fallows for proper resource utilisation.
5. Adoption of silvi-pastoral or horti-pastoral systems with emphasis on cover management for soil and water conservation.
6. Ensuring people participation in hill development and sharing of the forest resources through suitable legislations. Earlier programmes in this region have failed owing to non-participation of the people in resource management.

Fig. 15.6 (**a**) Very severe water erosion – Shivaliks near Panchkula. (**b**) Rock outcrops as a result of very severe soil erosion in Shivaliks

15.2.8.6 Very Severe Erosion Class

Land experiencing annual soil loss in the range of 40–80 t ha^{-1} is categorised under very severe erosion class. It covers an area of 0.32 M ha, which constitute about 6 % of the total state area. The annual soil loss from this class is 19.18 M tonnes, which is about 7.5 % of the total soil loss from the state (Fig. 15.6).

Very severe erosion covers the entire state except Bilaspur district where no area falls under this class of soil erosion. This class includes tehsils of Salooni, Dalhousie, Chamba and Bharmour of Chamba; Indora, Jawali, Fatehpur, Dehra, Jaisinghpur and Dharamsala of Kangra; Una and Bangana of Una; Arki, Solan and Kandaghat of Solan; Shillai and Ponta Sahib of Sirmaur; Barsar of Hamirpur; Jogindernagar, Mandi, Sunder Nagar, Karsog and Paddar of Mandi; Rampur, Kumharsain, Shimla, Jubbal, Theog, Chirgaon and Kotkhai of Shimla; Kullu, Nirmand, and Saraumga glacier area of Kullu; Spiti of Lahaul and Spiti; and Sangla, Kalpa and Puh of Kinnaur districts.

The average annual rainfall in this region is good enough to cause high magnitude of run-off. High run-off on steep slope carry heavy load of detached soil mass leading to the development of rainy season streams called 'torrent'. These rainy-season torrents, locally known as *choes*, spread in plains as their carrying capacities are reduced and deposit boulders and silts in the streambed. The torrents frequently change their course bringing new fertile land under degradation. These regions are deprived of canal irrigation facility because of the undulating topography and tube-well irrigation due to very deep groundwater tables.

Although the area under this class is very less, it causes havoc not only in the catchment area but also in the command area through flash floods in torrents passing through them. The region facing very severe erosion occurs in close association of region with extremely severe erosion, and the same set of factors is responsible for their cause. During heavy rains, the torrents overflow banks causing inundation of adjoining lands, cutting the *choe* banks, uprooting trees and electric and telephone poles and damaging roads. The areas facing very severe erosion generally occupy the top position of the landscape and will erode the middle and lower topographic positions if not controlled at the site.

The conservation measures in these regions should focus on solving the twin problems of high soil losses and water deficit during non-monsoon seasons. This can be achieved only through watershed approach, and the state government is making the earnest efforts through various phases of World Bank-funded Integrated Watershed Development Project (hills). The watershed programme should include the following components:

1. Soil conservation measures in the catchment for in situ moisture conservation to enhance vigorous plant growth to reduce run-off and sediment loss.
2. Inclusion of rainwater-harvesting component for recycling it for crop production and reducing peak flow to torrents.
3. Gully control structures in the upstream channel for stabilisation of torrent in hilly catchment.
4. Construction of engineering structures and vegetative measures to train the course of the water flow in the torrent.
5. Stabilisation of torrent bank and rehabilitated portion of the bed with plantation of suitable tree, bush and grass species.
6. Adoption of land and crop management practices for restoration for degraded fields around torrents.
7. All efforts should be concentrated on maintaining good cover of forest and grass. Only C value <0.01 can bring down soil loss of 40 t ha^{-1} or more to below tolerance limit.

15.2.8.7 Extremely Severe Erosion Class

Land experiencing annual soil loss >80 t ha^{-1} is categorised under extremely severe erosion class. The soil losses up to 683 t ha^{-1} year^{-1} can be seen at some places in the state. It covers an area of 0.56 M ha, which constitutes ~10 % of the total state area. The area under this class of soil erosion is only 10 %, but it contributes more than 80 % of soil loss 213.23 t ha^{-1} year^{-1}. The area with extremely severe erosion occurs in close proximity with the regions experiencing very severe erosion (Fig. 15.7).

Extremely severe soil erosion covers a major part of Kullu, Shimla, Kangra and Chamba districts, while no part of Hamirpur and Bilaspur districts faces the problem. Besides the four districts having a major problem, parts of Bangana and Una of Una; Kandaghat tehsil of Solan; Shillai and Ponta Sahib tehsils of Sirmaur; Karsog, Mandi and Paddar tehsils of Mandi; Spiti tehsil of Lahaul and Spiti; and Puh tehsil of Kinnaur districts have large chunk of areas under severe form of degradation.

The main factors responsible for torrential form of erosion in these hills are:

Fig. 15.7 A schematic representation of various erosion sites and different management practices for its control. (**a**) Extremely severe soil erosion (screes) in Spiti; (**b**) Extremely severe soil erosion due to landslide near Khadradhang in Lesser Himalayas; (**c**) Extremely severe soil erosion on sidehill slopes cultivation in stabilised area (Lahaul valley); (**d**) Wind erosion – sand remnants looking like sentinels; (**e**) Extremely severe wind erosion – Spiti valley; (**f**) Cultivation on levelled land Sangla valley (Regional Research Station HPKVV); (**g**) Soil erosion control measures in Greater Himalayas and soil conservation practices in Spiti valley; (**h**) Natural soil binders (*Ephedra* species) near Yang Thang – the highest village connected by roads in the world (**i**) Ogla (*Fagopyrum esculentum*) cultivation in Sangla valley

1. Cultivation of potato, paddy and other cereal crops on terraced fields in hills with slopes in the range of 8–50 %
2. Poor- to medium-managed forests with canopy less than 75 % on moderate to steep slopes in high-rainfall zones
3. Cultivation of crops like maize, potato, etc. on gentle-to-moderate slopes without conservation measures
4. Cultivated fallows on moderate slopes
5. Denuded hills without appreciable vegetation in low-rainfall areas in north-western part of the state

The factors responsible for erosion and management measures are almost similar as discussed under class very severe erosion, but 83 % of the total soil loss is caused by this class alone having only 10 % of state geographical area. So there is an urgent need to undertake appropriate soil conservation measures for the management of this area. There is need to implement integrated approach of watershed management and mechanical and engineering measures of advanced level in conjunction to scientific farming, which can be helpful in minimising the soil loss by controlling torrential erosion.

In addition to the area under above seven erosion classes, about 2.18 M ha area, (39 % of TGA of the state), is under rock outcrops and snow and water bodies. In this area either soil is absent or not available at the surface for detachment. So this area has not been discussed under soil erosion classes, but it also requires certain type of conservation measures.

15.3 Conclusions

It is inferred from the studies that state of Himachal Pradesh is suffering from very slightly to slight, moderate, moderately severe, severe, very severe and extremely severe erosion soil losses. Topography, slope, rainfall and vegetation canopy influence the soil erosion losses to a great extent. Gridded soil information is generated from the soil map of the state using universal soil loss equation. Suitable soil conservation measures were suggested for different types of soil erosion to reduce soil losses.

Acknowledgements The authors convey their thanks to the staff of ICAR-NBSS&LUP and ICAR-IISWC for carrying the soil resource survey and computation of soil loss of grid observations. The critical review of the MS by reviewers and editors is thankfully acknowledged.

References

Anonymous (2008) Harmonization of wastelands/degraded lands datasets of India. Published by National Rainfed Area Authority (NRAA). Ministry of Agriculture. Government of India, NASC Complex, DPS Marg, New Delhi – 110012, 5p

Anonymous (2014) District-wise area and production of crops. Directorate of Economics and Statistics, Department of Agriculture and Co-operation, Ministry of Agriculture, Govt. of India, New Delhi

Brown LR (1991a) The global competition for land. J Soil Water Conserv 46:394–397

Brown LR (1991b) Feeding six billion. The world watch reader on global environmental issues. World Watch Inst, Washington, DC, pp 147–164

Dhruva Narayana VV, Babu R (1983) Estimation of soil erosion in India. J Irrig Drain Eng 109:419–434

El-Swaify SA (1991) Land based limitation and threat to world food production. Outlook Agric 20:235–242

Ghosh BN, Meena VS, Alam NM, Dogra P, Bhattacharyya R, Sharma NK, Mishra PK (2016) Impact of conservation practices on soil aggregation and the carbon management index after

seven years of maize – wheat cropping system in the Indian Himalayas. Agr Ecosyst Environ 216:247–257
Gurung SM (1989) Beyond the myth of eco-crisis in Nepal: local response to pressure on land in the middle hills. Dissertation Abstracts Intl., A Humanities and Social Sciences 49, Thesis, University of Hawaii 1988, 312 pp
HPKKV (1991) Status report on Kinnaur and Spiti catchment of Satluj River in Himachal Pradesh. HPKVV, Palampur, 179 + i-XLII p
ICAR, NAAS (2010) Degraded and wastelands of India status and spatial distribution, ICAR and NAAS Publ, pp 1–158
Jacobson T, Adams RM (1958) Salt and silt in ancient Mesopotamian agriculture. Science 128:1251–1258
Kazgan G (1991) Population growth, land depletion and productivity. Medelingan van de FaculteitLandbouwwtenschappen. Rijksuniversiteit Gent 56:67–73
Khybri ML, Prasad SN, Ram S (1980) Effect of topsoil removal on the growth and yield of rainfed maize. Indian J Soil Conserv 8:164–170
Kumar A, Bahadur I, Maurya BR, Raghuwanshi R, Meena VS, Singh DK, Dixit J (2015) Does a plant growth promoting rhizobacteria enhance agricultural sustainability? J Pure Appl Microbiol 9(1):715–724
Lal R (1998) Soil erosion impact on agronomic productivity and environmental quality. Crit Rev Plant Sci 17(4):319–464
Lowdermilk W (1953) Conquest of land through 7000 years, SCS Agric. Information Bull.99. Govt. Printers, Washington, DC
Mahapatra SK, Rana KPC, Sidhu GS, Walia CS (2000) Assessment of soil degradation status of Jammu and Kashmir State for their amelioration. J Indian Soc Soil Sci 48(3):577–580
Mannering JV (1981) The use of soil tolerance as a strategy for soil conservation. In: Morgan RPC (ed) Soil conservation: problem and prospects. Wiley, Chichester, pp 337–349
McCracken RJ (1987) Soils, soil scientists and civilization. Soil Sci Soc Am J 51:395–1400
Meena RS, Yadav RS, Meena VS (2013) Heat unit efficiency of groundnut varieties in scattered planting with various fertility levels. Bioscan 8(4):1189–1192
Meena RS, Meena VS, Meena SK, Verma JP (2015a) The needs of healthy soils for a healthy world. J Clean Prod 102:560–561
Meena RS, Meena VS, Meena SK, Verma JP (2015b) Towards the plant stress mitigate the agricultural productivity: a book review. J Clean Prod 102:552–553
Meena VS, Meena SK, Verma JP, Meena RS, Ghosh BN (2015c) The needs of nutrient use efficiency for sustainable agriculture. J Clean Prod 102:562–563
Meena VS, Meena SK, Verma JP, Meena RS, Jat LK (2015d) Current trend to mitigate climate change and sustainable agriculture: a book review. J Clean Prod 102:548–549
Meena VS, Verma JP, Meena SK (2015e) Towards the current scenario of nutrient use efficiency in crop species. J Clean Prod 102:556–557
Meena RS, Bohra JS, Singh SP, Meena VS, Verma JP, Verma SK, Sihag SK (2016) Towards the prime response of manure to enhance nutrient use efficiency and soil sustainability a current need: a book review. J Clean Prod 112(1):1258–1260
Olson GW (1981) Archeology: lessons on future soil use. J Soil Water Conserv 36:261–264
Olson GW (1985) Pedological soil properties which indicate pre-historical chemical and physical changes in the soils. In: Farrington IS (ed) Prehistoric intensive agriculture in tropics, vol 2. BAR International Series, Oxford, pp 717–736
Pimentel D, Allen J, Beers A, Guinand L, Hawkins A, Linder R, McLaughlin P, Meer B, Musonda D, Perdue D, Poison S, Salazar R, Siebert S, Stoner K (1993) Soil erosion and agricultural productivity. In: World soil erosion and conservation. Cambridge Studies in Applied Ecology and Resource Management, Cambridge, pp 277–292
Pimentel D, Harvey C, Resodudarmo P, Sinclair K, Jurz D, McNair M, Crist S, Shpritz L, Fiton L, Saffouri R, Blair R (1995) Environmental and economic costs of soil erosion and conservation benefits. Science 267:1117–1123

Ram B, Tejwani KG, Agarwal MC, Subhash C (1978) Rainfall erosion potential and iso-erodent map of India. Central Soil and Water Conservation Research and Training Institute, Dehradun, Bulletin No. 2, p 47

Rao RVS, Mahapatra SK, Verma TP, Sidhu GS, Rana KPC (1997) Characterization and classification of some soils of Siwalik Hills of H.P. Agropedology 7(1):14–21

Renard KG, Laften JM, Foster GR, McCool DK (1994) The revised soil loss equation. In: Lal R (ed) Soil erosion – research methods. Soil and Water Conservation Society, Ankeny, pp 105–124

Sidhu GS, Rana KPC, Sehgal J, Velayutham M (1997) Soils of Himachal Pradesh for optimising land use. NBSS Publ. 57b (Soils of India Series). National Bureau of Soil Survey and Land Use Planning Nagpur, India. 44pp. + 2 sheets soil map (1:500 000 scale)

Sidhu GS, Walia CS, Sachdev CB, Rana KPC, Dhankar RP, Singh SP, Velayutham M (2000) Soil resource of N-W Siwaliks for perspective land use planning. In: Mittal SP, Aggarwal RK, Samra JS (eds) Fifty years of research on sustainable resource management in Shivaliks. CSWCRTI, Research Centre, Chandigarh, pp 23–34

Sidhu GS, Rana KPC, Tarsem L, Mahapatra SK, Verma TP, Rao RVS, Ram J, Singh SK, Singh SP (2007a) Soils of Himachal Pradesh: inventory for land use planning. J Indian Soc Soil Sci 55 (2):203–208

Sidhu GS, Rana KPC, Tarsem L, Mahapatra SK, Verma TP, Rao RVS (2007b) Soils of Himachal Pradesh-II: land capability classification and assessment of soil degradation status for suggested land use. J Indian Soc Soil Sci 55:335–339

Sidhu GS, Yadav RP, Sharma JP (2009) Soil erosion status in Himachal Pradesh for land use planning. In: Bhan S, Karale RL, Singh S, Bharti VK, Bali JS (eds) Conservation farming, post publication of 2nd international conference. Soil Conservation Society of India, New Delhi, pp 259–268

Sidhu GS, Yadav RP, Singh SP, Sharma JP, Aggarwal RK, Tiwari AK, Gajbhiye KS, Sarkar D, Sharda VN (2010) Soil erosion status of Himachal Pradesh. NBSS & LUP Publ. Bulletin No. 132. (ISBN -81-89043-11-0), 50 p

Singh G, Ram B, Subhash C (1981) Soil loss prediction research in India, ICAR Bulletin No.T-12/ D-9. Central Soil and Water Conservation Research and Training Institute, Dehradun, p 70

Singh DK, Singh SK, Singh AK, Meena VS (2014) Impact of long term cultivation of lemon grass (Cymbopogon citratus) on post-harvest electro-chemical properties of soil. Ann Agric Biol Res 19(1):45–48

Suri VK, Sidhu GS, Kumar A (2013) Physical attributes: soil and landscape characteristics of Western Himalayan region of India. In: Chopra VL (ed) Climate change and its ecological implications for the Western Himalaya. Scientific Publishers, Jodhpur, pp 1–48. ISBN 9788172338091

Tiwari DN (1998) Green accounting: natural resource depreciation, environmental costs and implications for environmental policy integration for Nepal. Draft paper. Kathmandu

Velayutham M, Bhattacharyya T (2000) Soil resource management. In: Yadav JSP, Singh GB (eds) Natural resource management for agricultural production in India. Secretary General, International Conference on Managing Natural Resources for Sustainable Agricultural Production in the 21st Century, New Delhi, pp 1–137

Verma JP, Jaiswa DK, Meena VS, Meena RS (2015a) Current need of organic farming for enhancing sustainable agriculture. J Clean Prod 102:545–547

Verma JP, Jaiswal DK, Meena VS, Kumar A, Meena RS (2015b) Issues and challenges about sustainable agriculture production for management of natural resources to sustain soil fertility and health. J Clean Prod 107:793–794

Wallace A (1994) Soil improvement to match 21st century agriculture. Commun Soil Sci Plant Anal 25:165–169

Wischmeier WH, Smith DD (1958) Rainfall energy and its relationship to soil loss. Trans Am Geophys Union 39:285–291

Wischmeier WH, Smith DD (1978) Predicting rainfall erosion losses – a guide for conservation planning, Agriculture Hand book No. 537. USDA, Washington, DC, p 58

Wolman MG (1985) Soil erosion and crop productivity: a worldwide perspective. In: Follet RF, Stewart BA (eds) Soil erosion and crop productivity. American Society of Agronomy, Madison, pp 9–21

Yadav RP, Sidhu GS (2010) Assessment of soil erosion in Himachal Pradesh. J Indian Soc Soil Sci 58(2):212–220

Towards the Natural Resource Management for Resilient Shifting Cultivation System in Eastern Himalayas

16

Rakesh Kumar, M.K. Patra, A. Thirugnanavel, D. Chatterjee, and Bidyut C. Deka

Abstract

Eastern Himalayan states in general and Nagaland in particular are known for their diversified cultural heritage and jhum cultivation as mainstay of livelihood. Inhabitants in this region are otherwise non-vegetarian and rice is the staple food. Traditional jhum cultivation is not subsistent to fulfil the requirement of the increasing population in the hilly states. Reduced jhum cycle (3–5 years), widespread water scarcity during the post-monsoon seasons, subsistence agriculture practice, lack of awareness about improved agriculture technologies, poor credit and marketing facilities, lack of infrastructures, pitiable socio-economic status of the people, etc. are the major issues confronted by the tribal people in the eastern Himalayan region. In this perspective, holistic approaches with modern agro-based interventions for augmenting production, productivity, income and employment generation in a sustainable manner are the need of the hour. This document depicts the agricultural scenario in several cluster villages in different districts particularly Mon, Wokha, Longleng, Peren, Kohima and Dimapur in Nagaland. The introduction and popularization of scientific intervention in agriculture and allied sector have changed the mindset of many tribal poors in adopted villages and became instrumental in the transformation of traditional jhum cultivation into a profitable agricultural enterprise. Moreover, the farmers have accepted the modern agro-techniques and high-yielding varieties (HYVs) of crops and livestock, cultivation practices, soil and water conservation measures and integration of livestock components in traditional agriculture. Increasing cropping intensity, productivity, gross returns from unit area and several-fold increase in employment generation were among the few several achievements following the implementation of agro-based intervention.

R. Kumar (✉) • M.K. Patra • A. Thirugnanavel • D. Chatterjee • B.C. Deka
ICAR-RC for NEH Region Nagaland Centre, Jharnapani, Medziphema 797 106, India
e-mail: rakeshbhu08@gmail.com

J.K. Bisht et al. (eds.), *Conservation Agriculture*,
DOI 10.1007/978-981-10-2558-7_16

Keywords
Shifting cultivation • Restoration of soil fertility • Integrated farming • Eastern Himalayan region • Livelihood security

16.1 Introduction

Shifting cultivation traditionally named as 'jhum' is considered as one of the oldest practices in the eastern Himalayan region of India particularly in Nagaland. In this form of agriculture, a piece of forest land is slashed, burnt and cropped without tilling the soil, and the cropped land is subsequently fallowed to attain pre-slashed forest status through natural succession (Rathore et al. 2012). Over the past two decades, due to increasing human population, the jhum cycle in the same land, which extended to 10–12 years in earlier days, has now been reduced to 3–6 years (Singh et al. 2003). A huge number of tribal farmers are involved in this primitive cultivation, and estimation reveals that more than 100 indigenous tribes and over 6.2 lakh families of north-eastern region depend on jhum cultivation (Patiram and Verma 2001; Anonymous 2002). ~85 % of the total cultivation area in north-east India is covered by shifting cultivation (Rathore et al. 2012). The existing shifting cultivation practice in north-east India is an injudicious form of land use. Cropping is continued for one or two cropping seasons only till all nutrients are mined up. Then the lands are left for again growing of forest and the same site is chosen after 10–15 years. Since complete eradication of shifting cultivation is practically impossible, research for prescribing resilient shifting cultivation for sustainable development is required.

The existing scenario of shifting cultivation practice and its impact on livelihood has been studied during the past several years to chalk out the strategies for developing the resilient shifting cultivation practice with a goal of sustainable development and food security. The phenomenon encircling shifting cultivation with soil-water-plant-society continuum approach has shown positive impact for holistic development of the tribal populations who are otherwise dependent on shifting cultivation and traditional animal husbandry for their livelihood security. This chapter describes the present scenario of shifting cultivation, its impact on soil fertility and agriculture production system and possible alternatives applied for augmenting the productivity of crops and livestock, opportunities in income and employment generation through livestock, horticulture and agroforestry-based farming system for gradual transformation towards settle cultivation in Nagaland. Adoption of improved agro-based technology has resulted in the restoration of degraded jhum land to a settled cultivation consisting of agriculture, horticulture and animal husbandry components in several cluster villages of different districts, namely, Mon, Longleng, Wokha, Peren, Dimapur, etc. This success in cluster villages was the eye-opener for many tribal farmers in the state.

16.2 Present Scenario of Shifting Cultivation

16.2.1 Impact on Soil Degradation

- Severe soil and nutrient losses accompany land clearing in the early stages of plantations.
- Shifting cultivation in north-east India has increased the problem of land degradation.
- Rapidly expanding population pressure has resulted in misuse of land resource.
- Estimates reveal 88.3–146 MT of the soil is lost annually as a result of shifting cultivation.

A study conducted by Arunachalam (2002) reported that soil organic carbon, available P, total Kjeldahl nitrogen, ammonium N and nitrate N decreased as the duration of cultivation increased under jhum cropping. However, the microbial biomasses C, N and P were high in forest stand. Microbial biomass C increased gradually as cultivation progressed, while microbial biomass N and P showed postburn decreasing trend. Bacterial and fungal populations were drastically reduced following the slash burning. Over the past two decades, due to increasing human population, jhum cycle in the same land, which extended to 20–30 years in older days, has now been reduced to 3–6 years (Borthakur 1992; Singh et al. 2003). Presently, it is estimated that the number of people practising shifting cultivation is around 367,000 tribal families, and the area affected by this practice is ~385,400 ha annually (Patiram and Verma 2001). The loss of 100–250 metric tonnes of topsoil ha^{-1} $year^{-1}$ is depleted due to jhum cultivation in Bangladesh hills (Karim and Mansor 2011). On an average, an area of 3869 km^2 is put under shifting cultivation every year. Excessive deforestation (net loss of 1577 km^2 forest cover between 1999 and 2001) coupled with jhum practice has resulted in tremendous soil loss. In general, shifting cultivation practices deteriorate soil fertility due to huge soil loss of ~2–200 t ha^{-1} $year^{-1}$ (Saha et al. 2011), and a minimum period of 10–15 years is very much essential to maintain the soil fertility for sustainable crop production (Singh et al. 2003; Meena et al. 2013, 2015a, b; Singh et al. 2014; Kumar et al. 2015a; Ghosh et al. 2016). Carbon and nitrogen in soil are the most limiting factors for plant growth after the forest is cut and then burned. Mishra and Saha (2003) reported that only fallow period under shifting cultivation is not enough for consideration of restoration capacity of soil.

16.2.2 Constraints in Jhum

Although jhum is practised since time immemorial, it faces several constraints which lead to a decline in the productivity and profitability. The major constraints listed below came into our knowledge while interacting with *jhumias* in several forums:

- Reduced jhum cycle: earlier cycle was 10–15 years which is now reduced to 3–5 years.
- Water scarcity during the post-monsoon/winter seasons particularly from Oct. to March.
- Lack of awareness about improved agriculture technologies and vegetable cultivation.
- Monocropping with traditional practices and management.
- Lack of high-yielding quality of seed and planting materials.
- Indigenous breed of livestock and poultry managed with zero to negligible input.
- Shortfall in animal protein availability.
- Lack of disease control facilities for crop and livestock.
- Poor socio-economic status of the people.
- Poor credit and marketing facilities.

16.3 Alternatives to Sustaining the Jhum Farming

Since a huge number of tribal farmers are involved in this cultivation, therefore the complete eradication of this method is practically impossible. Thus, only two ways are left to check the damage of the environment either by increasing the jhum cycle or taking some immediate reclamation process to the areas already affected by jhum through a wasteland development plan. Arunachalam and Arunachalam (2002) suggested the utilization of bamboos in eco-restoration of jhum fallows in Arunachal Pradesh. Soil pH increased after burning and decreased as the cultivation progressed in the jhum field. In different parts of north-east India, land is opt to be abandoned after the first year of jhum cropping, and second-year cropping is sometimes practised with plantations of banana and pineapple (Anonymous 2002). However, the crops, viz. maize (*Zea mays* L.), beans (*Phaseolus vulgaris* L.), potato (*Solanum tuberosum* L.), orange (*Citrus* spp.), colocasia (*Colocasia esculenta* L.), tapioca (*Manihot esculenta* L.), pineapple (*Ananas comosus* L.), etc., are popularly grown in a jhum field, while the wet rice cultivation on terrace (*panikheti*) is practised in Kohima, Phek, Peren and Dimapur district of Nagaland. Studies about soil fertility status under different jhum cycles at various elevations suggest that plots with short jhum cycles (5 years) have lower fertility than those with longer cycles (10 or more years) as reported by Mishra and Saha (2003). Saha et al. (2011) studied the soil erodibility characteristics under six land-use systems, i.e. agriculture, agri-horti-silvi-pasture, natural forest, livestock-based land use, natural fallow and shifting cultivation (*jhum*). They also observed that shifting cultivation showed the highest erosion ratio (12.5) followed by agriculture (~10.4), indicating the need to adopt tree-based land-use systems for resource conservation. They also reported that soil loss was significantly higher in shifting cultivation (30.2–170.2 t ha^{-1} $year^{-1}$), agriculture (5.1–68.2 t ha^{-1} $year^{-1}$) and livestock-based land-use systems (0.88–14.3 t ha^{-1} $year^{-1}$) as compared to other modified land-use systems. Rathore and Bhatt (2008) observed that rice-vegetable pea/bean cropping system was the most suitable under jhum land of Nagaland, and integration of fish, pig, dairy cattle, duck and crops such as rice, vegetable pea and French bean showed maximum system productivity followed by cultivation of rice, vegetable pea and beans along with dairy cattle (free grazing).

16.4 Scientific Intervention Towards Sustainable Agricultural Production

To address these constraints, various scientific interventions in soil, water and nutrient conservation, crops, livestock and subsidiary enterprises were undertaken:

- Terracing of existing *jhum* land for wet rice (*panikheti*)-based farming system
- Increasing of cropping intensity by introduction short-duration crops after rice fallow
- Introduction of high-yielding varieties of crop and improved agro-techniques
- Restoration of degraded *jhum* lands through agroforestry-based farming system
- Introduction of water-harvesting techniques and its multiple uses
- Horticulture-based farming system
- IFS through promoting scientific animal husbandry (e.g. pig-fish integration)
- Capacity-building programmes

16.4.1 Paddy-Based Farming System in Terrace

In Mon district of Nagaland under the NAIP project, an initiative was taken to introduce the settle cultivation where altogether ~210 nos. of terraces were constructed in degraded jhum land covering an area of ~7.5 ha. It has in turn resulted in settled cultivation of paddy and other crops. By seeing the success, several farmers of the region have started to make terraces themselves, thereby helping in significantly reducing the jhum area of the region. Cropping intensity which was ~104 % before the intervention has now increased to ~146 %. Moreover, the farmers have accepted the modern agro-techniques and high-yielding varieties of crops like paddy, maize, rapeseed and mustard, potato, beans, tapioca and other vegetables (Fig. 16.1). This has finally helped the farmers to increase the

Fig. 16.1 Paddy/*toria* cultivation on terrace

Table 16.1 Production and productivity of different crops in cluster villages of Mon district in Nagaland

Crops	Cumulative		Productivity (t/ha)	
	Total area covered (ha)	Household (no.)	Local	Improved
Rice	12.2	38	1.9	2.44
Maize	1.25	22	0.70	1.2
Vegetables	1.75	37	4.50	6.40
Colocasia	3.25	35	10.50	12.25

productivity up to two- to threefolds higher than traditional cultivars (Chatterjee et al. 2012; Sahoo et al. 2012a; Deka et al. 2013; Meena et al. 2015c, d).

In upland condition, paddy-maize-based cropping system was introduced with the use of high-yielding crops such as paddy (SARS-1, 2, 4, 5; Bhalum-1, 2, 3; Rakchu, Lampanah, Shasarang) and maize (RCM-76, Navjot composite), which became popular among the farmers due to its higher yield potential. Among the different crops, the productivity of paddy, maize, vegetables and colocasia was increased from 1.9 to 3.5 t ha^{-1}, 0.7 to 1.2 t ha^{-1}, 4.5 to 6.4 t ha^{-1} and 10.5 to 12.3 t ha^{-1}, respectively (Table 16.1). Due to increase in productivity, the net income for paddy, maize, vegetables and colocasia was enhanced to Rs. 20,000–79,056 (Deka et al. 2013; Chatterjee et al. 2012; Sahoo et al. 2012b).

Monocropping was practised in the village before the intervention in the form of terrace making and introduction of high-yielding upland paddy varieties (Munda et al. 2012). Indigenous landrace, foxtail millet, colocasia, tapioca and some seasonal vegetables were grown by the farmers in the jhum fields prior to the interventions. The productivity of local upland paddy was found to be lesser than 1 t ha^{-1}; likewise, the productivity of colocasia, tapioca and foxtail millet was recorded to be 5.97, 21.8 and 1.4 t ha^{-1}, respectively. Through wet land/terrace paddy cultivation, an attempt was made to increase not only the rice productivity but also to introduce second cropping, which otherwise was not followed by the stakeholders (Chatterjee et al. 2012; Sahoo et al. 2012a; Deka et al. 2013; Verma et al. 2015b; Meena et al. 2016).

Terracing was done in the lower part of the hillock in Lampong Sheanghah village having a slope lesser than ~30 %. Paddy varieties (Shasarang and Lampanah) developed by ICAR RC for NEH Region, Umiam, Meghalaya, were cultivated, and the productivity of the varieties was threefold higher than the indigenous landraces, i.e. Rakchu having the productivity of 1.27 t ha^{-1} in wetland or terrace condition. Vegetable crops, viz. tomato, French bean, carrot, radish, potato and coriander, were being cultivated during the winter season with limited assured irrigation facilities after harvesting paddy. Before the intervention, only upland paddy, colocasia, tapioca and foxtail millet were grown, indicating symptoms of malnutrition. However, the nutritional security was achieved at household level only after introduction of the terrace system (Munda et al. 2012).

16.4.1.1 Productivity Enhancement

For increasing crop production and family income, rice-maize cropping systems with improved varieties of paddy (SARS-1, 2, 4 and 5; Bhalum-1, 2, 3 for upland and Lampanah and Shasarang and RCM-9 for lowland) and maize (RCM-76, Vijay and Navjot composite) and modern agro-techniques were introduced (Table 16.2). The improved varieties and technologies so adopted by the farmers have increased the yield of various crops in the range of 57.2–90 % (Chatterjee et al. 2012; Sahoo et al. 2012b; Deka et al. 2013). Nowadays, the double cropping has become a normal practice in the adopted villages of Mon district of Nagaland in these days.

16.4.1.2 Income Enhancement Through Improved Agro-Based Intervention

With monocropping, indigenous landraces of paddy, foxtail millet, colocasia, tapioca and some seasonal vegetables were grown by the farmers in the jhum fields which was traditionally practised in villages before the initiation of the project. The productivity of all the crops was found to be very low than the improved variety available at ICAR, SAUs and research stations (Table 16.3). With the introduction of short-duration and fast-growing high-yielding varieties of paddy, maize and vegetable, cultivation became feasible as double cropping during the winter months. Thus, the overall productivity and profitability were enhanced by two- to threefolds (Munda et al. 2012; Chatterjee et al. 2012; Sahoo et al. 2012a, b; Deka et al. 2013; Meena et al. 2015e; Verma et al. 2015a).

The intervention adopted for restoration of abandoned jhum area through bench terraces helps the farmers for cultivation of wet rice with suitable high-yielding varieties. This practice has finally enhanced the productivity of paddy up to threefolds over the traditional practices. Thus, the gross income has been increased to Rs. 32,850 ha^{-1} (Table 16.4). It has been observed that once the irrigation facilities are assured, farmers are ready to take up the second crop. However, water harvesting offers opportunities for integration of livestock and fisheries, and the wetland cultivation shall be converted into farming system mode of food.

Table 16.2 Productivity of different local and improve varieties of crops at Mon district, Nagaland

Crop	Local variety	Yield (t ha^{-1})	Improved variety	Yield (t ha^{-1})	Percent increase
Upland rice	Rakchu	1.0–1.8	SARS-1, 2, 3,4 and Bhalum-1, 2, 3	2.0–2.4	57.14
Lowland rice	Local	2.0	Lampanah, Shasarang, RCM-9	3.6–4.0	90.00
Maize	Local	0.7	RCM-76, Navjot	1.2	71.43
Rapeseed and mustard	–	–	TS-36 and TS-38	0.8	–
Potato	–	–	Kufri Jyoti, Kufri Chamatkar	8.0–8.4	–

Table 16.3 Net income of the farmers adopting different farming system interventions

Farming system	Baseline yield (t ha^{-1})	Yield (t ha^{-1})	Annual income/ household
Wetland rice based	1.90	2.3	8518/–
Vegetable based	60	84	3200/–
Cardamom based	NIL	0.095	7600/–

Table 16.4 Economic evaluation of terracing and traditional cultivation practices

Technology (terracing)	Before	After
Household	01	09
Variety introduced	Rakchu	Shasarang
Productivity (t ha^{-1})	1.2	3.39
Increase in productivity	–	182.50
Gross income	18,000/–	50,850/–
Profit against baseline	–	32,850/–

Farmers are cultivating their second crop in large scale which was not followed earlier (Sahoo et al. 2012b).

16.4.2 Vegetable-Based Farming System

Cultivation of second crops including vegetables become feasible after adoption of water-harvesting facilities, viz. base-flow-harvesting structures along with irrigation channels, *jalkund*, modified Thai jar, rooftop water harvesting, check dams, ponds, etc. These facilities enabled farmers to grow winter vegetables and oilseed crops, viz. dwarf pea (Azad), potato (Kufri Jyoti), cabbage (Samrat), tomato (Bioseed 56), onion (Nasik red), coriander (Ramses), bhindi (Tokita), bean (Yard Long), bitter gourd (Champion), ridge gourd (F1 hybrid), cucumber (Garima Super) and rapeseed and mustard (TS-36, TS-38), as second crops after paddy.

The beneficiary farmers were trained for nursery raising and package of practices for vegetable cultivation at a demonstration unit established in the village (Fig. 16.2). The vegetables were cultivated in an area of ~1.65 ha. Among the vegetables, potato cultivation occupied the maximum area (0.75 ha). Since the shifting cultivation was the mainstay of economy of the villagers before the implementation of the project, the concept of water harvesting and its multiple uses was new to them. During the implementation of the project, the major thrust was given for water harvesting and its multiple use based on the lesson learnt from the past. The net monetary income of the farmers has increased significantly with the present intervention (Table 16.5). It has opened up a new avenue for increasing the production and productivity not only in the target area but in other regions of the states (Chatterjee et al. 2012; Sahoo et al. 2012a).

Fig. 16.2 Water-harvesting structure for vegetable farming at Mon

16.4.3 Agroforestry- and Horticulture-Based Farming System

A total of 44,950 nos. of multipurpose tree saplings of khokan, hollock, tita chap, phulsap, Himalayan alder, bonsum and puma were planted covering an area of ~112 ha for restoration of degraded *jhum* land. Further, in horticulture-based jhum farming system (~28 ha), ~1285 saplings of fruit crops comprised of Khasi mandarin, Assam lemon, peach, litchi, guava and mango in multi-storey cropping system; ~63,210 large cardamom suckers and ~3300 banana suckers were planted in 10.5 ha.

These interventions facilitated in improving the degraded *jhum* land into a productive zone by substantially increasing soil health (Fig. 16.3). A small demonstration on Himalayan alder- and large cardamom-based cropping system in an area of ~0.16 ha with standard package of practices opened up the eyes of the poor farmers, and this has become a regular practice in many of the abandoned jhum fields (Chatterjee et al. 2012; Deka et al. 2013).

16.4.3.1 Restoration of Jhum Land by Multipurpose Tree Species

In the vast abandoned *jhum* areas (86.1 ha), multipurpose tree species, namely, khokan (*Duabanga grandiflora* Roxb), Himalayan alder (*Alnus nepalensis*), Tita chap (*Michelia champaca*), phulsap (*Michelia oblonga*), bonsum (*Phoebe goalparensis*), hollock (*Terminalia*) and puma (*Chukrasia tabularis*) saplings (34,450 nos.), were planted for restoration. Pits of 2 × 2 × 2 ft for each plant were prepared at a spacing of 5 × 5 m before the onset of monsoon, and 15 g of lime

Table 16.5 Economics of vegetables cultivation at the project site

Vegetables	Production cost (Rs.)	Area (ha)	Yield (q ha^{-1})	Total prod. (q)	Gross income (Rs.)	Net return (Rs.)
French bean	4375	0.25	38	9.5	19,000	14,625
Potato	29,625	0.75	80.5	60.38	120,760	91,135
Radish	1675	0.25	124	31	3100	2940
Carrot	1145	0.15	7.8	1.17	5265	4120
Tomato	1550	0.20	36	7.2	36,000	34,450
Coriander	755	0.05	12	0.6	30,000	29,245

Fig. 16.3 Agroforestry intervention for improving the abandoned land at Mon

Fig. 16.4 Fruit crops under half-moon terrace at Mon

and 10 kg of farmyard manure (FYM) were mixed thoroughly with soil, and each pit was filled 2 months before plantation. Besides restoration of the land area, these trees enriched the soil nutrient status and soil microbial flora. The populations of beneficial insects and earthworms were also more in the restored land. As the alder is a nitrogen-fixing tree species, it has also increased the status of available nitrogen in the soil (Chatterjee et al. 2012; Sahoo et al. 2012b; Deka et al. 2013).

16.4.3.2 Half-Moon Terraces

The half-moon terraces were constructed for planting and maintaining saplings of fruit and fodder trees in horticulture and agroforestry land-use system. This type of terrace is made by earth cutting in half-moon shape to create circular level bed having 1–1.5 m diameter (Fig. 16.4). The bed may also have inward slope with an interval of planting spacing of the fruit and fodder trees. It helps in retaining soil fertility, moisture and added fertilizers and manures for healthy growth of the plant (Chatterjee et al. 2012; Deka et al. 2013).

16.4.4 Livestock-Based Farming System

Tribal communities inhabiting in the eastern Himalayas are mostly non-vegetarian; hence, the demand of animal protein is much more compared to other parts of the country. Pig, poultry, cattle, mithun and rabbit are popularly maintained in majority of households in the region. However, the low-input small-scale traditional livestock production system with indigenous variety could not produce enough meat, milk and egg to meet the nutritional requirement (Bhatt and Bujarbaruah 2005). Non-availability of quality germplasm and critical inputs like feeds, medicines and vaccines and lack of awareness are major stumbling blocks in the development of the livestock sector in the state. ICAR RC for NEH Region, Nagaland Centre, had taken initiative to augment livestock productivity through introduction of quality pig and poultry germplasm and creating awareness about scientific management and health-care practices. Further, integration of livestock component along with traditional agriculture and horticulture was also popularized to provide an opportunity of additional income and ensure optimum use of farm resources. There is a dynamic relationship between livestock and crops in this kind of farming systems. Integration of farming basically aims at enchasing upon the interdependencies of the systems. Livestock depend on crops and crop by-products for their feed and fodder requirements and return nutrients to the crops via manure for sustainability of system (Bhatt and Bujarbaruah 2005).

16.4.4.1 Production Enhancement

Livestock-based interventions were made by our groups in the remote villages of the eastern Himalayan region, particularly in Nagaland, where modern facilities of agriculture and animal husbandry were not available. For increasing livestock production and family income, improved technologies, viz. suitable varieties, scientific management practices, integration of agriculture and livestock components and health-care facilities, were introduced and popularized. The improved varieties and technologies were so adopted by the farmers that have increased the overall productivity (Table 16.6) in the range of 125–275 % (Patra et al. 2015a, b). The continuity of these technologies will further increase the production and productivity of agriculture and allied components as the farmers are by now fully empowered with the required skill, training and motivation.

16.4.4.2 Income Enhancement

Due to introduction and adoption of improved germplasm of livestock and poultry, the family income has increased several folds. Further, the intensive integrated farming system has benefited the farmers for regular income and employment

Table 16.6 Productivity enhancement of livestock and poultry with improved technologies

Attributes	Indigenous var./cultivar	Improved var. with scientific management	% increase productivity
Body weight gain of poultry at 5 months	0.9 kg	2.5 kg	177.78
Annual egg production of poultry (up to 72 weeks of age)	40–50 no.	140–150 no.	250.0
Body weight gain of pig up to 10 months	35–40 kg	80–90 kg	125.0
Overall production of integrated farming system	Rs. 15,000–20,000/ha	Rs. 60,000–70,000/ha	275.0

even from a small piece of land (<1 acre), which, otherwise, remain non-remunerative. The feedback analysis given below has shown an increasing trend in income following the adoption of improved technologies.

16.4.4.3 Fattening of Improved Variety of Pigs

The low-cost fattening unit with two to four pigs of improved variety is still very popular among the poor farmers and landless labourer. The performance of such low-cost fattening unit (six nos.) located at Jharnapani, Dimapur, was evaluated to understand its economic impact on the livelihood of farm labourers (Fig. 16.5). The piglets were procured from the ICAR farm and maintained at low-cost housing facilities with traditional feeding of crop residues and kitchen waste for ~10 to 11 months. Regular health care and vaccination were taken care of by ICAR (Table 16.7). The net profit from each pig was recorded at an average of Rs. 3925 to Rs. 16,520 (Patra et al. 2015a). In the survey study with the six beneficiaries in Jharnapani, the performance of the piggery unit owned by Shri Meren was the best remuneration followed by Mrs. T. Jami. Thus, low-cost fattening unit could generate additional income and employment for farm labourers and help in livelihood security.

16.4.4.4 Semi-intensive Poultry Farming (100–400 Nos.)

Rearing of improved varieties of Vanaraja and Gramapriya birds at lesser numbers (10–20) is very popular for backyard farming. However, many farmers have tried with medium- to large-scale semi-intensive poultry unit and generated additional income of rupees 10,000–40,000 from 100 birds (Table 16.8 and Fig. 16.6). Farmers who are interested in meat production can maintain at least three batches per year and depending on capacity could generate additional income in the range of rupees 1.0–1.5 lakh from 350 to 400 birds (Patra et al. 2015b).

Fig. 16.5 Low-cost pig fattening units at Jharnapani village

Table 16.7 Economic analysis of low-input pig fattening units at Jharnapani village

Beneficiaries	T. Jami	Elias	Manoj Kumar	Ganesh Jadav	Bikash	Meren
Number of piglet	3	2	4	2	4	3
Cost of piglet (Rs.)	7500	5000	9200	5000	10,000	7500
Survivability (%)	100	100	100	50	100	100
Cost of housing (Rs.)	3000	1500	5000	1000	2500	3000
Feed cost (Rs.)	19,200	10,000	29,000	3500	6000	15,000
Production cost (Rs.)	29,700	16,500	43,200	9500	18,500	25,500
Avg. body weight at 12 months (kg)	146	103	127	95	95	140
Gross income (Rs.)	78,840	36,900	75,000	15,000	34,200	75,060
Net profit (Rs.)	49,140	20,400	31,800	5500	15,700	49,560
B:C ratio	2.65	2.24	1.74	1.58	1.85	2.94

16.4.4.5 Nutritional Improvement

In most of the remote villages in the eastern Himalayan region, the availability of animal protein was meagre. The main source of animal protein was hunting of either wild animals or very few local animals available in the villages. However, after popularization of backyard poultry and pig varieties in the state, animal protein particularly egg and meat is now available round the year. As an entry point activity, Vanaraja and Gramapriya birds were distributed at 20–30 per household for backyard farming in many villages (Table 16.9). With the minimum scientific input in housing, feeding and health care, these improved backyard poultry varieties performed better than the indigenous birds available in the villages and attained an average body weight of 2.5–3.0 kg after 6 months of rearing and produced an average of 140–160 eggs per annum. At least 30–40 % of these birds were consumed at home; thus, the monthly consumption of animal protein has increased from a near negligible amount to 2–8 kg/person (Fig. 16.7). Besides nutritional security, the farmers earned an additional income of Rs. 140–160 per bird after rearing ~5 to 6 months (Kumar et al. 2015d).

Table 16.8 Economic analysis of semi-intensive poultry unit established at different districts in Nagaland

Farmer's name	No. of birds	Wt at 20 weeks (kg)	Cost of prod. (Rs)	Gross income (Rs)	Net benefit (Rs)	B:C ratio
Smt. Thezano, Sethikema, Dimapur	400	2.54	1,18,770	158,588	39,818	1.34
Shri Shekhoto D. Vadeo, Sakraba, Phek	400	3.01	1,16,000	1,79,625	63,625	1.55
Shri Supung Jamir, Watiym, Mokokchung	350	2.5–3.25	1,00,000	1,28,625	28,625	1.29
Shri Yuao, Lampong Sheanghah, Mon	400	2.28–2.75	60,000	105,000	45,000	1.75
Mr. Zhasavilie Kehie, Sirhima village, Dimapur	150	2.27	69,785	130,000	60,215	1.86

Fig. 16.6 Low-cost semi-intensive poultry units

16.4.4.6 Employment Generation

People living below the poverty line were mostly dependent on agriculture and livestock component for their livelihood and had minimum scope of employment for 3–4 months during the cropping season. However, after introduction of the second crop after rice, integrated farming system, medium- to large-scale pig-breeding unit and semi-intensive poultry farming have created employment for farm women and rural unemployed youth round the year. The semi-intensive poultry unit of 200–400 birds' capacity maintained for a dual purpose generated employment for farm women/youth round the year and gave an income of at least 5000–7000 per month.

An integrated farming system with agri + horti + livestock + fishery + vermicompost components created employment for more than 200 days for a family as compared to only 60 days while practising only rice cultivation in less than one acre of land. Similarly, pig-breeding unit with ten sows has the potential to generate employment for a person for ~200 days with monthly income more than rupees 10,000 per month (Table 16.10). The rural youth trained in animal health care and vaccination and those who are practising artificial insemination for pig

Table 16.9 Availability of poultry meat after introduction of improved variety

Village	No. of household	No. of birds reared	Survivability (%)	Meat produced (MT)	Availability of meat/household (kg)
Hukpang village, Longleng	100	1000	80	2.2	22.0
Kanching village, Tamlu town, Longleng	400	2000	90	4.95	12.38
Medziphema, Dimapur	135	2700	87	5.29	39.15
Songluh and Inbung village, Peren	50	1070	85	2.05	40.92
Porba village, Pfutsero, Phek	54	1000	87	2.35	43.50
Lampong Sheanghah, Mon	116	2130	85	4.0	34.48
Tumei, Mon	117	1500	80	2.7	23.08
Naginimora, Mon	135	2000	86	3.87	28.67
Total	**1107**	**13,400**	**85**	**31.32**	**28.29**

Fig. 16.7 Small-scale poultry unit maintained for household consumption and additional income

have got a tremendous opportunity to earn at least rupees 20,000–25,000 per month (Kumar et al. 2015d; Patra et al. 2015a, b).

16.4.5 Intensive Integration of Agri-horti-livestock-Based Farming System

To overcome the problems of resource-poor farmers, a holistic, resource-based, client-oriented and interacting approach popularly known as intensive integrated farming system (IIFS) was developed at the institute level (Fig. 16.8). The performance of each model consisting of agri-horti-livestock (pig/poultry) and fisheries was evaluated for three consecutive years and later on replicated at farmers' field. Rice/maize + *toria* + mung bean-based cropping system, along with fruits (mango,

Table 16.10 Employment generation potential of various interventions adopted

Interventions	Name of the beneficiary	Unit size	Employment (days)	Income (Rs./ month)
Semi-intensive poultry unit	Shri Libemo Ezung, Dimapur	700	365	30,000–50,000
	Mrs. Thezano, Sethikema A	400	365	15,000–18,000
	Mr. Vevoyi, Khushiabill village, Dimapur	200	365	10,000–12,000
Pig-breeding unit	Shri Kumar Thapa	4 sow	75	4000–5000
	Mrs. Nzano Ezung of Pongitong village, Wokha	10 sow	200	10,000–12,000
Integrated farming system	Mrs. Mhonlumi Murry, New Wokha village	2.5 ha	>200	6600–7000

Fig. 16.8 Integrated farming system model site at Mon and Wokha

lemon, banana, etc.), vegetables (year-round seasonal vegetables), livestock (pig/poultry), vermicompost, *Azolla* and mushroom, was integrated in less than 1 acre of land which has increased the cropping intensity for more than 300 %, with additional income of Rs. 11,000–32,000 and employment for 250–365 days as compared to only 60 days in rice-based (Table 16.11) monocropping system (Kumar et al. 2015b).

In farmers' field, integrated farming system models were developed in participatory mode by two approaches, first by utilizing the existing water body available with the farmers and second by constructing water-harvesting structure using *jalkund* or pond lining with UV-stabilized HDPE sheet. The agriculture land located at the periphery was used for crops, and the bunds were utilized for cultivation of vegetables and fruits and rearing livestock. Depending on the capacity of water-harvesting structures, labour force and topography, the agriculture plots were utilized for agri + horti + livestock-based farming system. The models developed at farmers' field with the adopted technology are presented in Table 16.12 (Annual Report, 2014–2015; Kumar et al. 2015a).

Table 16.11 Economics of farming system model developed at ICAR Nagaland Centre

Model component	Cost of production	Gross return	Net return	B:C ratio	Total system employment
Agriculture + horticulture + fishery + piggery + vermicomposting	17,891	42,285	24,394	2.36	350
Agriculture + horticulture + fishery + poultry + *Azolla* + mushroom + vermicomposting	29,395	61,600	32,040	2.09	365
Agriculture + horticulture + fishery + duckery + vermicomposting	16,600	28,320	11,720	1.70	250

16.4.5.1 Water Storage in Farms

In Nagaland, most of the hills are steep having a slope more than 50 % and are separated by deep river gorges. Despite the heavy monsoon rain, people face acute water problems every year in dry season. The geological formation does not permit water retention; run-off is quick and springs and small streams dry up when there is no rain. Rooftop harvesting structures for drinking purpose have been developed locally and now spread in the entire Mon district of Nagaland. It has proved to be quite successful. Most houses are built with sloping roofs with galvanized iron sheets which are conducive to rainwater harvesting. A common method of storing rainwater is to place horizontal rain gutters along the sides of a sloping roof, which is normally made of corrugated iron sheets (Fig. 16.9). Rainwater pours into a pipe connected to the tank which is mostly made from GCI sheets/galvanized plain sheets. But many people have started using reinforced cement concrete tanks, located in the courtyard or under the house (Chatterjee et al. 2012; Sahoo et al. 2012a; Deka et al. 2013).

16.4.6 Restoration of Degraded *Jhum* Land Through Conservation of Soil Fertility and Improving Water Productivity

North-eastern states receive heavy rain during peak monsoon from April to September. High rainfall intensity leads to more run-off. Such water has been traditionally conserved in situ by locally made structures. It may be lined/unlined and of various capacities depending on catchment size and farmers' need. Farmers construct storage tanks that may be in the form of tanks, ponds or small reservoirs. These tanks may be located on top/middle of the slope or foothills. This is preferably made of natural depressions to keep down the cost of excavations. Irrigation in the land below the pond is done through gravity method and land above by lifting of the water from the pond. If the pond is unlined, stored water recharges the soil profile, and the lowland or storage reservoirs at the lower reaches are benefited. Due to coarse-textured soils, water retention in these ponds is a great problem (Fig. 16.10).

Table 16.12 Details of integrated farming system model developed/supported under TSP

Name of the district	Name of the villages (total no.)	Model component	No. of units	Area (ha)	Capacity of water body (lakh lit)
Wokha	New Wokha, Ralan, Longsachung, Chukitong, Pongitong, Liphanyan, Koio and Sanis, Yanpha and Liphanyan	Duck + fish + agriculture + vegetables + fruits + pig/poultry, alder-based farming, tree bean plantation	17	9.0	8.75
Longleng	Lingtak, Hukpang, Orangkong, Shayong, Dungkhao, Pongo	Water-harvesting structure + agriculture + horticulture + poultry/pig, agriculture + horticulture + silviculture + pastoral; large cardamom-based agroforestry model	18	4.0	11.55
Dimapur	Medziphema, Ruzaphema	Agriculture + horticulture + fish + pig/poultry; Fruit-based farming system	4	4.65	8.75
Peren	Navala	Water harvesting structure + agriculture + horticulture + poultry/pig	1	2.0	8.75
Mon	Lampong Sheanghah, Naginimora	Water-harvesting structure + agri + horti + livestock	4	4.0	4.0
Mokokchung	Mokokchung	Agroforestry + horticulture + poultry based	1	2.0	4.0
Total	**22**	–	**45**	**25.65**	**45.8**

16.4.6.1 Mulching with Paddy Straw

During winter season after harvesting of the rice, these rice straw pieces are spread in the field as mulch for the succeeding maize crop. The crop residue left on the surface cushions raindrops' impact and reduces water movement, and hence soil erosion is checked. As run-off and evaporation are reduced, water infiltration is improved. Application of crop residues in the long run improves soil structure and fertility. However, farmers have expressed that the use of these materials for mulch often reduces the availability of fodder to cattle; therefore when fodder is in short supply, they do not practice mulching. Kumar (2015) conducted a field experiment at ICAR, Nagaland Centre, Jharnapani, to evaluate the best management practices,

Fig. 16.9 Rooftop water-harvesting structure at Mon

Fig. 16.10 Farm pond at Kolasib, Mizoram

i.e. mulching, liming and farmyard manures, for maximizing the productivity, profitability, nutrient uptake and quality on winter maize during the two consecutive rabi seasons of 2010–2012. Results revealed that application of straw mulches significantly increased growth and yield attributes and grain yield by rabi maize. This may be attributed to higher water regime and better water balance, which lead to vigorous growth and more yield attributes produced in mulch plot (Sharma et al. 2010).

Among the levels of lime, higher grain yield (3.91 t/ha) and stover yield (4.24 t/ha) were noted with application of lime at 0.6 t/ha. This might be due to the release of Ca^{2+} from lime, which meets the demands and creates favourable conditions for better uptake of essential nutrients particularly P (Kumar et al. 2012). Significantly, higher grain (3.79 t/ha) and stover yields (4.17 t/ha) were recorded with application of FYM at 12 t/ha. Improvement in yield of crop may be attributed to better nutrient availability resulting into a higher yield (Kumar 2014).

16.4.6.2 SRI: An Alternative Method of Rice Cultivation

In India, SRI technology started picking up. States like Andhra Pradesh and Tamil Nadu have done good progress with this technology. Even in north-east India, also a lot of works are undertaken in SRI. In Tripura after introduction of SRI techniques, an average of ~20 % higher yield is obtained compared to conventional practice. This state has covered ~15 % of its area under SRI. These practices can improve rice productivity by 15–20 % over conventional practices. The significant aspect of these practices is that the crop duration gets shortened by ~10 to 15 day. Kumar et al. (2015c) conducted a field experiment at the Agricultural Research Farm of

ICAR RC for NEH Region, Nagaland Centre, Jharnapani, in two consecutive kharif seasons of 2011–2013. This study compared the effect of crop establishment methods and nutrient management practices on production potential, nutrient uptake and energetics in transplanted rice in hill ecosystem of the eastern Himalayan region. Results showed that growth characteristics (plant height, tillers/m^2, dry matter production, root volume and root biomass) recorded significantly higher with SRI followed by ICM and CTR. Thus, maximum utilization of available plant nutrient resulted in ultimately higher grain filling (%), panicle length and weight, number of grains per panicle and test weight and finally increased the crop productivity.

Mirza et al. (2010) reported an increase in the number of tillers in rice plants due to integrated application of organic and inorganic nutrients. Similarly, the higher yield attributes (number of panicles/hill and panicle length) and yields were recorded under SRI as compared to ICM/CTR. Among the nutrient management practices, application of 100 % RDF + rice straw 5 t ha^{-1} produced higher grain yield (4.7 t/ha) followed by 100 % RDN (farmyard manure) + rice straw 5 t ha^{-1} (4.57 t ha^{-1}). This significant response might be due to enhanced nutrient availability to the crop by the application of organic manures in combination with inorganic fertilizers and the higher grain and straw yield of rice with integrated application (Das et al. 2013).

16.4.6.3 Surface Seeding/Zero Tillage

In low-lying poorly drained heavy rice soils, at the time of harvesting of paddy in the month of November, soil moisture is too high which does not allow timely tillage operation for sowing till the month of January. Thereby, farmers were forced to leave field vacant. So they have evolved practice of sowing mustard in the months of November and December just after harvesting of paddy by broadcasting in saturated soil surface without any tillage operation. It is widely practised by small and marginal farmers of Serchhip and Champhai district of Mizoram (Prasad et al. 2009). This practice is similar to the zero-tillage practice being advocated by the scientists nowadays. However, when seeds are sown on open soil surface, they are damaged/eaten by birds. To protect the seeds from bird's damage, farmers use thin layer of cow dung on seed surface before the seed. Similarly, different *toria* varieties during the rabi season were evaluated at ICAR Jharnapani under zero tillage, and it was found that the maximum yield was recorded by variety TS-38 (490 kg ha^{-1}). This showed that under the resource conservation technology, these varieties may be commercially exploited for farmers' fields (Annual Report 2013–2014).

16.4.6.4 Application of Common Salt for Weed Management in *Jhum* Rice

The farmers in north-east regions including Nagaland manage weeds manually, restricting its feasibility especially at peak weeding period. This might be owing to the non-availability of labourers during critical physiological stages and high expenditures on labour forces in upland rice cultivation which sometimes becomes unprofitable. However, extremely acidic soil condition in shifting cultivation areas helps in managing weeds with the use of salt application since the time immemorial. Jhum farmers of these regions traditionally used to apply common salt in upland rice as post-emergence spray to manage the annual broad-leaved weeds. Common salt is not a recommended herbicide to control broad-leaved weeds; however, alien weeds, e.g. *Ageratum conyzoides* and *Parthenium hysterophorus*, have been successfully controlled with application of 15–20 % common salt.

However, still time and levels of common salt application in rice have not been accredited under *jhum* field. Therefore, efforts have been made to validate the indigenous technical knowledge for weed management and improving the productivity, profitability, nutrient uptake and soil health of upland jhum rice to evolve a realistic weed management approach under shifting cultivation area of Nagaland. An experiment was conducted during the kharif season of 2012–2014 to assess the effect of common salt (NaCl) to manage the weed problem in upland rice with different doses of salt application, viz. 20–200 kg ha^{-1} (2–20 % NaCl) at 20 and 40 DAS along with a control/weedy check and weedy-free check. Result revealed that altogether 17 weed species were identified, out of which the broad-leaved weeds (BLWs), grasses and sedges comprised of 82.5 %, 12 % and 5.5 %, respectively. The BLWs *Borreria hispida*, *Urena lobata*, *Eupatorium odoratum*, *Bidens pilosa* and *Ageratum conyzoides*; the grasses *Cynodon dactylon*, *Digitaria sanguinalis*, *Echinochloa colonum* and *Cyperus rotundus*; and the sedges were among the noticed prominent weed flora. Application of increasing levels of common salt up to 20 % at 20 DAS recorded markedly lower weed population and dry matter as compared to the application of common salt at 40 DAS. Similarly, common salt applied at 20 % at 20 DAS recorded the highest weed control efficiency (WCE, 36.24 %), whereas comparatively lower WCE (32.67 %) was observed with salt applied at 40 DAS.

The better performance of growth characters of jhum rice in weed control treatments was due to enhanced crop growth attributes and effectiveness of chemicals in controlling weeds (Tabin and Singh 2008; Chatterjee et al. 2015). Application of 10 % common salt recorded significantly higher grain yield (2315.6 kg ha^{-1}) and straw yield (3589.1 kg ha^{-1}) as compared to their preceding levels (2–8 % NaCl), and beyond that, the effect of applied common salt was recorded at par (12–20 %). The increase in yield could attribute to an increase in growth and yield attributes, viz. the number of tillers, dry matter, number of panicles/m^2 and panicle length, along with a decrease in chaffy grains in a controlled plot (Kumar et al. 2016a; Chatterjee et al. 2015). Similarly, application of common salt up to 150 kg ha^{-1} recorded significantly higher grain yield due to better performance of growth and yield attributes of *jhum* rice (Tabin and Singh 2008). In respect of soil health, application of common salt (2–20 %) did not exert any marked influence on soil physico-chemical properties, viz. pH, SOC and available NPK; however, EC increased for a short period. The enormous rainfall coupled with steep slope reduced the salt concentration from the soil. The differences in timing and intensity of rainfall after salt application and amount and type of ground cover would affect the extent to which salt was available on the soil surface (Kumar et al. 2016a).

16.4.6.5 Bio-terracing

The shifting cultivation adversely affects the fertility status of *jhum* land, and the land becomes unsustainable for cultivation of the crops. The problem is very much intense in *jhum* lands where organic matter, soil nutrients and microbial biota are lost due to the burning of the forests. The bio-terracing models with growing of different suitable hedgerow species which act as restoration factory for maintaining the fertility status of the soil reduce soil erosion and also conserve the moisture in situ for sustainable production of agricultural crop in the hilly regions. The similar concept has been also tested under the lowland condition for in situ biomass production and fertility restoration of the soil and to reduce the dependence on external inputs like fertilizer (Sahoo et al. 2012a, b). The hedgerow species like *Tephrosia candida*, *Crotalaria juncea*, *Indigofera tinctoria*, *Flemingia macrophylla* and *Cajanus cajan* have been tried on alternate raised bunds in paddy fields by many researchers. A good amount of biomass has been produced to supplement the nutrient requirement for these crops. On an average, from pruning of the hedgerow species, an amount of 20–80, 3–4 and 8–38 kg NPK ha^{-1} $year^{-1}$, respectively, is added into the soil. The higher concentration of N in the foliage of N in the foliage of all the hedgerow species might be due to fixation of atmospheric N_2. In the mineralization process of organic matter and subsequent accumulation in the foliage, thus contour hedgerow provides an option for farming on the hill slopes on a sustainable basis. The growing of nitrogen-fixing hedge species on the field bunds helps in the fixation of atmospheric nitrogen and reduces the leaching losses of mineral nitrogen. Their vigorous root system mobilizes phosphorus, potassium and other trace elements. Decomposition of organic matter through leaf litter of hedge species improves the water-holding capacity of soils and other physical properties (Laxminarayan et al. 2006).

16.5 Promotion of Crop Varieties for Eastern Himalayan Agroclimatic Condition

In Nagaland, rainfed farming is prominent, and therefore second cropping is very much difficult under these circumstances. The soil is having very poor nutrient and residual moisture retention capacity. A field experiment was carried out in maize rabi season under moisture stress condition with agronomic management practices like straw mulching, where a total of 45 maize germplasms were screened to assess the best suitable line in terms of higher production potential and results revealed that maximum grain yield was recorded with maize cv. RCM-75 (3200 kg ha^{-1}). This conclusion is based on 2 years of the experimentation, so it may be recommended for commercial cultivation in moisture stress condition of Nagaland (Kumar et al. 2016a, b). Another experiment was carried out to assess the performance of rabi maize cultivars on production potential in climate change condition, and results revealed that maize cv. RCM-75 that was grown with application of RDF + FYM + lime + mulching recorded the maximum yield attributes and yield. This is due to the combined effect of the treatment which minimizes the moisture loss and provides better soil health and also due to genetic potential of the varieties (Annual Report 2012–2013).

Most of the interventions were made in remote villages of Nagaland, where modern facilities of agriculture are not available. Under such backdrop, increasing even 1 kg of yield might help these downtrodden farmers. For increasing crop production and family income, improved technologies, viz. suitable high-yielding varieties, scientific management practices and rainwater-harvesting facilities, were introduced and popularized. For increasing crop production and family income, rice-maize, rice-linseed/*toria* and rice-mustard-mung bean cropping systems with improved varieties of rice (SARS-1, 2, 5; Bhalum-1, 2, 3 for upland and Shasarang, RCM-9,11 and IET-16363 for lowland), maize (RCM-76, 75; DA-61), linseed (Sweta and Parvati) and soybean (JS-335 and Bragg] and modern agro-techniques were introduced. The improved varieties and technologies so adopted by the farmers have increased the overall productivity in the range of 85–280 % (Table 16.13). The continuity of these technologies will further increase the production and productivity of agriculture and allied components as the farmers are by now fully empowered with the required skill, training and motivation (Chatterjee et al. 2012; Deka et al. 2013).

16.6 Strategies to Adopt Improved Agricultural Technology for Augmenting Food Security

A number of awareness programmes on farm training, exposure visit, field day and method and result demonstration on improved technologies at progressive farmers' field were conducted to upgrade the knowledge and skill of the farmers. Further, several trainings on soft skill development for farm women were conducted to engage in income-generating activities. Besides regular technical inputs on agricultural operation, ~28 specific training-cum-demonstrations were organized on income-generating activities, i.e. weaving-cum-sewing, beekeeping and honey box making, for village women, unemployed youth and elders. A technology is blind if it is not adopted by the farming community. Around 20 improved technologies were introduced at adopted villages in the project site. Out of 20 technologies, eight technologies were well accepted by the villagers in an area of >500 ha, which they can continue even after completion of the project.

These technologies improved the farm income by~ 40 to 74.5 % and also generated annual employment of ~ 200 man-days per family. Bench terrace was constructed, where land was steeper than 16 % and slope and depth of soil was good enough (Fig. 16.11). The process consists of transforming relatively steep land into a series of nearly level steps across the slope, and the outward edges were supported by stones and wooden log. All the terraces were made at a vertical interval of 1 m keeping intact the topmost soil there. Irrigation channels were prepared to divert

Table 16.13 Productivity enhancement of various crops across the state

Crops	Local variety	Yield (t/ha)	Introduced HYVs with scientific management	Yield (t/ha)	Percent increase
Paddy					
Upland	Rakchu	1.0–1.5	SARS-3, 5; Bhalum-1, 2, 3; RCPL-1-300;	2.5–3.0	125
Lowland	Eangya	1.5–2.0	Shasarang, RCM-9,11; IET-16,363; Ranjit, RCM-5	3.–4.5	110
Maize					
Kharif	Local	0.7–1.2	RCM-76,75, VPQM, DMH-11, DA-1-61A	3.0–4.5	275
Rabi	Local	1.5–1.8	RCM-75, 76, DHM-117, VPQM, DA-1-61A	3.0–3.5	280
Mung bean					
Rabi	Local	0.5–0.7	Pratap; T-21; TRCM-8-2-1	1.0–1.2	85
Summer	Local	Nil	Pratap; T-21; TRCM-8-2-1	1.2–1.5	–
Rapeseed and mustard	Local	0.3–0.4	TS-36, 38,67 and PM-25,27,28	0.8–1.2	280
Soybean	Local	0.5–0.75	JS-35 and Bragg	1.2–1.5	120
Linseed	Neelam	0.3–0.4	Sweta and Parvati	0.8–1.0	150
Veg. French bean	Local	–	Anupam	40–45	–
Veg. pea	Local	–	Arkel and Azad pea	20–25	–
Baby corn	–	–	HM-4	2.0–2.5	–

Fig. 16.11 Terrace construction for demonstration of settled cultivation in Mon

water from the stream. For nutrient management, a thick row of hedgerow species like *Tephrosia candida* and *Crotalaria* spp. was planted, and green biomass was mulched into the terraces for soil fertility management (Sahoo et al. 2012a; Deka et al. 2013).

16.7 Future Prospective

- Jhum improvement through building upon indigenous conservation practices along with development of improved varieties of crops, fruits and vegetables and management practices to improve the productivity of jhum farming system at least by 50 % from the present level.
- Setting up of an independent regional resource centre on shifting cultivation involving various stakeholders to share the knowledge, information and dissemination of technology.
- Available high-yielding altitude-specific varieties of crops should be tested in all the districts of Nagaland.
- Intensification of improved jhum cultivation and improved fallow management involving physical and biological practices.
- Improvement of existing and indigenous farm implements and tools to reduce the drudgery and increased efficiency of *jhumias.*
- Demonstration of site-specific improved agroforestry-/horticulture-/livestock-based integrated farming system models for jhum improvement.
- Frequent interface meeting of the State Department of Agriculture/Horticulture/Veterinary/Soil and Water Conservation/Irrigation/Forestry/Fishery, KVKs, ATMA, Research Institutes and NGOs at district/block level to take stock of the situation and discuss various options/development programmes on jhum.
- Making available the required quantity of quality seeds, planting materials and bio-fertilizers in time and place to the farmers.
- Appropriate soil and water conservation measures including water harvesting should be encouraged in jhum areas on priority basis for sustainable production.
- Since jhum is indispensable, synergy among the departments/research institutes/universities for convergence of available programmes (RKVY, MNREGS, NWDPRA, HTM MM-2) in a holistic manner should be taken up for improvement of the existing systems.
- Strengthening the existing government policy and act on jhum regulation and forest policy involving local people/village development council without any prejudice.
- Organizing frequent awareness programmes in local language.

16.8 Conclusions

The shifting cultivation (*jhum*) is the mainstay of traditional agriculture in the eastern Himalayan region. Considering its association with tribal livelihood prospective, approaches were implemented to strengthen the existing cultivation practice instead of imposing modern intervention. Adoption of site-specific agro-based interventions has proved to be beneficial in augmenting productivity of major crops and livestock, thus ensuring more income, employment and food security. Therefore, the success achieved in our study areas located in the remote part of

Nagaland could be a model for future policy making for sustainable development in wider coverage and assured development in the vast eastern Himalayan region.

Acknowledgements The author is grateful to the NAIP/NICRA/TSP project for providing the financial support and all the other facilities to carry out the research and extension activities.

References

Annual Report (2012–2013) ICAR Research Complex for NEH Region, Nagaland Centre Jharnapani, Medziphema-797106, Nagaland

Annual Report (2013–2014) ICAR Research Complex for NEH Region for NEH Region Nagaland Centre Jharnapani, Medziphema-797106, Nagaland

Annual Report (2014–2015) ICAR Research Complex for NEH Region for NEH Region Nagaland Centre Jharnapani, Medziphema-797106, Nagaland

Anonymous (2002) Basic statistics of North Eastern Region, Government of India, NE Secretariat, Shillong, pp. 42

Arunachalam A (2002) Dynamics of soil nutrients and microbial biomass during first year cropping in an 8-year *jhum* cycle. Nutr Cycl Agroecosyst 64:283–291

Arunachalam A, Arunachalam K (2002) Evaluation of bamboos in eco-restoration of '*jhum*' fallows in Arunachal Pradesh: ground vegetation, soil and microbial biomass. For Ecol Manage 159:231–239

Bhatt BP, Bujarbaruah KM (2005) Intensive integrated farming system: a sustainable approach of land use in eastern Himalayas, technical bulletin no 46. ICAR Research Complex for NEH Region, India, Meghalaya

Borthakur DN (1992) Agriculture of the North Eastern region with special reference to hill agriculture. Beecee Prakashan, Guwahati

Chatterjee D, Deka BC, Patra MK, Thirugnanavel A, Sahoo B, Kumar R, Bhatt BP, Krose M (2012) Livelihood improvement of Konyaks through integrated farming system at Mon, Nagaland, pp 9–11

Chatterjee D, Kumar R, Kuotsu R, Deka BC (2015) Validation of traditional weed control method through common salt application in hill region of Nagaland. Curr Sci Accepted

Das A, Patel DP, Munda GC, Ramkrushna GI, Kumar M, Ngachan SV (2013) Improving productivity, water and energy use efficiency in lowland rice (*Oryza sativa*) through appropriate methods and nutrient management practices in the mid altitude of North-East India. Exp Agric. doi:10.1017/S0014479713000483

Deka BC, Chatterjee D, Sahoo B, Patra MK, Thirugnanavel A, Kumar R, Krose M, Bhatt BP, Ngachan SV (2013) Soil-water-plant-animal-society continuum approach for livelihood improvement of Konyaks tribes of Mon in Nagaland. Technical Bulletin, ICAR Research Complex for NEH Region, Nagaland Centre, Jharnapani, Nagaland, Published by Joint Director, ICAR RC NEHR, Nagaland Centre Jharnapani, Medziphema, Nagaland -797 106

Ghosh BN, Meena VS, Alam NM, Dogra P, Bhattacharyya R, Sharma NK, Mishra PK (2016) Impact of conservation practices on soil aggregation and the carbon management index after seven years of maize–wheat cropping system in the Indian Himalayas. Agr Ecosyst Environ 216:247–257

Karim SMR, Mansor M (2011) Impact of *jhum* cultivation on agro ecology of mountains and socio economy of tribal peoples. Asian J Agric Res 5:109–114

Kumar R (2014) Productivity, profitability and nutrient uptake of maize (*Zea mays* L.) as influenced by management practices in North East India. Indian J Agron 62(2):69–74

Kumar R (2015) Influence of mulching, liming and farm yard manures on production potential, economics and quality of maize (*Zea mays* L.) in rainfed condition of North Eastern India. Bangladesh J Bot 44(3):391–398

Kumar R, Deka BC, Baishya LK, Pandey A, Ngachan SV (2012) Effect of mulching, liming and integrated nutrient management on production potential of *rabi* maize under rainfed condition of Nagaland. (In) Extended summaries, Vol. 2, pp 50–51. International Agronomy Congress during 26–30 November, 2012 at New Delhi, India

Kumar A, Bahadur I, Maurya BR, Raghuwanshi R, Meena VS, Singh DK, Dixit J (2015a) Does a plant growth promoting rhizobacteria enhance agricultural sustainability? J Pure Appl Microbiol 9(1):715–724

Kumar R, Deka BC, Thirugnanavel A, Patra MK, Chatterjee D, Borah TR, Barman KK, Rajesha G, Talang HD, Kumar M, Ngachan SV (2015b) Comparative evaluation of different farming system models suitable for small and marginal farmers of Nagaland. Lead paper presented in the National Seminar on "Sustaining Hill Agriculture in Changing Climate" (SHACC) during December 5–7th 2015 in Agartala, Tripura, pp 8–10

Kumar R, Kumar A, Kumar A, Pandey A (2015c) Productivity, profitability, nutrient uptake and soil health as influenced by establishment methods and nutrient management practices in transplanted rice under hill ecosystem of North East India. *Indian Journal of Agricultural Sciences* 85 (5): 30-35

Kumar R, Patra MK, Deka BC, Ngullie E, Kumar M, Sangtam HM, Ngachan SV (2015d) Promotion of Agro-based interventions for sustainable livelihood in Nagaland through Tribal Sub Plan. Published by Joint Director, ICAR RC NEH Region, Nagaland Centre Jharnapani, Nagaland-797 106

Kumar R, Chatterjee D, Deka BC, Kumar M, Kuotsu R, Ao M, Ngachan SV (2016a) Weed management practices in upland direct seeded *jhum* rice under Eastern Himalaya. Res Crops, ROC-15/9365 Vol. 17, No. 2 (June), 2016

Kumar R, Hangsing N, Zeliang PK, Deka BC (2016b) Exploration, collection and conservation of local rice germplasm of Nagaland. Environ Ecol 34(4D):2514–2517

Laxminarayan K, Bhatt BP, Rai T (2006) Soil fertility buildup through hedgerow intercropping in integrated farming system: a case study. In: Bhatt BP, Bujarbaruah KM (eds) Agro-forestry in North East India: opportunities and challenges. ICAR RC NEHR Region, Umiam

Meena RS, Yadav RS, Meena VS (2013) Heat unit efficiency of groundnut varieties in scattered planting with various fertility levels. Bioscan 8(4):1189–1192

Meena RS, Meena VS, Meena SK, Verma JP (2015a) The needs of healthy soils for a healthy world. J Clean Prod 102:560–561

Meena RS, Meena VS, Meena SK, Verma JP (2015b) Towards the plant stress mitigate the agricultural productivity: a book review. J Clean Prod 102:552–553

Meena VS, Meena SK, Verma JP, Meena RS, Ghosh BN (2015c) The needs of nutrient use efficiency for sustainable agriculture. J Clean Prod 102:562–563

Meena VS, Meena SK, Verma JP, Meena RS, Jat LK (2015d) Current trend to mitigate climate change and sustainable agriculture: a book review. J Clean Prod 102:548–549

Meena VS, Verma JP, Meena SK (2015e) Towards the current scenario of nutrient use efficiency in crop species. J Clean Prod 102:556–557

Meena RS, Bohra JS, Singh SP, Meena VS, Verma JP, Verma SK, Sihag SK (2016) Towards the prime response of manure to enhance nutrient use efficiency and soil sustainability a current need: a book review. J Clean Prod 112(1):1258–1260

Mirza H, Ahamed KU, Rahmatullah NM, Akhetar KN, Rahman MJ (2010) Plant growth characters and productivity of wetland rice (*Oryza sativa* L.) as affected by application of different manures. Emirates J Food Agric 22(1):46–58

Mishra VK, Saha R (2003) Soil physico-chemical behaviour and recovery pattern under different phases of *jhum* cultivation in hilly regions of Meghalaya. J For. (Accepted)

Munda GC, Ngachan SV, Das A, Malngiang S, Chowdhury S (2012) Site specific farming system options for rural livelihood-success stories from NEH region. NAIP Bulletin No. 2 ICAR Research Complex for NEH Region, Umiam-797103, Meghalaya pp. 1–78

Patiram, Verma ND (2001) Reformed system to increase the productivity of shifting cultivation in north-eastern hills. J N East Counc 21:24–32

Patra MK, Deka BC, Das RK, Addison MP, Ngachan SV (2015) Promotion of rural poultry farming in Nagaland. Published by ICAR RC for NEH region, Nagaland Centre, Jharnapani, Medziphema, Nagaland

Patra MK, Deaka BC, Kent Y, Ngullie L, Das RK, Ngachan SV (2015) Promotion of pig breeding through quality piglet production and artificial insemination. Published by ICAR RC for NEH region, Nagaland Centre, Jharnapani, Medziphema, Nagaland

Prasad K, Kumar R, Prakash N, Sah AK, Verma RP (2009) Indigenous soil and water conservation practices prevailing among the tribal farmers of Mizoram. Indian J Soil Conserv 38(1):31–36

Rathore SS, Bhatt BP (2008) Productivity improvement in jhum fields through integrated farming system. Indian J Agron 53:167–171

Rathore SS, Krose N, Naro M, Shekhawat K, Bhatt BP (2012) Weed management through salt application: an indigenous method for shifting cultivation areas, Eastern Himalaya, India. Indian J Tradit Knowl 11:354–357

Saha R, Mishra VK, Khan SK (2011) Soil erodibility characteristics under modified land-use systems as against shifting cultivation in hilly ecosystems of Meghalaya, India. J Sustain For 30:301–312

Sahoo B, Bhatt BP, Moanaro, K (2012a) Terracing for *panikheti* based farming system for food security and livelihood in Mon, Nagaland, pp 30–32. Edn Munda GC, Ngachan SV, Das A, Malngiang S, Chowdhury S (2012) Site specific farming system options for rural livelihood-success stories from NEH region. NAIP Bulletin No. 2 ICAR Research Complex for NEH Region, Umiam-797103, Meghalaya, pp 1–78

Sahoo B, Moanaro, Krose M (2012b) Water harvesting and its multiple use for enhancing productivity and livelihood in Mon, Nagaland, pp 30–32. Edn Munda GC Ngachan SV, Das A, Malngiang S and Chowdhury S (2012) Site specific farming system options for rural livelihood-success stories from NEH region. NAIP Bulletin No. 2 ICAR Research Complex for NEH Region, Umiam-797103, Meghalaya, pp 1–78

Sharma AR, Singh R, Dhyani SK, Dube RK (2010) Effect of live mulching with annual legumes on performance of maize (*Zea mays*) and residual effect on following wheat (*Triticum aestivum*). Indian J Agron 55(3):177–184

Singh J, Bora IP, Baruah A (2003) Changes in the physico-chemical properties of soil under shifting cultivation with special references to Karbi Anglong district of Assam. Indian J For 26 (2):116–122

Singh DK, Singh SK, Singh AK, Meena VS (2014) Impact of long term cultivation of lemon grass (Cymbopogon citratus) on post-harvest electro-chemical properties of soil. Ann Agric Biol Res 19(1):45–48

Tabin D, Singh MK (2008) Effect of common salt and 2, 4-D Na salt application on weed growth and yield of upland direct seeded rice. Oryza 45:296–299

Verma JP, Jaiswa DK, Meena VS, Meena RS (2015a) Current need of organic farming for enhancing sustainable agriculture. J Clean Prod 102:545–547

Verma JP, Jaiswal DK, Meena VS, Kumar A, Meena RS (2015b) Issues and challenges about sustainable agriculture production for management of natural resources to sustain soil fertility and health. J Clean Prod 107:793–794

17 Impact, Adaptation Strategies and Vulnerability of Indian Agriculture Towards the Climate Change

Ved Prakash, Santosh Kumar, S.K. Dwivedi, K.K. Rao, and J.S. Mishra

Abstract

Change in climate is one of the most important worldwide environmental challenges, with implication for production of food, supply of water, health, energy resources, etc. Proper scientific understanding with coordinated action at global level is required for addressing the climate change. Climatic system of the earth has apparently changed on both regional as well as global scales since the pre-industrial era. The Intergovernmental Panel on Climate Change (IPCC) declared that the daily mean temperature (globally) may be increased between 1.4 and 5.8 °C by the twenty-first century. Historically, the responsibility for greenhouse gases (GHGs) emissions' increase lies largely with the developed world, though the developing nations are likely to be the major cause of a growing proportion of future emissions of GHGs. Agriculture sector is not only sensitive to climate change but also one of the key drivers for climate change. The sensitivity of climate towards agriculture is uncertain, as there is regional variation of climatic factors like rainfall, temperature (maximum and minimum), crop stands and different cropping system, soils properties and management practices. Overall, in the winter season (rabi), rise in the average temperature is likely to be much higher than in the monsoon season (kharif). Agriculture is one of the sectors where impact of climate change will be significant, and it will affect every part of agriculture including crops, fisheries, livestock, etc. Hence, it is necessary to identify the possible impacts of climate change on agricultural productivity and its allied sectors in order to recommend adaptation and mitigation strategies. Developing countries is reducing the vulnerability of their natural and socio-economic systems to the projected climate change that is the issue of the highest importance for developing countries. Developing countries will face the challenge of promoting mitigation and

V. Prakash (✉) • S. Kumar • S.K. Dwivedi • K.K. Rao • J.S. Mishra
ICAR-Research Complex for Eastern Region, Patna 800014, India
e-mail: jyanivedprakash486@gmail.com

J.K. Bisht et al. (eds.), *Conservation Agriculture*,
DOI 10.1007/978-981-10-2558-7_17

adaptation strategies for climate change, bearing the cost of such an effort, and its implications for social and economic development. We conclude by making interim recommendations on the practical strategies needed to develop a more resilient and dynamic world agriculture in the twenty-first century.

Keywords
Climate change • IPCC • GHGs • Adaptation • Mitigation and vulnerability

17.1 Introduction

Nowadays the whole world is concerned about the climate change and its impact. Climate change is the change in the behaviour of different weather parameters (temperature, humidity, sunshine, evaporation, rainfall, etc.) over an area during a given time. Climate change is a very serious and promising threat to food, feed and livelihood security. Climatic zones are changing due to global warming worldwide and will have a variety of effects on different parts of agriculture. The gases responsible for the global warming are known as GHGs, which are comprised of carbon dioxide, methane (CH_4), nitrous oxide (N_2O) and water vapours. These GHGs are producing by a number of anthropogenic activities (Mendelsohn and Sanghi et al. 2001). At present, climate change is the most important concern that affects the whole earth environment globally, and among them agriculture is the most vulnerable sector to climate change because it is highly dependent on surrounding weather and climate. Most of the population (~60 %) of this region is directly or indirectly dependent on agriculture as a source of livelihood.

Proper scientific understanding with coordinated action at global level is required for addressing the climate change. The proposed climate change under different scenarios is likely to have so many implications on food production, energy, water supply, forest ecosystems, health, etc. In the developing countries, the impact of changing climate on adaptive capacity of communities is lower than others. The efforts of the United Nations Framework Convention on Climate Change (UNFCCC) and the Kyoto Protocol provisions are clearly not enough to deal with the challenges of climate change. To address the climate change, the most effective way is by adapting a sustainable development pathway with the help of shifting of environmentally sustainable technologies and advertising of energy efficiency, forest conservation, renewable sources, reforestation, water conservation, water harvesting, etc. In the fourth assessment (AR) report of IPCC, it was observed that 'continuous warming of climatic system is now an unequivocal, as is now evident from different observations of increases in global average temperature (of air and ocean), extensive melting of snow and ice from polar region, and continuous rising of global sea level' (Solomon et al. 2007). It is also observed that synthetic (man-made) GHG emissions have increased ~70 % over the last three decades and will increase notwithstanding the present mitigation policies for climate change and associated sustainable expansion practices. The FAO of the

United Nations estimates that as a result of climate change 'India could lose approximately 125 MT cereal production in rain-fed areas, which is contribute ~18 % of the total world production'. The UNFCCC, which is an international apex body on climate change, refers to adaptation in the context of climate change only. In other words, there is no climate change without emissions of any GHGs, and due to this, there is no need to go for adaptation strategy. Stern (2006) reported that adaptation is essential to deal with the inescapable effects of climate change for which the entire world is already committed.

Agricultural sector is continuously releasing different GHGs (CO_2, CH_4, and N_2O) to the atmosphere in significant amounts (Cole et al. 1997; IPCC 2001; Paustian et al. 2004). Smith (2004b) and Janzen (2004) have reported that microbial decomposition or burning of plant litter and soil organic matter play a major role to release CO_2 in the atmosphere. Mosier and Kroeze (2000) observed that methane is produced during the decomposition of organic materials in oxygen-deprived conditions, notably from fermentative digestion by ruminant livestock, from old manures and from rice cultivation (under flooded conditions). Nitrous oxide (N_2O) is released by the microbial transformation of N_2 in manures and soil and is often enhanced where availability of N_2 exceeds more than the plant requires, especially under wetland conditions (Smith and Conen 2004; Oenema et al. 2005). Fluxes are complex and heterogeneous of all agricultural GHGs, but it can be mitigated by the active management of agricultural systems.

Cultivation of land with high management intensity is the most important human practices for the dynamics of the atmospheric biogeochemical and cycling of water in the Earth system. Global biogeochemical cycles mainly affect the production of food and are closely reliant on the input of fossil fuel energy. Now, farmers are using the different products of fossil fuels as fertiliser, pesticides and herbicides and machinery to boost the percentage of solar energy that is trapped by the green part of the crop to drive dry matter production and thereby harvested yield.

Resource conservation technologies (RCTs) of agriculture, which include reduced/zero tillage/surface seeding practices and crop residue retention, can reduce GHG emissions and curb global warming. Appropriate RCTs cover an innovative cropping system for higher production that combines the dramatic reductions in tillage operation with an ultimate goal to achieve zero till or controlled till seeding for all type of crops in a cropping system if possible. It also supports for rational retention of sufficient levels of different crop residues on the soil surface to seize run-off and control soil erosion, improve water use efficiency and reduce evaporation losses and increase organic matter in soil and other microbial activity to enhance land and water productivity on sustainable basis (Khan et al. 2000).

17.2 Indian Agriculture Scenario

Majority of Indian population depends on agriculture, where for the majority of population depends on crop cultivation, forestry, fishery, livestock, etc. for their livelihood. In India, agriculture sector is considerably dependent on rainfall which comes from south-west monsoon. Economy of India mostly depends on the onset of south-west monsoon and its further performance. The agriculture sector represents ~35 % of India's gross national products (GNP). It plays a critical role in the development of the country. It will continue to occupy an important place in the national economy (Meena et al. 2013, 2015a; Ved et al. 2015a; Singh et al. 2014; Kumar et al. 2015; Ghosh et al. 2016). It sustains livelihood ~75 % of the population. Future food security of mankind on earth will be mainly affected by the impact of climate change on agriculture. In Indian economy, the contribution of agriculture and associated activities in current years has been no less significant than that of industries and other services. 'If agriculture sector survives, India survives' is the best statement to show the importance of agriculture to the country.

Food and nutritional security is a constant challenge in the background of a burgeoning population where agriculture has emerged as a key factor for the growth of the Indian economy. According to the latest report with a contribution of ~16 % to India's GDP and 10.23 % (provisional) to the total exports in 2008 and 2009 and the fact that it provides employment to 58.2 % of the population, a constant growth and development of agricultural sector is very vital to meet other challenges as well. From a total net sown ~141 M ha area, out of this ~61 M ha area is under irrigated condition. Thus, maximum part of the net sown area is under rain-fed condition, thereby making the agriculture sector in India very sensitive to any changes in the amount and pattern of rainfall. For instance, the impact of the overall deficit of 23 % in rainfall during the south-west monsoon in 2009 and 2010 badly affected the crop production of kharif season, and it was reflected in the growth rate of agriculture GDP which shows a decline of 0.2 % as in opposition to the previous year's growth rate of 1.6 %. The agriculture is the only sector that has the potential to transcend from being a problem to becoming an essential part of the clarification to climate change provided there is a more holistic idea of agricultural mitigation approaches, food and nutrient security, climate change adaptation strategy and agriculture's pro-poor development contribution (Tables 17.1 and 17.2).

Table 17.1 Atmospheric concentration, lifetime and global warming potential (GWP) of major greenhouse gases

GHGs	Atmospheric concentration	Lifetime (years)	GWP (100 years)
CO_2	387 ppm	Not fix (variable)	1
CH_4	1780 ppb	12	25
N_2O	319 ppb	114	298

Source: IPCC (2007a)

Table 17.2 Adaptation techniques for different sectors

Temperature increase by 1 °C in the months of June to August	Farm management	Livestock production	Fishery	Development of agricultural biotechnologies	Improvement of agricultural infrastructure
>Use of drought and heat stress-tolerant crop varieties under water stress conditions	>Altered application of nutrients/fertiliser	>Breeding livestock for greater tolerance and productivity	>Breeding fish tolerant to high water temperature	>Development and distribution of more drought-, disease-, pest- and salt-tolerant crop varieties	>Improve pasture water supply. Improve irrigation systems and their efficiency
>Use of more disease- and pest-tolerant crop varieties	>Altered use of insecticide/ weedicide/pesticide	>Increase stocks of forages for unfavourable time periods	>Fisheries management capabilities to cope with impacts of climate change must be developed	>Develop improved processing and conservation technologies in livestock production	Improve use/store of rain and snow water
>Introduce high yielding and early maturing varieties of crops in cold regions	>Change date of planting to efficiently use the prolonged growing season and irrigation	>Improve pasture and grazing management including improved grasslands and pastures		>Improve crossbreeds of high productivity animals	>Improve information exchange system on new technologies at national as well as regional and international level
>Grow the salt-tolerant crop varieties in saline areas	>Develop adaptive management strategy at farm level	>Improve management of stocking rates and rotation of pastures			>Improve sea defence and flood management
		>Increase the quantity of forages used to graze animals			
		>Increase plant coverage per hectare			

Source: IPCC (2007b)

17.3 General Trend of Climate Change

Climate change and climate variability are a subject of great concern to humankind. According to the United Nations Framework Convention on Climate Change, if any change is made in climate, then it will directly and indirectly affect the human activities that alter the environmental composition. IPCC defined it as 'any change in the climatic conditions over time and given place whether due to natural cause or as a result of human activities'.

17.3.1 Current Projection

Indian Meteorology Department (IMD, Delhi) and the Indian Institute of Tropical Meteorology (IITM), Pune, did the analyses and found the same trends for temperature, heat waves, cold waves, droughts, glaciers, floods and sea-level rise as by the IPCC. The scale of the changes in environmental condition varies in some cases. At India level, there is no trend that has been observed in monsoon rainfall during the last 100 years, but some regional patterns are available. An increasing trend of monsoon rainfall are found along the west coast, north part of A.P and north-west India, and those of declining trends over east Madhya Pradesh and its adjoining areas and parts of Gujarat and Kerala (6–8 % of normal rainfall over the last 100 years). Air temperature of the surface from 1901 to 2000 indicates a considerable warming of 0.4 °C for 100 years. The spatial distribution in temperature changes indicated a significant warming trend which has been observed along the central part of India, west-coast, and over north-east India. Instrumental records of the last ~130 years do not show any considerable long-term trend in the frequencies of large-scale droughts or floods in the summer monsoon season. The total frequency of cyclonic storms that form over Bay Bengal has remained almost constant over the period from 1887 to 1997. All these are the evidences that glaciers in Himalayas are moving back at a rapid pace.

17.3.2 Future Projection

By the end of the twenty-first century, it is projected that rainfall will increase by 15–31 %, and the temperature (annual mean) will increase by 3–6 °C. The warming effect is more obvious over land areas, with the maximum increase over the northern part of India. Relatively greater warming is projected in winter and post-monsoon seasons.

17.4 Global Warming Through Emission of Greenhouse Gases (GHGs) Globally and in Indian Context

The GHGs are the key culprits of the global warming. The GHGs like CO_2, CH_4 and N_2O are in performing hazards in the resent times. These warming due to greenhouse effect are called global warming. The greenhouse effect is visible more significantly in the current years, with number of natural disaster on the rise in the global world. Although CO_2, CH_4 and N_2O take place naturally in the environment, their current significant atmospheric build-up is largely due to anthropogenic activities. The composition of the environment has been altered by this raise and will have an impact on future global climate. Out of all the GHGs, CO_2 is the most important GHG because it recorded 80 % increase from 21 to 38 gigatonnes (Gt) from 1970 to 2004 which constitutes ~77 % of the total GHG emissions in the year of 2004. According to IPCC 2007, it has been observed that CO_2 contributes 60 %, CH_4 15 % and N_2O 5 % to the global warming.

In India, CO_2 emissions in 2013 continued to increase by 4.4 % to ~2 billion tonnes; it helps India to become the fourth largest CO_2-emitting country in the world, near to the European Union (EU) and well ahead of the Russian association, which is the fifth largest emitting country. This high ranking is partly caused by the size of its population and economy; the workforce is expanding in the industry and services sectors, partially because of international outsourcing (World Bank 2014). An agricultural community is the principal supplier to global anthropogenic non-CO_2 greenhouse gases, accounting for ~56 % of emissions in the year of 2005 (U.S.EPA 2011). According to FAOSTAT (2013); U.S.EPA (2006); Tubiello (2013), the yearly total non-CO_2 GHG emissions from agricultural production in 2010 was estimated to be 5.2–5.8 Gt CO_2 eq/year and comprised ~10 to 12 % of the global anthropogenic emissions (Tables 17.3 and 17.4).

Nitrous oxide acts as a greenhouse gas in the troposphere and is the major source of ozone depletion (Graedel and Crutzen 1993). Nitrous oxide (N_2O) is emitting from forest covers, grasslands, nitrogenous fertilisers (urea, DAP, etc.), oceans, etc and burning of biomass fossil and fuels are the main sources of N_2O. The main source of total nitrous oxide emission is soil, with a contribution of ~65 %. In terms of global warming potential (GWP), it is approximately 310 times more effective than CO_2, the single largest greenhouse gas (Prather and Ehhalt 2001). The GWP is an index to compare the strengths of different greenhouse gases in increasing temperature on a common basis. CO_2 is used as the reference gas to compare the ability of a GHG to trap atmospheric heat.

The most significant activities from agriculture sector contributing to GHG emissions are:

1. Destruction of forests and other land-use changes as a source of CO_2.
2. Rice-wheat cropping systems are a source of methane and nitrous oxide (and also source of CO_2 due to burning of agricultural residues).
3. Methane emission from animal husbandry.

Table 17.3 Local coping strategies as adaptation tools to mitigate the impacts of climate change in agriculture

Subregion	Regional area	Natural disaster	Impacts	Adaptation measures
Himachal Pradesh	Erratic rainfall	Water shortage	Sustainable water management	Collection, distribution and utilisation of run-off water from glacier
Several North Eastern States	Erratic rainfall	Losses of crops	Proper crop selection	Domesticating indigenous varieties of cereals and fruit trees
Goa	Sea-level rise	Water logging	Integrated agriculture aquaculture system	Balancing agriculture and fisheries through sluice gates. Application of Khazan – traditional community managed integrated agri-aquaculture ecosystems
	Drought and/or aridity	Water shortage	Rainwater harvesting	Building small- and medium-sized dams to serve water reservoirs
Himalayas	Erratic rainfall	Losses of crops	Alternative cultivation methods	Growing apricots, walnuts, grapes and vegetables in the cold deserts
	Erratic rainfall	Losses of crops	Appropriate crop selection in cold deserts	Rotational cropping seed selection
Western Himalayas	Erratic rainfall	Losses of crops	Disaster risk management, appropriate cropping practices	Using meteorological indicators and animal behaviour to predict rain
Central Himalayas	Drought/ aridity	Losses of crops	Diet diversification	Use of wild foods and medicinal plants by Bhotiya tribes (Tolccha, Marchha, Jadhs)
Thar Desert	Drought/ aridity	Water shortage	Rainwater harvesting	Building underground tanks
North-East	Drought/ aridity	Losses of crops. Water shortage	Sustainable water management	Using bamboo to transport stream and spring water to irrigate plantations
Andaman and Nicobar	Drought/ aridity	Losses of crops	Alternative cultivation method	Intercropping with banana and using plant residues. Selecting and storing rice, pulse and vegetable seeds
Gujarat	Drought/ aridity	Water shortage	Rainwater harvesting	Cleaning, desilting and deepening of ponds to collect more rainwater at a time

(continued)

Table 17.3 (continued)

Subregion	Regional area	Natural disaster	Impacts	Adaptation measures
Maharashtra	Drought and/aridity	Water shortage/ soil erosion	Rainwater harvesting	Building ground barriers and shallow excavations through various barriers
Orissa	Drought/ aridity	Losses of crops	Appropriate crop selection	Storing and exchanging rice varieties and medicinal plants
Rajasthan	Drought/ aridity	Water shortage	Rainwater harvesting	Harvesting water and recharging groundwater with earthen check dams
	Drought/ aridity	Losses of crops	Appropriate crop selection	Pearl millet (Bajra) cultivating in arid regions
	Drought/ aridity	Losses of crops	Appropriate crop selection	Growing Sona Mukhi as medicinal cash crop
	Drought/ aridity	Land degradation	Nutrient management	Using earthworms to decompose organic waste
Tamil Nadu	Drought/ aridity	Water shortage	Sustainable water management	Improving wells and irrigation facilities
	Drought/ aridity	Losses of crops	Post-harvest management	Threshing, winnowing, cleaning and drying for dry land crops
Uttar Pradesh	Floods	Losses of crops	Appropriate crop selection	Breeding rice varieties for flood prone region
	Drought/ aridity	Land degradation	Nutrient management	Improving soil fertility through gypsum and manures

Source: Adapted from ADB and IFPRI (2009)

Table 17.4 Measurement scale for different categories of vulnerability

Categories	Vulnerability criteria	Different scales for measurement
Features of agricultural system	Sensitivity	Land resources
		Inputs and technology
		Irrigation share
		Production and productivity
Biophysical indicators	Exposure	Soil parameters and climatic factor
		Water availability and storage capacity
		Biomass/yield
		Crop calendar of every year
Socio-economic data	Adaptive capacity	Rural welfare
		Protection and trade
		Poverty and nutrition
		Crop insurance

Source: Adapted from Tubiello and Rosenzweig (2008)

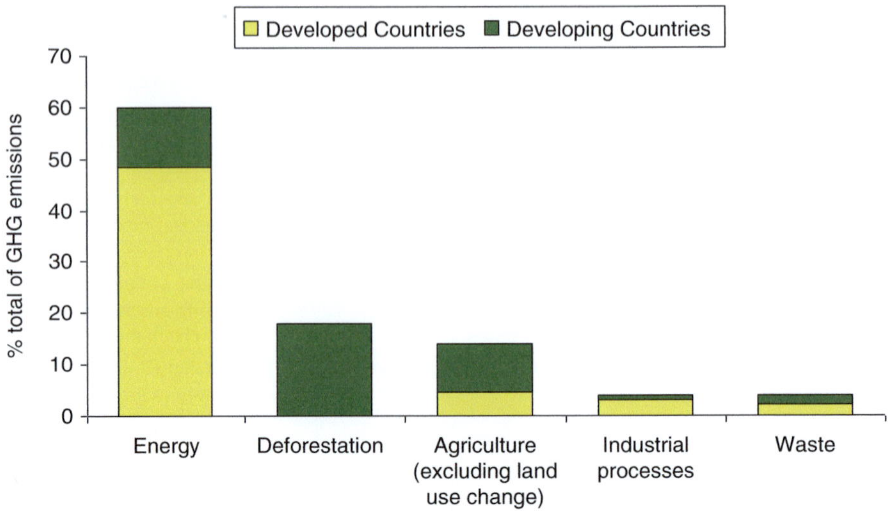

Fig. 17.1 Different sources of GHG emissions (Adapted from WRI 2007; WDR 2008)

Lybbert and Sumner (2010) reported that 13–15 % of global GHGs is emitted from agricultural sector (as agriculture contribution in global GDP is just ~4 %, this suggests that agriculture is very GHG intensive). This Fig. 17.1 is confined to direct GHG emissions at production level, not including agricultural production inputs and fixed capital equipment, processing and trade of agricultural products (in GHG inventory reports, these emissions appear under industries, energy and transport system).

The agriculture contribution to GHG emissions rises to ~30 to 32 % if deforestation, land-use changes and land degradation are included. Methane and nitrous oxide emissions through agricultural sector grew by 17 % in the period 1990–2005 (IPCC 2007a). This increase is more or less proportionate, for example, global cereals' production increase, but three times as increase in cereal productivity. By 2030, GHG emissions are expected to rise by 35–60 % in response to the increasing population and diet changes in developing countries, in addition to the further stretch of industrial and factory farming in developed and developing countries (IPCC 2007b).

These plant metabolic activities and production directly affected by change in atmosphere are caused by the anthropogenic GHGs. The rise in carbon dioxide concentration in the atmosphere may decrease the soil pH, which will affect both nutrient availability to plants and microbial activity. The change in CO_2 concentration may also change the yields of crops, for example, photosynthesis rate of C_3 plants such as wheat, rice and soybean may increase by as much as 30–100 % if the CO_2 concentration is doubled. In the livestock rearing, bovines and the small ruminants are the most dominant features of Indian agriculture scenario and act as major source of methane emissions.

17.5 Climate Factors Connected to Climate Change and Agricultural Productivity

17.5.1 Average Temperature Increase

It has been projected that in the next 100 years, the Earth's average temperature will increase between 1.8 and 4.0 °C. An increase in the average temperature can increase the growing season in regions with a relatively cool spring, and it adversely affects crops in regions where summer heat already limits production. Increasing average temperature also increases soil evaporation rates as well as increases the chances of severe droughts.

17.5.2 Changing Rainfall Amount and Its Patterns

Changes in precipitation can affect the rate of soil erosion and soil moisture status; both of them are important for crop yield. The IPCC forecasts that rainfall will increase at high latitudes and decrease in most subtropical parts. While according to IPCC (2007a) regional amount of precipitation will vary, the number of extreme precipitation events is expected to increase.

17.5.3 Increasing Atmospheric CO_2

Increasing level of atmospheric CO_2, which is emitted from different human activities, can be used as a fertiliser and enhance the growth of particular crops like wheat, rice and soybeans. CO_2 can be a limiting factor for some crop and, when increased, can enhance crop growth. Other limiting factors include water and nutrient availability. IPCC (2007b) reported that CO_2 fertilisation will have a positive impact on some crops; another side of climate change (e.g. changes in temperature and rainfall) may temper any positive CO_2 fertilisation effect.

17.5.4 Level Pollution Such as Tropospheric Ozone (O_3)

Buse et al. (2003) and the WHO (2003) have reported that besides being an effective greenhouse gas, O_3 is also toxic to human health, animals and plants and other living things. Factors that influence high ozone concentration at a given level: (1) Populated and highly industrialised areas have higher ozone levels because they have more traffic and more factories that produce ingredient gasses like VOCs and NO_2. (2) There are various products available today that are called 'ozone generators' which means that when utilised, these products emit the various pollutant (mainly ozone). Included in the list are office equipments such as zerox machine and laser printers, industrial manufacturing process of ozone treatment for bottled water and certain kinds of electric or ion-generating air purifiers for clear air rooms and spaces (3) and local and regional emissions. Since levels in the lower atmosphere are shaped by both emissions of GHGs and temperature, climate change will most likely increase O_3 concentrations. Those changes may also offset any favourable yield effects that result from high CO_2 levels.

17.5.5 Change in Climatic Variability and Extreme Events

Changes in the occurrence and severity of natural calamities (cold waves, heat waves, drought, floods, cyclones and hurricanes) remain a key uncertainty in future climate change. All those changes are anticipated by global climatic models; it is very difficult to forecast the regional changes and potential effects on agriculture.

17.6 Impact of Climate Change

The greenhouse gas concentration in the atmosphere would be doubled by 2035 as compared to pre-industrial level and eventually leads to the increase of temperature over 2 °C if immediate action has not been taken to control the GHGs emission. In the longer term, there would be more than a 50 % probability that the temperature rise would go beyond 5 °C. This rise would be very risky undeniably; it is alike to the change in the average temperature from the last ice age to today. Such a drastic change in the physical geography of the world must direct to key changes in the human geography – where and how people live their lives. All the evidence from the detailed studies of regional and sectoral bearings of changing weather patterns through to economic models of the global effects show that even at the more modest levels of warming climate, change will have severe impacts on the ecosystem, on world output and on human life. Despite the fact that the poorest countries and populations have contributed least to the causes of climate change, they will experience the effects earliest and most.

Climate change will have profound and extensive effects on the environment, ecosystems, natural resources, economy and human life. The 4th assessment report of the IPCC indicates that climate change is likely to alter risk patterns in numerous ways like raise the frequency and intensity, trim down the reliability and change the spatial distribution of extreme climatic hazards, such as extreme temperatures, heat waves, droughts and floods, fires and storms, with a variety of effects in diverse regions.

17.6.1 Impact of Climate Change on Indian Agriculture

It is projected that climate change will affect food security by the mid of the twenty-first century, with the largest numbers of food-insecure people located in South Asia. There are very limited data globally for observed impacts of climate change on food production systems and this is true also for Asia (IPCC 2014). The future food security is majorly decided by the impact of climate change on agriculture. Agriculture is one of the major drivers for climate change which is contributing and itself affected by the change in temperature and rainfall. To understand the impact of temperature, rainfall and CO_2 in crop growth and yield, several methods are used by researchers to assess the climatic variability impact ranging from the traditional method of historical data analyses to environmental studies by means of different statistical tools (Aggarwal 2008; Meena et al. 2015b, c, 2016; Verma et al. 2015b). The C_3 plants will get benefit by increase in ambient atmospheric temp by improving their photosynthetic efficiency. However, the food production may considerably be affected by the dual effect of temperature and rainfall. There are chances of 10–40 % crop production loss due to increase in temp by 2080–2100 (Aggarwal 2008). The modelling-based estimates done at Indian Agricultural Research Institute (IARI) show that with each degree increase in temperature throughout the growing period will cause a possible loss of 4–5 MT of wheat grain production after carbon fertilisation is taken into account.

All agricultural commodities even today are sensitive to the rising climatic variability associated with global warming resulting in substantial seasonal/annual fluctuations in food production. Droughts, floods, tropical cyclones, heavy precipitation events and heat waves are known to negatively influence agricultural production and livelihood of farmers. The anticipated increase in these events will result in greater wavering in food production and threaten livelihood security of farmers. Increasing glacier melt in Himalayas will affect irrigation water availability particularly in the Indo-Gangetic plains, which, in turn, has greater effect on our food production. Despite the above-outlined gravity of the climate-change impact on agriculture, according to the former head of the International Food Policy Research Institute (IFPRI), Joachim von Braun, governments 'underestimate the climate-related threats and there is little work on how the negative effects can be mitigated' (Braun 2008).

17.6.2 Impacts of Climate Change on Different Agricultural Sectors

17.6.2.1 Crop

On the global market, India is the major exporter of wheat, corn and rice. Crop yields highly depend on temp variations, content of atmospheric CO_2 and intensity of extreme weather conditions. Warmer conditions lead to rapid growth in most of the crops. Nevertheless, faster growth in grain crops will moderate the time for seeds to mature due to increased respiration, reduction in rainfall and number of irrigations. The rate of photosynthesis will be increased in several C_3 crops such as wheat and rice, with increase in ambient CO_2 concentration (Aggarwal and Swaroopa 2009). In the rain-fed areas, the quality of fruits, vegetables, tea, coffee, and aromatic and medicinal plants may be affected due to augmented crop water demand and altering rainfall during monsoon season. Prevalence of pest and diseases of crops may alter because of more enhanced pathogen and vector development, swift in pathogen transmission and increased host vulnerability.

17.6.2.2 Soil

Soil organic matter quality will be affected which in turn reduces the organic carbon content of soil. As Indian soils are already quite low in organic carbon content

(<0.5 %), it would become depleted further. The decomposition rate and nutrient supply of crop residues which have wider C:N ration may further be decreased under elevated CO2 concentrations. The increase in soil temp has positive effect on N mineralisation, but N losses will be increased due to processes like volatilisation and denitrification. The rate of soil erosion may be increased due to alterations in rainfall volume and severity of winds. Agricultural land area may be decreased due to salt water ingression in the coastal lands, tuning them to saline soils.

17.6.2.3 Water

The rise in temperature and evapotranspiration rate will increase the irrigation water demand of crops which may lead to depletion of groundwater table levels. In the short run, water availability in the Ganges and Brahmaputra and their tributaries will increase due to the melting down of Himalayan glaciers, but in the long run, a noticeable decrease in the available water may occur. A vast expansion of storage infrastructures is very beneficial as the increase in run-off is anticipated in the wet season. This additional water in the wet season may lead to increase in rate of recurrence and extent of floods. The intrusion of sea water along the coastal track will be affecting the quality of groundwater and cause the water imbalance in different parts of India.

17.6.2.4 Livestock

Climate change will affect fodder production and nutritional security of livestock by decreasing production of feed and fodder due to water scarcity and enhancing lignifications of plant tissue cell walls which may reduce the digestibility. Changes in rainfall pattern may influence expansion of vector populations in cooler areas, leading to disease outbreaks. Climate change could affect animals both directly and indirectly by increasing requirements of water and energy by mulching animals and by affecting their reproductive performance. Climate change is expected to aggravate heat stress in dairy animals which can increase disease vulnerability, infertility and reduction in milk production. Drought reduces the amount of quality forage and may threaten feed and pasture supplies.

17.6.2.5 Fishery

Many fish species are already facing numerous stresses, including overfishing, higher sea surface temperature and water pollution. Increasing surface sea and river water temperature is likely to affect breeding, migration and harvests of fishes and also causes coral bleaching. Temperature and the changes in season will control many stages of aquatic animal's life cycle which could affect the reproduction time and migration. For example, there is large decline in salmon population in the north-west warmer water temperatures due to changes in life cycle and increase in the likelihood of disease. Migration to new areas may cause these species into competition with other species over food and resources.

17.7 Adaptation Practices, Strategies, Opportunities and Mitigation Options for Climate Change

Taking steps to build resilience and curtail costs is essential towards the adaptation to climate change. It is still possible to protect our societies and economies from its impacts to some extent by improved planning, providing better information and more climate-resilient crops and infrastructure. Adaptation efforts should be accelerated, particularly in developing countries, but this will put further pressure on already scant resources. Intergovernmental panel on climate change (IPCC) in 2007 emphasised on climate change adaptations through adjustments to reduce vulnerability or increase resilience in response to observed or expected changes in climate and associated extreme weather events. Adaptation measures can be divided into (a) individual or autonomous adaptations: these are initiatives by private actors rather by governments due to actual or predictable climate change. (b) Policy driven or planned adaptation: it is the result of policy decision by public agency or governments based on an awareness that conditions are about to change or have changed and that action is necessary to minimise losses or benefit from opportunities (Rosegrant and Cline 2003).

IPCC (2007a) summarises more common adaptation measures intended to increase adaptive capacity by altering the farming practices crop and livestock improvement through breeding and investing in new technologies and infrastructure (Wang et al. 2004; IPCC 2007b). Upgrading of irrigation systems and breeding of novel rice varieties minimise the risk of severe productivity losses caused by climate change, and information, education and communication programmes enhance the level of consciousness and understanding of the vulnerable groups (IPCC 2007a). Changes in the management of agriculture and allied activities could also enhance adaptive capacity. One example is the coastal zone management by integration of aquaculture and fisheries management to increase the coping ability of small communities in East Asia, South Asia and South-East Asia to sea-level rise.

17.7.1 Local Coping Strategies as Adaptation Tools

The coping strategies to extreme weather variation of countries in the Asia-Pacific region would be effective to have long-term adaptation strategies. However, despite some commonalities, the coping strategies and aboriginal knowledge vary by subregion, country and provinces.

17.7.2 To Mitigate Climate Change, the Followings Are the Management Options

In the present scenario of changing climate with region to region, there is an urgent need for reassessment of all the agricultural practices. Conservation practices that conserve available natural resources and have the benefit to delay effect of stress should be given importance, and essential policy may be encouraged (Meena et al. 2015d, e; Ved et al. 2015b; Verma et al. 2015a). The most important thing, i.e. a reforestation programme, should be engaged by the communities, institutes, schools, governments and NGOs. Among them some of the resource conservation technologies (RCTs) suitable for the particular region are:

- Zero or reduced tillage practices with residue cover on the surface for improving soil prosperities and conserving soil moisture that save fuel and improve plot-level water productivity.
- Integrated systems of farming (IFS) and watershed development with livestock, fishery and hedge row to row cropping for moisture conservation and recycling of nutrient.
- Mulching, in situ residue management cover cropping and restoration of degraded lands for moisture conservation and improved carbon sequestration.
- Spread the technologies like system of rice intensification (SRI), zero till rice and aerobic rice cultivation for the maximum saving of water and mitigation of GHGs emissions.
- Selection and screening of short-duration varieties for their drought resistance.
- In situ biomass management in shifting cultivation (cover ~1.6 Mha in NER) instead of biomass burning to reduce CO_2 emission and improve hydrology.
- Encourage the technologies that increase biological nitrogen fixation and improve nutrient and water use efficiency to reduce N_2O emission and boost up dependence on renewable energy.
- Change in date of sowing and crop cultivars are another adaptive measure to overcome the impacts of climate change to some extent.
- Shift in rainfall pattern and rise in temperature in North-East India, it is important to recheck the present date of sowing and varieties, i.e. at mid-altitude of Meghalaya (950 m), where it was very difficult to take a second crop of rice after *kharif* rice because of early onset of winter, presently double cropping system is possible at least at the experimental land with some adjustment of varieties and sowing dates.

17.8 Vulnerability to Climate Change

Vulnerability of a system refers to its physical, social and economic aspects in which a system is unable to cope with adverse effects of climate change, including climate variability and weather extremes (IPCC 2007b). It is defined by IPCC that vulnerability is a function of exposure, sensitivity, adaptive, and vulnerability = f (exposure, sensitivity, adaptive capacity) IPCC 2001). In other words, the greater the exposure or sensitivity, the greater is the vulnerability. The estimation of key vulnerabilities in any system, and damage implied, will depend on the exposure or sensitivity, which is determined in part and where relevant by development status and adaptive capacity. However, adaptive capacity is inversely related to vulnerability. So the greater the adaptive capacity, the lesser is the vulnerability. Tubiello and Fischer (2007) provide framework for all the functions of vulnerability for assessing susceptibility in agricultural sector. Table 17.4 describes the indicators of vulnerability in agricultural region. As mentioned in the table, exposure refers to biophysical indicators. The measures of exposure can be soil (properties of soil) and climate (temperature/rainfall), water availability, crop calendar and yield. On exposure, the present and past climate fashion and variability in Asia show that the trend of surface air temperature is in increasing manner which is more obvious during the winter season rather than in summer season (IPCC 2007a) and that types of increasing trend have been observed across Asia and its subregion. The observed increase in temperature in some parts of Asia during the current decades varies between <10 and 30 °C every century.

Regarding rainfall patterns, interannual, interpersonal and spatial variability have been observed in the last few decades across the whole Asia. It may be noted that decreasing trends in annual mean rainfall are observed in North China and North-East, coastal belts and parts of North-East India, Philippines, Indonesia and some parts of Japan. On the other side, increasing trends of annual mean rainfall are observed in the western part of China, South-Eastern coast of China, Arabian Peninsula, Bangladesh and western part of the Philippines (IPCC 2007b). On changes in extreme events, usually, the frequency of occurrence of high intensive rainfall events in various parts of Asia has increased and resulted in severe floods, while the total number of rainy days and total annual rainfall has declined.

In the twenty-first century, temperature projections based on the IPCC's Fourth AR and Atmosphere-Ocean General Circulation Models (AOGCMs) suggest a considerable acceleration of warming over that observed in the twentieth century. The projections show that increase in temperature is not as much rapid as the global mean warming in South-East Asia, stronger over East Asia and South Asia and highest in Central part, North and West Asia. It may be noted that projected increase in mean temperature over Asia is higher during winter in the Northern Hemisphere than during summer for all seasons.

The maximum warming is projected in North Asia at high latitudes. The warming is considerably high in the Himalayan highlands including the arid parts of Asia. According to the Fourth AR of IPCC, sensitivity is the degree to which a system is affected, either positively or negatively, by climatic variability or climate change. The effect may be direct (e.g. a changing in the mean, range or variability of temperature affects the crop yield and its attributes) or indirect (e.g. damages caused by coastal flooding due to rise in sea level). The sensitivity indicators can be high moisture stress, high degradation of land, and high dependence of livelihoods on agriculture sector.

17.9 Concluding Remarks and Future Prospective

The following key elements of future international frameworks should include:

- *Technology support*: Effectiveness of investments in innovation can be boosted up by informal coordination as well as formal agreements around the world. Internationally, support for energy R&D should at least double, and support for the deployment of newly low carbon technologies should increase up to five times. To boost energy efficiency, the international cooperation on product standards is the only powerful way.
- *Cap and trade or emissions trading*: Emission trading is a government mandated, market-based approach to controlling emission by providing economic incentives for achieving reductions in the emissions of pollutants. Expanding and connecting the growing number of emission trading schemes around the globe is an efficient way to promote cost-effective reductions in emissions of GHGs and to take forward action for all the developing countries: strong objective in rich countries could drive flows amounting to tens of billions of dollars every year to maintain the transition to low carbon improvement paths.
- *Steps to reduce deforestation*: Deforestation around the whole world contributes further increase to global emissions every year than the transport sector. Limiting deforestation is a very much cost-effective way to ease the emissions of all GHGs; large-scale nation and international pilot programmes to explore the best ways to do this could get underway very rapidly.
- *Adaptation*: Vulnerability to climate change is more in poorest countries rather than developed countries. It is very essential that climate change be fully incorporated into developmental policy and that developed countries honour their pledges to increase support via overseas developmental assistance. Funding by international organisation should also support enhanced regional information on impacts of climate change and research into new crop cultivars that will be more resilient to flood and drought condition.

Acknowledgements We thank the editors and anonymous reviewers for giving constructive and beneficial comments, which helped us to improve the manuscript.

References

ADB and International Food Policy Research Institute (2009) Building climate resilience in the agriculture sector in Asia and the Pacific. ADB, Manila

Aggarwal PK (2008) Global climate change and Indian agriculture: impacts, adaptation and mitigation. Indian J Agric Sci 78(11):911–919

Aggarwal PK, Swaroopa DN (2009) Assessment of climate change impacts on wheat production in India. In: Aggarwal PK (ed) Global climate change and Indian agriculture-case studies from ICAR network project. ICAR, New Delhi, pp 5–12

Braun J (2008) Impact of climate change on food security in times of high food and energy prices. ICTSD, Land Management, 2(3), October. Available at: http://ictsd.org/i/environment/31511/

Buse A, Mill G, Harmens H, Cinderby S, Holland M, De Temmerman L (2003) Air pollution and vegetation. UNECE ICP Vegetation, Bangor. http://icpvegetation.ceh.ac.uk/annual report 2003 pdf.htm

Cole CV, Duxbury J, Freney J, Heinemeyer O, Minami K, Mosier A, Paustian K, Rosenberg N, Sampson N, Sauerbeck D, Zhao Q (1997) Global estimates of potential mitigation of greenhouse gas emissions by agriculture. Nutr Cycl Agroecosyst 49:221–228

FAOSTAT (2013) FAO statistics. Food and Agriculture Organization of the United Nations, Rome

Ghosh BN, Meena VS, Alam NM, Dogra P, Bhattacharyya R, Sharma NK, Mishra PK (2016) Impact of conservation practices on soil aggregation and the carbon management index after seven years of maize–wheat cropping system in the Indian Himalayas. Agr Ecosyst Environ 216:247–257

Graedel TE, Crutzen PJ (1993) Atmospheric change: an earth system perspective. WH Freeman, New York, p 446

Intergovernmental Panel on Climate Change (2007) The physical science basis. Summary for Policymakers. Contribution of Working Group I to the third assessment report of the IPCC. IPCC Secretariat, Geneva

IPCC Climate Change (2001) Impacts, adaptation and vulnerability. In: McCarthy JJ, Canziani OF, Leary NA, Dokken DJ, White KS (eds) Contribution of Working Group II to the third assessment report of the Intergovernmental Panel on Climate Change. Cambridge University Press, Cambridge, p 1032

IPCC Climate change (2007) Impacts, adaptation and vulnerability, technical summary of Working Group II. In: Parry ML, Paultikof OF, Hanon CE (eds) Fourth assessment report Intergovernmental Panel on Climate Change. Cambridge University Press, Cambridge, pp 23–78

IPCC Climate Change (2014) Mitigation of climate change. In: Edenhofer O, Pichs-Madruga, Sokona Y, Farahani E, Kadner S, Seyboth Brunner S, Eickemeier P, Kriemann B, Minx JC (eds) Contribution of Working Group III to the fifth assessment report of the Intergovernmental Panel on Climate Change. Cambridge University Press, Cambridge/New York

Janzen HH (2004) Carbon cycling in earth systems – a soil science perspective. Agric Ecosyst Environ 104:399–417

Khan TA, Singh OP, Rahman MD (2000) Recent sea level and sea surface temperature trends along the Bangladesh coast in relation to the frequency of intense cyclones. Mar Geod 23:103–116

Kumar A, Bahadur I, Maurya BR, Raghuwanshi R, Meena VS, Singh DK, Dixit J (2015) Does a plant growth promoting rhizobacteria enhance agricultural sustainability? J Pure Appl Microbiol 9(1):715–724

Lybbert T, Sumner D (2010) Agricultural technologies for climate change mitigation and adaptation in developing countries: policy options for innovation and technology diffusion. ICTSD and International Food & Agricultural Trade Policy Council, Issue Brief No. 6, Geneva. Available at: www.ictsd.org

Meena RS, Yadav RS, Meena VS (2013) Heat unit efficiency of groundnut varieties in scattered planting with various fertility levels. Bioscan 8(4):1189–1192

Meena RS, Meena VS, Meena SK, Verma JP (2015a) The needs of healthy soils for a healthy world. J Clean Prod 102:560–561

Meena RS, Meena VS, Meena SK, Verma JP (2015b) Towards the plant stress mitigate the agricultural productivity: a book review. J Clean Prod 102:552–553

Meena VS, Meena SK, Verma JP, Meena RS, Ghosh BN (2015c) The needs of nutrient use efficiency for sustainable agriculture. J Clean Prod 102:562–563

Meena VS, Meena SK, Verma JP, Meena RS, Jat LK (2015d) Current trend to mitigate climate change and sustainable agriculture: a book review. J Clean Prod 102:548–549

Meena VS, Verma JP, Meena SK (2015e) Towards the current scenario of nutrient use efficiency in crop species. J Clean Prod 102:556–557

Meena RS, Bohra JS, Singh SP, Meena VS, Verma JP, Verma SK, Sihag SK (2016) Towards the prime response of manure to enhance nutrient use efficiency and soil sustainability a current need: a book review. J Clean Prod 112(1):1258–1260

Mendelsoh R, Dinar A, Sanghi A (2001) The effect of development on the climate sensitivity of agriculture. Environ Dev Econ 6:85–101

Mosier A, Kroeze C (2000) Potential impact on the global atmospheric N_2O budget of the increased nitrogen input required to meet future global food demands. Chemosphere-Global Chang Sci 2:465–473

Oenema O, Wrage N, Velthof GL, Groenigen JW, Dolfing J, Kuikman PJ (2005) Trends in global nitrous oxide emissions from animal production systems. Nutr Cycl Agroecosyst 72:51–65

Paustian, K, Babcock BA, Hatfield J, Lal R, McCarl BA, McLaughlin S, Mosier A, Rosenberg NJ, Rosenzweig C, Schlesinger WH, Zilberman D (2004) Agricultural mitigation of greenhouse gases: science and policy options. CAST (Council on Agricultural Science and Technology) Report, R141 2004, ISBN 1-887383-26-3, p 120

Prather M, Ehhalt D (2001) Atmospheric chemistry and green house gases: the scientific basis. In: Houghton et al (eds) Third assessment report of the Intergovernmental Panel on Climate Change. Cambridge Univ Press, Cambridge, pp 239–287

Rosegrant MW, Cline SA (2003) Global food security: challenges and policies. Science 302:1917–1919

Singh DK, Singh SK, Singh AK, Meena VS (2014) Impact of long term cultivation of lemon grass (Cymbopogon citratus) on post-harvest electro-chemical properties of soil. Ann Agric Biol Res 19(1):45–48

Smith P (2004b) Engineered biological sinks on land. In The global carbon cycle. Integrating humans, climate, and the natural world, SCOPE 62. Island Press, Washington, pp 479–491

Smith KA, Conen F (2004) Impacts of land management on fluxes of trace greenhouse gases. Soil Use Manag 20:255–263

Solomon SD, Qin A, Manning M, Wratt D (2007) Technical summary. In: Solomon S, Qin D, Manning M, Chen Z, Marquis M, Averyt KB, Miller HL (eds) Climate change 2007: the physical science basis. Contribution of Working Group I to the Fourth Assessment Report of the Intergovernmental Panel on Climate Change. Cambridge University Press, Cambridge/New York

Stern Review: The Economics of Climate Change (2006)

Tubiello FN, Fischer G (2007) Reducing climate change impacts on agriculture: global and regional effects of mitigation, 1990–2080. Technol Forecast Soc 74:1030–1056

Tubiello FN, Rosenzweig C (2008) Developing climate change impact metrics for agriculture. Integ Assess J 8:165–184

Tubiello FNM, Salvatore S, Rossi A, Ferrara N Fitton, P Smith (2013) The FAOSTAT database of greenhouse gas emissions from agriculture. Environmental Research Letters 8, 1–11. doi: 10.1088/1748-9326/8/1/015009, ISSN: 1748-9326.

U.S.EPA (2006) Global mitigation of non-CO_2 Greenhouse Gases (EPA Report 43o-R-o6-oo5). Available at: http://www.epa.gov/climatechange/economics/international.html

U.S.EPA (2011) Draft: global antropogenic non-CO_2 greenhouse gas emissions: 1990–2030. Washington, DC. Available at: interned-pdf:/US-EPA Non CO_2 projection 2011 draft

Ved P, Ram N, Khichar ML, Dinesh S, Manmohan SB (2015a) Agrometeorological indices and intercepted photosynthetically active eradiation in cotton crop under different growing environments. J Cotton Res Dev 29(2):268–272

Ved P, Singh AK, Kumar S, Dwivedi SK (2015b) Effectiveness of cropping systems in mitigation of climate change through carbon sequestration. Indian Farm 65(1):31–34

Verma JP, Jaiswa DK, Meena VS, Meena RS (2015a) Current need of organic farming for enhancing sustainable agriculture. J Clean Prod 102:545–547

Verma JP, Jaiswal DK, Meena VS, Kumar A, Meena RS (2015b) Issues and challenges about sustainable agriculture production for management of natural resources to sustain soil fertility and health. J Clean Prod 107:793–794

Wang P, Saraswati S, Guan Z, Watkins CJ, Wurtman RJ, Littleton JT (2004) A Drosophila temperature-sensitive seizure mutant in phosphoglycerate kinase disrupts ATP generation and alters synaptic function. J Neurosci 24:4518–4529

WHO (2003) Health aspects of air pollution with particulate matter, ozone and nitrogen dioxide (http://www.euro.who.int/document/e79097.pdf), p 98, EUR/03/5042688, World Health Organisation, Bonn.

World Bank (2014) Data from database: World development indicators. Last Updated: 11/06/2014.internet: http://databank.worldbank.org/data/views/variableselection/selectvariables

Potential Impacts of Climate Change on Quality Seed Production: A Perspective of Hill Agriculture

18

Aniruddha Maity, Dunna Vijay, Anirban Mukherjee, and Amrit Lamichaney

Abstract

Indian agriculture has witnessed a paradigm shift in the agricultural productivity from the 1960s onward with the introduction of quality seed of high-yielding varieties which helped in achieving the goals and impacts of green revolution. However, in addition to the inherent bottlenecks in agriculture, yielding a plethora of emerging problems has slowed down the smooth pursuance of enhanced productivity. And that is why the need of a second green revolution in Indian agriculture has been emphasized. The second green revolution is possible through effective resource utilization. It is universal that without high-quality seed, utilization of other resources and better technologies remain ineffective. Seed being the foremost requirement, with its quality and availability, plays the most critical role in realizing the targeted production level of second/evergreen revolution. Among these, the burning issue of climate change and its probable consequences on quality of seed as well as on seed production itself has not received importance till recent past, but the problem is very real. Seed quality comprises of several parameters, viz., physical and genetic purity of seed, seed germination, viability, vigor, seed health, and appearance like size, shape, weight, and color. Each of these parameters depends on climatic variables prevailing during the entire crop growth period, harvest, subsequent seed processing, and storage. The weather aberrations particularly during seed setting lead to severe fluctuations in overall seed production and subsequently on its quality. If the climatic factors are adverse, resultant loss in quality and quantity

A. Maity (✉) • D. Vijay
ICAR-Indian Grassland and Fodder Research Institute, Jhansi, Uttar Pradesh 284003, India
e-mail: aniruddha.maity@icar.gov.in

A. Mukherjee
ICAR-Vivekananda Institute of Hill Agriculture, Almora 263601, Uttarakhand, India

A. Lamichaney
ICAR-Indian Institute of Pulses Research, Kanpur, Uttar Pradesh 208 024, India

J.K. Bisht et al. (eds.), *Conservation Agriculture*,
DOI 10.1007/978-981-10-2558-7_18

of seeds upshot to lower market value and nonavailability followed by decreased crop area and poor crop harvest. This in turn severely affects both food and nutritional security of the nation as well as the economy of farmers. On the other hand, according to the Intergovernmental Panel on Climate Change report, the Himalayan ecosystem is one of the highly vulnerable zones after coastal ecosystem to climate change in India. The socio-economic condition of hill depends mainly on agriculture. Therefore, owing to its importance and need of time, in this chapter we reviewed the potential impacts of climate change on seed setting and quality parameters, which ultimately alters seed production and performance, as well as the plausible strategies, combining the traditional and modern knowledge bases to combat those issues in a holistic way.

Keywords
Climate • Environment • Seed quality • Seed production • Seed germination

18.1 Introduction

With every sunrise, the world is being challenged to produce adequate food for its ever-growing population. Besides some inherent bottlenecks in agriculture, a basket of emerging problems has narrowed down the smooth pursuance of enhanced productivity. Among these, the burning issue of altering climate and its impact on agricultural production system attracted the interests of researchers very late, but the problem is very real (Maity and Pramanik 2013). If nothing is done to seize or adapt to climate change, the situation will be very grim for us in the coming years. Agricultural productivity is a function of better inputs, among which quality seed holds the utmost importance. As per an estimate, the use of quality seed with other inputs remaining the same enhanced the yield by more than 20 %. At present, Indian farmers use only 20–25 % quality seed, which makes it possible to produce about 264 MT of food grain annually (Rajendra Prasad et al. 2015). So we can think of the production potential of high-quality seeds. Seed quality depends on the environment where the seeds were developed and stored. Thus, seed quality is susceptible to vagaries of weather. The stages in seed setting and development will be influenced by the change in climatic conditions. Thus, a steady temporal fluctuation in climatic variables has put seed quality under threat.

18.1.1 Climate Change Projections in Himalayan Region

The basics of climate change have been well understood as it encompasses the basic principle of physics that keeps the earth habitable. The presence of greenhouse gases (GHGs) in the atmosphere (like water vapor and carbon dioxide, CO_2) keeps the atmosphere warmer by about 33 °C than it would be in its absence. With the industrial development, increased usage and burning of fossil fuels and deforestations have increased the level of CO_2 and other heat-trapping GHGs in the atmosphere and are thickening the GHG's blanket (AMS 2012). The Intergovernmental Panel on Climate Change (IPCC) has projected 1.8–4.0 °C increase in temperature by the end of the twenty-first century (IPCC 2007a). Intense tropical cyclones accompanied by increased wind speed and heavy precipitation have also been projected. An increase in the amount and intensity of precipitation is very likely in high altitudes, while a decrease is expected in subtropical regions. Himalayan glaciers and snow covers are projected to melt and contract with a projected sea level rise of 0.18–0.59 m by the end of the twenty-first century. For the Indian region, the rise in the aforesaid factors is not an exception (Kavi Kumar 2009).

India is confronted with the real issue of sustaining rapid economic growth and food security amidst the increasing global threat of climate change. As per the UN Human Development Report (UNDP 2007/2008), most of the poor people residing in rural India are completely dependent on natural resources for sustaining their livelihood. In India, ~60 % of people are engaged directly or indirectly in agriculture and its allied sectors. Evidence has shown that the changing climate will affect the quality and distribution of India's natural resources like the monsoon rainfall pattern, which will ultimately threaten the livelihood security of most poor and marginal sector of the population who are closely tied to India's natural resource base, i.e., agriculture. The highly fertile riverine delta lands highly contribute to the Indian food grain production. The Himalayan glaciers are the lifelines for the key rivers in Northern India which are the main water sources for the delta agriculture. Lamentably, the Himalayan region is one of the threatened places which are facing the consequences of climate change at the most. The projected change in temperature in Northeast and Northwest India, i.e., Himalayan belt, is much higher than that in Southeast and Southwest India (Table 18.1). Therefore, the impact of climate change is going to be very high in this region.

Himalaya is one of the 34 hotspots in the world. Rich biodiversity is the characteristic feature of this region. In recent past several climate change studies have been conducted in which several geographical areas have been focused

Table 18.1 Projected changes in climatic variables of India: 2070–2099 (Kavi Kumar 2009)

Region	Jan–March	April–June	July–Sep	Oct–Dec
Temperature change (°C)				
Northeast	4.95	4.11	2.88	4.05
Northwest	4.53	4.25	2.96	4.16
Southeast	4.16	3.21	2.53	3.29
Southwest	3.74	3.07	2.52	3.04
Precipitation change (%)				
Northeast	−9.3	20.3	21.0	7.5
Northwest	7.2	7.1	27.2	57.0
Southeast	−32.9	29.7	10.9	0.7
Southwest	22.3	32.3	8.8	8.5

Source: Cline (2007)

(globally) and various domains of knowledge (subject matter) have been explored in relation to climate change. A number of authentic, reliable, and peer-reviewed documents were assembled to generate some knowledge base about the agriculture in the Indian Himalayan region vis-a-vis climate change. Shrestha et al. (2012) reported a significant alteration in climatic variables (rainfall and temperature), vegetation, and its phenology in the Himalayas over a period since 1982–2006. During these 25 years, the average increase in annual mean temperature and mean annual precipitation was 0.06 °C year^{-1} and 6.52 mm year^{-1}, respectively. They also reported an advanced (by 4.7 days or 0.19 days year^{-1}) average start of the growing season (SOS) and advanced (4.7 days or 0.19 days year^{-1}) length of the growing season (LOS), with no change in the end of the growing season (EOS). The rate of warming in the Himalayas is greater and faster than the global average, confirming the Himalayas among the regions are most vulnerable to the consequences of climate change. The Ministry of Environment and Forests projected an increase in mean annual temperature from 0.9 to 2.6 °C in the 2030s, with a net temperature increase of 1.7–2.2 °C in the 1970s. Temperature increased in all the seasons. In this region, minimum and maximum temperatures are expected to increase by 1–4.5 °C and 0.5–2.5 °C, respectively. The rainy days are expected to increase on an average by 5–10 days in the Himalayan region in the 2030s. It will increase by $>$15 days in Jammu and Kashmir. The intensity of precipitation is expected to increase by 1–2 mm/day (INCCA 2010). These changes over a period may lead to reduced water resources and thereby the existence of lifeline rivers in the Northern part. Further, the enhanced intensity of rainfall resulted in cloud bursts and unintended floods devastating the delta agriculture as a whole.

18.1.2 Seed Setting and Quality

Agricultural production depends on a lot of variable factors. The basic catalyst of efficiency of all other factors is a seed. A seed is the basic and most vital input of agriculture. Modern crop production and the science of agriculture also confirm that without quality seed, we won't have a successful agricultural production. At adverse field conditions, good-quality seeds along with recommended doses of other inputs provide uniform and rapid germination, healthy crop establishment, and subsequently good crop harvest. Seed quality comprises of several parameters, viz., should be physically and genetically pure with good seed germination; should be viable, vigorous, and free from seed diseases; and must be uniform in size, shape, weight, and color. Each of these parameters depends on climatic variables prevailing during the entire crop growth period and subsequent seed processing.

18.2 Quality Seed Production Under Changing Climate

To become high quality, the seed has to pass through several well-defined sequential steps starting from timely sowing with required inputs in the field to the good and uniform germination and growth and timely flowering; complete development of flower primordia, pollen, and egg; pollination followed by double fertilization and triple fusion resulting in embryo formation and endosperm development; and seed maturation, desiccation, and timely harvest (Fig. 18.1). The postharvest steps, viz., seed processing, transportation, storage, and handling until next sowing, are also crucial for maintaining quality. Every stage of this long chain is vulnerable to stresses, may it be biotic or abiotic (Maity et al. 2012; Ellis 2011; Khanduri 2011; Dadlani et al. 2009; Meena et al. 2013, 2015a; Singh et al. 2014; Kumar et al. 2015; Ghosh et al. 2016). Alarming reports are being generated and disseminated frequently by various scientific groups and nongovernment organization in every aspect of plant growth (Kumar et al. 2009).

The paper, recently published in Annals of Botany, has warned that warming of climate may culminate in a shift in the season of germination from spring to autumn (Phys.org 2013). Seed dormancy has been naturally modified by the changing environment and promotes germination at undesirable time. Study shows that the temperature affects both crop phenology and pollen viability in wheat. Exposing crop to lower temperature at crop growth period increased crop duration, while high temperature hastens anthesis and reduced the crop duration. Lower temperature during anthesis increases pollen sterility resulting into poor pollen germination affecting good seed set (Chakrabarti et al. 2011). Various stresses due to climate aberration during the entire seed development stage alter seed quality at

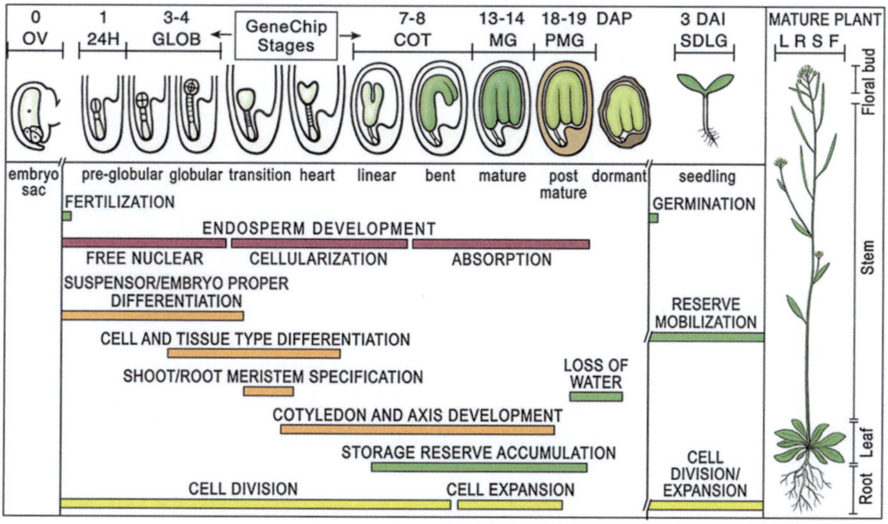

Fig. 18.1 Different steps of seed development (Le et al. 2010)

physiological and biochemical level by many means. Change in biochemical composition, oversynthesis of antiquality factors, imposition of unpredictable dormancy, expression of undesirable genes, and many more are some of the effects that could hamper the quality and quantity of seed production. In most of the tropical and subtropical areas, the seed yields are expected to reduce with the increase in temperature. Many researchers have reported that the direct effect of changing climate on "kharif" would be minor, but agricultural production during "kharif" will become vulnerable due to frequent and increased incidence of weather extremes like delayed or early onset of monsoon, intensity and frequency of wet and dry spell, asymmetry in high and low temperature, altered humidity, and incidence of pest (Krishnan et al. 2009). Agriculture in "rabi" season is relatively more risky because of shortened crop growth periods and increased terminal heat and water stress (Mall et al. 2006).

Among various climatic variable, high temperature and moisture stress not only affect seed yields but also affect seed quality and performance of resultant crops (Alqudah et al. 2011). Climate change imposes threat for existing crop species by altering the phenology and crop growth periods, promoting prevalence of new seedborne diseases, affecting pollinators' behavior (Hegland et al. 2009), shifting crop suitability areas, and reducing seed yields, storability, and longevity of seeds by altering the biochemical properties of the seed. Weather stress during seed production also lowers the duration of stigma receptivity (Maity and Chakrabarty 2013), which ultimately leads to the reduction of seed set and size. Decrease in seed size may also cause problems during seed processing. However, every adverse effect reflects on whole seed quality characteristics, which are directly related to lower market value and poor farm economy.

18.2.1 Climate Change Influences Reproductive Growth

Plants' flowering time depends greatly upon diverse environmental factors, including temperature and day length. Recent reports suggest that climate change could shift the flowering time of many plant species, in spite of the fact that flowering in specific season is important for their reproduction (Korner and Basler 2010; Kobayashi and Shimizu 2011). Long-term phonological data coupled with remote sensing measurements has indicated an altered plant phenology in spring and autumn. Plant phenology was advanced in spring by 2–3 days, whereas it was delayed by 0.3–1.6 days in autumn in the past 30–80 years, resulting in extended growing season causing increased production in terrestrial as well as in marine ecosystems (Edwards and Richardson 2004; Nemani et al. 2003) and prolonging the production of allergic pollens (Wan et al. 2002). Alteration in crop vegetative phenology could affect the functioning of ecosystem. Community and population dynamics depends upon the reproductive events, thus affecting the very process of evolution. As the reproductive behavior and phenology of a plant are very sensitive to the moisture, temperature, and photoperiod (Rathcke and Lacey 1985; Meena et al. 2015b, c, d), it is crucial that the consequences of the changing climate on the reproductive behavior and phenology be well understood. Recently, efforts were made to combine ecological and molecular functional studies to understand better the functioning of flowering time control in natural and complex environments (Meena et al. 2015e, 2016; Verma et al. 2015a, b). Warming can advance reproductive (flowering and fruiting) phenology for those species that begins flowering before the peak heat but can delay those species which begins to flower after the peak heat (Sherry et al. 2007). Such temperature-induced alteration of reproductive phenology in different growing seasons results in a gap in the staggered progression of flowering and fruiting in the community during the middle of the season. Difference among species in their response to increasing temperature alters the length of the reproductive periods, hence modifies the degree of intersection between the reproductive phases, and thus generates probabilities for a changed selection pressure to reshape communities in the projected warmer world.

Chronological variation in reproductive stages among species during the growing season may lessen the competition by widening the use of primary resource throughout the growing period (Korner and Basler 2010; Parmesan and Yohe 2003; Wilczek et al. 2010). Alteration in phenology and growth cycle among species responding to changing climate could result in the appearance of new species coexistence and even ultimately can alter the composition of species in the community (Kobayashi and Shimizu 2011; Nemani et al. 2003).

Pollen or male gametophyte remains highly sensitive throughout its developmental stages to heat stress, whereas the female gametophyte can fairly tolerate heat stress (Hedhly 2011). Heat stress fastens anthesis, forcing the genotype to initiate the reproductive phase prior to sufficient storage or accumulation of resources (Zinn et al. 2010). In angiosperm, stamen of male and pistil of female are the primary organs where reproductive development takes place (Pachauri 2007; Ahmad et al. 2010). The development of the pollen grain initiates as the reproductive tissue separates followed by a series of meiosis and mitosis cell division resulting into microspore, and its maturation results into functional male gametophyte. During this process, some nonreproductive yet vital organs are also developed, i.e., the tapetum for food or nutrient support and the stomium that helps in pollen dehiscence. For male fertility, development of microspore and tapetum is essential (Christensen and Christensen 2007; Tauber et al. 2007; Zinn et al. 2010). Most of the gene expression observed under heat stress in many plant species is often associated with faulty tapetum generation or pollen sterility (Oshino et al. 2007; Endo et al. 2009).

Sensitive crop plants can experience male sterility in response to climatic stress, and altered anatomy and physiology of their pollen development can lead to reduction in yield when exposed to heat stress (Sakata and Higashitani 2008; Wassmann et al. 2009). Exposure of anthers of barley and *Arabidopsis* to high temperature causes cessation of cell proliferation, alters chloroplast development, swells vacuoles, and causes mitochondrial aberrations (Sakata et al. 2010). Heat stress cuts the accumulation of carbohydrate in pollens and also in stigmatic tissue by changing partitioning of assimilates and altering the balance between apoplastic and symplastic loading in the phloem (Taiz and Zeiger 2006). In tomato, a widely used vegetable, the alteration of sink and source relationship even under slightly increased temperatures results in a diminution in available carbohydrates at critical stages of plant development, which leads to reduced yield parameters (Sato et al. 2006).

Higher temperature also causes premature abortion of the tapetum, which forces PMC to quickly enter into prophase of meiotic division and undergo PCD resulting into sterile pollen (Oshino et al. 2007; Parish et al. 2012). High temperature stress caused some structural anomalies in microspore of snap bean which was found to be linked with degeneration of tapetal cells caused by abnormal endoplasmic reticulum (Suzuki et al. 2001).

Like heat stress, the occurrence of dry spells at the time of flowering affects the reproductive growth or the pollination process. The occurrence of long dry spell is also a characteristic result of climate change. Such dry spell has the ability to terminate the reproductive development (Siddique et al. 2000; Turner et al. 2006) and such dry spell is known as terminal drought. Such period of dry spell is known as terminal drought. Fang et al. (2010) studied the effect of terminal drought on the reproductive growth in chickpea and to ascertain whether defective pistil or defective pollen is responsible for flower abortion caused by terminal drought. They hand-pollinated the water-stressed pistil with non-stressed pollen and vice versa. Pistils in water-stressed plants when pollinated with the pollen from non-stressed or water-stressed plant had comparatively less germinated pollen that could reach the ovary as compared to non-stressed plants pollinated with water-stressed pollen suggesting that the alteration of the function of the female reproductive part, i.e., pistil, is responsible for flower abortion than pollen viability (Fig. 18.2).

Climatic stress is also known to influence several phytohormones. The concentration of phytohormones like salicylic acid (SA), ethylene (ET), and abscisic acid (ABA) increases under climatic stress, while the concentration of cytokinin (CT), gibberellic acids (GAs), and auxin (AUX) decreases. Such changes result into premature senescence of the plant (Larkindale et al. 2005). Increased concentration of ET and ABA and reduced concentration of auxins cause the dehiscence of the reproductive organs (Binder and Patterson 2009), while altered auxin synthesis during anther development stages produces sterile pollen (Sakata et al. 2010). Alteration in cytokinin concentration caused poor seed filling in cereals.

18.2.2 Influence of Climate Variables on Pollination Dynamics

Insect-pollinated plants are pollinated by a wide variety of insects, i.e., honeybees, native bees, flies, bumblebees, etc., and maintaining the diversity of pollinator is crucial to increase yields (Fig. 18.3). Changing climate might disturb the activity of pollinator by altering their foraging periods and behavior like the duration of its stay in one flower and the areas or distance it covers. Temperature, light intensity, solar radiation, wind speed, and humidity are important climatic variables that influence the activity of pollinators (Abrol 2010). However, different pollinating agents could respond differently to these variables, as they are active at different times and at

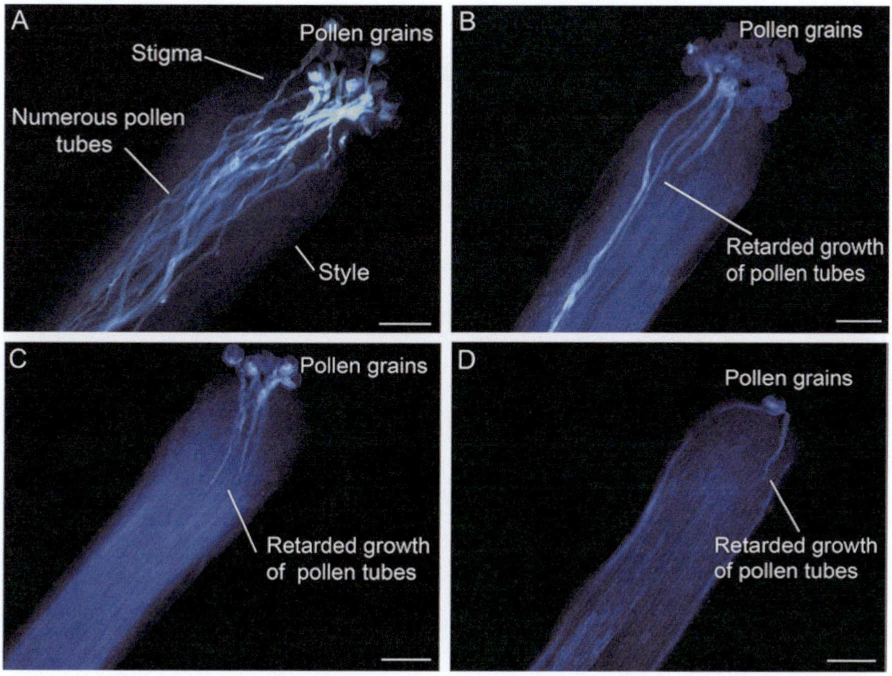

Fig. 18.2 Germinating pollen in the pistils of (**a**) non-stressed plant crossed with non-stressed plants (pollen), (**b**) non-stressed plant crossed with pollen produced in water-stressed plants, (**c**) water-stressed plants crossed with pollen from normal plant, and (**d**) water-stressed plants crossed with pollen from water-stressed plants (Fang et al. 2010)

different climatic conditions (activity windows). Bumblebees require comparatively lower temperature for foraging than honeybees (Zhao et al. 2011). Bumblebees have a large body size and mass which are insulated accompanied by long setae (Heinrich 1974) which allows them to keep their body warm when foraging at cooler places (Peat et al. 2005). But this may cause overheating under hot conditions (Peat et al. 2005). Honeybees on the other hand have the ability to maintain a constant thorax temperature (33.7–35.7 °C) when foraging at temperatures in the range of 10–27 °C (Kovac and Stabentheiner 2011). However, foraging continuously at temperature below 20 °C might lower the thoracic temperature below the minimum required for foraging. Apart from a few species-specific defaults, fluctuation in the variables of climate (particularly temperature) has been found to impact the pollinator activity windows resulting into the alteration in pollinator diversity, abundance, and behavior (Maity and Chakrabarty 2013). The data from several scientific reports confirm a direct association between pollinator activity and temperature. Apart from temperature, light intensity, RH, and wind speed also showed relationships with the pollinator belonging to some specific taxa.

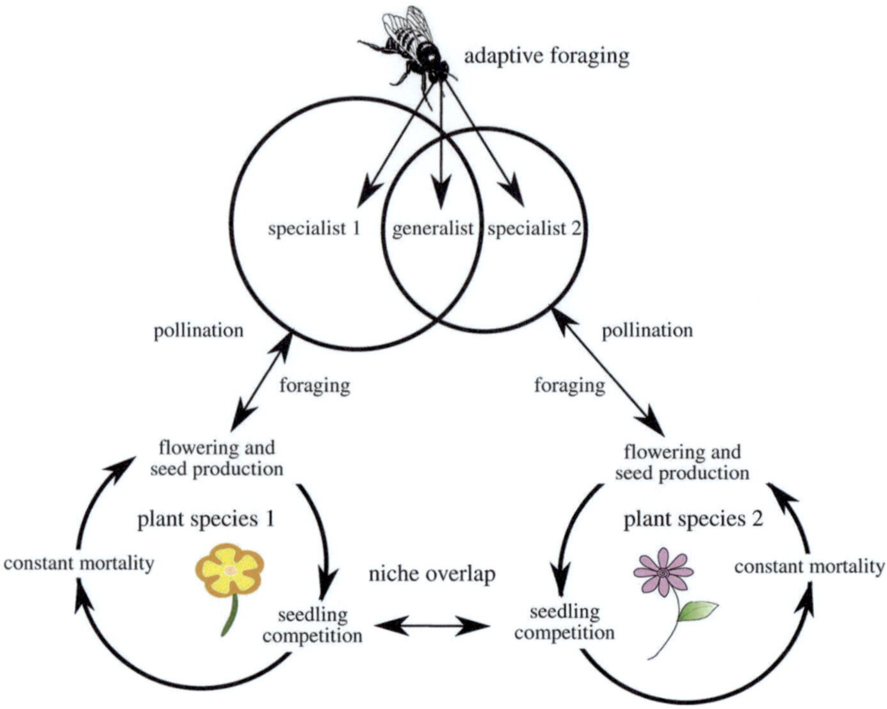

Fig. 18.3 Structure of the model that couples adaptive foraging by individual pollinators and plant population dynamics

Insect pollinators play an important role in carrying successful pollination of numerous crop plants (Cunningham et al. 2002; Westerkampe and Gottsberger 2000), which contributes for about 9.5 % to the world's total agricultural production (Gallai et al. 2009). Although they are considered very important for enhancing crop yields, and their activity directly controlled by climatic conditions, knowledge regarding the potential impact of climate change on the interaction between the pollinators and pollination process remains scarce (Kjohl et al. 2011). Bee foraging is often influenced by the high energy demands of endothermic flight, hive building, and progeny provisioning (Corbet et al. 1993). They may go for several kilometers from the hives to collect pollen and nectar from flower, and due to these causes, Diptera are main pollinators of plants in cooler regions (Gonzalez et al. 2009). Diptera often depend notably on solar heat for sustained flight by perching on the ground or on flowers (Heinrich 1993; Arroyo et al. 1982).

Other than abundance, climate variability may alter pollinator behavior by affecting their movement between inflorescences and flowers. Honeybee requires minimum thoracic temperatures during flight for successful foraging. Therefore, if temperature falls, they tend to spend more time shivering rather than foraging affecting pollination (Heinrich 1993). Honeybees are unable to produce sufficient heat for flight if their flight muscles fall down to 10 °C. Low temperatures also likely affect the behavior of pollinators especially of Dipterans as they may spend much of their time absorbing solar radiation so that they could be able to raise the thoracic temperatures above the minimum required for flight (Morgan and Heinrich 1987).

Climate change could affect the total foraging time of particular pollinators and also it could affect the pollinator diversity. Therefore, the variable responses of pollination agents to climatic variables might have an important implication for pollination under changing climate. In insect-pollinated crops, seed setting and development precisely depends on effective pollination by insects. Therefore, an effect of climate change is obvious on increased production of high-quality seeds.

18.2.3 Effect of Temperature on Seed Quality

Temperature influences germination in three ways, i.e., enzyme activity, hormone production, and moisture. For seeds to germinate, they need to imbibe water. For this to occur, sufficient moisture must be present. A warmer climate may increase evaporation and decrease moisture, which would negatively affect germination. Additionally, some climate change models predict greater variation in precipitation, which will directly affect the hydrologic cycle in many regions (IPCC 2007a). Two different hormones regulate germination: abscisic acid and gibberellins. Abscisic acid promotes dormancy and inhibits germination, while gibberellins advance germination. Through environmental cues, most importantly temperature, genes that control the production of gibberellins are upregulated, dormancy is released, and germination occurs. Germination occurs when the embryo elongates and the radical protrudes from the seed coat. For many species, enzymes are required to facilitate this process. For example, enzymes degrade endosperm tissue and rupture the seed coat. Chemical signaling regulates the production of enzymes, which is in turn regulated by temperature (Finch and Leubner 2006). If the temperature window is breached, then these enzymes may become inactive (Peterson et al. 2007). Due to the temperature dependency of hormones and enzymes, a drastic change in temperature will significantly affect germination.

Seed weight either remains unaffected or sometimes increases as temperature increases (Peltonen et al. 2011). Small fluctuation of temperature during the process of seed development and maturation could influence the ability of the seeds to germinate in many plant species (Gutterman 2000). During seed filling, high temperatures disturb normal development of seed resulting into increased proportion of shriveled seeds affecting its quality (Spears et al. 1997). However, it has been observed that the germinability of the seeds decreases when the temperature increases during seed filling (Khalil et al. 2010; Thomas et al. 2009). Reduction in the germinability of the seeds often exposed to higher temperature before physiological maturity (PM) may be due to poor accumulation of the reserve assimilates as the plants are unable to supply the seeds with the necessary assimilates required to synthesize storage assimilates (Dornbos and McDonald 1986), and/or the seeds suffer from weathering/physiological damage (Powell 2006) to such an extent that they fail to germinate. After attaining PM, high temperature sometimes causes reduced germination (Green et al. 1965) but more often reduces seed vigor. The association of temperature during seed development, maturation, and seed germination requires further investigation. In soybean, for example, temperatures (32–38 °C) that lowered the germination in controlled growth chamber did not change during seed filling, in contrast to field temperatures which can differ remarkably (Egli 2005), and the plants were kept at these temperatures from anthesis to seed harvest. However, during seed development, there may be critical stages which are particularly sensitive to temperature. This was investigated by Shinohara et al. (2006a), in the pea (*Pisum sativum* L.), which showed that when a day/night temperature of 30/20 °C for 4 days (=24 °C h above a base temperature (Tb) of 25 °C) was imposed to plants at the start of the seed development followed by reverting to field until seed harvest, germination was remarkably reduced in few genotypes. When, at later stages of seed development, plants were exposed to these conditions, germination was unaffected. Temperature stress even after successful fertilization can retard the embryo development (Barnabas et al. 2008), which can lead to poor and nonuniform germination and weaker crop stand in the next sowing (Akman 2009; Ren et al. 2009).

Seed weight, an important trait of quality seed, is reported to be affected by heat stress. Mahmood et al. (2010) reported decline in seed weight and number, which was in direct proportion to the temperature rise during flowering and seed filling stages. Heat stress during seed development in several crop species has been reported to reduce the quality of the produce in terms of total oil yield, protein, and starch. Wheat plants exposed to high temperature stress during reproductive development stages are reported to decrease sugar content of the seeds, which may be due to low rate of photosynthesis (Shah and Paulsen 2003). Altered or poor nutritional quality of the grain/seed is a consequence of climate change (Hedhly et al. 2009) and also reduced physiological quality of seed (Maity and Pramanik 2013). The electrical conductivity (EC) of the seed leachate is negatively correlated to the quality of the seeds (if EC is high, seed quality is poor and vice versa). It is reported that the EC of the single seed did not increase when the developing seeds were treated with high temperature stress before PM. However, exposing the seeds to high temperature at or after PM increased the EC of the seeds. Loss in seed vigor is related to seed deterioration (Hampton and Coolbear 1990), and lipid per oxidation is considered as the main cause for seed deterioration (McDonald 1999). Heat stress to the parent plant can cause mitochondrial degeneration and reduced ATP accumulation and declined rate of oxygen uptake in imbibing wheat embryos, showing a clear proof of physiological changes at the mitochondrial level during the early stages of seed germination under heat stress (Grass and Burris 1995). When the reproductive growth encounters high temperatures, an increase in seed cell membrane damage is common (Nilsen and Orcutt 1996; Shinohara et al. 2006b), and as a result, electrolyte leakage from seeds is amplified (Shinohara et al. 2006b; Spears et al. 1997).

Climate change, resulting from the anthropogenic release of carbon dioxide, another greenhouse gas, may globally increase temperatures between 2.5 and 4.5 °C

(IPCC 2007a). This increase in global temperatures will likely affect future plant species distribution and could influence the survivability of individual species because of the relationship between temperature, germination, and dormancy. Temperature has been shown to be the most important variable affecting germination (Milbau et al. 2009; Roberts 1988). For many species, the optimal range of temperature for germination is 10–20 °C (Baskin and Baskin 2001). Within this temperature range, the germination rate is fastest (Baskin and Baskin 2001). Each individual species has a base and ceiling temperature that represents the extremes at which germination can occur. Below and above these extremes, no germination can occur (Finch-Savage and Leubner-Metzger 2006). If climate change results in temperatures that exceed the ceiling for a species, then that species will not be able to germinate, thus affecting its survivability. For seeds to germinate, they need to imbibe water. For this to occur, sufficient moisture must be present. A warmer climate may increase evaporation and decrease moisture, which would negatively affect germination. Additionally, some climate change models predict greater variation in precipitation, which will directly affect the hydrologic cycle in many regions (IPCC 2007a). Shinohara et al. (2006b) correlated the association of temperature with seed vigor using 262 seed lots of garden pea. They observed wide variation in terms of vigor, and such variation was associated with temperature during seed maturation and development. Usually, the higher the temperature during seed development and maturation, the poorer is the seed vigor.

18.2.4 Seed Quality Under Elevated CO_2 Levels

Initial seed quality determines the success of the agricultural production. Seed size and weight are an important trait of seed quality. Seed weight, usually described as 1000 seed weight, reflects its density, whereas seed size reflects its volume (Castro et al. 2006). Seed weight or mass is the least variable yield-attributing traits due to the breeding for high uniformity of seed (Almekinders and Louwaars 1999) and separation and removal of undersized seeds during processing. Plant genetic makeup, nutrient availability, and water availability are some factors that influence seed weight (Castro et al. 2006). Enhanced concentration of CO_2 in the atmosphere increases availability and translocation of the food assimilates, thus may increase the seed weight (Jablonski et al. 2002). However, the effect of enhanced CO_2 on seed mass is variable as there are reports that the seed weight has decreased (Huxman et al. 1998; Smith et al. 2000; Wagner et al. 2001), increased (Musgrave et al. 1986; Dijkstra et al. 1999; Steinger et al. 2000; Quaderi and Reid 2005) and shown no change (Edwards et al. 2001; Prasad et al. 2002) under elevated CO_2.

Jablonski et al. (2002) concluded that enhanced CO_2 increased the seed mass by approximately 4 %, and such increase was more in legumes (+8 %) than nonlegumes (+3 %), while C_4 plants showed no response in an experiment they conducted using about 79 species. They also found significance within and between species seed mass variations. Also they reported a reduced seed nitrogen content (by 14 %) due to elevated CO_2, but such reduction in seed N was more in nonlegumes than the legumes. Increase in seed mass ratio, i.e., the ratio of seed mass in enhanced CO_2 to ambient CO_2, ranged from $0 \cdot 88$ to $2 \cdot 07$, $0 \cdot 93$ to $1 \cdot 87$, and $0 \cdot 75$ to $4 \cdot 45$ in wheat, soybean, and rice, respectively. However, the only possible way to increase seed mass is by increasing the availability of N at elevated CO_2 condition (Hikosaka et al. 2011). Greater increase in seed mass in legumes than nonlegumes as observed by Jablonski et al. (2002) may be due to higher utilization of carbon (C) gain under enhanced CO_2 for higher nitrogen fixation (Allen and Boote 2000), resulting into greater seed mass without affecting the seed N content. However, increase in temperature could nullify the beneficial effect of enhanced CO_2 (Prasad et al. 2002) and may result in reducing seed mass (Spears et al. 1997) because of dry matter accumulation and shortening in the seed filling duration (Young et al. 2004). A reduction in seed mass does not necessarily mean that they are of poor quality or loss in any other seed quality traits as no significant relationship was found in the relationship between germination and seed mass (Castro et al. 2006) or with seed vigor (Powell 1988).

To consider a seed lot to be of high quality, it must have germination as close as 100 %. Producing seed is a sequential process that starts with field selection and ends with harvesting and threshing followed by seed processing, packaging, and shipping. The very quality of the seeds can be affected due to adverse weather and climatic variables at all the stages of crop growth and also on the conditions the seeds are exposed during harvesting and processing activities (Dornbos 1995), especially rainfall, temperature, and RH (Egli 2005). However, there are variable reports in terms of the effect of elevated CO_2 on seed germinability. Seed germination is reported to decrease (Quaderi and Reid 2005), increase (Edwards et al. 2001), or remain unaffected (Steinger et al. 2000; Thomas et al. 2009) in elevated CO_2 conditions. Seed C/N ratio is reported to be increased by elevated CO_2 (Steinger et al. 2000) in legumes, and in nonlegumes, enhanced CO_2 results into reduced seed N content and increases seed (Jablonski et al. 2002). Greater content of seed N increases the rate of seed germination but may not affect final germination. But, alteration in seed C/N ratio may result into reduced protein content, which in turn may affect the germination due to unavailability of the amino acids which are essential for embryo growth. Therefore, alteration in seed C/N ratio could indirectly affect seed viability (Andalo et al. 1996). Another possible reason for enhanced germination due to elevated CO2 is by increasing ethylene production as it is known for its promontory role on germination.

18.2.5 Rainfall Effects Seed Setting and Quality

Emergence of seedling is affected by several variables, availability of water being an important variable (Herrera et al. 2008). Availability of moisture also determines the proportion of seeds with integumentary dormancy and nonviable seeds (Johnston et al. 1998a). Rainfall affects quality seed production in many ways. If rainfall occurs during peak pollination time, the preparedness of reproductive structures for effective pollination gets hampered and results in failure of seed setting. The anthesis doesn't take place, the stigma doesn't become receptive, and the pollinators don't visit the flowers. Even if the crop pollination escapes rainfall at early stages, the seed development and maturity stage are also critically vulnerable to rainfall. This hampers the physiological processes for successful development of quality seeds. Therefore, the erratic rainfalls, as a result of climate change, occurring during unexpected stages of crop growth, lower the crop seed yield in terms of both quantity and quality. Such erratic rainfall that occurs just prior to harvest will cause preharvest sprouting (PHS). The PHS is defined as the phenomenon of germination of physiologically mature grains in the ear or panicle or pod usually under wet conditions shortly prior to harvest. PHS is also known to occur in crops like wheat, barley, paddy, maize, sorghum, soybean, chickpea, green gram, black gram, and groundnut. PHS not only reduces the final yield but also the quality of the seeds making it unfit for sowing.

18.2.6 Enhanced UV Radiation vs. Seed Quality

Increased burning of fossil fuels and other anthropogenic activities have resulted in the thinning of the protective ozone layer that had led to the enhanced penetrance of UVB radiation reaching earth's surface (McKenzie et al. 1999). UVB radiation affects certain biochemical and physiological processes including abnormal carbon partitioning from source to sink (Bassman 2004) resulting into altered crop phenology, altered photosynthesis (Sunita and Guruprasad 2012), and aborted reproductive organs ultimately reducing the yield (Mohammed and Tarpley 2009, 2010).

Exposure of soybean to enhanced UVB caused dwarfness due to internode shortening (Barnes et al. 1990), decreased leaf number and area, altered flowering time, reduced total chlorophyll content (Feng et al. 2001a), reduced plant biomass and plant yield (Feng et al. 2001b), and reduced the concentration of flavones and phenolic compounds (Kim et al. 2011). Yield reduction in soybean due to enhanced UVB radiation is because of reduction in pod number plant^{-1} (Liu et al. 2009). This reveals that exposure of plants to higher UVB at early flowering stages affects the availability and translocation of assimilates to the developing and new reproductive parts and also affects flowering resulting into poor number of pods per plant during harvest. However, other important yield-attributing traits, i.e., the number of seeds per pod, were not affected by the enhanced UVB radiation suggesting that these traits are controlled more by the plant genotype and are not much influenced by the environment. However, the size of the seed is negatively influenced by UVB radiation. The cell number and cell volume of the cotyledon are two main factors that determine the size of the seed (Mathew et al. 2000). UVB radiation decreases the cotyledon cell number. This implies that reduction in seed size by UVB may be due to reduction in the cell number of cotyledon rather than its volume.

18.2.7 Impact of Climate Change on Seed Physiochemical Composition

Even if acceptable quantity of seed is produced, possible change in its chemical composition renders its quality. Polyphenolics and fatty acid composition of seed oil have significant effect on viability, germination, dormancy, seed coat permeability, and storability. Oleic-to-linoleic acid ratio or unsaturated-to-saturated fatty acid ratio plays a main role. But the fatty acid profile and ratio in crop plants are severely vulnerable to environmental stress. Oil-producing plant grown in hotter regions contains higher amount of saturated and monounsaturated fatty acids and has lower amount of polyunsaturated fatty acids as compared to the plant grown under cooler climates (McVetty 2009). Climate change is reported to have adverse effects on the biochemical composition of the seeds with far-reaching consequences on the seed quality. Recent proteomics work reveals some important insight about the molecular mechanism involved in the germinating seeds of many crops exposed to stresses like heavy metal stress especially copper and cadmium ions, drought, and high or low temperature. Over 600 environmental factor-responsive proteins have been identified in the germinating seeds that are having various expression patterns (Tan et al. 2013). The activity of some important enzymes like callase (ß-1, 3-D-glucanase), sucrose and starch synthase, hydrogenase, α- and β-amylase, invertase, protease, lipase, mannose, etc. and plant growth regulators (PGR) like ABA, GA, ET, cytokinins, auxin, brassinosteroids (BR), etc. is indispensible for development and maturation of high-quality seeds. These enzymes and PGR regulate and control the genetic processes affecting seed development, maturity, release of dormancy, and germination. Amino acids like proline confer stress tolerance in plant as well as in seeds, and the enzymes of its synthesis pathway like proline dehydrogenase (PDH), pyrroline-5-carboxylate synthetase (P5CS), and ornithine aminotransferase (OAT) are also modified by environmental stresses (Song et al. 2005). Many researchers summarize the diverse effect of climatic alterations on the control and metabolic mechanisms of seed germination, including induction of environmental factor-responsive signaling pathways, mobilization and utilization of seed storage reserve, improvement of DNA repair, regulation of gene and synthesis of protein, and control of cell structure and cell defense. The messenger RNAs accumulated in the course of maturation of embryo in them, proteostasis and DNA integrity, metabolism pathway of sulfur amino acid regulating the cell metabolism, and their close associations with hormone signaling pathways play the key role in control of germination (Rajjou et al. 2012). Temperature and/or climatic irregularities at any stage of crop growth may hamper the normal functioning of these essential components resulting in physiological abnormalities which ultimately worsen seed quality (Fig. 18.4).

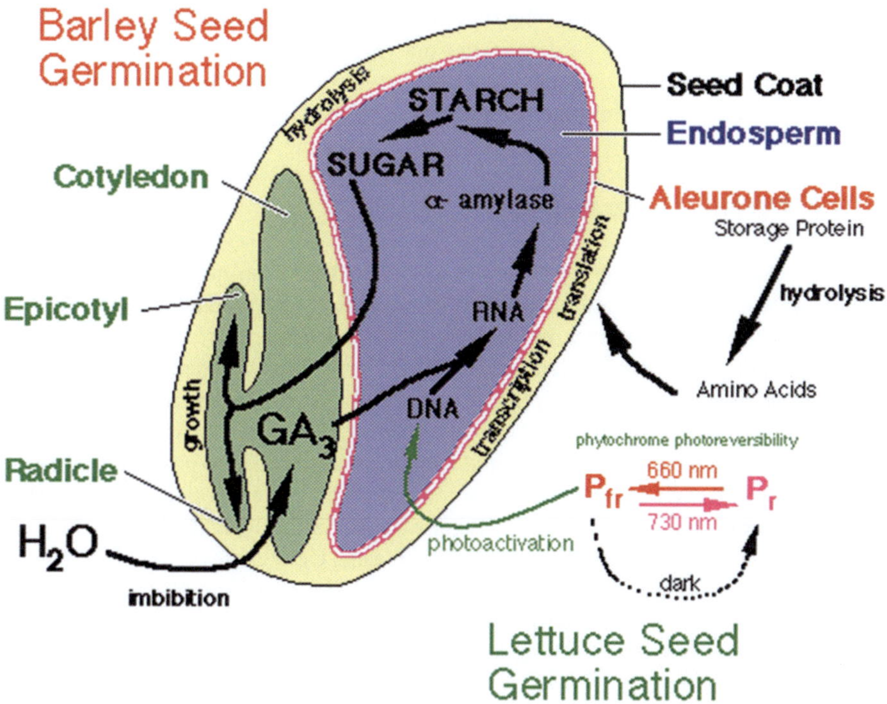

Fig. 18.4 Different steps involved in seed germination (http://plantphys.info/seedg/seed10.html)

18.3 Mitigation Strategy for Quality Seed Production

Plenty of strategic, basic, and hypothetical plans are being chalked out in every part of the world regarding the impact of climate change. The obvious effects can never be nullified ever by single permanent mean but it could be minimized. Therefore, strategy is being planned in a holistic approach that would include every aspect from crop breeding to its management at field as well as lab to use of indigenous technical knowledge (ITK).

18.3.1 Breeding Strategies

To survive in this changing environment, C3/C4 balance is changing with time in several ecosystems worldwide (Lattanzi 2010) like the way once C4 evolved from C3 by means of genetic, anatomical, and physiological changes for better utilization of enhanced temperature and CO_2 (Kalra et al. 2008). By modern research strategies, like C_4 engineering, a C_3 plant may be converted into C_4 plant (www.articles.timesofindia.indiatimes.com) to utilize increased level of atmospheric CO_2. Identifying and transforming the genes necessary to install C_4 photosynthesis are the key step for C_4 engineering. The functional C_4 cycle requires downregulation of some part of Calvin-Benson cycle in mesophyll cells. Another aspect of C_4 engineering is incorporation of transporters which are required to support fluxes of metabolites between subcellular compartments of the C_4 cycle.

18.3.2 Agronomic Strategies

Besides these research advances, there is a need to shift crop growing areas toward suitable growing locations, and accordingly, we may shift seed production areas. Studies have shown the steady and negative impact of changing temperature regime on productivity of mustard, chickpea, wheat, and barley and have recommended shifting the date of sowing as an adaptation strategy (Kalra et al. 2008). Maximum wheat yield reduction due to per degree increase in temperature was observed in Haryana (4.29 q per ha) and Rajasthan (2.49 q per ha). There are some temporal changes in seed production areas of India; in the south, erratic rainfall patterns are causing the coffee crops to fruit twice and sometimes thrice in a year resulting in inferior beans. Similarly, in the Kuttanad region of Kerala (state's Rice Bowl), heavy rains delay normal October sowing, pushing it up to December (http://www.upi.com). Because of late sowing, crop growth period coincides with the occurrence of insect pests and diseases. The additional cost of pest control increases the price of seeds. So, the combination of shifting the location and sowing time is needed to avoid these problems.

18.3.3 Value Addition

Improvement in microclimate around seeds may eliminate or curtail the risk of damage due to possible climatic calamities in field condition. Seed coating with thermostable polymer (http://www.ccimarketing.com) augmented with pesticides and growth supplements (e.g., plant growth regulators or hormones, etc.) is a promising option in high-value crops to cope up with harsh field condition. In this technique, seeds are allowed to germinate in the field only within a suitable range of temperature instead of a sudden unfavorable condition. Retaining high amount of moisture in soil for easy germination of seed is an alternative to cope with changing soil moisture pattern. Other modern technologies like seed priming and hardening, nanotechnology etc. are coming in the picture to aid in maintaining or improving seed quality. Recently nanotechnology is a promising tool to improve seed and seedling performance in harsh condition (Maity et al. 2016).

18.3.4 Low-Cost Indigenous Technical Knowledge (ITK) for Storage

Storage of grains was one of the most promising challenges that people were facing from time immemorial. The traditional knowledge has emerged out with solutions to protect grains for a long time. One of such promising grain storage structures is "mittikakotha" (mud grain storage structure) made through indigenous technological knowledge (ITK) for pest-restricted storage of wheat and maize. ITKs are acquired by the communities through generations of living in close proximity with nature. These are the time-tested practices. During the Field Experience Training of Foundation Course for Agricultural Research Services in Phalicheda-Khedi village of Udaipur district in February 2015, the ITK on grain storage was documented from the house of Shri Bhura Lal Jat. A structure called "mittikakotha" is made in the house for the storage of grains for home consumption purpose. It is made up of native soil, bamboo sticks (*Bambusa* spp.), and cow dung, around 30–45 cm above the ground supported with the wooden sticks all around the dimensions to keep the store away from rodents. Ash layer covered with the neem (*Azadirachta indica*) leaves at the bottom of this structure helps to restrict entry and attack of storage pests like *Sitophilus* spp., *Rhyzopertha* spp., *Trogoderma* spp., etc. Grains are stored in this indigenous structure up to 5 years. Application of inputs, may it be chemical fertilizer or pesticides in an extensive way, has curtailed the sustainability of agricultural production system. The ITKs are time-tested knowledge which belongs to farming community evolved as constant interaction with nature. These practices are well known to farmers and are helpful in maintaining and enhancing the quality of environment. Dissemination of modern agricultural practices leads to less emphasis on indigenous practices; as a result, these practices have started to lose their importance and eroded to a large extent. Farmers from the hilly region of Uttarakhand are still practicing farming by adopting the indigenous techniques. For leaf spot disease (*Cercospora* spp.), soybean farmers are employing ash made up of seeds and barks of reetha (*Sapindus mukorossi*). Karnal bunt (*Neovossia indica*) of wheat was managed by spraying the mixture of cow urine and leaf extract of bakain (*Melia azadirachta*). Seeds of cereals like rice, wheat, millets, etc. were stored by mixing the leaf of akhrot (*Juglans regia*) and bakain (*Melia azadirachta*). The oil of Indian mustard (*Brassica juncea*) and red roasted soil are coated to store pulses for a long time and to manage storage pest like *Sitophilus* spp. Several reports emphasized the use of neem products, viz., neem leaves, neem seed kernel, neem oil, etc., in enhancing the storability of various species. The use of eucalyptus leaves due to their inhibitory effect on storage pests is another cost-effective practice. The indigenous knowledge of ash application on crop in field as well as on seeds during storage deters the attack of various field and storage pests, respectively (Chakrabarty et al. 2012; Mehta et al. 2012) (Table 18.2).

In the changing climate era, to combat the major regional or global catastrophe, the Svalbard Global Seed Vault has been established on the Norwegian island of Spitsbergen in 2006. An Indian seed bank in the Himalayas in Chang La has been planned for the same.

18.4 Conclusions

Besides the above strategies, evaluation and screening, conservation, and seed multiplication of local landraces for their inherent resilience to climate change novel breeding for high germinable variety and hybrids having photo- and thermo-

Table 18.2 Plant parts and other material used for seed storage in Uttarakhand Himalaya

Material used for pest control	Crops
Bach (*Acorus calamus* L.) rhizome and leaves	Pulses and cereals
Peach (*Pinus persica* L.) leaves	Pulses and cereals
Neem (*Azadirachta indica* A. Juss) leaves	Pulses, oilseeds, and cereals
Timur (*Zanthoxylum armatum* DC.) leaves	Pulses, oilseeds, and cereals
Walnut (*Juglans regia* L.) leaves	Pulses and cereals
Bakayan (*Melia azadirachta* L.) leaves	Pulses and cereals
Turmeric (*Curcuma longa* L.) leaves	Pulses and cereals
Lemon (*Citrus limon* L.) leaves	Pulses and cereals
Wooden ash	Cereals
Cow dung + cow urine	Pulses and cereals
Kerosene oil	Pulses
Lime powder	Pulses
Mustard oil	Pulses
Red roasted soil	Pulses and cereals
Eucalyptus leaves	Pulses and cereals

Source: Modified after Mehta et al. (2012)

insensitive nature, with high concentration of advantageous chemical components (sugars, protein, lipids, and others) and enzymes, have been paving a smooth way toward a winning stroke against climate change. The indigenous knowledge for mitigation of climate change impact has to be linked with modern technologies. Moreover, the farmer who is the direct victim of climatic irregularities is to be educated with converged extension efforts so that they are less suffered by the climate change impact. The convergence of private and public extension system for creating awareness to farmers and revision of agricultural policy with respect to compensation, minimum support price and crop insurance are needed to tackle the situation from farmer's point of view (Mukherjee and Maity 2015). But human effort solely cannot overcome the natural consequences of the ever-increasing undesirable activities of human race on earth. Until and unless the causes of climate change are minimized to the maximum possible limit, the sufferings are going to increase by manifold for more but one count.

Acknowledgments We thank the editors and anonymous reviewers for their constructive comments, which helped us to improve the manuscript.

References

Abrol DP (2010) For aging behaviour of *Apis florea* F. an important pollinator of *Allium cepa* L. J Apic Res 49(4):318–325

Ahmad A, Diwan H, Abrol YP (2010) Global climate change, stress and plant productivity. In: Pareek A, Sopory SK, Bohnert HJ (eds) Abiotic stress adaptation in plants: physiological, molecular and genome foundation. Springer Science+Business Media, Dordrecht, pp 503–521

Akman Z (2009) Comparison of high temperature tolerance in maize, rice and sorghum seeds by plant growth regulators. J Anim Vet Adv 8:358–361

Allen LH Jr, Boote KJ (2000) Crop ecosystem responses to climatic change: soybean. In: Reddy KR, Hodges HF (eds) Climate change and global crop productivity. CAB International, Wallingford, pp 133–160

Almekinders CJM, Louwaars NP (1999) Farmer's seed production. New approaches and practices. ITDG Publishing, London

Alqudah AM, Samarah NH, Mullen RE (2011) Drought stress effect on crop pollination, seed set, yield and quality. In: Lichtfouse E (ed) Alternative farming systems, biotechnology, drought stress and ecological fertilisation. Sustainable agriculture reviews. Springer, Dordrecht, pp 193–213

AMS (2012) Climate change- an information statement of the American Meteorological Society (Adopted by AMS Council 20 August 2012)

Andalo C, Godelle B, Lefranc M, Mousseau M, Tillbottraud I (1996) Elevated CO2 decreases seed germination in Arabidopsis thaliana. Global Change Biol 2:129–135

Arroyo MTK, Primack R, Armesto J (1982) Community studies in pollination ecology in the high temperate Andes of central Chile. I. Pollination mechanisms and altitudinal variation. Am J Bot 69(1):82–97

Barnabas B, Jager K, Fehér A (2008) The effect of drought and heat stress on reproductive processes in cereals. Plant Cell Environ 31:11–38

Barnes PW, Flint SD, Caldwell MM (1990) Morphological response of crop and weed species of different growth forms to UV-B radiation. Am J Bot 77:1354–1360

Baskin CC, Baskin JM (2001) Seeds. Ecology, biogeography, and evolution of dormancy and germination. Academic Press, San Diego, p 666

Bassman JH (2004) Ecosystem consequences of enhanced solar ultraviolet radiation: secondary plant metabolites as mediators of multiple trophic interactions in terrestrial communities. Photochem Photobiol 79:382–398

Binder BM, Patterson SE (2009) Ethylene-dependent and -independent regulation of abscission. Stewart Postharvest Rev 5:1–10. doi:10.2212/spr.2009.1.1

Castro J, Hodar JA, Gomez JM (2006) Seed size. In: Basra AS (ed) Handbook of seed science and technology. Food Products Press, New York, pp 397–427

Chakrabarti B, Singh SD, Nagarajan S, Aggarwal PK (2011) Impact of temperature on phenology and pollen sterility of wheat varieties. Aust J Crop Sci 5:1039–1043

Chakrabarty SK, Joshi MA, Singh Y, Maity A, Vashisht V, Dadlani M (2012) Indian J Genet Plant Breed 72(2):136–142

Christensen JH, Christensen OB (2007) A summary of the prudence model projections of changes in European climate by the end of this century. Clim Change 81:7–30. doi:10.1007/s10584-006-9210-7

Cline W (2007) Global warming and agriculture: impact estimates by country. Peterson Institute, Washington, DC

Corbet SA, Fussell M, Ake R, Fraser A, Gunson C, Savage A, Smith K (1993) Temperature and the pollinating activity of social bees. Ecol Entomol 18(1):17–30

Cunningham SA, Fitz Gibbon F, Heard TA (2002) The future of pollinators for Australian agriculture. Aust J Agric Res 53(8):893–900

Dadlani M, Chakrabarty SK, Basu S (2009) Impact of climate change and IP regime on the production and availability of quality seed. Indain J Genet Plant Breed 69(4):325–330

Dijkstra P, Schapendonk HMC, Groenwold K, Jansen M, Van De Geijn SC (1999) Seasonal changes in the response of winter wheat to elevated atmospheric CO2 concentration grown in open-top chambers and field tracking enclosures. Glob Chang Biol 5:563–576

Dornbos DL Jr (1995) Production environment and seed quality. In: Basra AS (ed) Seed quality: basic mechanisms and agricultural implications. Food Products Press, New York, pp 119–152

Dornbos DL Jr, McDonald MB (1986) Mass and composition of developing soybean seeds at five reproductive growth stages. Crop Sci 26:624–630

Edwards M, Richardson AJ (2004) Impact of climate change on marine pelagic phenology and trophic mismatch. Nature 430:881–884

Edwards GR, Clark H, Newton PCD (2001) The effects of elevated CO2 on seed production and seedling recruitment in a sheep-grazed pasture. Oecologia 127:383–394

Egli DB (2005) Flowering, pod set and reproductive success in soya bean. J Agron Crop Sci 191:283–291

Ellis RH (2011) Rice seed quality development and temperature during late development and maturation. Seed Sci Res 21(2):95–101

Endo M, Tsuchiya T, Hamada K, Kawamura S, Yano K, Ohshima M, Higashitani A, Watanabe M, Kawagishi-Kobayashi M (2009) High temperatures cause male sterility in rice plants with transcriptional alterations during pollen development. Plant Cell Physiol 50:1911–1922. doi:10.1093/pcp/pcp135

Fang X, Turner NC, Yan G, Li F, Siddique KHM (2010) Flower numbers, pod production, pollen viability, and pistil function are reduced and flower and pod abortion increased in chickpea (*Cicerarietinum* L.) under terminal drought. J Exp Bot 61:335–345

Feng HY, An LZ, Xu SJ, Qiang WY, Chen T, Wang XL (2001a) Effect of enhanced ultraviolet-B radiation on growth, development, pigments and yield of soybean (*Glycine max* (L.) Merr.). Acta Agron Sin 27:319–323

Feng HY, Chen T, Xu SJ, An LZ, Qiang WY, Zhang MX, Wang XL (2001b) Effect of enhanced UV-B radiation on growth, yield and stable carbon isotope composition in *Glycine max* cultivars. Acta Bot Sin 43:709–713

Finch-Savage WE, Leubner-Metzger G (2006) Seed dormancy and the control of germination. New Phytol 171:501–523. doi:10.1111/j.1469-8137.2006.01787.x

Gallai N, Salles JM, Settele J, Vaissiere BE (2009) Economic valuation of the vulnerability of world agriculture confronted with pollinator decline. Ecol Econ 68(3):810–821

Ghosh BN, Meena VS, Alam NM, Dogra P, Bhattacharyya R, Sharma NK, Mishra PK (2016) Impact of conservation practices on soil aggregation and the carbon management index after seven years of maize–wheat cropping system in the Indian Himalayas. Agr Ecosyst Environ 216:247–257

Gonzalez AMM, Dalsgaard B, Ollerton J, Timmermann A, Olesen JM, Andersen L, Tossas AG (2009) Effects of climate on pollination networks in the West Indies. J Trop Ecol 25:493–506

Grass L, Burris JS (1995) Effect of heat stress during seed development and maturation on wheat (*Triticum durum*) seed quality. II. Mitochondrial respiration and nucleotide pools during early germination. Can J Plant Sci 75:831–839

Green DE, Pinnell EL, Cavanaugh LE, Williams LF (1965) Effect of planting date and maturity date on soybean seed quality. Agron J 57:165–168

Gutterman Y (2000) Maternal effects on seed during development. In: Fenner M (ed) Seeds: the ecology of regeneration in plant communities. CAB International, Wallingford, pp 59–84

Hampton JG, Coolbear P (1990) Potential versus actual seed performance – can vigour testing provide an answer? Seed Sci Technol 18:215–228

Hedhly A (2011) Sensitivity of flowering plant gametophytes to temperature fluctuations. Environ Exp Bot 74:9–16. doi:10.1016/j.envexpbot.2011.03.016

Hedhly A, Hormaza JI, Herrero M (2009) Global warming and sexual plant reproduction. Trends Plant Sci 14:30–36. doi:10.1016/j.tplants.2008.11.001

Hegland SJ, Nielsen A, Lazaro A, Bjerknes A, Totland O (2009) How does climate warming affect plant-pollinator interactions? Ecol Lett 12:184–195

Heinrich B (1974) Thermoregulation in endothermic insects. Science 185(4153):747–756

Heinrich B (1993) The hot-blooded insects: strategies and mechanisms of thermoregulation. Harvard University Press, Cambridge, MA, p 601

Herrera F, Ocumpaugh WR, Ortega JA, Lloyd-Reilley J, Rasmussen GA, Maher S (2008) Environmental influences on seed quality of windmill grass ecotypes in South Texas. Agron J 100:1205–1210

Hikosaka K, Kinugasa T, Oikawa S, Onoda Y, Hirose T (2011) Effects of elevated CO2 concentration on seed production in C3 annual plants. J Exp Bot 62:1523–1530

http://www.ccimarketing.com/CCINews/Landec/landec2.asp

http://www.upi.com/Business_News/Energy-Resources/2011/02/01/Indias-crops-affected-by-erratic-climate/UPI-58761296595012/

http://www.usanpn.org/

Huxman TE, Hamerlynck EP, Loik ME, Smith SD (1998) Gas exchange and chlorophyll fluorescence responses of three western *Yucca* species to elevated CO_2 and high temperature. Plant Cell Environ 21:1275–1283

INCCA (2010) India: greenhouse gas emissions 2007. Report of Indian Network for Climate Change Assessment, Ministry of Environment and Forests, Government of India, pp 64

IPCC (2007a) Climate change 2007. The physical science basis. Summery for policymakers. Intergovernmental panel on climate change

ISTA (2011) International rules for seed testing. International Seed Testing Association, Bassersdorf

Jablonski LM, Wang X, Curtis PS (2002) Plant reproduction under elevated CO2 conditions: a meta-analysis of reports. New Phytol 156:9–26

Johnston M, Olivares A, Contreras X (1998) El banco de semillasdelsuelo y surespuesta a regímenespluviométricossimulados. II. Géneros de interésforrajero. Avancesen Prod Anim 23:55–65

Kalra N, Chakraborty D, Sharma A, Rai HK, Jolly M, Chander S, Ramesh Kumar P, Bhadraray S, Barman D, Mittal RB, Lal M, Sehgal M (2008) Effect of increasing temperature on yield of some winter crops in northwest India. Curr Sci 94:82–88

Kavi Kumar KS (2009) Climate sensitivity of Indian agriculture. working paper 43/2009. Madras School of Economics. Gandhi Mandapam Road, Chennai, 600 025, India, April 2009

Khalil SK, Mexal JG, Rehman A, Khan AZ, Wahab S, Zubair M, Khalil IH, Mohammad F (2010) Soybean mother plant exposure to temperature stress and its effect on germination under osmotic stress. Pak J Bot 42:213–225

Khanduri VP (2011) Variation in anthesis and pollen production in plants. Am-Eur J Agric Environ Sci 11(6):834–839

Kim EH, Seguin P, Lee JE, Yoon CG, Song HK, Ahn JK, Chung IM (2011) Elevated ultraviolet-B radiation reduces concentrations of isoflavones and phenolic compounds in soybean seeds. J Agron Crop Sci 197:75–80

Kjohl M, Nielsen A, Stenseth NC (2011) Potential effects of climate change on crop pollination. Food and Agriculture Organization of the United Nations, Rome, p 38

Kobayashi MJ, Shimizu KK (2011) Challenges in the studies on flowering time: interfaces between phenological researches and the molecular network of flowering genes. Ecol Res. doi:10.1007/s11284-011-0835-2

Korner C, Basler D (2010) Phenology under global warming. Science 327:1461–1462

Kovac H, Stabentheiner A (2011) Thermoregulation of foraging honeybees on flowering plants: seasonal variability and influence of radiative heat gain. Ecol Entomol 36(6):686–699

Krishnan PB, Ramakrishnan B, Rao KS, Dash RN (2009) Simulation studies to characterize the impact of climate change on crop production and to identify strategies for adaptation and mitigation. In: Singh SN (ed) Climate change and crops. Environmental Science and Engineering. Springer, Berlin, pp 39–61

Kumar PR, Yadav SK, Sharma SR, Lal SK, Jha DN (2009) Impact of climate change on seed production of cabbage in north western Himalayas. World J Agric Sci 5(1):18–26

Kumar A, Bahadur I, Maurya BR, Raghuwanshi R, Meena VS, Singh DK, Dixit J (2015) Does a plant growth promoting rhizobacteria enhance agricultural sustainability? J Pur Appl Microbiol 9(1):715–724

Larkindale J, Hall JD, Knight MR, Vierling E (2005) Heat stress phenotypes of Arabidopsis mutants implicate multiple signaling pathways in the acquisition of thermotolerance. Plant Physiol 138:882–897. doi:10.1104/pp.105.062257

Lattanzi FA (2010) C3/C4 grasslands and climate change. In: Schnyder H (ed) Grassland in a changing world. Proceedings of 23rd general meeting of European Grassland Federation 15. Science in Europe, Kiel, pp 1–11

Le BH, Cheng C, Anhthu QB, Wagmaister JA, Henry KF, Pelletier J, Kwong L, Belmonte M, Kirkbride R, Horvath S, Drews GN, Fischer RL, Okamuro JK, Harada JJ, Goldberg RB, Le H (2010) Global analysis of gene activity during Arabidopsis seed development and identification of seed-specific transcription factors. PNAS 107:8063–8070

Liu B, Wang C, Jin J, Liu JD, Zhang QY, Liu XB (2009) Responses of soybean and other plants to enhanced UV-B radiation. Soybean Sci 28:1097–1102

Mahmood S, Wahid A, Javed F, Basra SMA (2010) Heat stress effects on forage quality characteristics of maize (*Zea mays*) cultivars. Int J Agric Biol 12:701–706

Maity A, Chakrabarty SK (2013) Effect of environmental factors on hybrid seed quality of Indian mustard (*Brassica juncea*). Afr J Agric Res 8(48):6213–6219

Maity A, Chakrabarty SK (2014) Foraging behaviour of honeybees (Apis spp.) (Hymenoptera: Apidae) in hybrid seed production of Indian mustard (*Brassica juncea*). Ind J Agric Sci 84(11):1389–1394

Maity A, Pramanik P (2013) Climate change and seed quality: an alarming issue in crop husbandry. Curr Sci 105(10):1336–1338

Maity A, Chakrabarty SK, Yadav JB (2012) Standardisation of hybrid seed production technology of first Indian mustard (*Brassica juncea*) hybrid NRCHB 506. Indian J Agric Sci 82(9):753–8

Maity A, Natarajan N, Vijay D, Srinivasan R, Pastor M, Malaviya DR (2016) Influence of metal nanoparticles (NPs) on seed germination and yield of forage oat (*Avena sativa*) and berseem (*Trofolium alexandrinum*). Proc Natl Acad Sci, India, Sect B Biol Sci, doi:10.1007/s40011-016-0796-x

Mall RK, Singh R, Gupta A, Srinivasan G, Rathore LS (2006) Impact of climate change on Indian agriculture: a review. Clim Chang 78(2):445–478

Mathew JP, Herbert SJ, Zhang SH, Rautenkranz AAF, Litchfield GV (2000) Differential response of soybean yield components to the timing of light enrichment. Agron J 92:1156–1161

McDonald MB (1999) Seed deterioration: physiology, repair and assessment. Seed Sci Technol 27:177–237

McKenzie RL, Conner B, Bodeker G (1999) Increased summertime UV radiation in four aquatic macrophytes. Chemosphere 74:642–647

McVetty PBE (2009) Plant composition and climate change- influence of climate and climate change on plant oil composition. Presentation at Green Crop Network Workshop

Meena RS, Yadav RS, Meena VS (2013) Heat unit efficiency of groundnut varieties in scattered planting with various fertility levels. Bioscan 8(4):1189–1192

Meena RS, Meena VS, Meena SK, Verma JP (2015a) The needs of healthy soils for a healthy world. J Clean Prod 102:560–561

Meena RS, Meena VS, Meena SK, Verma JP (2015b) Towards the plant stress mitigate the agricultural productivity: a book review. J Clean Prod 102:552–553

Meena VS, Meena SK, Verma JP, Meena RS, Ghosh BN (2015c) The needs of nutrient use efficiency for sustainable agriculture. J Clean Prod 102:562–563

Meena VS, Meena SK, Verma JP, Meena RS, Jat LK (2015d) Current trend to mitigate climate change and sustainable agriculture: a book review. J Clean Prod 102:548–549

Meena VS, Verma JP, Meena SK (2015e) Towards the current scenario of nutrient use efficiency in crop species. J Clean Prod 102:556–557

Meena RS, Bohra JS, Singh SP, Meena VS, Verma JP, Verma SK, Sihag SK (2016) Towards the prime response of manure to enhance nutrient use efficiency and soil sustainability a current need: a book review. J Clean Prod 112(1):1258–1260

Mehta PS, Negi KS, Rathi RS, Ojha SN (2012) Indigenous methods of seed conservation and protection in Uttarkashi and Himalaya. Indian J Tradit Knowl 11(2):279–282

Milbau A, Graae BJ, Shevtsova A, Nijs I (2009) Effects of a warmer climate on seed germination in the subarctic. Ann Bot 104:287–296. doi:10.1093/aob/mcp117

Mohammed AR, Tarpley L (2009) Effects of elevated ultraviolet-B radiation on productive tillers, spikelet sterility and grain characteristics of southern US rice (*Oryza sativa* L.) cultivars. J Agron Crop Sci 195:292–300

Mohammed AR, Tarpley L (2010) Differential response of Southern US rice (*Oryza sativa* L.) cultivars to ultraviolet-B radiation. J Agron Crop Sci 196:286–295

Morgan KR, Heinrich B (1987) Temperature regulation in bee- and wasp-mimicking syrphid flies. J Exp Biol 133:59–71

Mukherjee A, Maity A (2015) Public–private partnership for convergence of extension services in Indian agriculture. Curr Sci 109(9):1557

Musgrave ME, Strain BR, Siedow JN (1986) Response of two pea hybrids to CO2 enrichment: a test of the energy overflow hypothesis for alternative respiration. Proc Natl Acad Sci USA 83:8157–8161

Nemani RR, Keeling CD, Hashimoto H, Jolly WM, Piper SC, Tucker CJ, Myeni RB, Running SW (2003) Climate-driven increases in global terrestrial net primary production from 1982 to 1999. Science 300:1560–1563

Nilsen ET, Orcutt DM (1996) Plant membranes as environmental sensors. In: Nilsen ET, Orcutt DM (eds) The physiology of plants under stress: abiotic factors. Wiley, New York, pp 50–82

Oshino T, Abiko M, Saito R, Ichiishi E, Endo M, Kawagishi-Kobayashi M, Higashitani A (2007) Premature progression of anther early developmental programs accompanied by comprehensive alterations in transcription during high-temperature injury in barley plants. Mol Genet Genomics 278:31–42. doi:10.1007/s00438-007-0229-x

Pachauri RK (2007) Sustainable well-being. Science 315:913. doi:10.1126/science.1140975

Parish RW, Phan HA, Iacuone S, Li SF (2012) Tapetal development and abiotic stress: a centre of vulnerability. Funct Plant Biol 39:553–559. doi:10.1071/FP12090

Parmesan C, Yohe G (2003) A globally coherent fingerprint of climate change impacts across natural systems. Nature 421:37–42

Peat J, Darvill B, Ellis J, Goulson D (2005) Effects of climate on intra- and interspecific size variation in bumble-bees. Funct Ecol 19(1):145–151

Peltonen-Sainio P, Jauhiainen L, Hakala K (2011) Crop responses to temperature and precipitation according to long-term multi-location trials at high-latitude conditions. J Agric Sci (Camb) 149:49–62

Peterson ME, Daniel RM, Danson MJ, Eisenthal R (2007) The dependence of enzyme activity on temperature: determination and validation of parameters. Biochem J 402(2):331–337

Phys.org (2013) Germination of alpines: climate change could shift the timing of seed germination in alpine plants. www.phys.org/news/2013-03-germination-alpines-climate-shift-seed.html

Powell AA (1988) Seed vigour and field establishment. Adv Res Tech Seeds 11:29–80

Powell AA (2006) Seed vigour and its assessment. In: Basra AS (ed) Handbook of seed science and technology. Food Products Press, New York, pp 603–648

Prasad PVV, Boote KJ, Allen LH Jr, Thomas JMG (2002) Effects of elevated temperature and carbon dioxide on seed-set and yield of kidney bean (*Phaseolus vulgaris* L.). Glob Chang Biol 8:710–721

Quaderi MM, Reid DM (2005) Growth and physiological responses of canola (*Brassica napus*) to UV-B and CO2 under controlled environment conditions. Physiol Plant 125:247–259

Rajendra Prasad S, Uday BK, Umesh RK, Ramesh KV, Radhika C, Natarajan S, Sripathy KV, Rajiv KS, Govind P (2015) Decade of ICAR seed project: retrospect and prospects. ICAR-Directorate of Seed Research, Mau 275101, UP, India, 433pp. ISBN: 978-81-95128-6-6

Rajjou L, Duval M, Gallardo K, Catusse J, Bally J, Job C, Job D (2012) Seed germination and vigor. Annu Rev Plant Biol 63:507–533

Rathcke B, Lacey EP (1985) Phenological patterns of terrestrial plants. Annu Rev Ecol Syst 16:179–214

Ren C, Bilyeu KD, Beuselinck P (2009) Composition, vigor, and proteome of mature soybean seeds developed under high temperature. Crop Sci 49:1010–1022. doi:10.2135/cropsci2008.05.0247

Roberts EH (1988) Temperature and seed germination. Symp Soc Exp Biol 42:109–132
Sakata T, Higashitani A (2008) Male sterility accompanied with abnormal anther development in plants – genes and environmental stresses with special reference to high temperature injury. Int J Plant Dev Biol 2:42–51
Sakata T, Oshino T, Miura S, Tomabechi M, Tsunaga Y, Higashitani N, Miyazawa Y, Takahashi H, Watanabe M, Higashitani A (2010) Auxins reverse plant male sterility caused by high temperatures. Proc Natl Acad Sci USA 107:8569–8574. doi:10.1073/pnas.1000869107
Sato S, Kamiyama M, Iwata T, Makita N, Furukawa H, Ikeda H (2006) Moderate increase of mean daily temperature adversely affects fruit set of *Lycopersicon esculentum* by disrupting specific physiological processes in male reproductive development. Ann Bot 97:731–738. doi:10.1093/aob/mcl037
Shah N, Paulsen G (2003) Interaction of drought and high temperature on photosynthesis and grain-filling of wheat. Plant Soil 257:219–226. doi:10.1023/A:1026237816578
Sherry RA, Zhou XH, Gu SL, Arnone JA, Schimel DS, Verburg PS, Wallace LL, Luo YQ (2007) Divergence of reproductive phenology under climate warming. Proc Natl Acad Sci USA 104:198–202
Shinohara T, Hampton JG, Hill MJ (2006a) Effects of the field environment before and after seed physiological maturity on hollow heart occurrence in garden pea (*Pisum sativum*). N Z J Crop Hortic Sci 34:247–255
Shinohara T, Hampton JG, Hill MJ (2006b) Location of deterioration within garden pea (*Pisum sativum*) cotyledons is associated with the timing of exposure to high temperature. N Z J Crop Hortic Sci 34:299–309
Shrestha UB, Gautam S, Bawa KS (2012) Widespread climate change in the Himalayas and associated changes in local ecosystems. PloS ONE 7(5):e36741. doi:10.1371/journal.pone.0036741
Siddique KHM, Brinsmead RB, Knight R, Knights EJ, Paull JG, Rose IA (2000) Adaptation of chickpea (*Cicer arietinum* L.) and faba bean (*Vicia faba* L.) to Australia. In: Knight R (ed) Linking research and marketing opportunities for pulses in the 21st century. Kluwer Academic Publishers, Dordrecht, pp 289–303
Singh DK, Singh SK, Singh AK, Meena VS (2014) Impact of long term cultivation of lemon grass (Cymbopogon citratus) on post-harvest electro-chemical properties of soil. Ann Agri Bio Res 19(1):45–48
Smith SD, Huxman TE, Zitzer SF, Charlet TN, Housman DC, Coleman JS, Fenstermaker LK, Seemann JR, Nowak RW (2000) Elevated CO2 increases productivity and invasive species success in an arid ecosystem. Nature 408:79–82
Song SQ, Lei YB, Tian XR (2005) Proline metabolism and cross-tolerance to salinity and heat stress in germinating wheat seeds. Russ J Plant Physiol 52:793–800
Spears JF, Tekrony DM, Egli DB (1997) Temperature during seed filling and soybean seed germination and vigour. Seed Sci Technol 25:233–244
Steinger T, Gall R, Schmid B (2000) Maternal and direct effects of elevated CO2 on seed provisioning, germination and seedling growth in *Bromus erectus*. Oecologia 123:475–480
Sunita K, Guruprasad KN (2012) Solar UV-B and UV-A/B exclusion effects on intraspecific variations in crop growth and yield of wheat varieties. Field Crop Res 125:8–13
Suzuki K, Takeda H, Tsukaguchi T, Egawa Y (2001) Ultrastructural study on degeneration of tapetum in anther of snap bean (*Phaseolus vulgaris* L.) under heat stress. Sex Plant Reprod 13:293–299. doi:10.1007/s004970100071
Taiz L, Zeiger E (2006) Stress physiology. In: Taiz L, Zeiger E (eds) Plant physiology. Sinauer Associates, Sunderland, pp 671–681
Tan L, Chen S, Wang T, Dai S (2013) Proteomic insights into seed germination in response to environmental factors. Proteomics 13:1850–1870
Tauber E, Zordan M, Sandrelli F, Pegoraro M, Osterwalder N, Breda C, Andrea D, Alessandro S, Karen M, Clara B, Ezio R, Charalambos PK, Rodolfo C (2007) Natural selection favors a

newly derived timeless allele in *Drosophila melanogaster*. Science 316:1895–1898. doi:10.1126/science.1138412
Thomas JMG, Prasad PVV, Boote KJ, Allen LH (2009) Seed composition, seedling emergence and early seedling vigour of red kidney bean seed produced at elevated temperature and carbon dioxide. J Agron Crop Sci 195:148–156
Turner NC, Abbo S, Berger JD, Chaturvedi SK, French RJ, Ludwig C, Mannur DM, Singh SJ, Yadava HS (2006) Osmotic adjustment in chickpea (*Cicerarietinum* L.) results in no yield benefit under terminal drought. J Exp Bot 58:187–194
UNDP (2007/8) Fighting climate change: human solidarity in a divided world. Human development report
Verma JP, Jaiswa DK, Meena VS, Meena RS (2015a) Current need of organic farming for enhancing sustainable agriculture. J Clean Prod 102:545–547
Verma JP, Jaiswal DK, Meena VS, Kumar A, Meena RS (2015b) Issues and challenges about sustainable agriculture production for management of natural resources to sustain soil fertility and health. J Clean Prod 107:793–794
Wagner J, Luscher A, Hillebrand C, Kobald B, Spitaler N, Larcher W (2001) Sexual reproduction of *Lolium perenne* L. and *Trifolium repens* L. under free air CO2 effects of high CO2 and temperature on seed quality 161 enrichment (FACE) at two levels of nitrogen application. Plant Cell Environ 24:957–965
Wan S, Yuan T, Bowdish S, Wallace L, Russell SD, Luo Y (2002) Response of an allergenic species Ambrosia psilostachya (Asteraceae), to experimental warming and clipping: implications for public health. Am J Bot 89:1843–1846
Wassmann R, Jagadish S, Sumfleth K, Pathak H, Howell G, Ismail A, Serraj R, Redona E, Singh RK, Heuer S (2009) Regional vulnerability of climate change impacts on Asian rice production and scope for adaptation. Adv Agron 102:91–133. doi:10.1016/S0065-2113(09)01003-7
Westerkampe C, Gottsberger G (2000) Diversity pays in crop pollination. Crop Sci 40:1209–1222
Wilczek AM, Burghardt LT, Cobb AR, Cooper MD, Welch SM, Schmitt J (2010) Genetic and physiological bases for phenological responses to current and predicted climates. Phil Trans R Soc B 365:3129–3147
Young LW, Wilen RW, Bonham-Smith PC (2004) High temperature stress of Brassica napus during flowering reduces micro-and mega gametophyte fertility, induces fruit abortion, and disrupts seed production. J Exp Bot 55:485–495
Zhao Y, An J, Zhou Z, Dong J, Xing Y, Qin J (2011) Pollination behavior of *Apis melliferaligustica* and *Bombus hypocrita* (Hymenoptera, Apidae) and the influencing factors in peach greenhouse. Acta Entomol Sin 54(1):89–96
Zinn KE, Tunc-Ozdemir M, Harper JF (2010) Temperature stress and plant sexual reproduction: uncovering the weakest links. J Exp Bot 61:1959–1968, 10.1093/jxb/erq053

Impact of Climate Change on Water Resources in Indian Himalaya

19

Pragati Pramanik and Debarati Bhaduri

Abstract

The Indian Himalayan Region (IHR) which is a domicile ~4 % of the country's population is a rich depository of biodiversity, a natural habitat of ~8000 species of flowering plants, over 816 tree species, 675 edibles and nearly 1740 species of medicinal plants. The region encompasses over several states of India, i.e., JK, HP, UK, Sikkim, and Arunachal Pradesh and the hill regions of Assam and West Bengal. This vast region is important in global atmospheric circulation and is vulnerable due to its unique geology, rich natural resources, and socioeconomic milieu. Climate change may cause a significant impact on this region. Water resources in this region are one of the vital inputs for the survival of mankind and rich biodiversity. Climate change can affect significantly in the quantum of flow as snow and glacial melt is the major contributor of the rivers' flow in this region. The serious challenge is related to frequency and magnitude of extreme weather events like rainfall which may lead to flash floods, landslides and debris flow. There will be both short- and long-term implications due to climate change in this region. A large knowledge gap exists in the present scenario regarding the climate change implications on water resources and related hazards in the Himalayas and their downstream river basins. Primary data generation and its utilization in developing scenarios taking into account water demand and socio-economic development as a whole are required. Establishment of monitoring system for snow, ice, and water and use of latest hydrological model are the keys. Climate change may have a detrimental effect on the present socio-economic structure in the region also. Society will also have to adapt to

P. Pramanik (✉)
Center for Environment Science and Climate Resilient Agriculture, ICAR – Indian Agricultural Research Institute, New Delhi 110012, India
e-mail: pragati.iari@gmail.com

D. Bhaduri
ICAR-Directorate of Groundnut Research, Junagadh 362001, Gujarat, India

J.K. Bisht et al. (eds.), *Conservation Agriculture*,
DOI 10.1007/978-981-10-2558-7_19

the stresses of the climate change on the livelihood. Participation of people in their general welfare backed by institutional support and updated knowledge base will be important in the changed climate scenario.

Keywords
Indian Himalayan Regions (IHR) • Climate change • Hydrology • Hydropower • Indian rivers • Monsoon • Modeling

19.1 Preface: The Himalayan Region

Himalaya, 'the abode of snow', is a Sanskrit term. Literally the whole Himalayan region is a home of extensive mountain ranges like the Himalaya, the Karakoram, the Hindu Kush and many more minor ranges extending from the Pamir Knot. The Himalayan mountain range in South Asia separates the Indo-Gangetic Plain (IGP) from the Tibetan Plateau. The region extends to five nations which include India, Nepal, Bhutan, Pakistan, and China. Himalayan mountain system is the largest in the world, and to name a few, Mount Everest, Kanchenjunga, Makalu, Cho Oyu, Dhaulagiri, Manaslu, Annapurna, and Naga Parbat are among the significant peaks (Figs. 19.1 and 19.2).

Physically the Himalayas are divided into three parallel zones, i.e., the Great Himalayas, the inner Himalayas (also known as middle or lesser Himalaya), and the Sub-Himalayan foothills with the adjacent Terai and Duar plain. The Great Himalayas consists of a great line of snowy peaks with an average height of more than 6100 m. Most of the highest peaks are in the Nepal and Sikkim followed by Kumaon section and then Punjab and Bhutan section. Peaks in the Assam portion are of least heights. Ranges like Karakoram, Kailas, Ladakh, and Zanskar come under the Great Himalayan region. The inner Himalaya is characterized by uniform height, between 6000 and 10,000 ft. The significant ranges in this region include Nag Tibba, Dhauladhar, Pir Panjal, North Kashmir, Mahabharata, Mussoorie, and Ratanpur. In the outermost and lowest zone, the Sub-Himalaya is characterized by the large number of longitudinal flat-bottomed valleys and plains known as Terai and Duars.

This region is having the third largest deposit of ice and snow in the world after Antarctica and the Arctic. There are about 15,000 glaciers which prominently include Gangotri, Yamunotri, Khumbu, Langtang, Zemu and many more. These glaciers provide storage of about 12,000 km^3 of fresh water (IPCC 2007). The permanent snow line of the IHR is among the highest in the world at typically around 5500 m (Shi et al. 1978). The higher regions of the Himalayas are snowbound throughout the year and serve as sources of several large perennial rivers like Indus, Ganges, Brahmaputra, Yamuna, and Ayeyarwady (or Irrawaddy). Mountain lakes or tarns, formed due to glacial activity, are found mostly in the upper reaches of the Himalayas, i.e., above 5500 m. Below this altitude, snow-fed lakes are found. Tilicho Lake in Nepal in the Annapurna

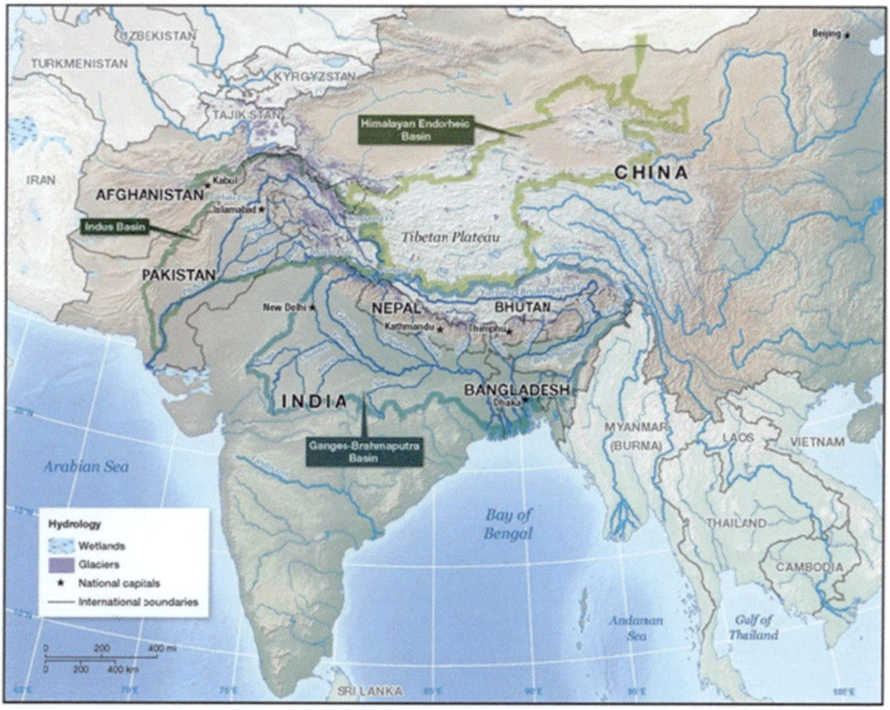

Fig. 19.1 Origin and expansion of major Indian river systems (Source: Publication titled 'Himalayan Glaciers: Climate Change, Water Resources, and Water Security', National Academic Press 2012)

Massif is one of the highest lakes in the world. Some of the important lakes are Yamdrok Tso in Tibet, Shey Phoksundo Lake in Nepal, Gurudongmar lake in North Sikkim, Gokyo Lakes in Nepal, and lake Tsomgo, near the Indo-China border in Sikkim (Figs. 19.2 and 19.3).

The unique flora and fauna is the wealth of the Himalayan region. Variations are basically due to wide range of climatic condition in different parts of this region. The region also supports a strong ethnicity of several cultures and religions. Based on the general spatial differentiation with associated geographic elements and broad pattern of human occupancies, the entire Himalayas was divided into three major realms, i.e., Western Himalayas, Central Himalayas, and Eastern Himalayas (Karan 1966). Each of these three realms possesses some unifying physical and cultural traits along with a certain measure of geographic homogeneity. The Western Himalayan realm may be divided into the Sub-Himalayan Kashmir (Poonch and Jammu), Pir Panjal, Vale of Kashmir, Ladakh and Baltistan, and the Kohistan and Gilgit regions. This realm is dominated by Afghan-Iranian culture and Hinduism and Lamaistic Tibetan cultures. The Central Himalayan realm between Kashmir and Sikkim may be divided into three major geographic regions, i.e., Himachal and Punjab Himalayas, Garhwal and Kumaon Himalayas, and Nepal Himalayas. The

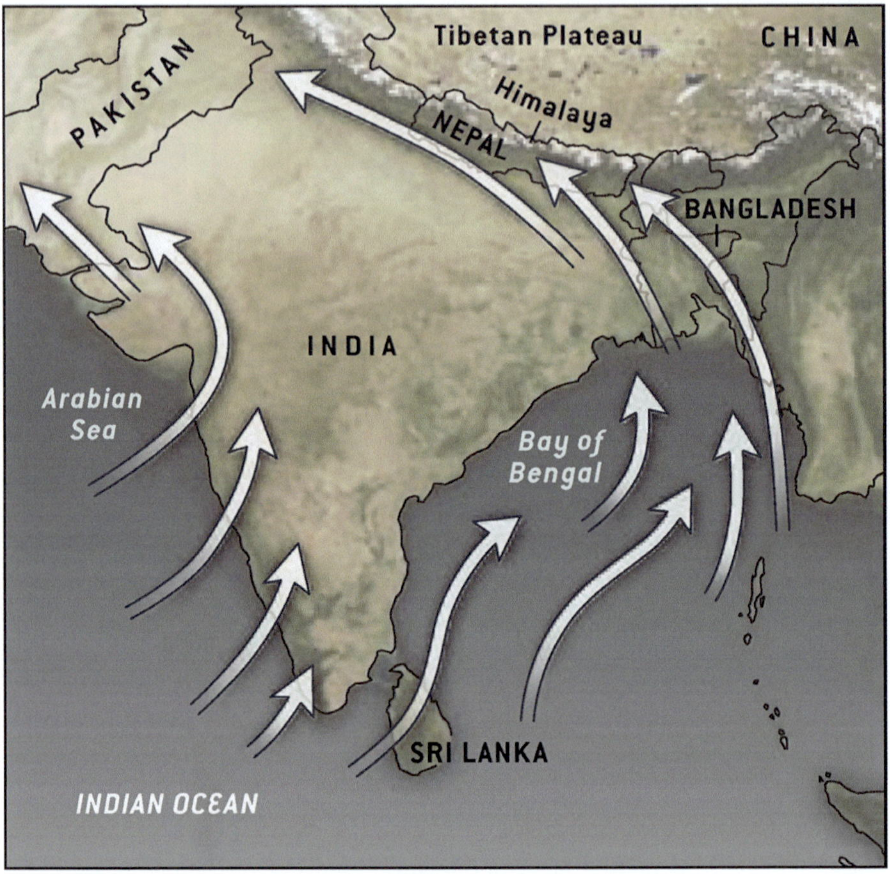

Fig. 19.2 The moist air currents that drive the South Asian Monsoon are indicated by *white arrows*. Monsoon flow transports moisture from the Arabian Sea to the Indian subcontinent, resulting in heavy monsoon rain over the Indo-Gangetic Plain and the Bay of Bengal (Source: Hodges 2006)

realm is dominated by Hinduism and Buddhism. Eastern Himalayan realm may be divided into three geographic regions, i.e., Darjeeling and Sikkim Himalayas and Bhutan and Assam Himalayas. The realm is dominated by Hindu and Tibetan culture.

Since this region is sensitive to its unique ecology, environment and culture and religion and climate change may have serious consequences altogether. Studies (Kothawale et al. 2010; Basistha et al. 2009; Bhutiyani et al. 2007; Arora et al. 2005; Meena et al. 2015d, e) have suggested for temperature variations in either direction (Figs. 19.1, 19.2, and 19.3). This will have an impact on the flora and fauna base, snow and glacial base and culture and environment base. The impact on the snow and glacial base and effect on the major river systems

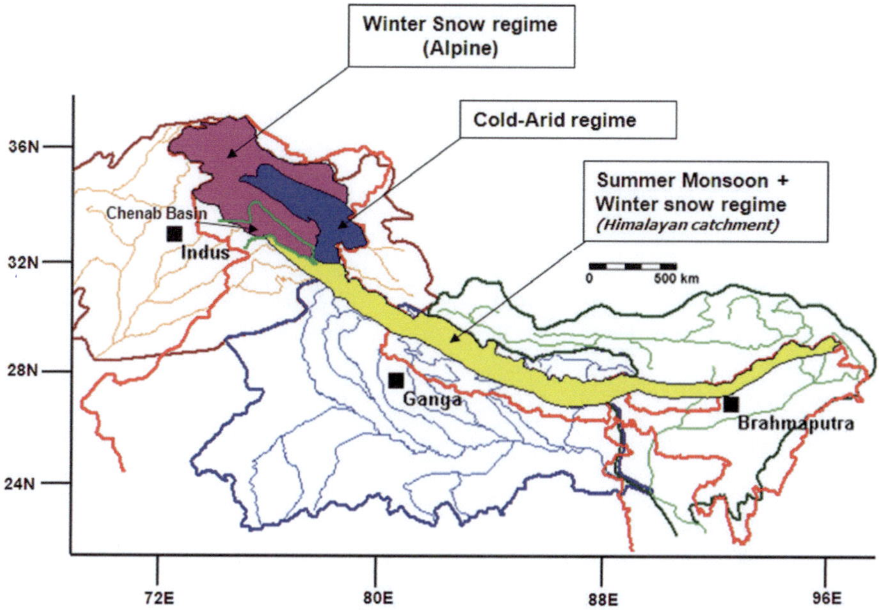

Fig. 19.3 The climate varies across the HKH region. In the west, indicated by *purple* in the figure, the climate is alpine and dominated by the mid-latitude westerlies. Most precipitation takes the form of winter snow. This area adjoins a cold arid climate regime, indicated by *blue* in the figure. In the east, indicated by *yellow* in the figure, the climate is dominated by the summer monsoon, with most of the precipitation coming during the summer months. The Indus, Ganges, and Brahmaputra watersheds are also shown in the figure (Source: Thayyen and Gergan 2010)

originating from this region, in the context of Indian subcontinent, would be the focus in the subsequent sections of this chapter.

19.2 Indian Himalayan Region (IHR)

The IHR extends from Kashmir in the west to Arunachal Pradesh in the east and is almost 2500 km long, and its breadth is about 80–300 km and rising from low-lying plains to over 8000 m above msl. The region spreads between 21° 57′–37° 5′ N latitudes and 72° 40′–97° 25′ E longitudes and covers 12 states of India, viz., Jammu and Kashmir, Himachal Pradesh, Uttaranchal, Sikkim, Arunachal Pradesh, Nagaland, Manipur, Mizoram, Tripura, Meghalaya, and hills of Assam and West Bengal of a total geographical area around 5, 33,604 km^2. The region is full of peaks and lakes, which feed the river system originating from this region (Table 19.1).

This region is a repository of diverse natural resources and habitat of about 40 million people of different cultures and beliefs. Multiple ethnic compositions are a striking feature of the region, where more than 171 out of 573 scheduled tribes of

Table 19.1 Major peaks and lakes of Indian Himalayan Region

Peaks	Altitude (m asl)	High altitude lakes	Altitude (m asl)
K2	8611	Suraj Tal	4950
Kangchenjunga	8598	Tso Moriri	4524
Nanga Parbat	8128	Panggon	4350
Masher Bram (East)	7821	Hemkund Sahib	4329
Nanda Devi	7817	Vasuki Tal	4300
Rakaposhi	7788	Chandra Tal	4300
Kamet	7756	Daitya Tal	4275
Saser Kangri	7672	Dashir lake	4270
Chaukhamba (Badrinath Shikhar)	7138	Manimahesh	4080
Trisul (west)	7138	Gandhi Sarovar	3970
Noonkonn	7135	Nako lake	3662
Phaunari	7128	Tsomgo (Changu)	3658
Kaunto	7090	Tulian lake	3353
Dunagiri	7066	Deoria Tal	3255

Source: ENVIS Monograph 3

India inhabit the region (Samal et al. 2000). Varied climatic conditions have been the reason for the rich flora and fauna base in the region. Three major Indian river systems, i.e., Indus, Ganges, and Brahmaputra (Fig. 19.1), provide for the economic and cultural development of the whole region in particular and for the country in general (Fig. 19.1).

19.2.1 Climate and Meteorology

The climate of the IHR ranges from subtropical (<800 m) at the foothills to arctic zone (>3500 m) at the highest altitude (Tibetan Plateau). During the late spring and early summer, the plateau surface heats up quickly and serves as an elevated heat source, which draws warm and moist air from the Indian Ocean toward the Himalayas and Tibetan Plateau region. As the monsoon flow transports moisture from the Arabian Sea to the Indian subcontinent, it causes heavy monsoon rain over the Indo-Gangetic Plain (IGP) and the Bay of Bengal (Fig. 19.2). During the winter, the low-level monsoon flow reverses to northeasterly, with prevailing large-scale subsidence and relative dry conditions over India (NAP 2012).

The climatic gradient is strong not only across but also along the arc of the Himalayas. The Sutlej valley serves as a rough dividing line between the climate regimes of the Western and Eastern Himalayas (Bookhagen and Burbank 2010). In the west (Fig. 19.3), the climate is alpine and dominated by the mid-latitude westerlies (indicated by purple) where most of the precipitation takes the form of winter snow. This area adjoins a cold arid climate regime (indicated by blue). However in the east, the climate is dominated by the summer monsoon, and most of the precipitation comes during the summer months (indicated by yellow) (NAP 2012).

The areas in Indus basin have a relatively dry climate, with annual precipitation of 400–600 mm, primarily from winter time storms associated with the mid-latitude westerlies. In the cold arid regions of Ladakh, though the precipitation is somewhat higher in summer, but the mean annual precipitation is as low as 115 mm (Thayyen and Gergan 2010). Rainfall is higher in the east, mostly from summer monsoon rain. Based on 10 years data (1998–2007), Bookhagen and Burbank (2010) showed that mean annual rainfall ranges from ~1 to more than 4 m (1000–4000 mm) in the monsoon-precipitation-dominated portions of the region (Figs. 19.2 and 19.3).

19.2.2 Glaciers

Glaciers play an important role in maintaining hydrological cycle and ecosystem stability as they act as buffers and regulate the runoff water supply from mountains to the plains in both dry and wet spells. On the basis of the mode of occurrence and dimensions, glaciers have broadly been classified as valley glaciers, piedmont glaciers and continental glaciers. Himalayan glaciers which amount to about 70 % of world non-polar glaciers (Nandy et al. 2006) fall in the category of valley glaciers. It has been estimated that an area of about 32,000 km^2 is under permanent cover of ice and snow in the Himalaya (Negi 1991). Some of the important glaciers are enlisted in Table 19.2.

The committee on Himalayan glaciers, hydrology, climate change, and implications for water security has calculated the proportion of glacier area in

Table 19.2 Important glaciers of Indian Himalayan Region

Glaciers	Location
Siachen	Indus basin, Karakoram
Rulung	Indus basin, Trans-Himalaya
Neh-Nar	Sind basin, Great Himalayan range
Thanak-Lungpa	Suru basin, Zanskar range
Braham Sar	Pir Panjal range
Harmukh	Sind basin, North Kashmir range
Gara	Tirung Khad basin
Gor Garang	Baspa basin
Bara Shigri	Chenab basin, Great Himalayan range
Shaune Garang	Baspa basin
Gangotri	Alaknanda basin, Kumaon Himalaya
Pindari	Alaknanda basin, Kumaon Himalaya
Chorabari	Alaknanda basin
Dunagiri	Alaknanda basin
Changme-Khangpu	Sikkim Himalaya
Zemu	Sikkim Himalaya

Source: ENVIS Monograph 3

different elevation bands for the Indus, Ganges, and Brahmaputra basin (NAP 2012). In all the three basins, the majority of glacier area is in the 5000–6000 m band, and a significant amount lies in the 4000 to 5000 m band. The Indus basin has a slightly greater proportion of its glacier area below 4000 m than the Ganges/ Brahmaputra basin, whereas the latter has a slightly greater proportion of its glacier area above 6000 m. The calculation suggests that glacial retreat would be more significant to changes in climate in the Indus basin than in the Ganges/Brahmaputra basin. This also qualifies the evidence that glaciers are more stable in the Western Himalayas.

Based on the model study, it was reported that as much as 70 % of the summer flow in the river Ganges and 50–60 % of the flow in other major rivers are contributed from melting glaciers (Barnett et al. 2005). Although a large variation in the reported percentage figures was observed (Alford and Armstrong 2010; Racoviteanu 2011; Armstrong 2010), it has been generally accepted that the percent contribution increases from east to west across the IHR (Immerzeel et al. 2010). However, there is an ambiguity over the glacial and snow contribution. It is worth to mention that snowmelt is a renewable resource that is replenished every year, in contrast to the glacial melt (Barnett et al. 2005). Considering the widely varied climatic regime alone in the IHR, it can be anticipated that there will be variation in the relative contributions of rain, snowmelt, and melt of glacier ice toward the discharge of different rivers of the IHR.

19.2.3 Water Resources

The main sources of water in this mountain chain are primarily rainfall from southwest monsoon, winter rains and heavy snowfall in the winter season. A significant contribution of the water resources in this region comes from snow and ice melt. Even a part of the groundwater and natural springs in the IHR are also recharged from the snowmelt water. An estimated water flow of about 8634 million m^3 occurs down the Himalayan rivers every year (Negi 2003). The catchment of Indus river system in Western Himalaya forms part of Jammu and Kashmir and Himachal Pradesh, while Ganges river rising from the snout of Bhagirathi drains very large catchment of Himachal Pradesh and entire Uttarakhand. In the eastern most part, the Brahmaputra river system drains a considerable area of the North Bengal, Sikkim, and Arunachal Pradesh in the IHR (Table 19.3).

There are total 19 rivers in the Indus basin, 67 in the Ganges basin and 31 in the Brahmaputra basin besides many small rivulets, streams and tributaries which altogether contribute in a large way to the sustenance of diversified flora and fauna of this region. More than 50 % of average water resources potential of our country is shared by various tributaries of these river systems (Table 19.4).

There is a wide variation of precipitation pattern in different parts of this region. In the eastern and central part of the region experiences more rainfall than snow falls, whereas in the western part snowfall is predominant. Therefore the nature of river flows in the western part is different from the eastern and central part. Due to a

Table 19.3 Principal glacial-fed river systems of the Himalaya

River	Mountain area (km^2)	Glacier area (km^2)
Indus	268,824	7890
Jhelum	33,670	170
Chenab	27,195	2944
Ravi	8092	206
Sutlej	47,915	1295
Beas	12,504	638
Yamuna	11,655	125
Ganges	23,051	2312
Ramganga	6734	3
kali	16,317	997
Karnali	53,354	1543
Gandak	37,814	1845
Kosi	61,901	1281
Tista	12,432	495
Raikad	26,418	195
Manas	31,080	528
Subansiri	81,130	725
Brahmaputra	256,928	108
Dibang	12,950	90
Lohit	20,720	425

Source: ENVIS Monograph 3

Table 19.4 Potential of major water resources in India

River basin	Catchment area (km^2)	Average water resources potential (BCM)
Indus (up to border)	321,289	73.31
Ganges	861,452	525.02
Brahmaputra	194,413	537.24
Total (20 river basins of India)		1869.35

Source: CWC Publication 1993

lack of data and the inaccessibility of the region, the contribution of glacier to the river flow is not well established in the IHR.

Glacial melt and snowmelt are reported to play an extremely important role in the Indus basin followed by the Brahmaputra basin, but play only a modest role for the Ganges rivers (Immerzeel et al. 2010). Wulf et al. (2011) quantified the water resources and discharge components for the Sutlej river, a major tributary of the Indus River, for the period 2004–2009. Results indicated that the discharge of the Sutlej river at Bhakra located at low elevation and situated at the base of the mountains is sourced predominately by snowmelt (48 %) followed by an effective share of rainfall (evapotranspiration, 39 %) and glacial melt (13 %).

There is a strong seasonality observed in annual precipitation as well. Annual hydrographs for the Ganges and Indus rivers clearly show strong seasonality pattern in the amount of discharge, leading to seasonal differences in the water availability. The Ganges river exhibits a significant discharge during the summer months, resulting in a water surplus that can maintain streamflows and recharge groundwater storage in some areas. However, water consumption exceeds natural runoff in the late winter months (February and March) seems to have some reliance on groundwater and/or storage (Hoekstra and Mekonnen 2011). The Indus river has a lower peak discharge and lower annual discharge than the Ganges river. The Indus river discharge also varies seasonally and interannually (NAP 2012).

19.2.4 Land Use and Land Cover

In the IHR, about 52 % of the total area is covered with forest, followed by wastelands, whereas arable land cover is about 11 % (Nandy et al. 2006).

The land use/land cover changed from forest to other land uses in the past several decades in the region (Rai et al. 1994); this may lead to environmental degradation through massive soil erosion and nutrient loss. For example, Jhum cultivation in the Eastern Himalayan region showed to disrupt ecological balance due to soil erosion resulting from reduction of Jhum cycle (Nandy et al. 2006). Agricultural land area has also found to increase considerably over the past four decades in the Himalaya replacing the other land uses, particularly forests (Sharma et al. 1992). Understanding the relationship between land use/land cover and hydrology is critical for the prediction of soil fertility, nutrient budgets and local water recharges for the functioning of watershed (Sharma et al. 1992; Meena et al. 2013, 2015a, 2016; Singh et al. 2014; Kumar et al. 2015; Ghosh et al. 2016). Geographical distribution of states covering IHR is given in Table 19.5.

19.2.5 Agriculture

The Himalayan people have traditionally practiced integrated agriculture, balancing cultivation of agroforestry, animal husbandry, and forestry. Mountain geography and inaccessibility have helped to maintain agro-biodiversity; however commercial agriculture is not as high yielding and profitable like plains (Nandy et al. 2006). Dependency on its limited arable land is marginally higher in the IHR as cultivators and agricultural laborers together comprise about 59 % of total workforce in the region (Nandy and Samal 2005; Meena et al. 2015b, c; Verma et al. 2015a, b).

Table 19.5 Geographical distribution of states covering Indian Himalayan Region

States/regions	Geographical area (km^2)	% of area under Agricultural land	Wastelands	Forest cover
Jammu and Kashmir	222,236[a]	4.7	64.6	9.6
Himachal Pradesh	55,673	14.5	56.9	25.8
Uttaranchal	53,483	12.5	30.1	44.8
Sikkim	7096	16.1	50.3	45.0
West Bengal hills	3149	43.5	2.2	69.7
Meghalaya	22,429	48.2	44.2	69.5
Assam hills	15,322	10.5	56.6	79.8
Tripura	10,486	29.6	12.2	67.4
Mizoram	21,081	21.2	19.3	83.0
Manipur	22,327	7.3	58.0	75.8
Nagaland	16,579	38.4	50.7	80.5
Arunachal Pradesh	83,743	3.5	21.9	81.3
India	3,287,263	55.8	20.2	20.6

[a]Included 78,114 and 37,555 km^2 occupied by Pakistan and China, respectively, and 5180 km^2 handed over by Pakistan to china (Source: Wastelands Atlas of India and FSI 2000)

19.2.6 Agroclimatic Zonation

Considering the commonality of factors like soil type, rainfall, temperature, water resources, etc., Indian subcontinent has been divided into 15 agroclimatic regions (zones). The IHR represents two prominent agroclimatic zones (Table 19.6), viz., Zone I, the Western Himalayan region, and Zone II, the Eastern Himalayan region. Zone I characteristically covers the low altitude subtropical region of south to mid to high temperate region in the mid-hills and extended to high hills on the north. The northern part of Zone I belongs to cold arid regions of Himachal Pradesh and Jammu and Kashmir. Zone II is the hills and mountains of folded topography (Table 19.6) with near tropical to alpine climatic conditions (Nandy et al. 2006).

19.2.7 Forest Types and Forest Ecosystems of IHR

Forests in Western and Eastern Himalayas are typically different depending on the factors like altitude, climate, topography, soil, etc. The Western Himalayan forests are diverse both in content and composition, whereas Eastern Himalayan forests are very rich both in flora and fauna (Dhar et al. 1997).

Table 19.6 Agroclimatic zones of Indian Himalayan Region

Agroclimatic zone	Climate	Rainfall (in mm)	State/regions
Zone I	High altitude temperate (humid to cold arid)	<1200	Jammu and Kashmir
Zone I	Hill temperate (per humid to subhumid)	1200–1800	Himachal Pradesh and Uttaranchal
Zone II	Per humid to humid	1800–2200	Nagaland, Mizoram, Manipur, and Tripura
Zone II	Per humid to humid	2200–2800	Sub-Himalayan West Bengal, Sikkim, Assam, and Meghalaya
Zone II	Per humid to humid	>2800	Arunachal Pradesh

Source: Agro-climatic Regional Planning, Planning Commission (1989)

19.2.8 Biodiversity

IHR, as a whole, supports nearly 50 % of the total flowering plants of India, of which 30 % are endemic to the region. The IHR supports about 8000 species of angiosperms (40 % endemics), 44 species of gymnosperms (16 % endemics), 600 species of pteridophytes (25 % endemics), 1737 species of bryophytes (33 % endemics), 1159 species of lichens (11 % endemics) and 6900 species of fungi (27 % endemics) (Singh and Hajra 1996). There are over 816 identified tree species found in the IHR, of which 675 are edible and nearly 1743 species are of medicinal value (Samant et al. 1998). The diversity of plant species used in various ailments is dispersed all across the IHR (Samal et al. 2002, 2004).

19.2.9 Hydropower

The IHR is suitable for hydroelectric power development because of availability of perennial sources of water in rivers. At present this region has been estimated to produce electricity more than 1.10 lakh MW (Table 19.7) (Slariya 2013). Brahmaputra subbasin has the largest hydropower potential among three basins with Indus basin and Ganges basin in the second and third positions. Considering the completed and under construction projects, Brahmaputra subbasin, Indus, and Ganges have harnessed about 11 %, 48 %, and 31 % of the assessed potential, respectively (http://india-wris.nrsc.gov.in). Therefore, a large chunk of the hydropower potential of the basin is yet to be exploited. Some of the major running hydropower stations in this area are Tehri; Koteshwar; Rihand; Raj Ghat; Bhakra; Pong; Dehar; Ranjit Sagar; Chamera stages I, II, and III; Nathpa Jhakri; Uri; Salal; etc. (Table 19.7).

Table 19.7 Status of hydropower projects situated in Indian Himalayan Region

Indian Himalayan State	Status		Capacity (MW)
Himachal Pradesh	Projects under operation		6370.12
	Projects under execution/allotted		5744.10
	Projects under the process of allotment		5615.50
	Projects to be re advertised		1481.00
	Abandoned projects		0435.00
	Projects under investigation		0046.50
	Himurja projects proposed/under execution		0723.40
	Total		20,415.00
Jammu and Kashmir	Existing projects	Jhelum River basin (total = 6)	0252.60
		Chenab basin (total = 5)	0483.80
		Ravi basin (total = 1)	0009.00
		Indus basin (total = 8)	0013.03
		Central sector	1560.00
	Under construction	State sector	0453.61
		Central sector	0449.00
		Total	3221.31
Uttarakhand	Power projects under operation		3164.75
	Power projects under construction		5509.40
	Projects under development		1,7540.93
	Total		2,6214.68
Sikkim	Upper part of Sikkim (total = 9 projects)		2823.00
	Lower part of Sikkim (total = 11 projects)		1402.00
	Total		4225.00
Arunachal Pradesh	Central (total = 4)		05870.00
	State (total = 8)		1,4785.00
	Private (total = 74)		2,7512.00
	Total (total = 86 projects)		4,8167.00
Nagaland	State		0075.00
Manipur	To be implemented by NHPC (total = 7)		0037.75
	Under capital subsidy program (total = 3)		0800.00
	Under state sector (total = 3)		0560.84
	Under center sector (total = 2)		1590.00
	Mini/micro projects (total = 5)		3900.00
	Total (total = 20 projects)		6888.59
Mizoram	Central sector (total = 3)		0876.00
Tripura	State/center (total = 2)		0184.00
Meghalaya	State (total = 1)		0090.00
	Grand total		110,281.25

Source: Slariya (2013)

19.3 Climate Change in IHR

The increasing human activities due to industrial revolution have led to unprecedented changes in the earth's delicate climate system. There is now established evidence that the earth's surface has warmed during the past 100 years, which is mainly attributed to the anthropogenic activity. Changes in many components in the climate system, like precipitation, snow cover, sea ice, extreme weather events, etc., have also been observed. These changes, however, showed remarkable regional variations around the globe.

19.3.1 Climate Change Models

Studies on climate change in mountain areas remained incomplete and scattered (IPCC 2007; Nogués-Bravo et al. 2007), although certain studies from the Hindu Kush Himalayas (HKH) (Shrestha et al. 1999) do indicate that climate change has an undesirable impact on Himalayan biodiversity and its services. The Himalayan region, including the Tibetan Plateau, has shown consistent warming trends during the past 100 years (Yao et al. 2006). Specific knowledge on IHR with respect to climate change indicators are lacking due to both inaccessibility of the location and the insufficient theoretical attention given to the complex interaction of spatial scales in weather and climate phenomena in mountain areas. The analysis in the Eastern Himalayan region estimated for widespread warming, and the rate is generally greater than 0.01 °C per year or more (Shrestha and Devkota 2010). The highest rates of warming are in the winter and the lowest, or even cooling, are in summer. It has also been reported that more warming may be the reality in the higher elevation.

The spatial distributions of maximum temperature trends in Nepal (Central Himalayan region) showed high warming in most of the Himalayan region and the middle mountains, while low warming or even cooling trends are observed in most of the Terai and the Siwalik regions (Shrestha et al. 1999). Analyses of maximum temperature data from 49 stations in Nepal over a period of 1971–1994 revealed annual warming trends after 1977 varied from 0.068 to 0.128 °C in most of the middle mountain and Himalayan regions, while the Siwalik and Terai (southern plains) regions seemed to be less affected and showed warming trends of less than 0.038 °C $year^{-1}$.

19.3.2 Climate Change Impact on Water Resources in IHR

19.3.2.1 Glaciers

Since glaciers are contributing significantly in river flow, the impact of climate change on glacier is the most important from the point of view of water availability and ecosystem balance. Data that has been generated from the glacier studies, in the Himalayas, over the last 100 years or so, indicates that the glaciers, in the

Himalayas, have been, by and large, shrinking and retreating continuously, barring a flip here and there, but the rate of retreat cannot be considered as alarming/abnormal, especially in the last decade or so (Raina, MoEF discussion paper 2009).

Glacier monitoring in the Indian Himalayas started in the early twentieth century, when 20 odd glaciers in the Himalayas, located across the Indian Himalayas, from Jammu and Kashmir in northwest to Sikkim in northeast, began to be monitored by the Geological Survey of India (GSI). Observations had revealed that the degeneration of the glacial mass has been highest in the northwest side and declined toward the northeast side (Raina, MoEF discussion paper 2009).

The Ministry of Environment & Forest (MoEF) discussion paper has revealed that glaciers in the Himalayas, over a period of the last 100 years, behaved in contrasting ways. As an example, Sonapani glacier has retreated by about 500 m during the last hundred years. On the other hand, Kangriz glacier has practically not retreated even an inch in the same period. Siachen glacier is believed to have shown an advance of about 700 m between 1862 and 1909, followed by an equally rapid retreat of around 400 m between 1929 and 1958, and hardly any retreat during the last 50 years. Gangotri glacier, which had hitherto been showing a rather rapid retreat, along its glacier front, at an average of around 20 m per year till up to 2000 AD, has since slowed down considerably and between September 2007 and June 2009 is practically at a standstill (NAP 2012). The same is true of the Bhagirath Kharak and Zemu glaciers. The paper highlighted 'It is premature to make a statement that glaciers in the Himalayas are retreating abnormally because of the global warming. A glacier is affected by a range of physical features and a complex interplay of climatic factors'.

19.3.2.2 Major Rivers

Given the difference in climate between the drier western and monsoonal eastern ends of the region, future warming is unlikely to affect river flow uniformly throughout. Impacts of declining glacier area on river flow will be greater in smaller and more highly glacierized basins in both the west and east and in the west, where precipitation is scarce, for considerable distances downstream (Rees and Collins 2006). Glacier shrinkage and the relative contribution of glacier melt to the region's river discharge are lower than reported in the Fourth Assessment Report of IPCC, 2007 (Miller et al. 2012). There has been some research on the water availability in three basins of IHR.

Indus

Snow and glacier melt both are considered as the primary drivers of the hydrological regime (Mukhopadhyay and Dutta 2010). Modeling of the upper Indus indicated that some 34 % of total streamflow in this area is generated by snowmelt, while 26 % is from glacier melt (Immerzeel et al. 2010; Bookhagen and Burbank 2010). A decrease in the observed average annual and summer monsoon discharge data from the river Sutlej was also reported (Bhutiyani et al. 2008; Tahir et al. 2011). Impacts of declining glacial mass on river discharge as a result of climate change will be more substantial in the Indus basin because of the high proportion of discharge derived from melt water (Rees and Collins 2006). Increased glacier melt as a result of climate change will provide short-term increases in the contribution to discharge of the Indus river, but is likely to lead to decrease in the future (Miller et al. 2012).

Ganges

Considering the changes to the meteorological variable, Arora and Boer (2001) has reported 5 % increases in mean annual runoff for the Ganges by 2070–2100. Immerzeel et al. (2010) projected a decrease in mean runoff of 17.6 % at higher elevation (>2000 mean sea level) with a likely scenario of declining precipitation and glacier melt. However, in small high-altitude catchments a contrasting pattern was projected (Immerzeel et al. 2012). Glacier retreat may not have a drastic influence on the flow of Ganges river (Jain 2008). Toward downstream its influence will be reducing but it is unlikely that the Ganges will become a seasonal river (Miller et al. 2012).

Brahmaputra

Comparatively a lesser studies were carried out regarding the impacts of climate change on the glacier hydrology, and discharge exists for the Brahmaputra basin compared to the Indus or Ganges (Miller et al. 2012). Glacier melt is not a significant component of downstream discharge in the Brahmaputra. There is no visible spring contribution from glacier and snowmelt (Immerzeel 2008). Despite the fact that there is considerable uncertainty in predicting changes to the South Asian monsoon, all global climate models (GCM) were found to indicate an increase in discharge in the lower Brahmaputra river owing to a projected increase in precipitation downstream (Miller et al. 2012).

19.3.2.3 Extreme Events

In recent decades, the hydrological characteristics of the watersheds in the Himalayan region seem to have undergone substantial changes as a result of extensive land use (e.g., deforestation, agricultural practices, and urbanization), leading to frequent hydrological disasters, enhanced variability in rainfall and runoff, extensive reservoir sedimentation, and pollution of lakes. Global warming and its impact on the hydrological cycle and nature of hydrological events have posed an additional threat to this mountainous region of the Indian subcontinent (Mall et al. 2006).

Climate change may induce physical hazards like flash flooding due to extreme precipitation, flooding due to monsoon rainfall, lake outbursts, landslides and avalanches (NAP 2012). Although the main reason behind the monsoon flood is heavy rain, other factors like snowmelt and change in land use may increase the flood volume, and the outcome may be disastrous. Major Himalayan rivers being intercountry rivers, such disasters may have ramifications on the political scenario of the neighboring countries. As an example, the flooding of the Kosi river led to a dispute between Nepal and India in 2008 (Malhotra 2010).

A glacial lake outburst flood (GLOF) occurs when water dammed by a glacier or a moraine is rapidly released by failure of the dam (Bajracharya et al. 2007; Hewitt 1982; Xin et al. 2008). Failure of the confining dam can have a variety of causes, including earthquakes, catastrophic failure of slopes into the lake (avalanches, rockslides, icefall from a glacier into the lake), a buildup of water pressure, or even simple erosion of the confining dam over time.

Twenty-four GLOF events have been occurred in Nepal in the recent past, causing considerable losses of life and property. For example, the Sun Koshi GLOF in 1981 damaged the only road link to China and disrupted transportation for several months, and the Dig Tsho GLOF in 1985 destroyed the nearly completed Namche small hydroelectric project, in addition to causing other damages farther downstream (Bajracharya et al. 2007).

Landslide lake outburst flood (LLOF) is a catastrophic release of impounded water from behind a natural dam formed by a landslide. In the steep mountainous Himalayas, landslides are a common event, whether they are triggered by normal weathering and erosion processes, extreme rainfall events, or earthquakes. Gupta and Sah (2008) presented an example of a LLOF in the Sutlej catchment in Himachal Pradesh, India, well below the termini of any glaciers above it. Like GLOFs, LLOFs also pose a serious hazard to people, property, and infrastructure downstream from the landslide dams. LLOFs are even less predictable than GLOFs.

19.3.2.4 Hydropower

The response of hydrological systems, erosion processes and sedimentation of IHR could alter considerably due to climate change. Therefore the hydropower stations may face some kind of insecurity either due to uncertain water availability or the water quality issues. Increased sedimentation may certainly have implications on the available head to the power turbines. Therefore, the planning of hydropower stations should consider the climate change scenarios as the Himalayan regions are the available powerhouse for Indian subcontinent from which a significant share is yet to be exploited.

19.4 Conclusions

It has been ample clear that, although the outcome of the climate change scenarios from various circulation models varies in quantitative terms, there is an overall agreement for the rise in temperature and rainfall events in the Indian Himalayan Region. Realizing the importance of this region's high forest cover which acts as sink for carbon dioxide and provides major ecosystem services, the impacts of global warming, due to both natural causes and anthropogenic emissions, may cause serious consequences. Since the water availability is directly related to the forest cover, the impact on the water resources, sediment load in the rivers and its seasonal variability may have the potential to change the environmental, social and economic structure of this region. Climate change studies of IHR on several aspects have raised the awareness about the threats and possibilities. Depending on the model assumptions, the developed scenarios sometimes rule out the alarming forecast and sometimes make the comfort level disturbed. The most important areas in the context of water resources in the IHR, the glaciers, are either accessible with utmost difficulty or completely inaccessible, and thereby the ground data collection in itself is many times not possible. Therefore mapping of the present and future scenarios mostly depends on the secondary data and assumptions. So, efforts are needed to standardize the assumptions and setting up observation stations for collection of primary data to encourage the scientific community to reduce the uncertainty associated with the research outcomes. Many organizations are putting their efforts to understand the possible effects of climate change in this region. Coordination among those organizations is also needed. This will help to formulate schemes which will be beneficial to the people of this region, to the preservation of forest land and to control of soil erosion. More emphasis is needed toward the adaptation strategies as the phenomenon of global climate change involves the solution from both local and global perspectives.

Acknowledgments We thank the editors and anonymous reviewers for their constructive comments, which helped us to improve the manuscript.

References

Alford D, Armstrong R (2010) The role of glaciers in stream flow from the Nepal Himalaya. Cryosphere Discuss 4:469–494

Armstrong RL (2010) The glaciers of the Hindu-Kush-Himalayan region: a summary of the science regarding glacier melt/retreat in the Himalayan, Hindu Kush, Karakoram, Pamir, and Tien Shan Mountain Ranges. Technical Paper. International Centre for Integrated Mountain Development, Kathmandu, Nepal

Arora VK, Boer GJ (2001) Effects of simulated climate change on the hydrology of major river basins. J Geophys Res 106:3335–3348

Arora M, Goel NK, Singh P (2005) Evaluation of temperature trends over India. Hydrol Sci J 50:81–93

Bajracharya S, Mool PK, Shrestha B (2007) Impact of climate change on Himalayan glaciers and glacial lakes: case studies on GLOF and Associated Hazards in Nepal and Bhutan. Technical Paper. International Centre for Integrated Mountain Development, Kathmandu, Nepal

Barnett TP, Adam JC, Lettenmaier DP (2005) Potential impacts of a warming climate on water availability in snow-dominated regions. Nature 438(7066):303–309

Basistha A, Arya DS, Goel NK (2009) Analysis of historical changes in rainfall in the Indian Himalayas. Int J Climatol 29:555–572

Bhutiyani MR, Kale VS, Pawar NJ (2007) Long-term trends in maximum, minimum and mean annual air temperatures across the Northwestern Himalaya during the twentieth century. Clim Chang 85:159–177

Bhutiyani MR, Kale VS, Pawar NJ (2008) Changing streamflow patterns in the rivers of northwestern Himalaya: implications of global warming in the 20th century. Curr Sci 95(5):618–626

Bookhagen B, Burbank DW (2010) Toward a complete Himalayan hydrological budget: spatiotemporal distribution of snowmelt and rainfall and their impact on river discharge. J Geophys Res 115, F03019. doi:10.1029/2009JF001426

Dhar TN, Gupta SP, Johri SC (1997) Forestry in the Himalayan states of India – a comparative study. SHERPA, Lucknow, p 272

ENVIS, resource information database of the Indian Himalaya, Monograph 3, ISSN 0972–1819

Ghosh BN, Meena VS, Alam NM, Dogra P, Bhattacharyya R, Sharma NK, Mishra PK (2016) Impact of conservation practices on soil aggregation and the carbon management index after

seven years of maize – wheat cropping system in the Indian Himalayas. Agric Ecosyst Environ 216:247–257

Gupta V, Sah MP (2008) Impact of the trans-Himalayan landslide lake outburst flood (LLOF) in the Satluj catchment, Himachal Pradesh, India. Nat Hazards 45(3):379–390

Hewitt K (1982) Natural dams and outburst floods of the Karakoram Himalaya. Hydrol Sci J 27 (2):266–267

Hodges K (2006) Climate and the evolution of mountains. Sci Am 295:72–79

Hoekstra AY, Mekonnen MM (2011) Global water scarcity: the monthly blue water footprint compared to blue water availability for the world's major river basins, Value of Water Research Report Series No. 53. UNESCO-IHE, Delft

Immerzeel W (2008) Historical trends and future predictions of climate variability in the Brahmaputra basin. Int J Climatol 28:243–254

Immerzeel WW, Van Beek LPH, Bierkens MFP (2010) Climate change will affect the Asian water towers. Science 328(5984):1382–1385

Immerzeel WW, Van Beek LPH, Konz M, Shrestha AB, Bierkens MFP (2012) Hydrological response to climate change in a glacierized catchment in the Himalayas. Clim Chang 110:721–736, http://dx.doi.org/10.1007/s10584-011-0143-4

India-WRIS website. http://india-wris.nrsc.gov.in

IPPC (2007) The Himalayan glaciers. Fourth assessment report on climate change

Jain SK (2008) Impact of retreat of Gangotri glacier on the flow of Ganga River. Curr Sci 95 (8):1012–1014

Karan PP (1966) Geographic regions of the Himalayas. Namgyal Institute of the Tibetology

Kothawale DR, Revadekar JV, Rupa Kumar K (2010) Recent trends in pre-monsoon daily temperature extremes over India. J Earth Syst Sci 119:51–65

Kumar A, Bahadur I, Maurya BR, Raghuwanshi R, Meena VS, Singh DK, Dixit J (2015) Does a plant growth promoting rhizobacteria enhance agricultural sustainability? J Pure Appl Microbiol 9(1):715–724

Malhotra P (2010) Water issues between Nepal, India, and Bangladesh. Institute of Peace and Conflict Studies, New Delhi

Mall RK, Gupta A, Singh R, Singh RS, Rathore LS (2006) Water resources and climate change: an Indian perspective. Curr Sci 90(12):1610–1625

Meena RS, Yadav RS, Meena VS (2013) Heat unit efficiency of groundnut varieties in scattered planting with various fertility levels. Bioscan 8(4):1189–1192

Meena RS, Meena VS, Meena SK, Verma JP (2015a) The needs of healthy soils for a healthy world. J Clean Prod 102:560–561

Meena RS, Meena VS, Meena SK, Verma JP (2015b) Towards the plant stress mitigate the agricultural productivity: a book review. J Clean Prod 102:552–553

Meena VS, Meena SK, Verma JP, Meena RS, Ghosh BN (2015c) The needs of nutrient use efficiency for sustainable agriculture. J Clean Prod 102:562–563

Meena VS, Meena SK, Verma JP, Meena RS, Jat LK (2015d) Current trend to mitigate climate change and sustainable agriculture: a book review. J Clean Prod 102:548–549

Meena VS, Verma JP, Meena SK (2015e) Towards the current scenario of nutrient use efficiency in crop species. J Clean Prod 102:556–557

Meena RS, Bohra JS, Singh SP, Meena VS, Verma JP, Verma SK, Sihag SK (2016) Towards the prime response of manure to enhance nutrient use efficiency and soil sustainability a current need: a book review. J Clean Prod 112(1):1258–1260

Miller JD, Immerzeel WW, Rees G (2012) Climate change impacts on glacier hydrology and river discharge in the Hindu Kush-Himalayas: a synthesis of the scientific basis. Mt Res Dev 32 (4):461–467

MoEF discussion paper titled Himalayan glaciers a state-of-art review of glacial studies, glacial retreat and climate change – V.K. Raina, Ex. Deputy Director General, Geological Survey of India

Mukhopadhyay B, Dutta A (2010) A stream water availability model of upper Indus Basin based on a topologic model and global climatic datasets. Water Res Manag 24(15):4403–4443
Nandy SN, Samal PK (2005) An outlook of agricultural dependency in the IHR. ENVIS Newsl Himal Ecol 2:4–5
Nandy SN, Dhyani PP, Samal PK (2006) Resource information database of the Indian Himalaya. ENVIS Monograph 3. 108 p. http://gbpihed.gov.in/envis/HTML/monograph3/Contents.html. Accessed on 25 Dec 2009
National Academic Press (2012) Himalayan glaciers: climate change, water resources, and water security
Negi SS (1991) Himalayan rivers, lakes and glaciers. Indus Publishing Co, New Delhi, 182 p
Negi SS (2003) Natural resource management in the Himalaya. Vol. I: land and water environmental management. A.P.H. Publishing Corporation, New Delhi, 239 p
Nogués-Bravo D, Araújo MB, Errea MP, Martinez-Rica JP (2007) Exposure of global mountain systems to climate warming during the 21st century. Glob Environ Chang 17(3):420–428
Racoviteanu A (2011) Himalayan glaciers: combining remote sensing, field techniques and indigenous knowledge to understand spatio-temporal patterns of glacier changes and their impact on water resources. Ph.D. dissertation. University of Colorado, Boulder
Rai SC, Sharma E, Sundriyal RC (1994) Conservation in the Sikkim Himalaya: traditional knowledge and land-use of the Mamlay watershed. Environ Conserv 21(01):30–34
Raina VK (2009) Himalayan glaciers: a state-of-art review of glacial studies, glacial retreat and climate change. Himalayan glaciers: a state-of-art review of glacial studies, glacial retreat and climate change. MoEF discussion paper
Rees HG, Collins DN (2006) Regional differences in response of flow in glacier-fed Himalayan rivers to climatic warming. Hydrol Processes 20(10):2157–2169
Samal PK, Fernando R, Rawat DS (2000) Influences of economy and culture in development among mountain tribes of Indian central Himalaya. Int J Sustain Dev World Ecol 7:41–49
Samant SS, Dhar U, Palni LMS (1998) Medicinal plants of Indian Himalaya: diversity distribution potential values. GB Pant Institute of Himalayan Environment and Development, Almora
Samal PK, Shah A, Tiwari SC, Agrawal DK (2002) Indigenous animal health care practices and their relevance to bioresource conservation in Indian central Himalaya. Int J Sust Develop World Ecol 8:167–178
Samal PK, Shah A, Tiwari SC, Agrawal DK (2004) Indigenous medicinal practices and their linkages in resource conservation and physical well being of the locals in central Himalayan region of India. Ind J Tradit Knowl 3(1):12–26
Sharma E, Sundriyal RC, Raj SC, Bhatt YK, Rai LK, Sharma R, Rai YK (1992) Integrated watershed management: a case study in Sikkim Himalaya. Gyanodaya Prakashan, Nainital, 120p
Shi Y, Xie Z, Zheng B, Li Q (1978) Distribution, feature and variations of glaciers in China. World Glacier Inventory, Riederalp Workshop
Shrestha AB, Devkota LP (2010) Climate change in the Eastern Himalayas: observed trends and model projections; climate change impact and vulnerability in the Eastern Himalayas – technical report 1. ICIMOD, Kathmandu
Shrestha AB, Wake CP, Mayewski PA, Dibb JE (1999) Maximum temperature trends in the Himalaya and its vicinity: an analysis based on temperature records from Nepal for the period 1971–94. Climate Change Research Center, Institute for the Study of the Earth, Oceans, and Space, University of New Hampshire, Durham
Singh DK, Hajra PK (1996) Floristic diversity. Changing perspectives of biodiversity status in the Himalaya. In: Gujral (ed) Biodiversity status in Himalaya. British Council, New Delhi, pp 23–38
Singh DK, Singh SK, Singh AK, Meena VS (2014) Impact of long term cultivation of lemon grass (Cymbopogon citratus) on post-harvest electro-chemical properties of soil. Ann Agric Biol Res 19(1):45–48

Slariya MK (2013) Development and water resources in Indian Himalayan region: an invitation to disaster- a study of hydroelectric power developmental projects in Himachal Pradesh. Sch Res J Interdiscip Stud 1(1)

Tahir AA, Chevallier P, Arnaud Y, Ahmad B (2011) Snow cover dynamics and hydrological regime of the Hunza River basin, Karakoram Range, Northern Pakistan. Hydrol Earth Syst Sci 15(7):2275–2290

Thayyen RJ, Gergan JT (2010) Role of glaciers in watershed hydrology: a preliminary study of a 'Himalayan Catchment'. Cryosphere 4:115–128

Verma JP, Jaiswa DK, Meena VS, Meena RS (2015a) Current need of organic farming for enhancing sustainable agriculture. J Clean Prod 102:545–547

Verma JP, Jaiswal DK, Meena VS, Kumar A, Meena RS (2015b) Issues and challenges about sustainable agriculture production for management of natural resources to sustain soil fertility and health. J Clean Prod 107:793–794

Wulf H, Bookhagen B, Scherler D (2011) Differentiating rainfall, snow and glacial melt in the Sutlej Valley (western Himalaya) by distributed hydrological modeling. Geophys Res Abs 13 (EGU2011-4494)

Xin W, Shiyin L, Wanqin G, Junli X (2008) Assessment and simulation of glacier lake outburst floods for Longbasaba and Pida lakes, China. Mt Res Dev 28(3/4):310–317

Yao TD, Guo XJ, Lonnie T, Duan KQ, Wang NL, Pu JC, Xu BQ, Yang XX, Sun WZ (2006) δ18O record and temperature change over the past 100 years in ice cores on the Tibetan Plateau. Sci China Ser D Earth Sci 49(1):1–9

Diversity Among Rice Landraces Under Static (Ex Situ) and Dynamic (On-Farm) Management: A Case from North-Western Indian Himalayas

20

Santosh Kumar, S.K. Dwivedi, Ved Prakash, K.K. Rao, S.K. Samal, and J.S. Mishra

Abstract

Crop genetic diversity is the building block of sustainable agricultural development. Crop landraces contain enormous genetic diversity. The genetic structure of rice landraces is an evolutionary approach to existence and performance, especially under rainfed and to some extent under irrigated conditions in valleys and organic inputs in Himalayan agroecosystems. The combined effects of farmer and natural selection led to the building of genotypes representing diverse combinations of traits. The better understanding into the dynamics of genetic resources of rice is needed in order to identify detrimental evolutionary patterns and draw up conservation priorities. During the last two to three decades, the introduction of high-yielding varieties as well as important changes in rice farming systems has led to the loss of genetic diversity particularly from valleys in lower elevation ranges. In order to develop a rational conservation plan, in global climate change scenario, a conservation concept is required that goes far beyond ex situ conservation. In situ conservation on farm has been reflected as a backup and complementary strategy to ex situ conservation. In this scenario, the current study demonstrated farmer management and temporal evolution of rice genetic diversity in traditional production systems. The study also compared gene bank-conserved (ex situ) populations and on-farm-managed (in situ) landrace populations of same landraces Jaulia and Thapachini and revealed a greater number of alleles per locus under on-farm-managed populations as compared to static management. The marker diversity by using STMS indicates the genetic diversity among populations resulting from combined effects of many evolutionary forces operating within the biological and historical context of the landrace. The results indicated the low diversity of the populations under static management. On the other hand, the variations in adaptations indicate the degree

S. Kumar (✉) • S.K. Dwivedi • V. Prakash • K.K. Rao • S.K. Samal • J.S. Mishra
ICAR-Research Complex for Eastern Region, Patna 800014, India
e-mail: santosh9239@gmail.com

J.K. Bisht et al. (eds.), *Conservation Agriculture*,
DOI 10.1007/978-981-10-2558-7_20

to which populations are adapted to their environments and their potential for continued performance or as donors of characters in plant breeding. It also provides particular information on loss of diversity over time and space at allelic and genotype level. This piece of writing is a step towards understanding the impact of traditional farmer management on rice landrace populations in Himalayan agroecosystems of India.

Keywords
Crop diversity • In situ conservation on farm • Ex situ conservation • Landrace

20.1 Introduction

Crop genetic diversity is the building block of sustainable agricultural development both in subsistence and scientifically advanced societies. Genetic diversity permits plant breeders and farmers to adapt a crop to diverse and changing environment. The genetic diversity erosion in many food crops that accompanies the spread of commercial agriculture is a major concern for farmers and consumers throughout the world (FAO 1996). Massive genetic erosion was also caused by the farming of high-yielding and photoperiod-insensitive crop varieties, even though they helped in agricultural production increase (Harlan 1992). Although the cultivation of high-yielding and photoperiod-insensitive crop varieties has helped in the increase of agricultural production at aggregate levels worldwide, it has also caused extensive genetic erosion (Harlan 1992). This has had the effect of increasing the genetic vulnerability of some crops. Prior to the development of high-yielding varieties (HYVs) and in certain parts of the world, where HYVs are not yet ubiquitous, farmers relied on and continue to rely on a diverse array of traditional varieties through agriculture and human values. These traditional varieties are well adapted to local situation and exhibit flexible responses to environmental stresses (Harlan 1992; Bellon 1996). The conservation of traditional varieties will be important for the development of more sustainable modes of agricultural production and may provide the genetic diversity necessary for the development of plants better adapted to marginal conditions, evolving pests, changing climates and soils (Cleveland and Murray 1997). The Carpenter (2005) study on the static management of rice plant genetic diversity by resource-poor farmers from the Philippine island of Bohol demonstrates that farmers tend to favour genetically heterogeneous varieties that adapt to varied environments. Despite the decline in genetic diversity of rice plant that occurred during the green revolution, the study proven that local in situ conservation programmes can significantly increase the genetic resources which are available to resource-poor farmers (Meena et al. 2015e, 2016; Verma et al. 2015a, b; Ghosh et al. 2016).

Characterization information on plant genetic resources is the principle management tool for their efficient maintenance and utilization. These data are used to discriminate between accessions, detect redundancies and monitor genetic changes during maintenance. The characterization and diversity estimates are made from analysis of markers that vary between the accessions analysed. Several methods have been introduced by many workers to identify, characterize and differentiate the genotypes of interest. These extend from age-old morphological traits to modern molecular techniques. As science is advancing day by day, germplasm characterization at the DNA level is gaining steady popularity because of their several advantages over the morphological and biochemical markers. An ideal marker would be highly heritable, easy to measure and evaluate discrimination between individuals.

20.1.1 Rice as a Crop Plant

Rice is one of the major food crops of the world, providing ~20 % of the human per capita energy and 15 % of protein globally. The crop is mainly grown in the South, Southwest and East Asian regions, consisting of 17 countries of the 25 major rice-producing countries. Globally, rice occupies one-tenth (147.14 M ha) of arable land with India in the lead ~43 M ha (Siddiq et al. 2005). India is the largest rice-growing country; it is grown in nearly all the states of India, covering ~30 % of the total cultivated area. The rice cultivation is generally concentrated in the river valleys, deltas and low-lying coastal areas of Northeast, Northwest and Southern India, especially in the states of Andhra Pradesh, Chhattisgarh, Bihar, Assam, Karnataka, Odisha, Kerala, Maharashtra, Tamil Nadu, Uttarakhand, Uttar Pradesh and West Bengal, which together contribute ~97 % of the country rice production. The producing rice not only forms the mainstay of diet for majority of its people (>55 %), but also it is the livelihood for more than 70 % of population in the traditional rice-growing regions (Siddiq 2000; Meena et al. 2013; Singh et al. 2014).

Rice crop belongs to the genus *Oryza*, family Gramineae and tribe Oryzeae. The genus *Oryza* is distributed throughout the tropical and subtropicals of the world. The genus *Oryza* comprises of 23 wild species and 2 cultivated species, viz. the African *O. glaberrima* and Asian *O. sativa. Oryza sativa* is domesticated in Asia but now spread to almost all the rice-growing areas of the world, whereas *Oryza glaberrima*, domesticated in western tropical Africa, is limited to that part of the world alone. A total of nine distinct genomes (A, B, C, D, E, F, G, H, J) have been identified in the genus, based on which the species have been grouped into five complexes, namely, *sativa*, *officinalis*, *ridleyi*, *schlechteri* and *meyeriana* (Khush and Virk 2000).

20.1.2 India-Rich Source of Rice Variability

India possesses rich diversity in many crop plants and their wild relatives. Indian gene centre has 166 species of agri-horticultural crop plants and ~320 species of wild relatives. Over 60 species are relatively of rare occurrence being endemic/rare/

endangered or threatened due to over-exploitation (Srivastava 2001). Indian gene centre is appreciably rich in the genetic wealth of wild rice, particularly Indica types (Arora and Nayar 1984). India's rice possesses wide diversity in their morphological and physiological characteristics. They vary in duration from 60 to over 200 days and can be grown in varied elevations from sea level to 2,300 m. They are well adapted to different seasons of the year under both upland and deep water conditions (Hore and Sharma 1991).

India is rich in rice diversity by virtue of northeastern region being on the periphery of the centre of origin of rice on the one hand and having many centres of diversity (secondary centres) on the other. It exists in terms of thousands of landraces of Asian cultivar at the centre of origin/centre of diversity; the hundred of improved cultivars adapted to varied agro-ecologies and closely related wild/weedy species constituting rice germplasm in India. Natural and human selection of cultivars over centuries for varied agro-ecological conditions, viz. altitude, temperature, rainfall soil type, etc., has substantially contributed to the diversity.

In Uttar Pradesh, rice production is confined to the valley and mountainous zones under rainfed/irrigated conditions. Rice cultivation is more widespread and more intensive. The local names of the varieties and traditional upland rice are many with striking variability. Around one hundred twenty rice landraces are having different morpho types i.e. plant height, spike arrangement and awn length was identified so far (Pant and Negi 1997). Due to the diverse situation of cultivation in the Uttarakhand, the local cultivars of rice have a wide variation in the agro-morphological attributes. Realizing the need for conserving this diversity, several explorations were conducted in this region to tap the native diversity of rice and maintain them ex situ (Gupta et al. 1999). These variations in duration (maturity time), dwarf to moderately tall types, taste/palatability, yield drought tolerance, pest/disease resistance, glutinous type sticky/non-sticky, scented/non-scented types, growth habit (non-lodging/lodging, cold adaptable) and farmer's preferences for particular landraces were observed in the rice germplasm collected from Gangetic plains of Uttar Pradesh (Pant and Negi 1997).

20.2 Characterization of Rice Germplasm

The characterization of rice germplasm through different qualitative and quantitative traits is an important step for assessment of its genetic potential. This information is the principle management tool for their efficient maintenance and utilization. These data are used to discriminate between accessions, detect redundancies and monitor genetic changes during maintenance. Several methods have been introduced by many workers to identify, characterize and differentiate the genotypes of interest.

20.2.1 Morphological Characterization

The oldest and conventional method of identification of germplasm is based on morphology of different characters among several species including rice. These are the most widely used traits to describe the observable characters of an organism. The uses of morphological or agronomic traits as markers are irreplaceable because of their omnipresence and easy availability. Kato et al. (1928) is the pioneer worker in this regard who identified two main varietal types in rice, viz. Indica and Japonica through morphological, serological and inter-varietal hybrid fertility test. Later on, a third type of rice was identified by Matusuo (1952) on the basis of its morphology. Within a couple of years, these three types were referred to as Indica, Japonica and Javanica by Moringa (1954). Cheng et al. (1984) showed that morphological characters such as grain shape, leaf pubescence, glume colour at heading, glume hairiness, phenol reaction and interval between first and second nodes of panicle axis, etc., which are used to classify Hsien (Indica) and Keng (Japonica) in China, can also be used to identify Asian rices (Indica, Japonica and three morphological types of Javanica).

Many indigenous red accessions were known to be a useful source for diverse resistance genes. Useful genes for yield and other agronomic characteristics have been identified from landraces and wild relative of rice (Xiao et al. 1996). Red rice plants are difficult to differentiate from other rice cultivars at the seedling stage; however at tillering, flowering or maturity stage, these plants are recognizable by the following characteristics: (i) hispid, light green leaves, (ii) profuse tillering, (iii) longer and more slender panicles and (iv) heavy shattering of grains (Cohn and Hughes 1981). The use of morphological characters to identify genotypes has several limitations. According to Gilliland (1989), those morphological characters that are unambiguous, repeatable and indifferent to environmental influences can only be used to differentiate germplasm. It is very difficult to nullify the influence of environment because phenotype is the interaction between the gene and the environment and a particular variety can show variation in morphological traits in different test centres because of environmental variation. Another drawback of morphological markers lies in the fact that newer varieties may have similar characteristics, particularly in seed and seedling characters (Payne 1987). This problem is particularly prominent in those crops where genetic diversity is limited or when convergent selection towards similar morphology is practised (Guiard 1997). For cereals, grasses and forages, the frequency of variation between varieties and cultivars is limited for conventional characters. Therefore, the use of morphological markers alone is insufficient to distinguish closely related genotypes.

Ugo et al. (2003) observed increasing diversity of the grain size of rice landraces during varietal identification, appeared to have results from speciation or human selection in five geographical region of Yunnan. Joshi and Bauer (2007) studied the farming and loss of rice landraces in *Terai* region of Nepal and identified the factors which are influencing the possibility of cultivating the most dominant landraces by using a logistic regression model. They observed that the area under landraces has declined considerably following the development and the extension of high-yielding modern varieties in Nepal. The number of landraces under cultivation is declining, though the varietal diversity appeared too high.

20.2.2 Biochemical Markers for Characterization of Rice Germplasm

To identify a crop variety, the use of biochemical markers, especially isozymes or protein profiles, can be effectively used. Larsen (1969) has stated “all inherent morphological manifestation of cultivar difference must ultimately have a biochemical basis but not all biochemical differences are reflected morphologically”. This indicates that there are sufficient number of biochemical markers which show difference in cultivars, and these include (a) phenol reaction; (b) total soluble proteins, albumins and globulins; (c) isozymes like esterases, malate dehydrogenase, peroxidases, alcohol dehydrogenase, glutamate dehydrogenase, catalases, etc.; and (d) free amines (putrescine, spermidine, spermine, tyramine) or amine conjugates (di-feruloylputrescine, di-feruloylspermidine, etc.). Electrophoresis is employed to analyse biochemical markers like isozymes or seed proteins DNA or RNA. Here, voltage gradients are generated across a medium (usually starch, polyacrylamide or agarose) for the separation of charged biomolecules like DNA, RNA or proteins. The separation occurs due to the difference in size and charge of the molecules (Davis 1986). Analysis of isozymes through starch gel and agarose gel was reported by Hunte and Markert (1957) and Wieme (1959), respectively. In the year 1959, Raymond and Weintrauh described the isozyme analysis through acrylamide gel electrophoresis. Through the electrophoresis technique, banding pattern of proteins is generated. These protein profiles are unique for a particular cultivar which subsequently is used to generate data for identifying the cultivars. Numerous reports suggest the use of protein profiles for cultivar identification.

The use of total soluble proteins for genotype identification in rice has been successfully employed by several workers (Siddiq et al. 1972; Guo et al. 1986). Sarkar and Bose (1984)) used variations in protein for distinguishing and characterizing eight rice varieties. Isozymes are defined as multiple molecular forms of an enzyme, with similar or identical catalytic activities (Markert and Moller 1959), which can be demonstrated by electrophoresis followed by in situ biochemical staining for specific activity. The polymorphism observed is due to the presence of different structural genes and/or alleles at a single locus. Since isozymes may differ in their molecular weight and/or configuration, they can be detected and differentiated by appropriate histochemical methods (Hunte and Markert 1957). In cultivated rice (*O. sativa*), isozyme polymorphism has received much attention for the last few years. This is largely due to the work of Second and Tronslot (1980) who demonstrated that considerable variation could be revealed by starch gel electrophoresis. As a result, more than 40 loci encoding isozymes have been proposed in rice. Most of the genetic studies on the isozymes of rice deal with linkage relationships, heterosis in hybrid rice and evolution and classification of rice cultivars. More recent studies use isozymes as tool for hastening the breeding procedures. Isozymes have been extensively used with regard to the evolution of cultivated rice. The differentiation of rice ecotypes has been found to be related to a peroxidase band (AC band). Chu (1967) found that this band exists in most Indica varieties but is absent in Japonica rice. Nakagahara et al. (1975) found remarkable geographic clines upon comparing the esterase banding patterns to four varietal types: Indica (from India), Sinica (from South China), Japonica (from Japan, Korea and North China) and Javanica (from Indonesia).

Although seed proteins and isozymes can be used for the differentiation of varieties, sometimes polymorphic storage proteins in cereals or isozyme variations do not provide enough information to distinguish germplasm unequivocally; especially closely related varieties may be impossible to distinguish (Ainsworth and Sharp 1989). Other than this, biochemical markers have some disadvantages since their expression may be influenced by environmental factors or expression may be stage specific. Moreover, biochemical markers like isozymes are not as numerous as molecular markers and also epistatic effect prevails. Further, biochemical data alone cannot provide a basis of unique varietal identification since the genome coverage provided by these markers is limited. Varietal identification through DNA profiling can overcome the limitations associated with morphological and biochemical methods and can provide more detailed scrutiny of genetic variations among genotypes (Helentjaris et al. 1985).

20.2.3 Molecular Characterization and Identification of Rice Germplasm

The molecular markers are specific sequences of DNA, the variations of which can be used to identify organisms. The variations in the molecular markers may occur due to changes in the sequences of bases in the DNA. These changes may lead to the formation or deletion of a restriction site or a random change in the base sequences or even changes within the simple repeat regions of a DNA or in its flanking regions. The occurrence of these variations may be due to deletion, duplication, point mutation, insertion, inversion or spontaneous mutation of a DNA segment.

A range of techniques that can detect polymorphism at the DNA level are available. Molecular markers have significant value in breeding programmes to characterize and evaluate genetic variability in germplasm and to identify varieties. They offer several advantages over biochemical markers. Their expressions are neither the stage nor organ specific, nor they vary in any part of the plant organ (Morell et al. 1995) and being unlimited in number, a thorough sampling of genome is possible (Helentjaris et al. 1985). The uses of such modern analytical techniques to solve the problem of varietal characterization as well as their protection have been demonstrated (Smith and Helentjaris 1993). The commonly adopted approaches in the use of molecular markers are probe based, i.e. screening of restriction fragment length polymorphisms (RFLPs), polymerase chain reaction (PCR) based like random amplified polymorphic DNAs (RAPDs) or a combination of both like Amplified Fragment Length Polymorphisms (AFLP).

Simple sequence repeats (SSRs), usually mentioned to as microsatellites, are tandemly repeated mono- to hexa-nucleotide motifs; these are ubiquitous in eukaryotic genomes and exhibit highly variable numbers of repeats at a locus (Weber and May 1989). These SSRs consists of two to six nucleotide core units such as (CA), (ATT) or (ATGT) that are tandemly repeated in the genome. Length polymorphisms is created when PCR products from different alleles vary in length as a result of variation in the number of repeat units in the SSR and then can be analysed by acrylamide or agarose gel electrophoresis to resolve contrasting alleles. A significant number of microsatellite markers have been developed and used to construct genetic maps in rice, which will provide important co-dominant landmarks that are well spread all through the rice genome. These markers provide important co-dominant landmarks that are well distributed throughout the rice genome (McCouch et al. 2001).

Ravi et al. (2003) assessed the genetic diversity among the 40 cultivated varieties and 5 wild relatives of rice by using SSR and RAPD markers. This SSR analysis resulted in a more definitive separation of clusters of genotypes signifying a higher level of competence of SSR markers for the precise determination of relationships between accessions that are too close to be accurately differentiated 295 by RAPD markers. Microsatellite markers were used for fingerprinting of hybrids, assessing variation within parental lines and testing the genetic purity of hybrid seed lot in rice. Fingerprinting of 11 rice hybrids and their parental lines was done by using ten STMS markers by Nandakumar et al. (2004). Out of ten, nine STMS markers were found polymorphic across the hybrids and produced unique fingerprints for the 11 hybrids. A set of four markers (RM 206, RM 216, RM 258 and RM 263) differentiated all the hybrids from each other, which can be further used as referral markers for the identification and protection of these hybrids.

Microsatellite marker analysis was also used to distinguish premium traditional Basmati rice varieties from other cheaper cross-bred Basmati/long-grain rice varieties and monitor the cases of adulteration in milled rice samples. Thirteen rice cultivars were evaluated for allelic diversity using 35 SSR markers. A total of 123 alleles were detected, 25 of these were present in Basmati rice varieties only. SSR analysis generated polymorphism sufficient to differentiate all 13 rice genotypes. Rice varieties were clustered in three groups (Indica, Japonica and Basmati groups), which correspond well to their known pedigree data (Pal et al. 2004). The allelic variation detected by the STMS markers enables differential of the homozygotes from the heterozygotes and therefore is most useful in understanding varietal uniformity. STMS markers have been used widely in rice for varietal characterization and for conservation of genome map, mapping and tagging of genes (Kholi et al. 2004; Ganesh et al. 2007).

Tu et al. (2007) estimated genetic diversity of rice germplasm in Yunnan Province of China, by using microsatellite (SSR) fingerprints. Although they found high genetic diversity, it was not evenly distributed across the regions. They have suggested the need of strategic conservation of rice germplasm in Yunnan by collecting varieties across geographic regions with sufficient individuals within the same varieties. Thomson et al. (2007) also used 30 microsatellite markers to characterize the genetic diversity in 300 local Indonesian rice varieties. They got 394 alleles at the 30 simple sequence repeat loci, with an average number of 13 alleles per locus across all accessions and an average polymorphism information content value of 0.66.

20.3 In Situ Conservation

In situ conservation states to the conservation of genetic resources within their ecosystem and natural habitats. It can be grown outside their habitat, such as member of complex ecosystem, species with highly specialized breeding systems and that are endangered, wild relatives of crop species, etc.

20.3.1 In Situ Conservation On Farm

In situ conservation on farm, sometimes referred to as "on-farm conservation", has been defined as "the continuous cultivation and management of a diverse set of populations by farmers in the agro ecosystem where a crop has evolved" (Bellon et al. 1997; Kumar et al. 2015; Meena et al. 2015a, b, c, d). On-farm conservation is carried out by the farmers who are interested and willing to do so. It cannot be imposed on them. Therefore, the basis of on-farm conservation should be the participation of farmers that maintain crop-intraspecific diversity. It is very important to understand what is needed to conserve genetic diversity on farm, because it can be helpful to identify the specific needs of an on-farm conservation programme. In addition to preserve plant genetic resources, on-farm conservation has six major benefits which make it unique among the options available to conservationists:

- It helps in conserving the processes of evolution and adaptation of crops to their environments.
- Conservation of diversity at three levels – ecosystem, species and within species.
- Participation of farmers into a national plant genetic resources system.
- It helps in conserving ecosystem services critical to the functioning of the earth's life support system.
- It helps in improving the livelihood of resource-poor farmers through economic and social development.
- Maintain farmer's control over and access to crop genetic resources.

20.3.2 Management of Rice Diversity by Farmers

Management of crop diversity by farmers refers to the cultivation of a diverse set of more or less specialized crop population. These populations are named and recognized as units by the farmers: they are "farmers' varieties" as opposed to the improved varieties. The structure and level of crop genetic diversity in the farmers field are a result of a number of activity and external influence extending beyond the farmers' choice of how many and which varieties to plant. Farmers' seed selection decisions are based on the range of agro-morphological characteristics that their crop exhibits. Farmers influence the population structure of a crop by determining the proximity of the crop population to potential breeding partners and thus how genetic material is exchanged between and within fields. New genetic material may be introduced through the system of seed flows, in which seeds are acquired from a variety of a channel or stored on farm for later use.

20.3.3 Need for On-Farm Conservation of Rice Germplasm

Conservation of rice germplasm is urgently needed because some species are going to be extinct and others are threatened and endangered due to biotic and abiotic factors. We need to conserve these germplasm because of human dependence on them for many different uses like food and feed and ethnic and traditional uses and for future food and nutritional security. So far, only few species of this crop have been evaluated for their agricultural and medicinal potential. The conservation of rice crop diversity at all levels within local environments helps to ensure the ongoing process of evolution and adaptation of this crop to their local environments. The ongoing process of evolution and adaptation confirms that new germplasm is generated over time, rather than restrictive conservation of genetic resources conserved in gene banks. For farmers, on-farm conservation could serve to support cultural traditions; fit household labour and budget constraints; mitigate the effect of pests, disease and other environmental stresses; and provide insurance of new genetic material in the face of future environmental or economic change (Table 20.1).

20.4 Ex Situ Conservation of Rice Diversity

The ex situ conservation refers to the conservation of genetic materials of biological diversity outside their natural habitat in facilities supporting either storage or perpetuation with conditions suited to maintain their viability and genetic constitution (Dhillon et al. 2004). It virtually safeguards and provides required supply of germplasm for research and breeding. The various possible approaches can be grouped in:

Plant conservation: (i) botanical garden, (ii) arboreta, (iii) herbal garden, (iv) field gene bank and (v) clonal repositories.

Seed conservation: (i) low-temperature storage of seeds (seeds gene bank) and (ii) cryopreservation (storage of orthodox, intermediate and recalcitrant seeds in liquid nitrogen at −150 to −196 °C).

In vitro conservation: (i) conservation of cells, tissue and organs in glass or plastic containers under aseptic condition through slow growth of cultures and (ii) cryopreservation of cultures at −150 to −196 °C in liquid nitrogen.

Table 20.1 Some possible benefits accruing from on-farm conservation

Items	Economics and sociocultural benefits	Ecological benefits	Genetic benefits
Farmers household	Manage risk and uncertainty	Minimize the use of chemical inputs	Insurance against environmental and socioeconomic change
	Fit different budget constraints	Soil structure amelioration	
	Avoid or minimize labour bottlenecks	Manage pest and disease	
	Fulfil rituals or forge social ties		
	Fill nutritional needs		
Society	Global food security	Reduction of chemical pollution	Insurance against environmental change, pest and diseases
	Empowerment of local communities	Restriction of plant diseases	Use for the agricultural industry
	Social sustainability	Regulation of hydrological flows	

Adapted from Jarvis (1999)

DNA conservation: conservation of genetic diversity in the form of functional unit called "gene". The whole genome in the form of genomic library or a sequence of DNA in the form of DNA library may be conserved following the appropriate DNA conservation methods.

For ex situ conservation of rice diversity, a gene pool approach has to be followed for safe and effective conservation. The rice gene pool consists of self-pollinated cultigens and a range of wild *Oryza* spp. habitat to the range of climatic conditions with breeding system ranging from obligate vegetative propagation to facultative and obligate self-pollination. In such a situation, it is quite logical to have an approach, which is an appropriate and balanced application of both in situ and ex situ conservation methods (Dhillon et al. 2004). At the national level, there have been significant efforts for ex situ conservation of rice diversity and various stakeholders have taken several initiatives at different levels.

20.5 Rice Diversity Under Static (Ex Situ) and Dynamic (In Situ) Management

In situ conservation has now been considered as a backup and complementary strategy to ex situ conservation (Maxted et al. 2002). It has been considered as enhanced PGR utilization at the local level and consistent with agricultural development (FAO 1998). As in situ conservation stands now, there seems to be a de facto conservation of landraces that farmers have been practising for centuries as part of their farming system (Seboka and Hintum 2006). An exploratory study was conducted for comparing diversity and adaptation in rice varieties under static and dynamic management by the help of morphological and biochemical (isozyme) markers and significant genetic changes observed including loss of adaptation. The population which is managed by farmers showed a common trend of later flowering and maturity time, more uniformity in grain quality, less off types and reduced drought stress tolerance compared with the corresponding ex situ population. The observed adaptability is at risk under on-farm conservation due to the natural and intentional selection pressure for rice (Tin et al. 2001). Soleri and Smith (1995) have studied the consequence of in situ and ex situ conservation in maize accessions by measuring morphological and genetic structure variation among accession of the same populations reserved under static and dynamic situation. They observed that genetic drift and shift have occurred under ex situ.

Hammer et al. (1996) studied the genetic erosion in landraces in Albania and South Italy and observed ~72 % genetic erosion, during 52 years in Albania and in South Italy, and 75 % during ~35 years in most of the field crops. Tsegaye and Berg (2006) demonstrated the overall loss of the once wide spread tetraploid wheat landrace (East Shewa) in the study area accounted for 77 % of the formerly available diversity. Treuren and van Hintum (2001) investigated intra-accession variation in two landraces, and wild population of ex situ conserved barley germplasm by using AFLP markers. Parzies et al. (2000) have reported that the genetic diversity of barley landraces accession conserved in ex situ gene banks was less compared to that in in situ conservation due to genetic drift during regeneration.

Traditional crop landraces play dynamic roles in the expression of native biological and cultural diversity via their central position in the genetic resource base, agroecosystems and social heritage of aboriginal peoples. Farmer's varieties meet local cultural practices and environmental constraints and play an essential role in cultural survival. Pfeiffer et al. (2006) studied the research efforts of rice farmers pertaining to the Tado clan, a Kempo Manggarai community on Flores Island. Research results demonstrated a complex suite of upland rice-based ethnobotanical traditions, significant and dynamic regional flux and dissemination of "old" and "new" landraces, community-level maintenance of distinct genotypes across a range of microenvironments, localized "extinctions" of ancestral landraces within one to two generations and a simultaneous loss of related traditions and the contributions of a two-way (indigenous and academic) approach to ethnographic and agronomic research.

20.6 A Case Study for Diversity Analysis of Rice Landraces from the Uttarakhand State in North-Western Himalayas

The experimental materials comprised 11 rice landrace populations. These are collected from different parts of Uttarakhand state in North-western Himalayas. Out of 11 rice landrace population, five populations of rainfed common landrace Jaulia, three populations of a common irrigated landrace Thapachini and three

Table 20.2 Passport information of the rice landrace populations from different parts of Uttarakhand

Landraces	Population/accession ID	Origin (district)
Jaulia	IC 100051	Pithoragarh
Jaulia	IC 548358	Champawat
Jaulia	IC 548363	Champawat
Jaulia	IC 548639	Rudraprayag
Jaulia	IC 548668	Rudraprayag
Thapachini	IC 113843	Chamoli
Thapachini	IC 548386	Almora
Thapachini	IC 548396	Bageshwar
Rare landrace (Punjabi)	IC 548401	Chamoli
Rare landrace (Sal)	IC 548390	Almora
Rare landrace (Syao)	IC 548398	Bageshwar

different locally had grown rare landrace populations. One population each of common landrace Jaulia and Thapachini represented static (ex situ) conservation since 1991 and were taken from the National Genebank of National Bureau of Plant Genetic Resources (NBPGR), whereas the remaining populations were collected directly from farmers' fields during 2006 cropping season for the present study and represented dynamic conservation (Table 20.2) on farm (Kumar 2009).

The rice landrace populations were also grown for morphological characterization in an on-farm field experiment at NBPGR Regional Station, Bhowali (Uttarakhand), during rainy season 2007. Observations on 15 quantitative and 7 qualitative characters were recorded. The range of variation for rice landrace populations revealed that maximum variations were recorded for grain yield, straw yield per plant, number of tillers per plant, panicle length and grain length. Least variation was observed for the traits such as days to ~75 % panicle emergence, days to maturity and flag leaf width. Not many variations were recorded for various qualitative traits, viz. panicle exertion, presence of awn, husk colour, seed coat colour, presence of aroma and thresh ability.

Highly significant differences for all quantitative traits among landraces were recorded. The quantitative traits such as number of tillers, grain yield and grain length revealed greater diversity than other traits. The population quantitative parameters revealed high grain yield per plant for population IC 548668 of Jaulia followed by IC 548390 (rare landrace) and IC 548396 (Thapachini). Population IC 548668 of Jaulia also showed maximum 100-grain weight followed by IC 548390 (rare landrace). Among the nine on-farm-managed rice landrace populations, the maximum straw yield per plant was recorded for population IC 548639 followed by IC 548668 of Jaulia. The population IC 548386 of Thapachini was early maturing. The maximum number of grain per panicle was recorded for population IC 548668 of Jaulia followed by IC 548398 and IC 548398 (rare landrace). These morphological variations indicated that landraces significantly varied for yield-related traits under farmer management. Even different populations of the same landrace Jaulia and Thapachini varied for yield traits.

The molecular (STMS) analysis of rice landrace populations indicated sufficient polymorphism to differentiate inter- and intrapopulation diversity. Greater genetic diversity with more number of effective alleles, greater Shannon's information index and greater expected heterozygosity were observed in on-farm-managed populations as compared to populations under static management. Out of the 11 landrace populations, the maximum number of 41 alleles was recorded in IC 548358followed by the population IC 548398. The total number of alleles ranged from 21 to 41 with maximum number of alleles for population IC 548358 and minimum for IC 100051, both of common landrace Jaulia. The population genetic parameters revealed high allelic richness for populations IC 548358 and IC 548668 of Jaulia, population IC 548396 of Thapachini and the rare landrace population IC 548398, together with high Shannon information index, in different categories of landraces. The expected heterozygosity was highest for rare landrace population IC 548398 followed by population IC 548358 of common landrace Jaulia.

The population differentiation under static management was greater ($F_{ST} = 0.84$) as compared to dynamic management ($F_{ST} = 0.65$) (Kumar et al. 2010). Differences of genetic differentiation may have been associated with differences in accession handling and sampling method. Genetic diversity between samples conserved in a gene bank and samples collected from farmer field could be explained by differences in the way of collection methods during the periods compared. The frequency of rare alleles was relatively more for on-farm-managed populations. The comparison of dynamic and static conservation made in the present study could only reveal the overall increase in intra-accession genetic diversity over time under dynamic management. This study also provided particular information on the loss of diversity over time and space at genotype, and allelic level is essential by comparing the landraces conserved under both static and dynamic conservation to develop a rational conservation plan. The results must be considered as a signal of low diversity of the populations under static management. The increased intra-accessions diversity observed in on-farm-managed populations may be described by the rise in the number of landraces cultivated during the last two to three decades. This research works also as a step towards understanding the impact of traditional farmer management on rice landrace populations in Himalayan agroecosystems of India. Investigating the population genetic structure is therefore helpful in monitoring change in diversity over time and space and also for devising a rational plan for management of farmer landraces on farm.

20.7 Conclusions

Crop landraces are valuable genetic resources. They are considered to contain huge genetic diversity and therefore it should complement and diversify the gene pool of advanced cultivars. The genetic structure of rice landraces is an evolutionary approach to existence and performance, particularly under rainfed condition and to some extent under irrigated conditions in valleys, and organic inputs in Himalayan agroecosystems. The combined effects of farmer selection and natural selection have led to the design of genotypes representing diverse combinations of traits. Better understanding into the dynamics of genetic resources of rice crop is required in order to identify harmful evolutionary patterns and draw up conservation priorities. Diversity at genetic level allows plant breeders and farmers to adapt a crop to diverse and changing environments. The introduction of high-yielding varieties (HYVs), and the basic changes in rice farming systems, has led to the loss of genetic diversity particularly from valleys in lower elevation ranges. In order to devise a rational conservation plan, in global climate change scenario, a conservation concept is required that goes far beyond ex situ (static) conservation. As in situ (dynamic) conservation stands now, there seems to be a de facto conservation of landraces that farmers have been practising for centuries as part of their farming system. In situ conservation on farm of agricultural biodiversity needs to be made an integral part of agricultural development and be supplemented by ex situ conservation. In this scenario, our study also established farmer management of crop population structure and temporal evolution of rice genetic diversity in traditional production systems. It provides precise information on the loss of diversity over time and space at allelic, and genotype level is necessary by comparing the landraces conserved under both static and dynamic conservation to devise a rational conservation plan. Our research works also as a step towards understanding the impact of traditional farmer management on rice landrace populations in Himalayan agroecosystems of India. Investigating the population genetic structure is therefore helpful in monitoring change in diversity over time and space and developing a rational plan for managing of farmer landraces on farm.

Acknowledgement The authors are thankful to the Director, NBPGR, for providing the facilities for the study. The contingency provided to the senior author (SK) by the CSIR is gratefully acknowledged.

References

Ainsworth CC, Sharp PJ (1989) The potential role of DNA probes in plant variety identification. Plant Var Seeds 2:27–34

Arora RK, Nayar ER (1984) Wild relativs of crop plants in India. NBPGR. ScientificMonograp 7:1–90

Bellon MR (1996) The dynamics of crop infraspecific diversity: a conceptual framework at the farmer level. Econ Bot 50:26–39

Bellon MR, Pham JL, Jackon MT (1997) Genetic conservation: a role for farmers. In: Plant genetic conservation: the in-situ approach. Chapman and Hall, London, pp 261–289

Carpenter D (2005) The in-situ conservation of rice plant genetic diversity: a case study from a Philippine barangay. Agric Hum Welf Value 22:421–434

Cheng KS, Wang XK, Zhou JW, Lu YX, Lou J, Huang HW (1984) Studies on indigenous rices in Yunan and their utilization. 2. A revised classification of Asian cultivated rice (in Chinese). Acta Agron Sci 10:271–280

Chu Y (1967) Variation in peroxidase isozymes of *Oryza perennis* and *O. sativa*. Jpn J Genet 42:233–244
Cleveland DA, Murray SC (1997) The world's crop genetic resources and right of indigenous farmers. Curr Anthropol 38:477–515
Cohn MA, Hughes JA (1981) Seed dormancy in red rice I. Effect of temperature on dry-after ripening. Weed Sci 29:402–404
Davis MG (1986) Electrophoretic technique. In: Wilson K, Goulding KH (eds) A biologists guide to principles and techniques of practical biochemists, 3rd edn. Edward Arnold, London, pp 245–268
Dhillon BS, Tyagi RK, Lal A, Saxena S (2004) Ex situ approaches for conservation of plant genetic resources: achievements and challenges. In: Plant genetic resources management. pp 179–193
FAO (1996) Global plan of action for the conservation and sustainable utilization of plant genetic resources for food and agriculture. FAO, Rome
FAO (1998) The state of the world's plant genetic resources for food and agriculture. FAO, Rome
Ganesh SR, Thiruvengadam V, Vinod K (2007) Genetic diversity among cultivars, landraces and wild relatives of rice as revealed by microsatellite markers. Theor Appl Genet 48(4):337–345
Ghosh BN, Meena VS, Alam NM, Dogra P, Bhattacharyya R, Sharma NK, Mishra PK (2016) Impact of conservation practices on soil aggregation and the carbon management index after seven years of maize–wheat cropping system in the Indian Himalayas. Agric Ecosyst Environ 216:247–257
Gilliland TJ (1989) Electrophoresis of sexually and vegetatively propagated cultivars of allogamous species. Plant Var Seeds 2:15–25
Guiard J (1997) Application of molecular techniques to DUS testing. Regional seminar on variety testing for UPOV convention. UPOV and Ministry of Agriculture, Indonesia and MAAFF, Japan, Dec 5–7
Guo YJ, Bishop R, Ferhnstorm H, Yu GZ, Lian YN, Huang SD (1986) Classification of Chinese rice varieties by electrofocussing. Cereal Chem 63:1–3
Gupta A, Sharma RK, Mani VP, Chauhan VS (1999) Pattern of genetic diversity and variability in rice germplasm of U.P. hills. Crop Improv 26(1):81–87
Hammer K, Knupffer H, Xhuveli L, Perrino P (1996) Estimating genetic erosion in landraces-two case studies. Genet Resour Crop Evol 43:329–336
Harlan JR (1992) Crops and man, 2nd edn. American Society of Agronomy, Madison
Helentjaris T, King G, Slocum M, Siedenstrang C, Wegmen S (1985) RFLPs as probes for plant diversity and their development as tools for applied plant breeding. Plant Mol Biol 5:90–115
Hore DK, Sharma BD (1991) Collection, evaluation and conservation of rices of South Asia with special reference to North East India. Indian J Plant Genet Res 4(1):61–72
Hunte RL, Markert CL (1957) Histochemical demonstration of enzymes separated by zone electrophoresis in starch gels. Science 12:1294–1295
Jarvis D (1999) Strengthening the scientific basis of in-situ conservation of agricultural biodiversity on-farm. Bot Lith Suppl 2:79–90
Joshi GR, Bauer S (2007) Cultivation and the loss of rice landraces in the Terai region of Nepal. Plant Genet Res Charact Utilisation 5(1):1–6
Kato S, Kosaka H, Hava S (1928) On the affinity of rice varieties as shown by the fertility of rice plants. Centre Agric Inst Kyushu Imp Univ 2:241–276
Kholi S, Mohapatra T, Das SR, Singh AK, Sharma RP (2004) Composite genetic structure of rice land races revealed by STMS markers. Curr Sci 86:850–854
Khush GS, Virk PS (2000) Rice breeding achievements and future strategies. Crop Improv 27 (2):115–125
Kumar S (2009) Studies on genetic structure of rice landraces and dynamics of their on-farm management in traditional agro-ecosystems of North-western Himalayas in Uttarakhand state. Ph.D. thesis, Indian Agricultural Research Institute, New Delhi
Kumar S, Bisht IS, Bhatt KV (2010) Population structure of rice (*Oryza sativa* L.) landraces under farmer management. Ann Appl Biol 156:137–146

Kumar A, Bahadur I, Maurya BR, Raghuwanshi R, Meena VS, Singh DK, Dixit J (2015) Does a plant growth promoting rhizobacteria enhance agricultural sustainability? J Pure Appl Microbiol 9(1):715–724

Larsen AL (1969) Isozymes and varietal identification. Seed World 140:5–6

Markert CL, Moller F (1959) Multiple forms of enzymes, tissue ontogeny and species specific patterns. Proc Natl Acad Sci U S A 45:753–766

Matusuo T (1952) Genecological studies on cultivated rice (in Japanese). Bull Natl Inst Agric Sci 3:1–11

Maxted N, Guarino L, Myer L, Chiwona EA (2002) Towards a methodology for on-farm conservation of plant genetic resources. Genet Resour Crop Evol 49:31–46

McCouch SR, Temnykh S, Lukashova A, Coburn J, De Clerk G, Cartinhour S, Harrington S, Thomson M, Septiningsih E, Semon M, Moncada P, Jiming Li (2001) In: Khush GS, Brar DS, Hardy B (eds) Rice genetics IV. Proceedings of the Fourth International Rice Genetics Symposium, 22–27 October. 2000, Los Baños, Philippines, New Delhi (India). Science Publishers and Los Baños (Philippines) International Rice Research Institute. pp 117–136

Meena RS, Yadav RS, Meena VS (2013) Heat unit efficiency of groundnut varieties in scattered planting with various fertility levels. Bioscan 8(4):1189–1192

Meena RS, Meena VS, Meena SK, Verma JP (2015a) The needs of healthy soils for a healthy world. J Clean Prod 102:560–561

Meena RS, Meena VS, Meena SK, Verma JP (2015b) Towards the plant stress mitigate the agricultural productivity: a book review. J Clean Prod 102:552–553

Meena VS, Meena SK, Verma JP, Meena RS, Ghosh BN (2015c) The needs of nutrient use efficiency for sustainable agriculture. J Clean Prod 102:562–563

Meena VS, Meena SK, Verma JP, Meena RS, Jat LK (2015d) Current trend to mitigate climate change and sustainable agriculture: a book review. J Clean Prod 102:548–549

Meena VS, Verma JP, Meena SK (2015e) Towards the current scenario of nutrient use efficiency in crop species. J Clean Prod 102:556–557

Meena RS, Bohra JS, Singh SP, Meena VS, Verma JP, Verma SK, Sihag SK (2016) Towards the prime response of manure to enhance nutrient use efficiency and soil sustainability a current need: a book review. J Clean Prod 112(1):1258–1260

Morell MK, Peakal R, Appels R, Preston CR, Liloyd HL (1995) DNA profiling techniques for plant variety identification. Aus J Exp Agric 35:807–819

Moringa T (1954) Classification of rice varieties on the basis of affinity. In: International rice commission. Working party on rice breeding. Rep. 5th Meeting. pp 1–4

Nakagahara M, Akhiana T, Hagushi K (1975) Genetic variation and geographic line of esterase isozymes in nature rice varieties. Jpn J Genet 50:373–382

Nandakumar N, Singh AK, Sharma RK, Mohapatra T, Prabhu KV, Zaman FU (2004) Molecular fingerprinting of hybrids and assessment of genetic purity of hybrid seed in rice using microsatellite markers. Euphytica 136:257–264

Pal S, Jain S, Saini N, Aarti, Jain RK (2004) Identification of microsatellite markers for differentiating some basmati and non-basmati rice varieties. Indian J Biotechnol 3:519–5236

Pant KC, Negi KS (1997) Exploration and collection of rice (*Oryza sativa*). Germplasm during 1985–1994 from Uttar Pradesh Himalaya and Gangetic Plains area. Indian J For 20(1):89–92

Parzies HK, Spoor W, Ennos RA (2000) Genetic diversity of barley landrace accessions (*Hordeum vulgare* ssp.*Vulgare*) conserved for different lengths of time in ex situ gene banks. Heredity 84:476–486

Payne RC (1987) Seed and cultivar identification. Seed Sci Technol 15:641–644

Pfeiffer JM, Rice KJ, Dun S, Mulawarman B (2006) Bicultural diversity in traditional rice-based agroecosystems: indigenous research and conservation of mavo (*Oryza sativa* L.) upland rice landraces of eastern Indonesia. Environ Dev Sustain 8:609–625

Ravi M, Geethanjali S, Sameeya F, Maheswaran M (2003) Molecular marker based genetic diversity analysis in rice (*Oryza sativa* L.) using RAPD and SSR markers. Euphytica 133:243–252

Raymond L, Weintrauh A (1959) In: Brewer GJ, Singh CF (eds) An introduction to isozyme techniques. Academic, New York, p 1

Sarkar R, Bose S (1984) Electrophoretic characterization of rice varieties. Singh seed (salt-soluble) proteins. Theor Appl Genet 68:415–419
Seboka B, Hintum TV (2006) The dynamics of on-farm management of sorghum in Ethiopia: implication of the conservation and improvement of plant genetic resources. Genet Resour Crop Evol 53:1385–1403
Second G, Tronslot P (1980) Polymorphism de treizezymograms observes d' enzyme de riz (*Oryza* spp.) Travaux et documents. ORSTOM, Paris
Siddiq EA (2000) Rice. In: Chopra VL (ed) Breeding field crops. Oxford and IBH Publishing Co. Pvt. Ltd, New Delhi, pp 1–85
Siddiq EA, Nerkar YS, Mehta SL (1972) Intra and inter sub-specific variation in soluble proteins of *Oryza sativa* L. Theor Appl Genet 42:351–356
Siddiq EA, SaxenaS, Malik SS (2005) In: BS Dhillon, S Saxena, A Agrawal, RK Tyagi (eds) Plant genetic resources: food grain crops. Indian Society of Plant Genetic resources, New Delhi, Narosa Publishing Houise, New Delhi, India pp 27–57
Singh DK, Singh SK, Singh AK, Meena VS (2014) Impact of long term cultivation of lemon grass (Cymbopogon citratus) on post-harvest electro-chemical properties of soil. Ann Agric Biol Res 19(1):45–48
Smith S, Helentjaris T (1993) DNA fingerprinting and plant variety protection. In: Paterson A (ed) Genome mapping in plants. R.G. Landes Company, Austin, pp 95–107
Soleri D, Smith SE (1995) Morphological and phenological comparisons of two Hopi maize varieties conserved in-situ and ex situ. Econ Bot 49:56–77
Srivastava UC (2001) Conservation of endangered plant genetic resources of economic importance in India. J Econ Taxon Bot 25(2):501–508
Thomson MJ, Septiningsih EM, Suwardjo F, Santoso TJ, Silitonga TS, McCouch SR (2007) Genetic diversity analysis of traditional and improved Indonesian rice (*Oryza sativa* L.) germplasm using microsatellite markers. Theor Appl Genet 114:559–568
Tin HQ, Berg T, Bjornstad A (2001) Diversity and adaptation in rice varieties under static (ex situ) and dynamic (in-situ) management. Euphytica 122:491–502
Treuren RV, Van Hintum JJL (2001) Identification of intra-accession genetic diversity in selfing crops using AFLP markers: implication for collection managements. Gen Resou Crop Evol 48:287–295
Tsegaye B, Berg T (2006) Genetic erosion of Ethiopian tetraploid wheat land races in Easter Shewa, Central Ethiopia. Genet Resour Crop Evol 715–726
Tu M, Lu B, Zhu Y, Wang Y (2007) Abundant within-varietal genetic diversity. Biochem Genet 45(11–12):789–801
Ugo Y, Fukuta Y, Ohsawa R, Fujimura T (2003) Variation of floral traits in Asian cultivated (*O. sativa* L.) and its wild relatives (*O. rufipogon* Griff.). Breed Sci 53:345–352
Verma JP, Jaiswa DK, Meena VS, Meena RS (2015a) Current need of organic farming for enhancing sustainable agriculture. J Clean Prod 102:545–547
Verma JP, Jaiswal DK, Meena VS, Kumar A, Meena RS (2015b) Issues and challenges about sustainable agriculture production for management of natural resources to sustain soil fertility and health. J Clean Prod 107:793–794
Weber JL, May PE (1989) Abundant class of human DNA polymorphisms which can be typed using the polymerase chain reaction. Am J Hum Genet 44:338–396
Wieme A (1959) In: Brewer GJ, Singh CF (eds) An introduction to isozyme techniques. Academic, New York, pp 51–62
Xiao J, Li J, Grandillo S, McCouch SR, Tanksley SD, Yuan L (1996) A wild species contain genes that may significantly increase the yield of rice. Nature 384:223–224

Effect of Climate Change on Growth and Physiology of Rice-Wheat Genotypes

21

S.K. Dwivedi, Santosh Kumar, Ved Prakash, and J.S. Mishra

Abstract

Under the changing climatic condition, both CO_2 and temperature are the key variables that may cause significant changes in crop productivity. To understand the effect of these key variables of climate change on rice and wheat genotypes, a study was conducted under different sets of varying environments inside open top chambers (OTCs). The sets of conditions were ambient condition, ~25 % higher CO_2 than ambient, 25 % higher $CO_2 + 2\ ^\circ C >$ ambient temperature, and 2 °C higher than the ambient temperature. Finding of the study showed that C_3 crops (rice and wheat) respond positively toward increasing atmospheric CO_2 in the absence of other stressful conditions, but the beneficial direct impact of elevated CO_2 can be offset by other effects of climate change, such as elevated temperature. Climate changes affect the development, growth, and productivity of plants through alterations in their biochemical, physiological, and morphogenetic processes. The rising level of atmospheric CO_2 led to the fertilization effect on C_3 crops which in turn improved their growth and productivity. Increasing CO_2 concentration in the atmosphere could lead to higher crop yields. Increased temperatures during the growing period may also reduce CO_2 effects indirectly, by increasing water demand. Among different rice genotypes, IR83376-B-B-24-2 was highly responsive, while IR84896-B-127-CRA-5-1-1 was least responsive toward elevated CO_2. Moreover, the response of wheat genotypes HD 2967 (4.18 t ha^{-1}) and HD 2733 (4.17 t ha^{-1}) was more positive toward elevated CO_2 as compared to other genotypes. In terms of tolerance to heat, the wheat genotype Halna followed by DBW 17 was least affected due to elevated temperature as compared to other genotypes. Finding also suggests that various physiological traits, viz., content of sugar, stability of membrane (MSI %), plant leaf water status (RWC %), and photosynthetic rate, were improved under elevated

S.K. Dwivedi (✉) • S. Kumar • V. Prakash • J.S. Mishra
ICAR-Research Complex for Eastern Region, Patna 800014, India
e-mail: sharad.dwivedi9736@gmail.com

J.K. Bisht et al. (eds.), *Conservation Agriculture*,
DOI 10.1007/978-981-10-2558-7_21

level of CO_2. However, rising temperature led to the negative response in terms of physiological traits. Advance knowledge of the future climate change scenarios in the zone on agroclimatic zone level would help in determining future climate risks and in identifying vulnerable areas to serve as the basis for crop planning and identification of suitable genotypes.

Keywords
Climate change • Open top chambers • Physiological traits • Yield attributes

21.1 Introduction

Climate change plays a very important role in the plant growth and physiology. The Nobel Laureate Svante Arrhenius pointed out that cause of global warming is the rising level of atmospheric CO_2 liberated from fossil fuels. Further, he reported that if present concentration of CO_2 (~260 ppm) will double, it would lead to the warming of earth atmosphere by about 2–6 °C. In the current context, the atmospheric level of CO_2 is ~400 ppm, arising largely from anthropogenic causes. Taking other GHGs along with CO_2 form, "CO_2 equivalent" is estimated to be ~470–480 ppm. The Intergovernmental Panel on Climate Change (IPCC) indicated that that the Earth atmosphere will warm by 1.4–5.8 °C till the end of twenty-first century (Swaminathan and Kesavan 2012). Rising of global atmospheric temperature led to the rise in mean sea level, change in precipitation pattern, and frequent occurrence of drought and heat waves, and heavy rainfall would take place. Further, due to shifting of temperature regimes, the acidification of ocean and species extinction happened. The negative consequences of climate change led to the decrease in crop yield which in turn affects food security and the loss of habitat from inundation. The GHG emissions between 2000 and 2010 grew 2.2 % $year^{-1}$, compared to 1970–2000 with 1.3 % $year^{-1}$. From the period 1906–2010, the surface temperature of earth rose by 0.74 ± 0.18 °C.

The main consequence of global climate change is the rising of atmospheric CO_2 level. The pace of increase in the level of atmospheric CO_2 is 1.9 ppm yr^{-1} and is forecasted to be reached at the level of 570 ppm by the middle of this century (IPCC 2007). Various studies on plant species have indicated that rising temperature and CO2 level will affect the morphophysiological processes of plant which in turn affect the growth and productivity (Bazzaz and Fajer 1992; Dwivedi et al. 2015a). The varieties developed through various means have contributed significantly in achieving higher yield. However, there is still need to identify and evolve new genotypes which can better withstand new environmental risks and possess some potential to produce high grain yield (Riaz-ud-din et al. 2010; Meena et al. 2013, 2015a, b; Singh et al. 2014; Kumar et al. 2015; Ghosh et al. 2016).

The emerging technology, i.e., conservation agriculture, is a kind of farming that ensures food security and helped in the conservation of biodiversity along with safeguarding the ecosystem services. Apart from these CA practices also contribute to combat climate change thus making agricultural system more resilient. The emission of GHGs tends to reduce due to practice of conservation agriculture approach and enhance its role as carbon sink. The main pillars of conservation agriculture are very less disturbance of soil by reducing tillage, increasing organic cover of soil, and crop diversification (Meena et al. 2015c, d, 2016; Verma et al. 2015b). Therefore, there is a need to develop suitable practices under conservation agriculture and identify suitable rice-wheat genotypes that are resilient to changing climate and produce good yield under changing climatic conditions.

21.1.1 Impact of Temperature on Crop Production and Productivity

Rising temperature severely hampered the production of food which in turn affects availability of food (Wassmann et al. 2009). Warmer temperatures have negative impact on yields of many crops. According to Mall and Singh (2000), the changes in temperature across the growing season/years seem to be the main factor of weather for fluctuations in the wheat yield. However, elevation in temperature could also make some areas favorable for crop production (Lioubimtseva and Henebry 2009). Tickell (1993) reported that till the year 2050, the mean temperature will increase by 1 °C and by 3 °C at the end of the twenty-second century. Gahukar (2009) also pointed that the production of cereals in India would go down up to 125 MT, and an overall increase of 2 °C in temperature might cause ~8 % loss in farm-level net revenue and ~5 % in GDP (FAO and IPCC). With such trends, by 2030, the maize production in Southern Africa could decrease by 30 %, while in South Asia the production of rice, millet, and maize could decrease by 10 %. Moreover, by 2080, the yield reduction in developing countries will be 10–25 %, while India could see a reduction of 30–40 %. The increment in temperature above 3 °C could result in yield drop in temperate regions that led to the reduction in total food production at global level. Elevated temperature also affects crop morphophysiology by altering crop duration, photosynthate partitioning, and rise in respiration rates. The elevation in temperature also affects the spatial distribution and survival of pest populations, fertilizer use efficiency, and rise in evapotranspiration rate. Moreover, plant biomass and crop yield also declined with increasing temperature (Rawson et al. 1995; Meena et al. 2015e; Verma et al. 2015a).

Aggarwal (2003) reported that 2–5 % reduction in yield potential of wheat and maize with a 0.5–1.5 °C increase in temperature in India. Further, rice yield was declined by 10 % for each 1 °C rise in minimum temperature above 32 °C (Pathak et al. 2003). Rice productivity in Punjab with temperature increases of 1, 2, and 3 °C would reduce the yield by 5.4 %, 7.4 %, and 25.1 %, respectively (Aggarwal and Swaroopa Rani 2009). Declining trend was observed in crop yield with rise in temperature in different parts of India. For example, a 2 °C elevation in mean air temperature decreases the rice yield by about 0.75 t/ha in the high-productivity areas and about 0.06 t/ha in the low-productivity coastal regions. A declining trend of potential yield of rice may be related to the negative trend in solar radiation, and an increase in minimum temperature was observed in Indo-Gangetic Plains (IGP) of India (Pathak et al. 2003).

21.1.2 Impact of CO_2 on Crop Production and Productivity

The rise in atmospheric CO_2 concentration has some positive impact in terms of plant phenology and physiology of crops. It was observed that an increase of the atmospheric CO_2 concentration would raise the production capacity of most of the herbaceous plants by utilizing physiological pathways (C_3, C_4, CAM). Continual increase in the level of atmospheric CO_2 since the 1950s led to the change in regional and global climate features, such as precipitation and temperature (Yu et al. 2002). Findings suggest that doubling of atmospheric CO_2 would lead to increase in the temperature by 1.5–4.5 °C (Houghton et al. 1990). Tao and Zhang (2013) also reported that rising atmospheric CO_2 has a fertilization effect on crops like rice and wheat (C_3 photosynthetic pathway). Thus increment in CO_2 concentration in the atmosphere could lead to higher crop yields.

A study showed that by increasing the 300 ppm atmospheric CO_2, it will result in yield advantage of ~49 % for C_3 cereals like wheat and rice, 20 % for C_4 cereals like maize, 15 % for CAM crops like pineapple, 44 % for legumes, and 24 % for fruits and melons. The elevated atmospheric CO_2 level led to the stimulation of photosynthetic process, leading to increase in plant biomass, productivity, and modified cycling process of nutrient (Kimball et al. 2002; Nowak et al. 2004). Ainsworth et al. (2004) reported that doubling the atmospheric CO_2 concentration (keeping other factor constant) have increased photosynthetic rate by 30–50 % in C_3 plants and 10–25 % in case of C_4 plant species. On average by increasing 150 ppm of CO_2 compared with current atmospheric CO_2 concentrations of 400 ppm, the crop yields increase in the range of 10–20 % for C_3 crops and 0–10 % for C_4 crops across several species under unstressed conditions (Gifford 2004; Long et al. 2005). The aboveground biomass of trees was also increased at 550 ppm CO_2 with the range of 0–30 % (Norby et al. 2003; Nowak 2004).

Furthermore, the increases in aboveground production of C_3 pasture grasses and legumes were 10 % and 20 %, respectively (Nowak 2004). Erda et al. (2005) pointed out that without carbon dioxide fertilization, the climate change could reduce the wheat, rice, and maize yields by up to 37 % in the next 20–80 years. Clifford et al. (1993) reported that the yield per plant of groundnut was higher as compared to sorghum with increase of atmospheric CO_2 from 350 to 700 ppm. Yao et al. (2007) with the CERES-rice model analyzed that CO_2 level impacts on rice yield in China and showed that the yield of rice will increase with CO_2 fertilization; otherwise, it will negatively be affected. Recently, Dwivedi et al. (2015) reported that the rice yield was increased with elevated CO_2 (500 ppm) in compression to ambient condition (400 ppm).

Various researchers indicated that the predicted yield increase in response to atmospheric CO_2 (700 ppm) was ~30 %. The yield increase of wheat was ~31 % (Amthor 2001), 29–35 % in rice (Cock and Yoshida 1973; Imai et al. 1985), ~55 %

in case of soybean crop (Rogers 1983), and in maize it was 50 % increase. However, some researchers argued that responses of crops toward elevated CO_2 might be lower than previously it was thought (Long et al. 2006).

21.1.3 Combined Impact of CO_2 and Temperature on Crop Growth and Yield

The rising level of atmospheric CO_2 also increased earth atmospheric temperature. It was found that CO_2 and temperature interact together and have caused significant effect on vegetative growth of plant. Moreover, the effect of CO_2 fertilization was more at warmer temperature than at cooler temperature. Both rising temperature and CO_2 could increase productivity of vegetative crops (pastures and forages) by increasing overall biomass by extending the length of the growing season in temperate regions. The interactive effects of temperature, CO_2, and rainfall can be best studied through the use of various crop growth simulation models. In India also several crop models have been used for assessment of impact arising due to climatic variability and climate change. Crop models like ORYZA for rice and WTGROWS for wheat have been used in many studies. The positive effect of CO_2 was negated by high temperature (HT) during the critical flowering period of a crop in terms of yield by reducing grain size, number, and quality (Thomas et al. 2003; Baker 2004; Caldwell et al. 2005). Further, the rising of atmospheric temperature during the critical growing period of crops may also reduce positive effect of CO_2, indirectly by increasing water demand. Xiao et al. (2005) pointed out that rainfed wheat yield at 450 ppm was increased up to 0.8 °C warming and then declined at higher temperature; additional irrigation was needed to counterbalance these negative effects of temperature.

Results showed that elevated temperature, CO_2, precipitation, and N deposition resulted in increased primary production of pastures, with changes in species distribution and litter composition (Shaw et al. 2002; Zavaleta et al. 2003). As per the fourth assessment report of the IPCC (2007), the net production of cereals in South Asia could decrease by 4–10 % by 2100 even under the most conservative climate change scenario (Cruz and Roger 2007). Rosenzweig and Iglesias (1994) noted that up to 4 °C warming and with elevated CO_2, the yield in low-latitude countries has been declined, but yield in mid- and high-latitude countries might be increased.

21.2 Climatic Condition for Evaluation of Rice and Wheat Genotypes

In the current perspective of climate change, two key factors play a major role, i.e., increasing CO_2 concentration and rising temperature. To understand the positive and negative effect of elevated CO_2 and rising temperature on rice and wheat genotypes, it is of utmost importance to study their effect in isolation as well as in combination. Thus, for the above purpose, four different sets of conditions were designed to evaluate the performance of four different rice and wheat genotypes.

21.2.1 Ambient Condition (C0T0)

Ambient condition is the situation of current atmospheric concentration of CO_2 (400 ppm) and temperature.

21.2.2 Elevated CO_2 (25 % Higher > Ambient CO_2) (C1T0)

One of the key factors in the present context of climate change is the rising level of atmospheric CO_2. Various agencies have predicted that due to the rising pace of human and nonhuman activities, the level of CO_2 will increase further. Last 10-year data showed that the pace of increase in the level of atmospheric CO_2 was 1.9 ppm yr^{-1}, and as per IPCC (2007) report, it will reach to 570 ppm by the middle of this century. Thus, such set of condition was designed to study the impact of elevated CO_2 (500 ppm) on plant growth, physiology, and yield of rice and wheat genotypes.

21.2.3 2 °C > Ambient Temperature (C0T1)

As the concentration of CO_2 rises, the temperature of earth atmosphere also increases. It was predicted that doubling of atmospheric CO_2 would raise the earth temperatures by about 2 °C. Thus such set of condition was designed to study the effect of elevated temperature (2 °C > ambient temperature) on rice and wheat genotypes.

21.2.4 25 % Higher CO_2 + 2 °C > Ambient (C1T1)

In the current scenario of climate change, both CO_2 and temperature rise in parallel. Thus it is very important to design a set of condition to study the effect of ~25 % higher CO_2 + 2 °C > ambient temperature on yield and yield attributes of rice and wheat genotypes.

21.3 Characterization of Rice and Wheat Genotypes Under Different Climatic Conditions

21.3.1 Impact of Elevated CO_2 and Temperature on Physiological and Biochemical Traits of Rice Genotypes

The rice is the staple food for the ~65 % of the world's population. The frequent occurrence of abiotic stresses due to climatic variation has been recognized as the key factor to the low productivity of rice in rainfed ecosystem. Pandey and Bhandari (2009) reported that a total of ~21 M ha of rainfed rice area is present in India, and out of this total area, ~16 M ha lies in the eastern belt of India, of which

6.3 M ha of upland and 7.3 M ha of lowland area are highly drought (water shortage) prone. The variation in degree and duration of changing climate during the crop growth in rainfed lowland rice is in need of identification and development of climate-ready crop of rice cultivars, which must survive under this climate change situation and will perform better in the future, quickly recover, and grow rapidly. Thus, keeping such views in mind, it's important to evaluate different rice genotypes under varying climatic conditions so that a suitable genotype will be identified which will ensure optimum yield under changing climatic condition. Effect of elevated CO_2 and temperature was enormous on the rice genotypes in terms of their growth, physiology, and economic yield.

21.3.1.1 Chlorophyll Content

The variation in content of chlorophyll is a very important indicator for suitability of genotype to perform under changing climatic condition. It is the trait which is directly linked with the photosynthetic performance of genotypes for production of optimum yield under such a varying environment.

Rice genotypes IR83376-B-B-24-2 and R. Bhagwati were more responsive to elevated CO_2, while there was sharp decline in pigment content under the influence of elevated temperature. The chlorophyll content of rice genotypes (IR83376-B-B-24-2 and R. Bhagwati) showed reduction by ~16 and 15 % when grown under elevated temperature (2 °C higher than ambient). Further, the decline in chlorophyll content under elevated temperature condition was 27 % and 18 % in genotypes IR84895-B-127-CRA-5-1-1 and IR64, respectively. Xie et al. (2012) reported that heading stage of rice was negatively influenced by high air temperature and estimated in terms of SPAD value (relative content of chlorophyll) in flag leaves of rice. A significant reduction occurred if the air temperature increased continuously.

21.3.1.2 Photosynthetic Rate

The positive impact of elevated CO_2 was also observed in terms of higher photosynthetic rate. The genotype which has higher photosynthetic rate under such a varying environment (elevated CO_2 and temperature) has superiority in terms of yield. Drake et al. (1997) reported that the rising of atmospheric CO_2 level might improve the potential of net photosynthetic rate of C_3 plants (rice and wheat) due to involvement of ribulose-1,5-bisphophate carboxylase/oxygenase (RuBisCO), in the process of CO_2 fixation and photorespiration. It was proven that ~60 % of assimilates demanded by rice during the process of grain filling were derived from photosynthetic activity of flag leaf during post-anthesis period, which significantly contributes to grain filling. Therefore, the photosynthetic characteristic of flag leaves is very crucial for the determination of grain yield and biomass. As depicted in Fig. 21.1, the photosynthetic rate (μmol m^{-2} s^{-1}) of rice genotypes grown under elevated CO_2 condition was significantly ($p < 0.05$) greater than the ambient condition.

Among rice genotypes, the maximum photosynthetic rate (i.e., 28.3) was achieved by genotype IR83376-B-B-24-2 under elevated CO_2 condition and

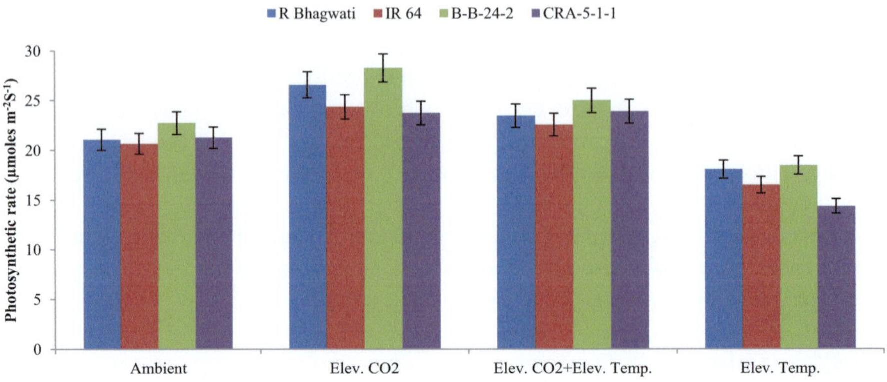

Fig. 21.1 Photosynthetic rate of rice genotypes grown under different climatic conditions inside open top chambers

reduced with elevated temperature condition. The elevation in temperature have pronounced negative effect and led to the reduction (~32 %) in mean photosynthetic rate in IR84895-B-127-CRA-5-1-1 as compared to ambient condition (Dwivedi et al. 2015). Nowak et al. (2004) also reported that elevation in atmospheric CO_2 concentration have stimulatory effect on photosynthesis process, leading to increased plant growth and production via modified water and nutrient cycles. Zhang et al. (2007) showed that high temperature caused the impairment of net photosynthetic rate mainly by reducing the content of chlorophyll as well as activities of RuBisCO and carboxylase (RuBP) involved in the process of photosynthesis in flag leaves.

21.3.2 Yield and Yield Attributes

To judge the performance of any genotype under varying environments, the grain yield is an important indicator if a particular genotype will fit in changed climatic system or not. As Pathak et al. (2003) reported that 1 °C increase in minimum temperature above 32 °C led to the reduction of 10 % in the grains yield of rice. Finding of Dwivedi et al. (2015) also suggests that yield and yield attribute traits (grain number panicle^{-1}, panicle length, test weight) of rice genotypes showed significant positive response with elevation of atmospheric CO_2, while response was negative with elevated temperature (Table 21.1 and Fig. 21.2).

Among rice genotypes, IR83376-B-B-24-2 (4.81 t ha^{-1}) and Rajendra Bhagwati (4.52 t ha^{-1}) gave higher yield in comparison to other genotypes under higher CO_2 condition ~500 ppm, although reduction in yield was highest in genotype IR84895-B-127-CRA-5-1-1 under elevated temperature (Tao and Zhang 2013). Krishnan et al. (2007) used ORYZA and info crop-rice models for data analysis to access the impacts of elevated CO_2 and temperature through an irrigated rice yield in eastern India, and the finding suggests that elevated CO_2 could increase the rice yield,

Table 21.1 Effect of varying climatic conditions (elevated CO_2 and temperature) on growth and yield attributes of rice genotypes grown inside open top chambers (O, OTCs; V, Varieties; OxV, Interaction)

Cultivar	Treatment						PH (cm)			PL (cm)	Grain panicle^{-1}		1000 GW (g)	GY (t ha^{-1})	
R. Bhagwati	Ambient CO_2						129			28.12	198		22.19	3.91	
	Elevated CO_2 (25 % higher > ambient)						137			30.48	215		23.90	4.52	
	Elevated CO_2 + elevated temp. (2 °C)						157			29.94	207		22.91	4.26	
	Elevated temp. (2 °C) > ambient						149			27.96	194		17.48	3.19	
IR64	Ambient CO_2						132			28.56	218		23.68	3.78	
	Elevated CO_2 (25 % higher > ambient)						142			30.74	242		25.03	4.23	
	Elevated CO_2 + elevated temp. (2 °C)						160			29.46	219		24.14	4.02	
	Elevated temp. (2 °C) > ambient						155			28.6	198		18.69	3.42	
IR83376-B-B-24-2	Ambient CO_2						153			31.32	228		24.09	4.18	
	Elevated CO_2 (25 % higher > ambient)						166			35.14	245		27.01	4.81	
	Elevated CO_2 + elevated temp. (2 °C)						177			32.16	233		25.08	4.49	
	Elevated temp. (2 °C) > ambient						170			28.96	231		20.52	3.9	
IR84895-B-127-CRA-5-1-1	Ambient CO_2						93			19.88	109		21.77	3.56	
	Elevated CO_2 (25 % higher > ambient)						95			20.42	115		22.07	3.88	
	Elevated CO_2 + elevated temp. (2 °C)						104			20.14	113		22.04	3.70	
	Elevated temp. (2 °C) > ambient						103			19	104		17.10	2.91	
Factors	O	V	OxV	O	V	OxV	O	V	OxV	O	V	OxV	O	V	OxV
LSD (p=0.05)	3.57	3.57	7.14	1.14	1.14	2.29	8.97	8.97	17.94	0.53	0.53	1.06	0.10	0.10	0.20
SEm+	1.78	1.78	3.57	0.57	0.57	1.15	4.49	4.49	8.98	0.24	0.24	0.49	0.04	0.04	0.09
CV (%)							4.06		6.60		7.39		2.23	2.47	

Fig. 21.2 Performance of rice genotypes inside different open top chambers

which is compensated with the sterility of rice spikelets due to higher temperature, the time of sowing, and the selection of genotypes. Thus, various findings suggest that elevated CO_2 and rising temperature have both positive and negative consequences of plant growth, physiology, and ultimately yield.

21.4 Effect of Elevated CO_2 on Wheat Crop Growth and Yield

Wheat is one of the leading cereal crops having worldwide importance and it is grown under a varying climatic condition. Among various abiotic stresses, high-temperature stress is one of the most detrimental environmental factor that limits the production and productivity of wheat crop (Dwivedi et al. 2015), with destructive economical and sociological impact. Moreover, the faster-than-predicted changes in global climate (IPCC 2007) indicated that there will be more frequent occurrence of heat waves as a consequence of the long-term effects of global warming. Wheat being a winter season crop is grown in the tropics and subtropics with high temperature (HT) that occur during the growth cycle. Performance of wheat genotypes under elevated CO_2 inside OTC were presented in Fig. 21.3.

Fig. 21.3 Performance of wheat genotypes (HD 2967, HD 2733, Halna, and DBW 17) under elevated CO_2 (~25 % higher than ambient) condition

High-temperature stress is an important constraint for wheat cultivation affecting its different growth stages especially grain filling duration which in turn affect crop productivity. It has been an already proven fact that the heat stress can be a significant factor for yield and quality reduction of wheat. A considerable portion of the wheat grown in an area in South Asia is considered to be highly affected by high-temperature stress, of which the majority is present in India (Joshi et al. 2007a). Transitory or constantly high temperatures cause a series of morphophysiological, anatomical, and biochemical changes inside plant system, which negatively affect growth and development of plant and led to a drastic reduction in grain yield, although elevated CO_2 has a positive impact on wheat production in terms of morphophysiological traits and yield attributes. Mitigation of heat stress due to imminent climate change and to enhance wheat productivity is the key to improve wheat productivity.

21.4.1 Impact of Elevated CO_2 and Temperature on Physiological and Biochemical Traits of Wheat Genotypes

Elevated CO_2 has positive effect in terms of physiological changes: RWC (%), MSI (%), photosynthetic rate, and TSS content (Fig. 21.4). Response of wheat genotype HD 2967 (4.18 t ha^{-1}) followed by HD 2733 (4.17 t ha^{-1}) was more positive toward elevated CO_2 as compared to other genotypes (Fig. 21.5). Halna followed by DBW 17 was least affected due to elevated temperature as compared to other genotypes (Fig. 21.5). The wheat yield was higher under elevated CO_2; however, the mean yield was declined with elevated temperature as compared to ambient condition. When along with elevated temperature the elevated CO_2 was also given, it was observed that the decline in mean yield was less as compared to elevated temperature condition alone. Elevated CO_2 condition leads to a significant yield advantage in terms of 1000 grain weight, grain number per year, year length, and harvest index.

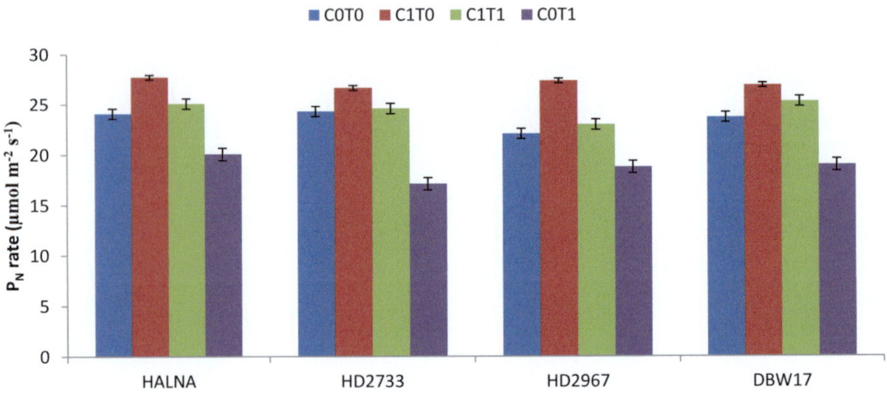

Fig. 21.4 Photosynthetic rate of wheat genotypes under varying climatic conditions

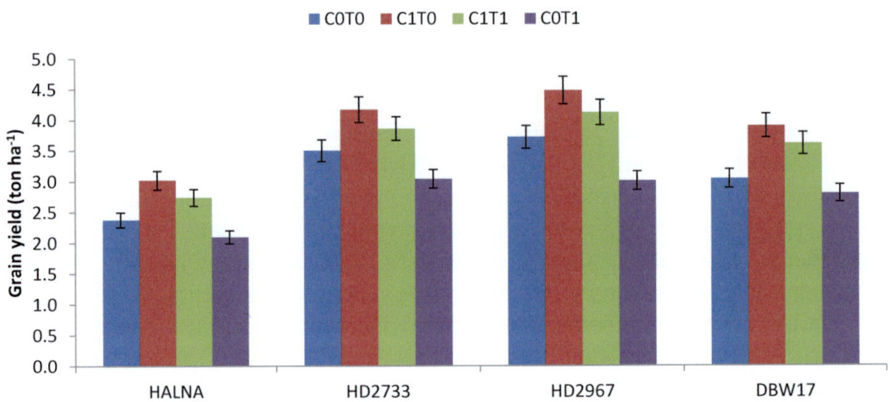

Fig. 21.5 Grain yield of wheat genotypes under varying climatic conditions

21.4.2 Mitigation and Adoption Strategies for Sustaining Productivity Under Changing Climate

Although the technology has been improved tremendously, yield potential of crops and food production remain highly dependent on climatic factors, because solar radiation, temperature, and precipitation pattern are the key drivers for the crop growth and development. The occurrence of plant diseases and pest infestations, as well as the demand for irrigation water supply are still very much influenced by climatic conditions. The change in the atmospheric temperature, precipitation pattern, and troposphere O_3 projected for 2050 is spatial and temporally variable, poorly constrained, and occurring in parallel. However, rise in CO_2 concentration is uniform at the global level and, unfortunately, certain. Even in the present context the levels of CO_2 are stabilized, atmospheric CO_2 would still be >500 μm per mol by 2050 (IPCC 2007). Therefore, attempts to design crop that will perform better under the situation of increasing environmental patterns should be considered against the backdrop of a guaranteed and ubiquitous rise in atmospheric CO_2 level (Ainsworth et al. 2004).

Vulnerability to changed climatic condition could be conceived as a function of exposure to hazards (variation in temperature or precipitation pattern from normal climate), sensitivity to such extreme events (e.g., yield of maize is highly vulnerable to water stress; yield of wheat and rice is sensitive to heat stress), and finally adoption mechanism to face the hazard (Ericksen et al. 2011). McCarthy et al. (2001) mentioned that if the people have sufficient strategies to manage a shock without suffering any harm, then they will not be vulnerable any more. Food insecurity depends upon many kinds of stressors, and change in climatic conditions will add more stress or potentially increase the chances of food insecurity if people or geographic areas are highly vulnerable to the climate hazard and not having sufficient coping capacity.

One of the key strategies that farmers can adapt under changed climatic condition is by shifting the date of sowing and selecting varieties with duration suitable for a given climatic conditions. Farmers can also change their crop rotation. Developing adaptation strategies to minimize the negative consequences of climate change might be risky with the number of uncertainties associated.

Possible adaptation option includes:

- Improving crop production by improved crop management practices under adverse climatic condition.
- Use of conservation agriculture practices can provide a new sight for combating climate change adverse impacts
- Better risk management through a forecasting system and policies that persuade crop insurance and provide better protection to the farmers. A forecasting system for disease and pest occurrence will be greatly helpful.
- Improved post harvest management techniques for minimizing the losses due to adverse climatic conditions.

21.5 Conclusions and Implications

Countries like India are highly vulnerable to adverse impacts of climate change in terms of agriculture. The anticipated changes are unavailability of water, rising of temperature, erosion of soil, and increase in extreme events like heat, droughts, and floods affecting agriculture severely. Productivity of wheat and rice genotypes is greatly influenced by changing climate. Climate change scenario emphasized low-cost adaptation strategies which included conservation agriculture practices, improving variety of crops with management, shifting in sowing time, and efficient utilization of water and fertilizer. Moreover, the higher grain yield in wheat and rice genotypes crop was observed under elevated CO_2. Elevated CO_2 has positive impact on rice and wheat genotypes in terms of plant growth, physiology, and grain yield; however, elevated temperature nullifies the effect of elevated atmospheric CO_2 in relation to crop growth and yield. Since rice and wheat are a staple food crop, thus long-term insight studies may provide better understanding on the efficiency of such genotypes under varying climatic conditions which are increasingly associated with food and nutritional security.

Acknowledgments We thank the editors and anonymous reviewers for their constructive comments, which helped us to improve the manuscript.

References

Aggarwal PK (2003) Impact of climate change on Indian agriculture. J Plant Biol 30:189–198

Aggarwal PK, Swaroopa Rani DN (2009) Assessment of climate change impacts on wheat production in India. In: Aggarwal PK (ed) Global climate change and Indian agriculture-case studies from ICAR Network Project. ICAR, New Delhi, pp 5–12

Ainsworth EA, Rogers A, Nelson R, Long SP (2004) Testing the 'source-sink' hypothesis of downregulation of photosynthesis in elevated [CO_2] in the field with single gene substitutions in *Glycine max*. Agric For Meteorol 122:85–94. doi:10.1016/j.agrformet.2003.09.002

Amthor JF (2001) Effects of atmospheric CO_2 concentration on wheat yield: review of results from experiments using various approaches to control CO_2 concentration. Field Crops Res 73:1–34

Baker JT (2004) Yield response of southern US rice cultivars to CO_2 and temperature. Agri Forest Mgt 122:129–137

Bazzaz FA, Fajer ED (1992) Plant life in a CO_2 rich world. Sci Am 266:68–74

Caldwell CR, Britz SJ, Mirecki RM (2005) Effect of temperature, elevated carbon dioxide, and drought during seed development on the isoflavone content of dwarf soybean grown in controlled environments. J Agri Food Chem 53(4):1125–1129

Clifford SC, Stronach IM, Mohamed AD, Azam-Ali SN, Crout NMJ (1993) The effects of elevated atmospheric carbon dioxide and water stress on light interception, dry matter production and yield in stands of groundnut (*Arachis hypogaea* L.). J Exp Bot 44:1763–1770

Cock JH, Yoshida S (1973) Changing sink and source relations in rice (*Oryza sativa* L.) using carbon dioxide enrichment in the field. Soil Sci Plant Nutri 19(3):229–234

Cruz RV, Roger VN (2007) Asia. Climate change 2007: impacts, adaptation and vulnerability. In: Contribution of Working Group 2 to the 4th Assessment Report of the Intergovernmental Panel on Climate Change, Parry, M. Lth edn. Cambridge University Press, Cambridge

Drake BG, Gonza'lez-Meler MA, Long SP (1997) More efficient plants: a consequence of rising atmospheric CO2? Annu Rev Plant Physiol Plant Mol Biol 48:609–639

Dwivedi SK, Kumar S, Prakash V, Mondal S, Mishra JS (2015a) Influence of rising atmospheric CO_2 concentrations and temperature on morpho-physiological traits and yield of rice genotypes in sub humid climate of Eastern India. Am J Plant Sci 6:2339–2349

Dwivedi SK, Kumar S, Singh SS, Mukharjee J, Haris A, Prakash V, Bhatt BP (2015b) Impact of elevated CO_2 concentrations and temperature on rice/wheat genotypes of Eastern India under predicted climate change scenario. Technical bulletin No. R-52/Patna–32, ICAR-RCER

Dwivedi SK, Kumar S, Prakash V (2015b) Effect of late sowing on yield and yield attributes of wheat genotypes in Eastern Indo Gangetic Plains (EGIP). J AgriSearch 2(4):304–306

Erda L, Wei X, Hui J, Yinlong X, Yue L, Liping B, Liyong X (2005) Climate change impacts on crop yield and quality with CO_2 fertilization in China. Philos Trans R Soc B 360:2149–2154

Ericksen P, Thornton P, Notenbaert A, Herrero M (2011) Mapping hotspot of climate change and food insecurity in the global tropics. CCAFS. REPORT no. 5. CGIAR research programme on climate change, Agricultural and Food security (CCAFS). Copenhagen, Denmark

Gahukar RT (2009) Food security: the challenges of climate change and bio-energy. Current Sci 96(1):26–28

Ghosh BN, Meena VS, Alam NM, Dogra P, Bhattacharyya R, Sharma NK, Mishra PK (2016) Impact of conservation practices on soil aggregation and the carbon management index after seven years of maize–wheat cropping system in the Indian Himalayas. Agr Ecosyst Environ 216:247–257

Gifford RM (2004) "The CO_2 fertilising effect—does it occur in the real world?" The international Free Air CO_2 Enrichment (FACE) workshop: short and long-term effects of elevated atmospheric CO_2 on managed ecosystems, Ascona, Switzerland, March 2004. New Phytol 163 (2):221–225

Houghton JT, Jenkins GJ, Ephraums JJ (1990) IPCC: climate change: the IPCC scientific assessment. Cambridge University Press, Cambridge

Imai K, Coleman DF, Yanagisawa T (1985) Increase in atmospheric partial pressure of carbon dioxide and growth and yield of rice (*Oryza sativa* L.). Jpn J Crop Sci 54(4):413–418

IPCC (Intergovernmental Panel on Climate Change) (2007) Intergovernmental Panel on Climate Change fourth assessment report: climate change 2007. Synthesis Report. World Meteorological Organization, Geneva

Joshi AK, Chand R, Arun B, Singh RP, Ortiz Ferrara G (2007) Breeding crops for reduced-tillage management in the intensive, rice–wheat systems of South Asia. Euphytica 153:135–151

Kimball BA, Kobayashi K, Bindi M (2002) Responses of agricultural crops to free-air CO_2 enrichment. Adv Agron 77:293–368

Krishnan P, Swain DK, Baskar C, Nayak SK, Dash RN (2007) Impact of elevated CO_2 and temperature on rice yield and methods of adaptation as evaluated by crop simulation studies. Agri Ecosyst Environ 122(2):233–242

Kumar A, Bahadur I, Maurya BR, Raghuwanshi R, Meena VS, Singh DK, Dixit J (2015) Does a plant growth promoting rhizobacteria enhance agricultural sustainability? J Pure Appl Microbiol 9(1):715–724

Lioubimtseva E, Henebry GM (2009) Climate and environmental change in arid central Asia: impacts, vulnerability and adaptations. J Arid Environ 73(11):963–977

Long SP, Ainsworth EA, Leakey AB, Morgan PB (2005) Global food insecurity. Treatment of major food crops with elevated carbon dioxide and ozone under large-scale fully open-air conditions suggests models may seriously over-estimate future yields. Philos Trans R Soc Lond B 360:2011–2020

Long SP, Ainsworth EA, Leakey ADB, Nosberger J, Ort DR (2006) Food for thought: lower than expected crop yield stimulation with rising CO_2 concentrations. Science 312:1918–1921

Mall RK, Singh KK (2000) Climate variability and wheat yield progress in Punjab using CERES-wheat and WTGROWS models. Vayu Mandal 30(3–4):35–41

McCarthy J, Canziani OF, White KS (2001) Climate change 2001: impacts, adaptation and vulnerability. Contribution of Working Group 2 to the third assessment report of the Intergovernmental Panel on Climate Change. Cambridge University Press, Cambridge

Meena RS, Yadav RS, Meena VS (2013) Heat unit efficiency of groundnut varieties in scattered planting with various fertility levels. The Bioscan 8(4):1189–1192

Meena RS, Meena VS, Meena SK, Verma JP (2015a) The needs of healthy soils for a healthy world. J Clean Prod 102:560–561

Meena RS, Meena VS, Meena SK, Verma JP (2015b) Towards the plant stress mitigate the agricultural productivity: a book review. J Clean Prod 102:552–553

Meena VS, Meena SK, Verma JP, Meena RS, Ghosh BN (2015c) The needs of nutrient use efficiency for sustainable agriculture. J Clean Prod 102:562–563

Meena VS, Meena SK, Verma JP, Meena RS, Jat LK (2015d) Current trend to mitigate climate change and sustainable agriculture: a book review. J Clean Prod 102:548–549

Meena VS, Verma JP, Meena SK (2015e) Towards the current scenario of nutrient use efficiency in crop species. J Clean Prod 102:556–557

Meena RS, Bohra JS, Singh SP, Meena VS, Verma JP, Verma SK, Sihag SK (2016) Towards the prime response of manure to enhance nutrient use efficiency and soil sustainability a current need: a book review. J Clean Prod 112(1):1258–1260

Norby RJ, Sholtis JD, Gunderson CA, Jawdy SS (2003) Leaf dynamics of a deciduous forest canopy: no response to elevated CO_2. Oecologia 136:574–584

Nowak RS, Ellsworth DS, Smith SD (2004) Functional responses of plants to elevated atmospheric CO_2– do photosynthetic and productivity data from FACE experiments support early predictions. New Phytol 162:253–280

Pandey S, Bhandari H (2009) Drought: economic costs and research implications. In: Serraj R, Bennett J, Hardy B (eds) Drought frontiers in rice: crop improvement for increased rainfed production. IRRI/World Scientific Publishing Company, Philippines/Hackensack

Pathak H, Ladha JK, Aggarwal P (2003) Trends of climatic potential and on-farm yields of rice and wheat in the Indo-Gangetic Plains. Field Crops Res 80:223–234

Rawson HM, Gifford RM, Condon BN (1995) Temperature gradient chambers for research on global environment change. I. Portable chambers for research on short-stature vegetation. Plant, Cell and Environ 18(9):1048–1054

Riaz-ud-din GM, Subhani N, Hussain M, Rehman A (2010) Effect of temperature on development and grain formation in spring wheat. Pak J Bot 42(2):899–906

Rogers HH (1983) Responses of selected plant species to elevated carbon dioxide in the field. J Environ Qual 12(4):569–574, ISSN 0047–2425

Rosenzweig C, Iglesias A (eds) (1994) Implications of climate change for international agriculture: crop modeling study. EPA 230-B-94–003. U.S. EPA Office of Policy, Planning and Evaluation, Climate Change Division, Adaptation Branch, Washington, DC

Shaw MR, Zavaleta ES, Chiariello NR, Cleland EE, Mooney HA (2002) Grassland responses to global environmental changes suppressed by elevated CO_2. Science 298:1987–1990

Singh DK, Singh SK, Singh AK, Meena VS (2014) Impact of long term cultivation of lemon grass (Cymbopogon citratus) on post-harvest electro-chemical properties of soil. Annals of Agri Bio Res 19(1):45–48

Swaminathan MS, Kesavan PC (2012) Agricultural research in an Era of climate change. Agri Res 1:3–11

Tao F, Zhang Z (2013) Climate change, high-temperature stress, rice productivity, and water use in Eastern China: a new super ensemble-based probabilistic projection. J App Meteorol Climatol 52(3):531–551

Thomas JMG, Boote KJ, Allen LH, Gallo-Meagher M, Davis JM (2003) Seed physiology and metabolism: elevated temperature and carbon dioxide. Crop Sci 43:1548–1557

Tickell C (1993) Global warming and its effects, Engineering for climate change. Institute of Civil Engineers, London

Verma JP, Jaiswa DK, Meena VS, Meena RS (2015a) Current need of organic farming for enhancing sustainable agriculture. J Clean Prod 102:545–547

Verma JP, Jaiswal DK, Meena VS, Kumar A, Meena RS (2015b) Issues and challenges about sustainable agriculture production for management of natural resources to sustain soil fertility and health. J Clean Prod 107:793–794

Wassmann R, Hosen Y, Sumfleth K (2009) Reducing methane emissions from irrigated rice. Focus 16, brief 3, a agenda for negotiation in Copenhagen 2020 vision for food, agriculture and the environment. International Food Policy Research Institute, Washington, DC

Xiao G, Liu W, Xu Q, Sun Z, Wang J (2005) Effects of temperature increase and elevated CO_2 concentration, with supplemental irrigation, on the yield of rain-fed spring wheat in a semiarid region of China. Agri Water Manag 74:243–255

Xie XJ, Li BB, Zhu HX, Yang SB, Shen SH (2012) Impact of high temperature at heading stage on rice photosynthetic characteristic and dry matter accumulation. Chin J Agromet 33:457–461

Yao F, Xu Y, Lin E (2007) Assessing the impacts of climate change on rice yields in the main rice areas of China. Clim Chang 80:395–409

Yu PS, Yang TC, Chou CC (2002) Effects of climate on evapotranspiration from paddy fields in southern Taiwan. Clim Chang 54(1–2):165–179

Zavaleta ES, Shaw MR, Chiariello NR, Thomas BD, Cleland EE, Field CB, Mooney HA (2003) Grassland responses to three years of elevated temperature, CO2, precipitation, and N deposition. Ecol Monogr 73:585–604

Zhang GL, Chen LY, Zhang ST, Liu GH, Tang WB, He ZZ, Wang M (2007) Effects of high temperature on physiological and biochemical characteristics in flag leaf of rice during heading and flowering period. Sci Agri Sinica 40:1345–1352

Rice Breeding for Drought Tolerance Under the Changing Climate Scenario

22

Santosh Kumar, S.K. Dwivedi, Ved Prakash, K.K. Rao, S.K. Samal, J.S. Mishra, and A. Kumar

Abstract

Rice (*Oryza sativa* L.) is one of the most important staple food crops for about two third of the world's population. Rice is cultivated in diverse ecosystems extending from rainfed upland to deep water. The rainfed rice covers ~45 % of the world's rice area. In a rainfed ecosystem, the recurrent event of drought has been attributed as the major reason to the low productivity. Present speculations regarding higher frequency of drought event along with a 1.1–6.4 °C rise in average global soil surface temperature by the end of this century pose an alarming threat to the production and productivity of rice posing a question mark to the food security of Asia. Drought is the greatest single yield-reducing factor among all other stresses influencing more than 23 M ha area of South and Southeast Asia. Out of the total 20.70 M ha located in India, ~16 M ha of area falls in eastern India including 6.3 M ha of upland and 7.3 M ha of lowland areas, which are highly susceptible to drought condition. Rice crop is very susceptible to soil moisture deficit and high-/low-temperature stresses, particularly at the reproductive stage. The majority of the existing high-yielding and traditional varieties of rice in the eastern part of India are very susceptible to moisture stress (drought). The majority of the high-yielding as well as traditional varieties of rice cultivated in the eastern part of India are very susceptible to moisture stress (drought). Farmers of drought-prone areas require such type of rice varieties that give them with high yield in years of normal rainfall and sustainable good yield in drought years. In this scenario, 42 advanced rice breeding lines were evaluated under drought stress at the reproductive stage with the objective of identifying drought stress-tolerant genotypes. The effect of drought stress on

S. Kumar (✉) • S.K. Dwivedi • V. Prakash • K.K. Rao • S.K. Samal • J.S. Mishra
ICAR-Research Complex for Eastern Region, Patna 800014, India
e-mail: santosh9239@gmail.com

A. Kumar
International Rice Research Institute, Los Baños, Philippines

J.K. Bisht et al. (eds.), *Conservation Agriculture*,
DOI 10.1007/978-981-10-2558-7_22

morphophysiological and biochemical traits was also studied. In the study, significant yield reduction was noticed almost in all rice genotypes under drought stress condition in comparison to control (non-stress). The varying responses of genotypes to the applied drought stress condition indicate its drought tolerance capacity. Grain yield was varied from 1.26 to 4.76 t/ha and 2.47–7.48 t/ha under stress and control condition, respectively. Based on preliminary screening, rice genotypes, viz., IR88867-4-1-1-4 (4.76 t/ha), IR88964-24-2-1-4 (4.73 t/ha), and IR88867-9-1-1-4 (4.55 t/ha), showed tolerance to drought at the reproductive stage as compared to check varieties Lalat (2.42 t/ha), IR64 (2.04 t/ha), and Sahbhagi Dhan (2.87 t/ha). Reproductive-stage drought also caused a decline in relative water content (RWC), membrane stability index (MSI), and plant biomass and an increase in grain sterility. These drought stress-tolerant rice genotypes may be cultivated in large areas of rainfed ecology where the occurrence of reproductive-stage drought is very frequent, especially in eastern India.

Keywords
Rice • Drought • Reproductive-stage stress • RWC • Grain yield • Physiological traits • Climate change

22.1 Introduction

Rice (*Oryza sativa* L.) is one of the most important food crops for human consumption. According to the USDA estimates, rice is cultivated in 159.46 M ha, and production was 463.8 MT in 2011–2012 (Table 22.1). The rice is the significant source of calories for ~50 % of the world's >7 billion people and has fundamental implication for food security and livelihood. Rice is cultivated in diverse ecosystems ranging from rainfed upland to deep water. Irrigated rice covers 55 % of the world's rice area, whereas rainfed ecosystem represents ~38 % of total rice area, accounting ~21 % of the world's rice production. These, because of semi-aquatic phylogenetic origin as well as diversity of rice ecosystems and growing situations, present rice production systems depend on plenty of water supply and, thus, are more susceptible to drought stress than other cropping systems. In terms of soil characteristics and water availability, rainfed rice systems are greatly variable, and they experience higher levels of abiotic stresses because of water scarcity or submergence than do irrigated ecosystems. In India, upland and rainfed lowland occupies 6.3 and 14.4 M ha, respectively, of the total rice area (Singh 2009). The recurrent occurrence of drought stress has been attributed as key factor of low rice productivity in rainfed ecosystems, especially in eastern India. In rainfed areas, drought can come at any crop stage (seedling, vegetative, and reproductive stage and physiological maturity) of the rice (Fig. 22.1).

Present speculations regarding higher frequency of drought event along with a 1.1–6.4 °C rise in average global soil surface temperature by the end of this century

Table 22.1 Rice area, yield, and production in different countries of the world during 2011–2012

Countries	Area (M ha)	Yield (Mt/ha)	Production (MT)
China	30.06	6.69	140.7
India	44.1	3.58	105.31
Japan	1.58	6.66	7.65
South Korea	0.85	6.58	4.22
North Korea	0.57	4.32	1.60
Pakistan	2.75	3.38	6.20
Bangladesh	11.72	4.31	33.70
Indonesia	12.16	4.73	36.50
Vietnam	7.74	5.61	27.15
Thailand	11.0	2.82	20.46
Burma	6.50	2.6	10.82
Philippines	4.58	3.71	10.71
Cambodia	2.77	2.41	4.27
Laos	0.82	2.71	1.40
Malaysia	0.68	3.85	1.90
World	159.46	4.36	463.8

Source: http://www.fas.usda.gov/psdonline and Kumar et al. (2013a, b, 2015a, b)

Fig. 22.1 Drought-affected rice field at vegetative and reproductive growth stages

pose an alarming threat to production and productivity of rice posing a question mark to the food security of Asia. As compared to the present condition, we will see much more land where the water stress will aggravate and only a small portion of the land where the water stress situation will be alleviated (Meena et al. 2015c, d, e, 2016; Verma et al. 2015a, b). In spite of an increased total water supply, the effects of seasonal runoff shifts, increased rainfall variability, water quality, and flood risks are likely to prevail in their impact on food production. A current estimation on climate change predicts the water scarcity to decline further in coming years (Wassmann et al. 2009), and the frequency and intensity of drought are predicted to become worse in years to come (Bates et al. 2008). Severe water scarcities lead to hunger and poverty through production losses in years of total crop failure with dramatic socioeconomic consequences on the affected population. Simulation studies have shown that increase in temperature in future time periods in different agroclimatic zones of Bihar may cause decline in rice productivity due to spikelet sterility (Haris et al. 2013).

Drought is one of the most serious constraints resulting in decline of rice grain yield in ~23 M ha of rainfed ecosystem in South and Southeast Asia (Huke and Huke 1997). Out of 20.7 M ha of rainfed rice-cultivated area in India, nearly ~16.2 M ha falls in eastern India, of which 4.2 M ha are severe drought prone (Singh and Singh 2000).

In areas where rainfed situation prevails, due to long spell between two rains or scarcity of rains or failure of rain, drought stress can occur at the different growth stages, i.e., seedling, vegetative, and reproductive stages of the rice. It can be an intermittent drought depending upon the rainfall pattern and distribution (Kumar 2011). The reproductive-stage drought has been identified as the most harmful to grain yield (O'Toole et al. 1982). Moreover, in most rainfed areas, the possibility of occurrence of terminal reproductive-stage drought is high due to the early withdrawal of rains (Kumar et al. 2008). Due to this reason, most of the drought breeding programs have focused on large-scale screening for reproductive-stage drought stress.

Drought stress is more difficult and complex than any other abiotic stresses. Successful improvement in this trait depends primarily on the selection of parents, screening criteria, and effective phenotyping protocols. From the beginning of the green revolution era in rice in 1965 to 2009, on 14 occasions, rice production in India failed to attain the estimated production, and it was in fact lower than the previous year's production. Drought was the main reason for the decreased production on 11 occasions. Severe drought stress in the rainy season not only created an adverse effect on production of rice but also reduced the area sown under pulses, wheat, and oilseeds in the succeeding dry season due to nonavailability of adequate moisture in the soil; it leads to reducing the production of these crops and forming food security in the country (Meena et al. 2013, 2015a, b; Singh et al. 2014; Kumar et al. 2015a, b; Ghosh et al. 2016). In the recent time, severe droughts in 2002 and 2009 observed a significant decline in rice as well as total food production. In 2002, in India, rice production fell by 21.5 MT, cereals by 7.0 MT, pulses by 2.2 MT, and wheat by 7.0 MT. Similarly, in 2009, 16.27 MT food production declined because of a 10.02 MT decrease in rice production and a significant decrease in the production of other crops such as cereals, oilseeds, and pulses (Kumar 2011).

Table 22.2 Drought-prone rice areas in eastern states of India

States	Geographical areas (M ha)	Area % of total India	% Area irrigated	Rice area (M ha)	Drought-prone rice area (M ha)	Rice areas (%)
Bihar	9.41	2.86	60.6	3.20	0.725	23
Eastern UP	8.64	2.62	74.9	5.92	0.985	17
West Bengal	8.87	2.69	48.4	5.94	0.956	16
Assam	7.84	2.38	20.3	2.50	0.221	9
Odisha	15.57	4.73	36.7	4.35	0.631	14
Jharkhand	7.97	2.43	12.0	1.67	0.243	15
Chhattisgarh	13.51	4.10	26.5	3.66	0.521	14
Eastern India	71.84	21.85	46.88	27.26	4.281	16

Source: IRRI (2013) and Kumar et al. (2015a, b)

22.2 Rice Areas Under Drought Prone in Eastern India

The eastern region includes eastern Uttar Pradesh, Bihar, Odisha, Chhattisgarh, West Bengal, Jharkhand, and plains of Assam. It represents ~22 % geographical area of the country and contributes to ~34 % of the country's production (Bhatt et al. 2011). In spite of the eastern region being bestowed with rich natural resources, the production level remained low. Rice production in eastern India has a direct effect on regional and national food security. The constraints like increasing water scarcity and high frequency of drought endanger the food security in the eastern region. Out of 27.26 M ha of rice area of the eastern states, nearly 4.28 M ha of area is highly prone to frequent drought (Table 22.2).

The present high-yielding varieties of rice growing in rainfed areas are bred for irrigated ecosystems, and most of them are highly susceptible to water scarcity (drought) conditions. The intensity and frequency of drought spells demand the development of rice cultivars which are able to survive under water deficit stress at the reproductive stage and quickly recover upon improved availability of soil moisture (De Datta et al. 1988; Kamoshita et al. 2008). There is an urgent necessity to disseminate and adopt drought-tolerant high-yielding varieties to achieve food self-sufficiency at the national level.

22.3 Productivity Trends in Different Rice Ecologies

The reasons for lower rice productivity and production vary among states and areas in eastern India. The rice productivity in Odisha has shown a fluctuating trend with the highest ever productivity of 1589 kg/ha in 2001–2002 as against the country's average of 2079 kg/ha. Jharkhand contributes ~2 M ha area under rice cultivation.

Table 22.3 Area, production, and productivity of different rice ecologies

Rice ecology	Area (M ha)	Production (MT)	Grain yield (t/ha)
Irrigated	21.6	59.8	2.9
Uplands	6	6.0	1.0
Favorable	2	2.8	1.4
Drought prone	4	3.2	0.8
Rainfed lowlands	13	24.0	1.8
Drought prone	5	10.0	2.0
Shallow favorable	2.5	7.5	3.0
Submergence prone	5.1	5.2	1.0

Source: Adopted from DRR (2011)

The average rice productivity in the state of Jharkhand is 1832 kg/ha. The average productivity of eastern Uttar Pradesh almost matches the national average. The average productivity of conducive irrigated area is more than 3.0 t/ha, whereas the average productivity of rainfed lowland and upland areas is 1.50 t/ha and 1.0 t/ha, respectively (Table 22.3). Rice productivity in Bihar is very poor. Out of the 38 districts of Bihar states, 25 districts come under the low-productivity group with the average yield of 1–1.5 t/ha (DRR 2011).

22.4 Present Status of Drought-Tolerant Rice Varieties in Eastern India

Most of the presently growing high-yielding varieties, viz., Swarna (MTU7029), Sambha Mahsoori (BPT 5204), IR64, IR36, Rajendra Bhagwati, Savitri, Sarjoo-52, Rajendra Subhasani, and Rajendra Sweta of eastern region, are highly prone to drought stress condition. They are originally bred for irrigated ecosystems, and they were never selected for drought stress tolerance. In drought years, these varieties are subjected to more grain yield losses, leading to a rapid decline in rice production of the country. Due to nonavailability of good-quality, high-yielding, drought-tolerant varieties, farmers in the rainfed areas continue to grow these varieties. Farmers of drought-prone areas require varieties that provide them with high yield in years of normal good rainfall and sustainable good yield in drought years.

In recent years, several promising drought-tolerant breeding lines for rainfed lowlands and uplands have been identified by employing direct selection for grain yield under drought stress condition. Similarly, some drought-tolerant varieties, i.e., Sahbhagi Dhan, CR Dhan 40, Shusk Samrat, Anjali, Vandana, Hazaridhan, NDR 97, NDR118, Abhishek, Naveen, Indira Barani Dhan, and DRR 42, are already released for eastern India (Table 22.4). Growing of these drought-tolerant rice varieties will be helpful for sustaining food security in the eastern states.

Table 22.4 Rice varieties for drought-prone areas of eastern states of India (DFF: days to 50 % flowering)

Varieties	DFF (days)	Plant height (cm)	Duration (days)	Yield (q/ha)		Grain type	Favorable land for cultivation	Recommended state for cultivation
				Stress	Non-stress			
Sahbhagi Dhan	85–90	100–110	110–115	25–30	40–45	Long-bold	Rainfed upland and medium lowland	Odisha, Bihar, Jharkhand, eastern UP, and West Bengal
ShushkSamrat	80–85	95–100	110–115	25–30	35–40	Long-slender	Rainfed upland and medium lowland	Eastern UP, Bihar, Jharkhand, and Odisha
CR Dhan 40	70–75	105–115	95–100	25–30	35–40	Short-bold	Rainfed upland (direct seeded)	Jharkhand and Bihar
NarendraDhan 97	80–85	85–90	95–100	20–25	30–40	Long-slender	Rainfed upland and medium lowland	UP, West Bengal, Bihar, and Chhattisgarh
NarendraDhan 118	70–75	95–100	90–95	20–25	30–35	Medium-slender	Rainfed upland and medium lowland	Uttar Pradesh and Bihar
Anjali	65–70	90–100	90–95	25–30	35–40	Short-bold	Rainfed upland	Bihar, Odisha, Jharkhand, Assam, and Chatishgarh
Vandana	70–75	100–105	95–100	25–30	35–45	Long-bold	Rainfed upland	Jharkhand, Bihar, Odisha, and Chatishgarh
Hazaridhan	80–85	90–95	115–120	20–25	35–45	Long-slender	Rainfed upland and lowland	Jharkhand and Bihar
Indira BaraniDhan	82–87	100–105	111–115	25–30	40–45	Long-slender	Rainfed upland and medium lowland	Chhattisgarh and MP
Naveen	90–95	95–100	120–125	20–25	35–40	Long-slender	Rainfed upland and medium lowland	Odisha and Jharkhand
Lalat	80–85	100–105	125–130	25–30	35–40	Long-slender	Rainfed medium upland and lowland	Odisha and West Bengal
Abhishek	90–95	95–100	120–125	25–30	35–40	Short-bold	Medium upland and lowland	Bihar, Odisha, Uttar Pradesh, Jharkhand, and Assam
BirsaVikas Dhan-109	60–65	90–95	85–90	20–25	30–35	Medium-slender	Rainfed upland	Jharkhand

22.5 Current Status of Rice Breeding

Latest result of experiment at IRRI revealed a moderate to high heritability of grain yield under drought (Bernier et al. 2007; Kumar et al. 2008; Venuprasad et al. 2007), thus creating scope for direct selection for grain yield instead of secondary traits. Kumar et al. (2008) and Venuprasad et al. (2008) have reported direct selection for grain yield under drought stress condition. Recently, several promising drought-tolerant breeding lines for rainfed lowlands and uplands have been identified by using direct selection for grain yield under drought (Mandal et al. 2010; Verulkar et al. 2010). As per Lafitte (2003), in order to select a drought-tolerant line, most breeding programs aim to reduce yield of the stress trial by 50 %. However, Kumar et al. (2008) did not observe any response to the selection in screens that had a yield reduction ~56 %. Most of the drought tolerance breeding programs fail to impose an adequately severe stress in their drought trials and, as a result, unable to identify true drought-tolerant lines.

In the recent past, many large-effect QTLs for grain yield under drought have been reported. These QTLs have allowed the development of drought-tolerant versions of popular mega-high-yielding varieties. At the molecular level, Kumar et al. (2007) reported a major QTL for grain yield under lowland drought stress in the CT9993/IR62266 population on chromosome 1 explaining 32 % of the genetic variance. Bernier et al. (2007) also reported a QTL on the 12th chromosome in the Vandana/Way Rarem population explaining ~51 % of the genetic variance for yield under severe drought stress. Venuprasad et al. (2009) also identified a major grain yield QTL on the 3rd chromosome which explained ~36 % of the genetic variance. The identification of several grain yield QTLs raises new expectations of improving yield under drought stress condition through marker-assisted breeding. Bernier et al. (2009) also tested the effect of this QTL in eastern India and at IRRI and showed consistent performance at both environments with increased effect as severity of stress increased. There was no effect of this QTL that was observed under irrigated (non-stress) condition. The QTL is reported to increase water uptake, which leads to increased grain yield under drought stress. Dixit et al. (2012) reported that qDTY12.1 was contributed by the susceptible parent Way Rarem; the QTL showed digenic interactions with two other QTLs (qDTY2.3 and qDTY3.2) contributed by the tolerant parent Vandana. Kumar et al. (2013a) identified drought yield (DTY) QTLs that showed a yield advantage of 300–500 kg/ha under moderate to severe drought conditions. This grain yield advantage needs to be more to make an impact on a commercial level in farmer's field.

22.6 Weed Management in Upland Rice

In India, rice is cultivated in a very diverse range of ecosystems from irrigated to shallow lowlands and mid-deep lowlands and deep water to uplands. Upland rice, which is mostly dry seeded, is found in parts of Chhattisgarh, Bihar, Assam,

Table 22.5 Grasses, sedges, and broad leaves weeds in upland rice of Bihar

	Botanical name	Common name	Family
Grasses			
1	*Echinochloa colona*	Bansawan	Gramineae
2	*Echinochloa crus-galli*	Bansawan	Gramineae
3	*Cynodon dactylon*	DoobGhas	Gramineae
4	*Eleusine indica*	Bankodo	Gramineae
5	*Dactyloctenium aegyptium*	Makra	Gramineae
6	*Setaria glauca*	Bottle grass	Gramineae
Sedges			
1	*Cyperus rotundus*	Motha	Cyperaceae
2	*Cyperus iria*	Motha	Cyperaceae
Broad leaves			
1	*Caesulia axillaris*	Thukaha (gurguza)	Compositeae
2	*Eclipta alba*	Bhangaria	Compositeae
3	*Euphorbia hirta*	Bari dudhi	Euphorbiaceae
4	*Solanum nigrum*	Ban makoy	Solanaceae
5	*Leucas aspera*	Gumma	Labiatae
6	*Phyllanthus niruri*	Hazardana	Euphorbiaceae
7	*Lippia nodiflora*	Mokana	Verbenaceae

Source: Anon (2005)

Jharkhand, Gujarat, Kerala, Karnataka, Madhya Pradesh, Odisha, Uttar Pradesh, and West Bengal. The upland rice is generally grown in well-drained soils in sloppy, undulated, or terraced land without water accumulation. The upland rice area is around 5.5 M ha which accounts for 12.33 % of the total rice area of the country. Weeds are the important constraint after moisture stress, and weed problem is most severe in dry-seeded upland rice. Weeds were reported to reduce rice yields by 12–98 %, depending on rice establishment methods. Rice yield losses due to weed growth and competition were least in transplanted rice and highest in dry-seeded rice sown without tillage (Singh et al. 2001a). The important grasses, sedges, and broad-leaved weeds are given in Table 22.5. These adoptions of proper and effective weed management technologies will result in an extra rice production.

22.7 A Case Study of Evaluation and Identification of Rice Genotypes for Drought Stress Tolerance

In spite of the significance of drought as a key constraint to rice productivity, fewer attempts have been done for developing drought-tolerant rice varieties. Most of the presently growing high-yielding varieties are susceptible to drought. Due to non-availability of drought-tolerant high-yielding varieties with good grain quality, farmers in the rainfed ecosystem continue to grow these varieties. Keeping this in view, under STRASA project (phase 2), a field drought screening was done during *kharif* season in 2013–2014 at the ICAR Research Complex for Eastern Region, Patna, India, with the aim of identifying drought-tolerant rice genotypes/donor lines. A set of 42 advanced rice breeding lines were evaluated under drought stress (reproductive-stage stress) and non-stress conditions with the objective of identifying drought stress-tolerant rice genotypes. In non-stress (irrigated) experiments, standing water was maintained from transplanting to 20 days before maturity. The stress experimental field was irrigated like the non-stress (control) experiments by keeping standing water up to 4 weeks after transplanting. Thereafter, the stress field was dried to develop stress by draining of water from the field. No supplemental irrigation was provided to the stress experiments after drainage till permanent wilting was observed in the susceptible checks. The stress experiments were not provided any supplemental irrigation after drainage till the susceptible checks showed permanent wilting. The effect of water stress on various morphophysiological and biochemical traits associated with drought tolerance was also studied at the reproductive stage in promising genotypes selected on the basis of yield performance under stress condition. Result revealed that significant yield decline was observed almost in all rice genotypes grown under drought stress condition as compared to non-stress condition. Kumar et al. (2014) also observed similar results. Grain yield was varied from 1.26 to 4.76 t/ha and 2.47–7.48 t/ha under stress and non-stress condition, respectively. Based on preliminary screening, some rice genotypes, viz., IR88867-4-1-1-4 (4.76 t/ha), IR88964-24-2-1-4 (4.73 t/ha), IR88867-9-1-1-4 (4.55 t/ha), IR88966-43-1-1-4 (4.33 t/ha), and IR88964-11-2-2-3 (4.07 t/ha), showed tolerance to reproductive-stage drought stress condition as compared to national and regional check varieties Lalat (2.42 t/ha), IR64 (2.04 t/ha), Swarna (1.26 t/ha), Rajendra Sweta (1.47 t/ha), and Sahbhagi (2.87 t/ha). Drought stress at the reproductive stage also caused significant reduction in relative water content (RWC %), membrane stability index (MSI%), the number of effective tillers, and plant biomass and an increase in grain sterility. Substantial variations were also observed among genotypes for leaf drying, leaf rolling, and stress recovery. Drought stress-tolerant rice genotypes had lesser leaf drying and leaf rolling and better stress recovery as compared to susceptible genotypes.

22.8 Incidence of Major Insect Pests and Diseases in Rice Crop Under Drought

Incidences of major insect pests like stem borer (*Scirpophaga incertulas*), grasshopper, leaf folder (*Cnaphalocrosis medinalis*), yellow rice bug (*Leptocorisa oratorius*), brown planthopper (*Nilaparvata lugens*), and green leaf hopper (*Nephotettix virescens*) were observed in different crop stages of the rice genotypes under drought stress and non-stress (irrigated) condition. It revealed that no significant variation in the incidence of insect pests was recorded within the genotypes. No significant difference in the percentage of dead hearts and white ear heads was observed in both conditions though the percentages of white ear head were recorded more in stress condition. Likewise, termite damage was also recorded more in stress condition. At vegetative stage, mainly grasshopper (GH), leaf folder (LF), and yellow stem borer (YSB) were observed in both stress and control conditions in all the genotypes. However, the stem borer activities were recorded more in stress condition as compared to non-stress. Incidence of rice bug (RB) was observed at flowering and milking stages of the crop. Frequency of occurrence of green leaf hopper (GLH) was recorded less in stress condition than control. No brown planthopper (BPH) was recorded in both conditions of drought-tolerant genotypes.

Under drought stress condition, incidence of blast disease and brown spot diseases was increased. Rice blast caused by *Pyricularia oryzae* is one of the most damaging and, therefore, important diseases of rice in many parts, particularly in drought-prone areas of the world. Brown spot is caused by *Helminthosporium oryzae*. Rice blast occurs erratically in most rainfed lowland environments, being severe in some seasons and absent in others. It is most severe where plants have been subjected to drought stress (Bouman and Mackill 1988). This disease accounts for yield loss ~65 % in susceptible cultivar of rice (Bhatt 1988). Drought-tolerant lines should have tolerance of blast and brown spot disease. In this endeavor, it is important to identify blast-resistant genotypes under drought stress condition.

Significant differences were observed for resistance to blast and brown spot diseases among genotypes. The screening of rice germplasm against rice blast and brown spot disease revealed that none of the genotypes was immune toward these diseases. The screening revealed that only few entries showed moderate to high resistance against blast and brown spot diseases. Promising drought-tolerant rice genotypes, viz., IR88867-4-1-1-4, IR88964-24-2-1-4, IR88867-9-1-1-4, IR88966-43-1-1-4, and IR88964-11-2-2-3, were found to be moderately to highly resistant to leaf blast and brown spot under rainfed condition which may be further utilized as the genetic sources for multiple resistance crop improvement program. Promising genotypes against blast disease also showed tolerance to drought stress condition.

22.9 Constraints and Gaps in Drought Rice Breeding

Drought has been regarded as one of the most serious abiotic constraints reducing rice yield in rainfed areas. Frequency of droughts along with an increase in global average surface temperature poses a serious threat to rice production in Asia. Simultaneously, its complexity has long been discussed. However, to tackle this situation, necessary efforts have not been made by either plant breeders or research leaders. Concentrated effort from scientific and management levels can only resolve the complex research problems. There is a scarcity of skilled drought breeders and physiologists dedicated to work on this problem, and an essential link among breeders and physiologists with molecular biologists is missing (Kumar 2011). Limited breeding programs placed in the drought-prone rainfed ecosystem have a systematic drought breeding program and screening. Even at institutes having drought breeding program, little activity is carried out to systematically screen drought-tolerant donors, study their combining abilities, and develop pre-breeding lines (the lines carrying new traits or trait combination which can be used as donor to improve existing cultivar) for use in a drought breeding program. In the national system, in majority of the countries, yield testing in rainfed environments is still done under irrigated conditions, providing favorable conditions for a drought-susceptible variety to excel (Kumar 2011). Lack of appropriate donor is one of the main constraints in rice breeding program for drought tolerance.

There is a necessity of improved donor parents with improved plant type (medium height, higher tillering, and lodging resistance), desirable grain type (medium-slender to long-slender), and also tolerance of various diseases and insect pests which should be selected for direct use in the rice improvement program. Most of the existing popular high-yielding varieties of rice grown in rainfed areas are originally bred for irrigated ecosystems. They are highly vulnerable to water scarcity (drought) conditions. Incorporating drought stress tolerance into high-yielding varieties will be a very effective approach. Adoption of high-yielding rice varieties having drought tolerance will play a significant role in developing sustainable food production which leads to food security in the rainfed areas.

22.10 Conclusions

Drought is one of the single most important constraints that affect rice production and productivity in rainfed areas. The challenge of increasing water scarcity and high frequency of drought has been identified as a major responsible factor in low rice productivity in rainfed ecosystems of the eastern region, endangering food security. Due to nonavailability of drought-tolerant high-yielding varieties with good grain quality, farmers in the rainfed ecosystem continue to grow traditional varieties, leading to severe grain yield losses. Severe drought stress in the rainy season not only adversely affected the production of rice but also reduced the area sown under pulses, wheat, and oilseeds in the succeeding dry season due to nonavailability of adequate moisture in the soil; it leads to reducing the production of these crops and forming food insecurity in the country. Several rice varieties have been released in South and Southeast Asia in recent years. Adoption of high-yielding drought-tolerant rice varieties with good grain quality will play a proactive and significant role in developing sustainable food production and lead to food security among farm families in the stress (water scarcity)-prone areas of the eastern region of India. Farmers of drought-prone areas need such type of rice varieties that provide them with high yield in years of good rainfall and sustainable good grain yield in drought years. Promising drought-tolerant rice genotypes (IR88867-4-1-1-4, IR88964-24-2-1-4, IR88867-9-1-1-4, IR88966-43-1-1-4, and IR88964-11-2-2-3) may be adopted (after multilocation testing and release) in a large area of rainfed ecosystem where drought is frequent, particularly at the reproductive stage.

Acknowledgments Authors profoundly acknowledge the International Rice Research Institute (IRRI), Philippines, for providing seed material for this study.

References

Anon (2005) Annual report of AICRIP on weed control, Rajendra Agricultural University, Pusa, Samastipur, Bihar

Bates BC, Kundzewicz ZW, Wu S, Palutikof JP Eds (2008) Climate change and water. Technical Paper of the Intergovernmental Panel on Climate Change, IPCC Secretariat, Geneva, 210 pp. www.ipcc.ch/ipccreports/tp-climate-change-water.htm

Bernier J, Kumar A, Ramaiah V, Spaner D, Atlin G (2007) A large-effect QTL for grain yield under reproductive-stage drought stress in upland rice. Crop Sci 47:505–516

Bernier J, Serraj R, Kumar A, Venuprasad R, Impa S, Gowda RPV, Oane R, Spaner D, Atlin G (2009) The large-effect drought-resistance QTL qtl12.1 increases water uptake in upland rice. Field Crops Res 110:139–146

Bhatt SC (1988) Yield loss in five rice varieties due to blast. J Hill Res 1:115–118

Bhatt BP, Haris AA, Islam A, Dey A, Mukherjee J, Brari SK, Das B, Kaushal DK (2011) Agriculture in Eastern States: opportunities and challenges. Technical Bulletin No.R-31/PAT-20

Bouman JM, Mackill DJ (1988) Durable resistance to rice blast disease. Oryza 25:103–110

De Datta SK, Malabuyoc JA, Aragon EL (1988) A field screening techniques for evaluating rice germplasm for drought tolerance during the vegetative stage. Field Crop Res 19:123–134

Dixit S, Swamy BPM, Vikram P, Bernier J, Sta Cruz MT, Amante M, Atri D, Kumar A (2012) Increased drought tolerance and wider adaptability of qDTY12.1 conferred by its interaction with qDTY 2.3 and qDTY3.2. Molecular Breed 30:1767–1779

Ghosh BN, Meena VS, Alam NM, Dogra P, Bhattacharyya R, Sharma NK, Mishra PK (2016) Impact of conservation practices on soil aggregation and the carbon management index after seven years of maize–wheat cropping system in the Indian Himalayas. Agric Ecosyst Environ 216:247–257

Haris AA, Bhatt BP, Chhabra V, Biswas S, Elanchezhian R (2013) Climate change impacts on yields of phenologically different rice varieties over a sub humid climatic environment. Agricul Res 2:319–329

Huke RE, Huke EH (1997) Rice area by type of culture: South, Southeast, and East Asia. International Rice Research Institute, Los Banos

IRRI (2013) Cluster demonstrations of stress tolerant rice varieties in stress prone parts of India (2012–13). Interim report, IRRI. www.irri.org/www.strasa.org

Kamoshita A, Babu RC, Boopathi NM, Fukai S (2008) Phenotypic and genotypic analysis of drought-resistance traits for development of rice cultivars adapted to rainfed environments. Field Crop Res 109:1–23

Kumar A (2011) Breeding rice for drought tolerance and adaptation to climate change. Rice knowledge Management Portal. http://www.rkmp.co.in

Kumar R, Venuprasad R, Atlin GN (2007) Genetic analysis of rainfed lowland rice drought tolerance under naturally-occurring stress in eastern India: heritability and QTL effects. Field Crops Res 103:42–52

Kumar A, Bernier J, Verulkar S, Lafitte HR, Atlin GN (2008) Breeding for drought tolerance: direct selection for yield, response to selection and use of drought-tolerant donors in upland and lowland-adapted populations. Field Crops Res 107:221–231

Kumar A, Dixit S, Henry A (2013a) Marker-assisted introgression of major QTLs for grain yield under drought in rice, in translational genomics for crop breeding: abiotic stress, yield and quality, volume 2, R. K. Varshney, R. Tuberosa, eds, 1st edition

Kumar M, Ganggaiah RB, Sreedevi B, Mahajan G, Rao PR, Senguttuvel P, Nirmala B, Rao KV, Viraktamath BC (2013b) Water saving technologies in rice. Technical bulletin No. 75/2013. Directorate of Rice Research (ICAR)

Kumar S, Dwivedi SK, Singh SS, Elanchezhian R, Mehta P, Singh BP, Singh ON, Bhatt BP (2014) Morpho-physiological traits associated with reproductive stage drought tolerance of rice

(*Oryza sativa* L.) genotypes under rain-fed condition of eastern Indo-Gangetic plain. Indian J Plant Physiol 19(2):87–93

Kumar A, Bahadur I, Maurya BR, Raghuwanshi R, Meena VS, Singh DK, Dixit J (2015a) Does a plant growth promoting rhizobacteria enhance agricultural sustainability? J Pure Appl Microbiol 9(1):715–724

Kumar S, Dwivedi SK, Haris AA, Singh SS, IdrisMd, Sangle UR, Monda, S, Kumar S, Singh SK, Kumar A, Bhatt BP (2015b) Drought tolerant rice for rainfed ecosystems of eastern India. Technical bulletin no. R-51/Patna-31. ICAR Research Complex for Eastern Region

Lafitte R (2003) Managing water for controlled drought in breeding plots. In: Fischer KS, Lafitte R, Fukai S, Atlin G, Hardy B (eds) Breeding rice for drought-prone environments. International Rice Research Institute, Los Banos, pp 23–26

Mandal NP, Sinha PK, Variar M, Shukla VD, Perraju P, Mehta A, Pathak AR, Dwivedi JL, Rathi SPS, Bhandarkar S, Singh BN, Singh DN, Panda S, Mishra NC, Singh YV, Pandya R, Singh MK, Sanger RBS, Bhatt JC, Sharma RK, Raman A, Kumar A, Atlin G (2010) Implications of genotype x input interactions in breeding superior genotypes for favorable and unfavorable rainfed upland environments. Field Crops Res 118:135–144

Meena RS, Yadav RS, Meena VS (2013) Heat unit efficiency of groundnut varieties in scattered planting with various fertility levels. Biogeosciences 8(4):1189–1192

Meena RS, Meena VS, Meena SK, Verma JP (2015a) The needs of healthy soils for a healthy world. J Clean Prod 102:560–561

Meena RS, Meena VS, Meena SK, Verma JP (2015b) Towards the plant stress mitigate the agricultural productivity: a book review. J Clean Prod 102:552–553

Meena VS, Meena SK, Verma JP, Meena RS, Ghosh BN (2015c) The needs of nutrient use efficiency for sustainable agriculture. J Clean Prod 102:562–563

Meena VS, Meena SK, Verma JP, Meena RS, Jat LK (2015d) Current trend to mitigate climate change and sustainable agriculture: a book review. J Clean Prod 102:548–549

Meena VS, Verma JP, Meena SK (2015e) Towards the current scenario of nutrient use efficiency in crop species. J Clean Prod 102:556–557

Meena RS, Bohra JS, Singh SP, Meena VS, Verma JP, Verma SK, Sihag SK (2016) Towards the prime response of manure to enhance nutrient use efficiency and soil sustainability a current need: a book review. J Clean Prod 112(1):1258–1260

O'Toole JC (1982) Adaptation of rice to drought prone environments. In: Drought resistance in crops with emphasis on rice. International Rice Research Institute, Los Banos, pp 195–213

Singh MP (2009) Rice productivity in India under variable climates. www.niaes.affrc.go.jp/marco/marco2009/english/. . ./W2-02_Singh_P.pdf

Singh S, Singh TN (2000) Morphological, chemical and environmental factor affecting leaf rolling in rice during water stress. Indian J Plant Physiol 5:136–141

Singh VP, Singh G, Singh RK (2001a) Integrated weed management in direct seeded spring sown rice under rainfed low valley situation of Uttaranchal. Indian J Weed Sci 33:63–66

Singh DK, Singh SK, Singh AK, Meena VS (2014) Impact of long term cultivation of lemon grass (Cymbopogon citratus) on post-harvest electro-chemical properties of soil. Annals Agri Bio Res 19(1):45–48

Venuprasad R, Lafitte HR, Atlin GN (2007) Response to direct selection for grain yield under drought stress in rice. Crop Sci 47:285–293

Venuprasad R, Sta. Cruz MT, Amante M, Magbanua R, Kumar A, Atlin GN (2008) Response to two cycles of divergent selection for grain yield under drought stress in four rice breeding populations. Field Crops Res 107:232–244

Venuprasad R, Dalid CO, Del Valle M, Bool ME, Zhao L, Espiritu M, Sta. Cruz MT, Amante M, Atlin G, Kumar A (2009) Identification and characterization of large-effect quantitative trait loci (QTL) for grain yield under lowland drought stress and aerobic conditions in rice using bulk-segregant analysis (BSA). Theor Appl Genet 120:177–190

Verma JP, Jaiswa DK, Meena VS, Meena RS (2015a) Current need of organic farming for enhancing sustainable agriculture. J Clean Prod 102:545–547

Verma JP, Jaiswal DK, Meena VS, Kumar A, Meena RS (2015b) Issues and challenges about sustainable agriculture production for management of natural resources to sustain soil fertility and health. J Clean Prod 107:793–794

Verulkar SB, Mandal NP, Dwivedi JL, Singh BN, Sinha PK, Mahato RN, Swain P, Dongre P, Payasi D, Singh ON, Bose LK, Robin S, Chandrababu R, Senthil S, Jain A, Shashidhar HE, Hittalmani S, Vera Cruz C, Paris T, Hijmans R, Raman A, Haefele S, Serraj R, Atlin G, Kumar A (2010) Breeding resilient and productive rice genotypes adapted to drought-prone rainfed ecosystems of India. Field Crops Res 117:197–208

Wassmann R, Jagadish SVK, Sumfleth K, Pathak H, Howell G, Ismail A, Serraj R, Redona E, Singh RK, Heuer S (2009) Regional vulnerability of climate change impacts on Asian rice production and scope for adaptation. Adv Agro 102:91–133

MIX
Papier aus verantwortungsvollen Quellen
Paper from responsible sources
FSC® C105338

If you have any concerns about our products,
you can contact us on
ProductSafety@springernature.com

In case Publisher is established outside the EU,
the EU authorized representative is:
Springer Nature Customer Service Center GmbH
Europaplatz 3, 69115 Heidelberg, Germany

Printed by Libri Plureos GmbH
in Hamburg, Germany